continued on back

Statistics for Experimenters

An Introduction to Design, Data Analysis, and Model Building

GEORGE E. P. BOX

WILLIAM G. HUNTER

J. STUART HUNTER

John Wiley & Sons
New York • Chichester • Brisbane • Toronto • Singapore

Library of Congress Cataloging in Publication Data

Box, George E. P.
 Statistics for experimenters.

 (Wiley series in probability and mathematical
statistics)
 Includes index.
 1. Experimental design. 2. Analysis of
variance. I. Hunter, William Gordon, 1937–
joint author. II. Hunter, J. Stuart, 1923–
joint author. III. Title.

QA279.B68   001.4'24   77-15087
ISBN 0-471-09315-7

Printed in the United States of America

22   23   24   25   26   27   28   29   30

To the memory of R.A.F.

to Judy

and to G.M.C.

Experiment!
Make it your motto day and night.
Experiment,
And it will lead you to the light.

The apple on the top of the tree
        Is never too high to achieve,
        So take an example from Eve . . .
                          Experiment!

Be curious,
           Though interfering friends may frown.
Get furious
           At each attempt to hold you down.

If this advice you only employ,
The future can offer you infinite joy
                And merriment . . .
                Experiment
                      And you'll see!

                COLE PORTER*

When the Lord created the world and people to live in it—an enterprise which, according to modern science, took a very long time—I could well imagine that He reasoned with Himself as follows: "If I make everything predictable, these human beings, whom I have endowed with pretty good brains, will undoubtedly learn to predict everything, and they will thereupon have no motive to do anything at all, because they will recognize that the future is totally determined and cannot be influenced by any human action. On the other hand, if I make everything unpredictable, they will gradually discover that there is no rational basis for any decision whatsoever and, as in the first case, they will thereupon have no motive to do anything at all. Neither scheme would make sense. I must therefore create a mixture of the two. Let some things be predictable and let others be unpredictable. They will then, amongst many other things, have the very important task of finding out which is which."

E. F. SCHUMACHER*

* From *Small Is Beautiful*. Used by permission.

Preface

Collaboration with research workers has been a source of satisfaction to us as practicing statisticians. It has put us in touch with a wide variety of investigations and has stimulated us by intellectual contact with many investigators. We have certainly learned from them, and they have sometimes learned from us. This book owes much to such interaction. Written for those who collect and try to make sense of data, it is an introduction to those ideas and techniques that we have found especially useful.

For a number of years we have used preliminary versions (in the form of notes) of this book in teaching engineers, chemists, biologists, statisticians, and other scientists at the University of Wisconsin and Princeton University. We have also used this material in courses for professional societies, industry, and government. It is neither a cookbook nor a textbook on mathematical statistics. It is an introduction to the philosophy of experimentation and the part that statistics plays in experimentation.

Actual problems are never so straightforward that they can safely be solved mechanically. Therefore we emphasize the necessity for thinking about the real nature of the scientific problem itself, for mulling over data plots and other graphical displays, as well as for understanding potentially useful statistical principles and their practical consequences. We point out possible difficulties caused by the violation of assumptions, such as the lack of independence of data. Furthermore we discuss ways by which such difficulties may be overcome. To illustrate principles we use both real and constructed data. For example, when the analysis of variance is first introduced, we simplify a real set of data so that all the values required in the analysis are whole numbers. Special emphasis is placed on the design of experiments because this is the most valuable aspect of statistical method. Frequently conclusions are easily drawn from a well-designed experiment, even when rather elementary methods of analysis are employed. Conversely, even the most sophisticated statistical analysis cannot salvage a badly designed experiment.

Statistical theory is introduced as it becomes necessary. Readers are assumed to have no previous knowledge of the subject; and although the book presents a properly sophisticated view of the philosophy of scientific investigation and the part played by statistical design and analysis, the mathematics needed is elementary. In particular, calculus does not appear in the main body of the text and is not needed to understand and use the book. Even more important than learning about statistical techniques is the development of what might be called a capability for *statistical thinking*. We hope this book will serve as a vehicle for this purpose. To help convey this facet of statistics, which is a challenge to the instructor, we have included several different types of questions and problems. In addition to exercises throughout the text, there are review questions at the end of each chapter, and problems at the end of each part of the book.

In addition to experimenters, others will find this book useful, for example, those who assess reported findings (Do the data really support these claims?), managers who direct research projects (What is the best way to approach this investigation?), and statisticians who work with experimenters (What is the best method of design and analysis for this problem?).

A typical one-semester college course might include fairly complete coverage of the first three parts of the book (Chapters 1–13) and a selection of topics from the fourth part (Chapters 14–18). Although this book is primarily intended for use as a text, we have also kept in mind the reader who will use it for self-instruction and/or as a reference.

Special Notes to the Reader

The questions at the end of each chapter reflect its contents and can be used in two ways: You can consider them for review *after* reading the chapter, or you can study them *before* reading the chapter to help identify key points. You may also find it helpful to use the problems listed at the end of each part of the book to guide your reading.

As you apply the ideas in this book, and especially if you meet with spectacular success or failure, we shall be interested to learn of your experience. We have tried to write a book that is useful and clear. If you have suggestions about how it can be improved, please write to us. As Chapter 1 suggests, we believe in iteration.

Acknowledgments

The manuscript was begun at Princeton University and continued at the University of Wisconsin (Department of Statistics, College of Engineering, and the Mathematics Research Center). We thank Bovas Abraham, Steve Bailey, Gina Chen, Art Fries, Johannes Ledolter, Greta Ljung, and Lars Pallesen for helping with the calculations and for checking the final manuscript. We are also indebted to the hundreds of students who used the notes on which this book is based and their proposed improvements for the valuable feedback they provided.

We are grateful to Professor E. S. Pearson and the Biometrika Trustees for permission to reprint condensed and adapted forms of various tables from *Biometrika Tables for Statisticians*, Vol. 1, 3rd ed., 1966. These are listed at the end of this book as Tables A, B1, C, D, F, and G.

We are grateful also to the Literary Executor of the late Sir Ronald A. Fisher, F.R.S., to Dr. Frank Yates, F.R.S., and to Longman Group Ltd., London, for permission to partially reprint Table III from their book *Statistical Tables for Biological, Agricultural and Medical Research*, 6th ed., 1974.

Our thanks are due to William G. Cochran and Gertrude M. Cox for allowing us to reproduce material on incomplete blocks designs from their classic book, *Experimental Designs*.

Finally, we are glad to acknowledge our debt to Mary Esser, Mary Arthur, and Doris A. Whitmore, who carefully typed the final manuscript.

G. E. P. Box
W. G. Hunter
J. S. Hunter

Madison, Wisconsin
Princeton, New Jersey
March 1978

Greek Alphabet

A α	alpha		N ν	nu
B β	beta		Ξ ξ	xi
Γ γ	gamma		O o	omicron
Δ δ	delta		Π π	pi
E ϵ	epsilon		P ρ	rho
Z ζ	zeta		Σ σ	sigma
H η	eta		T τ	tau
Θ θ	theta		Υ υ	upsilon
I ι	iota		Φ ϕ	phi
K κ	kappa		X χ	chi
Λ λ	lambda		Ψ ψ	psi
M μ	mu		Ω ω	omega

Contents

PART II COMPARING MORE THAN TWO TREATMENTS

PART III MEASURING THE EFFECTS OF VARIABLES

CHAPTER 1

Science and Statistics

1.1. THE LEARNING PROCESS

Scientific research is a process of guided learning. The object of statistical methods is to make that process as efficient as possible.

Learning is advanced by the iteration illustrated in Figure 1.1. An initial hypothesis leads by a process of *deduction* to certain necessary consequences that may be compared with data. When consequences and data fail to agree, the discrepancy can lead, by a process called *induction*, to modification of the hypothesis. A second cycle in the iteration is thus initiated. The consequences of the modified hypothesis are worked out and again compared with data (old or newly acquired*) that in turn can lead to further modification and gain of knowledge.

This process of learning can also be depicted as a feedback loop (Figure 1.2) in which the discrepancy between the data and the consequences of hypothesis H_1 leads to the modified hypothesis H_2, H_2 leads to H_3, and so on.

Nursery School Example

In a certain nursery school the teacher was pouring juice into cups for the children, who were sitting around a table. The teacher asked what would happen if, rather than stopping, she just kept pouring the juice into one of the cups. One child said the juice would fly *up* to the sky, and all the other children, except one, agreed. The one dissenting child (very likely drawing on her own past experience) said the juice would overflow and run *down* onto the table. The teacher took the opportunity to perform the experiment and demonstrate what actually happens. Learning took place that morning.

---

* In most of our applications the data acquiring process is scientific experimentation, but it could be a walk to the library to unearth already existing information or the conduct of a sample survey.

1

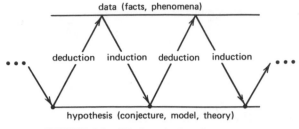

FIGURE 1.1. The iterative learning process.

Chemical Example

A chemist might have the following learning experience.

Hypothesis 1 and its deduced consequences	Because of certain properties of a newly discovered catalyst, its presence in a particular reaction mixture would probably cause chemical A to combine with chemical B to form, in high yield, a valuable product C.

The chemist has a tentative hypothesis and deduces its consequences, but he has no data on which to verify or deny its truth because, as far as he can tell from discussion with colleagues and careful examination of the literature, no one has ever performed the operation in question. He therefore decides to run some experiments.

Experimental design 1	A run is made at carefully selected reaction conditions. In particular, the chosen reaction temperature is 600°C.

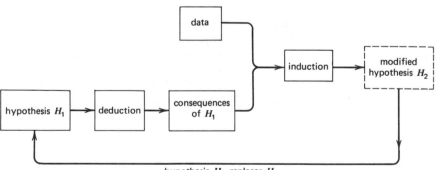

FIGURE 1.2. The learning process as a feedback loop.

Suppose, as often happens, that the result of the first experiment is disappointing. The desired product C is a colorless, odorless liquid, but what is obtained is a black, sticky, smelly mess containing only a very small trace of C.

Data 1   A black, tarry product containing less than 1 % of C is obtained.
(facts)

At this point hypothesis 1 and data 1 do not agree. The chemist considers the problem, is somewhat short with his wife at dinner that night, and the next morning, while taking a shower, begins to think along the following lines:

Induction   Product C might first have been formed in high yield, but it could then have been decomposed by the severe reaction conditions employed.

Hypothesis 2   Reaction conditions were too severe—a lower tempera-
and its   ture might produce a satisfactory yield of C.
deduced
consequences

As a result, during the course of that day:

Experimental   Two further runs are made with the reaction temperature
design 2   in the first run reduced to 550°C, and in the second to
               500°C.

Data 2   The product obtained for both runs is less black and tarry. The
(facts)   run at 550°C yields 4 % of the desired product C, and the run
             at 500°C yields 17 %.

The subsequent history of such an investigation can readily be imagined, with modification of the conjecture at each stage leading to further experiments calculated to best illuminate the current state of knowledge. Eventually, after a series of ups and downs, success would be celebrated or failure admitted.

Exercise 1.1. Describe a real or hypothetical example of iterative experimentation from a field such as agriculture, biology, education, engineering, or psychology.

1.2. THE ROLE OF EXPERIMENTAL DESIGN

Usually it is most efficient to estimate the effects of several variables simultaneously. Each experimental design will then contain a *group* of experimental runs. A new design is not necessarily employed for each iterative
cycle. Sometimes a sequence of cycles will occur in which the *same* data are
confronted by successive hypotheses. However, when it is not clear what
modification should be made to an unsatisfactory hypothesis, or, alternatively, when further confirmation of an apparently satisfactory hypothesis
is needed, additional data will be required. These are generated by further
experimental runs arranged in a new experimental design. Thus in Figure 1.3
additional data from the jth experimental design D_j is confronting deduced

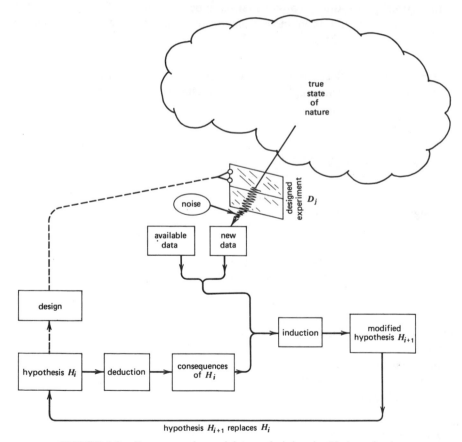

FIGURE 1.3. Data generation and data analysis in scientific investigation.

consequences of the ith hypothesis H_i. As the figure indicates, the *choice* of design depends on the current hypothesis. The chosen design should explore the shadowy regions of present knowledge whose illumination is currently believed to be important to progress. The design is represented by a movable window through which certain aspects of the true state of nature, more or less distorted by noise, may be observed.

The diagram emphasizes that, although the conjectured state of nature may be false or at least inexact, the data themselves are generated by the true state of nature. This is the reason why the process of continually updating the hypothesis and comparing the deduced states of nature with actual data can lead to convergence on the truth.

Notice that, on this view of scientific investigation, we are not dealing with a *unique* route to problem solution. Two equally competent investigators presented with the same problem would typically begin from different starting points, proceed by different routes, and yet could reach the same answer. What is sought is not uniformity but convergence. As a familiar illustration consider the game of "twenty questions," also called "animal, vegetable, or mineral." Here the object is to pose no more than twenty questions, each having only two distinct answers, and from this information to determine the object in the mind of the question answerer. Suppose that the object to be guessed is *Abraham Lincoln's stovepipe hat*. The initial clue is: the object is animal and vegetable with animal connections. For two competent teams of players, independently presented with the initial clue, the game might go as follows.

TEAM *A*

Question	Answer
1. Are the animal associations human?	Yes
2. Male or female?	Male
3. Famous or not?	Famous
4. Connected with the arts?	No
5. Politician?	Yes
6. U.S.A. or other?	U.S.A.
7. This century or not?	Not
8. 19th or 18th century?	19th
9. Connected with the Civil War?	Yes
10. Lincoln?	Yes
11. Is the object Lincoln's hat?	Yes

TEAM *B*

Question	Answer
1. Is the object useful?	Yes
2. Is it an item of dress?	Yes
3. Male or female?	Male
4. Worn above or below the belt?	Above
5. Worn on the head or not?	Head
6. Is it a famous hat?	Yes
7. Winston Churchill's hat?	No
8. Abraham Lincoln's?	Yes

The game follows the iterative pattern of Figure 1.3, although in this particular case a new design (the choice of question) is formulated with each cycle and if, the game is played fairly, the answers to the questions are honest and free of distractions. At each stage conjecture, progressively refined, leads to an appropriate choice of question that elicits the data which lead to appropriate modification of the conjecture. Teams *A* and *B* have followed alternative routes, but each is led to the correct answer because the data are generated by the truth.

The qualities needed to play this game well are (*a*) subject-matter knowledge and intellect, and (*b*) knowledge of strategy. Concerning strategy, it is well known that at each stage a question should be asked that divides the objects not previously eliminated into approximately equiprobable halves. Teams *A* and *B* evidently used such a strategy at least some of the time.

Knowledge of strategy in this example parallels knowledge of statistical methods in scientific investigation. Notice that without knowledge of the appropriate strategy it is possible to play the game, although perhaps not very well, whereas without subject-matter knowledge it cannot be played at all. However, notice also that one does best by using both subject-matter knowledge and strategy.

The conclusion carries over to scientific investigation and statistics. It is possible for a scientist to conduct an investigation without statistics, and it is impossible for a statistician to do so without scientific knowledge. However, a good scientist becomes a much better one if he uses statistical methods. Even if scientific data contained no noise (that is, the data were free of all disturbances caused by measurement and incomplete control of the experimental environment) induction of the true nature of complex systems could be hard enough. The existence of misleading experimental error makes the task even more difficult. In these circumstances the intellect and subject-matter knowledge of the investigator are put to best use when effective statistical tools are at his disposal. Convergence will occur most quickly and surely if he has available:

1. Efficient methods of experimental design, which enable him to obtain answers to his questions that are as unequivocal and as little affected by experimental error as possible.
2. Sensitive data analysis, which can indicate what can legitimately be concluded about current hypotheses and can suggest new ideas that should be considered.

Of these two resources design is more important. If the experimental design is poorly chosen, so that the resultant data do not contain much information, not much can be extracted, no matter how thorough or sophisticated the analysis. On the other hand, if the experimental design is wisely chosen, a great deal of information in a readily extractable form is usually available, and no elaborate analysis may be necessary. In fact, in many happy situations all the important conclusions are evident from visual examination of the data.

1.3. DIFFICULTIES MITIGATED BY STATISTICAL METHODS

Three sources of difficulty typically confronting the investigator are:

1. Experimental error (or noise).
2. Confusion of correlation with causation.
3. Complexity of the effects studied.

Experimental Error

Variation produced by disturbing factors, both known and unknown, is called *experimental error*. Usually only a small part of it is directly attributable to error in measurement. Important effects may be wholly or partially obscured by experimental error. Conversely, through experimental error, the experimenter may be misled into believing in effects that do not exist.

The confusing effects of experimental error can be greatly reduced by adequate experimental design and analysis. Furthermore, statistical analysis yields *measures of precision* of estimated quantities under study (such as differences in means or rates of change) and in particular makes it possible to judge whether there is any solid evidence of the existence of nonzero values for such quantities. The net effect is to increase greatly the probability that the investigator will be led along a true rather than a false path.

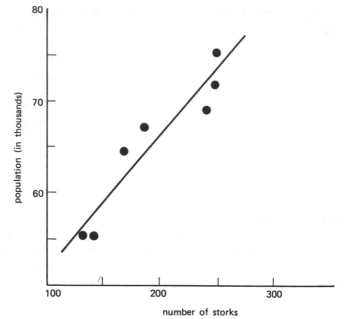

FIGURE 1.4.   A plot of the population of Oldenburg at the end of each year against the number of storks observed in that year, 1930–1936.

Confusion of Correlation with Causation

Figure 1.4 shows the population of Oldenburg at the end of each of 7 years plotted against the number of storks observed in that year.* Although in this example few would be led to hypothesize that the increased number of storks *caused* the observed increase in population, investigators are sometimes guilty of this kind of mistake in other contexts. Correlation between two variables Y and X often occurs because they are *both* associated with a third factor W. In the stork example, since the human population Y and the number of storks X both increased with time W over this 7-year period, it is readily understandable that a correlation appears when they are plotted together as Y versus X.

Exercise 1.2.   Give other examples where correlation exists but causation does not. (See, e.g., Huff, 1954.)

* These data cover the years 1930–1936. See *Ornithologische Monatsberichte*, **44**, No. 2, Jahrgang, 1936, Berlin, and **48**, No. 1, Jahrgang, 1940, Berlin, and *Statistiches Jahrbuch Deutscher Gemeinden*, 27–33, Jahrgang, 1932–1938, Gustav Fischer, Jena. We are grateful to Lars Pallesen for these references.

By using sound principles of experimental design and, in particular, randomization, data can be generated that provide a more sound basis for deducing causality.

Complexity of Effects

Consider an experimental study of the effects of alcohol and coffee on the reaction times of drivers operating a simulator. Suppose it was found that (a) if no coffee was taken, one shot of whiskey increased the reaction time by an average of 0.45 second, and (b) if no alcohol was taken, one cup of coffee reduced the reaction time by an average of 0.20 second.

In assessing the effects of several shots of whiskey and several cups of coffee and their combined effect, a great simplification would result if the effects were *linear* and *additive*. If they were linear, two shots of whiskey would increase the reaction time by 0.90 second [2(0.45) = 0.90], and three cups of coffee would reduce it by 0.60 second [3(−0.20) = −0.60]. If the effects were additive, one shot of whiskey and one cup of coffee would increase the reaction time by 0.25 second (0.45 − 0.20 = 0.25). Finally, if they were linear and additive, 10 shots of whiskey and 23 cups of coffee would reduce the reaction time by 0.10 second [10(0.45) + 23(−0.20)] = −0.10.

It is much more likely, however, that the effect of one additional shot of whiskey depends on (a) the number of shots of whiskey already consumed (the whiskey effect is *nonlinear*) and (b) the number of cups of coffee already consumed (there is an *interaction* effect between whiskey and coffee). Experimental designs are available that generate data in such a way that not only linear and additive effects but also effects of the interactive and nonlinear kind may be estimated with the smallest possible transmission of experimental error.

1.4. A TYPICAL INVESTIGATION

To illustrate what we have said about iterative learning and simultaneously provide a preview of what is discussed in this book, we now present an imaginary but reasonably realistic description of an investigation aimed at improving the quality of drinking water. Our imaginary investigator is a chemist, but when you read what follows consider how identical *statistical* problems could confront an investigator in, for example, medicine, education, engineering, psychology, agriculture, or any other experimental science.

Treatment of Water

There is, of course, only a limited amount of usable water on this planet, and what we have must be used and reused. The cleaning of water for reuse usually involves (1) primary treatment, such as sedimentation; (2) secondary treatment, such as biological oxidation; and (3) tertiary treatment, such as removal of refractory chemicals by adsorption. Let us study a (very full) year in the scientific life of an imaginary investigator named Peter Minerex. He knows that preliminary sedimentation and biological treatment of a polluted water source can produce an effluent of drinking water quality, except that it contains an unacceptably high level of sulfate. For tertiary treatment, Minerex has developed a new kind of ion exchange resin that, he hopes, may adsorb the offending sulfate. The resin is regenerated and reused as in a domestic water softener.

Usually this type of tertiary treatment is expensive. The attractive feature of Minerex's new resin is that it is more specific for sulfate and potentially much cheaper to use and regenerate than currently available resins. Although Minerex has taken care to reduce known sources of experimental variation by careful experimental technique, tests made under apparently similar conditions can vary quite markedly. The following outline shows how, as the investigation proceeds, he is led to consider different questions and employ statistical techniques of varying degrees of sophistication.

Iterative Cycles of Investigation

The description indicates the way the investigation might go* and shows where the relevant techniques are to be found in this book.

ITERATION I

> Question: Is there evidence that the new resin reduces sulfate levels in the partially treated water?
>
> Design and analysis: Chapter 3. Comparison of an average with a mean.
>
> Findings: Statistically significant reduction is obtained, but an unacceptably high level of sulfate remains.

COMMENTARY

Minerex thinks that using a very pure (hence more expensive) version of his new resin will improve its performance.

---

* This imaginary investigation has the remarkable property that in successive iterations it uses most of the techniques we discuss, approximately in the order in which they appear in the book. This is, of course, merely a pedagogical device. However, many investigations do go through various phases of sophistication in somewhat the manner illustrated here.

ITERATION II

Question: How does the "ordinary" new resin compare with the expensive, high-purity version?
Design and analysis: Chapters 3, 4, and 5. Comparison of two averages.
Findings: The expensive, high-purity version is about equal in performance to the "ordinary" new resin.

COMMENTARY

Minerex learns he was wrong about the high-purity version of the new resin, but the "ordinary" new resin still looks promising. He decides to compare it with five standard commercially available resins, all more expensive than his own.

ITERATION III

Question: How does the new resin compare with five commercially available resins?
Design and analysis: Chapters 6, 7, and 8. Comparison of more than two treatments.
Findings: Minerex's (cheaper) new resin is as good as any of the commercially available alternatives and perhaps somewhat superior.

COMMENTARY

The new resin has been shown to do at least as well as its more expensive competitors. Under the conditions being studied, however, the removal of sulfate is still insufficient to achieve the standard required for drinking water.

ITERATION IV

Question: Could modifications in the equipment affecting flow rate, bed depth, regenerant quantity, and so forth lead to improved sulfate removal?
Design and analysis: Chapters 9, 10, and 11. Empirical studies with factorial designs.
Findings: With suitable equipment modifications a sufficiently low sulfate level can be achieved.

COMMENTARY

Before the ion exchange process can be recommended as providing a satisfactory basis for tertiary treatment of water containing sulfates, its behavior must be considered under a variety of conditions that might be encountered at different locations and at different times. Questions now arise about the effect of such factors as the pH and hardness of the water to be treated, and the possibly inhibiting effect of trace amounts of certain chemicals $(A, B, C, \ldots, K)$ that may accompany the sulfate. Experimental runs can now be made in which deliberate changes are introduced in the water to be treated. Thus known amounts of various chemicals can be added, and the pH and hardness adjusted to desired levels. To test all combinations of these potentially disturbing factors would involve an impossibly large experimental program. Fortunately, *fractional* factorial designs, which employ carefully selected subsets of all possible combinations, can provide the desired information economically.

ITERATION V

> Question: How is the sulfate adsorption process affected by changes in pH, changes in hardness, and the presence of small amounts of chemicals $A, B, C, \ldots, K$?
>
> Design and analysis: Chapters 12 and 13. Fractional factorial designs.
>
> Findings: The adsorption of sulfate is insensitive to changes in pH, changes in water hardness, and the presence of small amounts of chemicals $A, B, C, \ldots, K$ over the ranges of the variables studied.

COMMENTARY

Minerex's company now concludes that it will probably be profitable to market this new resin. A pilot plant is built.

ITERATION VI

> Question: How do the settings of the process variables affect the quality and manufacturing cost of the resin?
>
> Design and analysis: Chapters 14 and 15. Response surface methods and the method of least squares.
>
> Findings: The pilot plant investigations indicate the feasibility of a process for the full-scale manufacture of the resin with satisfactory quality at reasonable cost.

COMMENTARY

A full-scale plant is built. Minerex now becomes involved in investigations to achieve efficient and smooth operation of this plant. He uses control charts (Chapter 17), which indicate that the measured product quality is excessively variable. He thinks that this problem may be caused by faulty analytical or sampling procedures. After variance component analysis and error transmission studies (Chapter 17) he finds that most of the trouble arises from real changes in the quality of the product.

ITERATION VII

> Question: Can the causes for the excessive variability of the product be identified and eliminated?
> Design and analysis: Chapter 11. Evolutionary Operation.
> Findings: Major causes of the variability are identified and eliminated.

COMMENTARY

Having solved the most urgent problems of resin production, Minerex now wants to study how the resin is actually being used in large-scale water treatment. Arrangements are made for him to investigate one customer's water purification unit in the field. He employs time series analysis (Chapter 18) to study the variation with time of the influent and effluent streams of this unit. This analysis suggests better ways to operate the unit and also leads to modifications in the engineering design. With a creditable history of success Minerex now returns to the research laboratory. Here he attempts to gain further insight into the detailed mechanism by which the new resin works. Much is already known about the mechanism of ion exchange.

ITERATION VIII

> Question: Can more be learned about why this new resin works the way it does?
> Design and analysis: Chapter 16. Mechanistic model building.
> Findings: A greater understanding is achieved of the mechanism whereby the new resin removes sulfate, suggesting new directions for further research.

Conclusions

Although the account of Peter Minerex is imaginary, it provides a realistic composite picture of the way statistical methods are used. One lesson to learn is that at successive stages of an investigation different problems confront the investigator and different techniques are needed to solve them. One characteristic of the good experimenter is that he chooses the appropriate tools for each stage of the job.* One of our objects in this book is to indicate in what circumstances the various methods we describe are useful.

Exercise 1.3.   Identify experimental problems, preferably in your own field, that are similar to those encountered by Minerex.

1.5   HOW TO USE STATISTICAL TECHNIQUES

All real problems have idiosyncrasies that must be appreciated before effective methods of tackling them can be adopted. Consequently each new problem should be treated on its own merits and with a certain amount of respect. Being too hasty can result in mistakes, such as obtaining the right answer to the wrong problem.†

Find Out as Much as You Can about the Problem

When an investigator or a consultant must rely on others for information, he should probe until he is satisfied with what is being presented to him and fully understands it. Here are some of the questions that ought to be asked and answered: What is the object of this investigation? I am going to describe your problem. Am I correct? Do you have any data? How were these data collected? In what order? On what days? By whom? How? How does the equipment work? What does it look like? May I see it? May I see it work? Do you have other data like these? How much physical theory is known about this phenomenon?

Don't Forget Nonstatistical Knowledge

Statisticians are often stunned by the over-zealous use of some particular statistical tool or methodology on the part of an experimenter, and we offer

---

* In certain circumstances he may realize that none of the tools presently available is appropriate, and he may have to fashion a new one himself (or have someone else do it)—see the final subsection in this chapter: "Learn from Each Other: The Interplay between Theory and Practice."
† Kimball (1957) has called this mistake "an error of the third kind."

the following caveat. Experimenters, when you are doing "statistics," do not forget what you know about your subject-matter field! Statistical techniques are most effective when combined with appropriate subject-matter knowledge. The methods are an important adjunct to, not a replacement for, the natural skill of the experimenter.

Define Objectives

In any investigation it is of utmost importance (1) to *define clearly the objectives* of the study to be undertaken, (2) to be sure that all interested parties agree on these objectives, (3) to agree on what criteria will determine that the objectives have been reached, and (4) to arrange that, if the objectives change, all interested parties will be made aware of this fact and will agree on the new objectives and criteria. It is surprising how often these steps are either ignored or not given the careful attention they deserve, a circumstance that often leads to difficulties and sometimes to disaster.

Learn from Each Other: The Interplay between Theory and Practice

While experimenters can benefit by learning about statistics, the converse is equally true: statisticians can benefit by learning about experimentation. Good statistical work seems to result from a genuine interest in practical problems. Sir Ronald Fisher, who was the originator of most of the ideas in this book, worked closely with experimenters and was one himself. For him there was no greater pleasure than discussing problems with scientists over a glass of beer. The same was true of his friend William S. Gosset (better known as "Student"), of whom a colleague commented, "To many in the statistical world 'Student' was regarded as a statistical adviser to Guinness's brewery; to others he appeared to be a brewer devoting his spare time to statistics. . . . Though there is some truth in both of these ideas they miss the central point, which was the intimate connexion between his statistical research and the practical problems on which he was engaged."* The work of Gosset and Fisher reflects the hallmark of good science, the interplay between theory and practice. Their success as scientists and their ability to develop useful statistical techniques were highly dependent on their deep involvement in experimental work.

---

* Launce McMullen in the foreword to *Student's Collected Papers*, edited by E. S. Pearson and John Wishart, Cambridge University Press, 1942, issued by the Biometrika Office, University College, London.

REFERENCES AND FURTHER READING

Two important books about the use of statistical methods in scientific investigations are:

Fisher, R. A. (1935). *Statistical Methods for Research Workers*, Oliver and Boyd.
Fisher, R. A. (1935). *The Design of Experiments*, Oliver and Boyd.

Other examples of "nonsense correlation" similar to the one in this chapter about storks and birth rate are contained in the book:

Huff, D. (1954). *How to Lie with Statistics*, Norton.

For a discussion and illustration of the iterative nature of experimentation see the following article and also the references given there:

Box, G. E. P. (1976). Science and statistics, *J. Am. Stat. Assoc.*, **71**, 791.

For further information on statistical consulting see the following articles and the references listed therein:

Boen, James R. (1972). The teaching of personal interaction in statistical consulting, *Am. Stat.*, **26**, No. 1, 30.
Cameron, J. M. (1969). The statistical consultant in a scientific laboratory, *Technometrics*, **11**, 247.
Daniel, C. (1969). Some general remarks on consultancy in statistics, *Technometrics*, **11**, 241.
Hyams, L. (1970). The practical psychology of biostatistical consultation, *Biometrics*, **27**, 201.
Kimball, A. W. (1957). Errors of the third kind in statistical consulting, *J. Am. Stat. Assoc.*, **57**, 133.
Watts, D. G. (1970). A program for training statistical consultants, *Technometrics*, **4**, 737.

QUESTIONS FOR CHAPTER 1

1. What is meant by the iterative nature of learning?
2. In what ways can statistics be useful to experimenters?
3. What is achieved by good statistical analysis? What by good statistical design? Which do you believe is more important?
4. What are three common difficulties encountered in experimental investigations?
5. Can you give examples (preferably from your own field) of real confusion (perhaps controversy) that has arisen because of one or more of these difficulties?
6. Which techniques in this book do you expect will be most useful to you?
7. How should you use the techniques in this book?
8. Can you think of an experimental investigation that was not iterative?

9. Read an account of the developments in a particular field of science over a period of time (e.g., the book *The Double Helix*, by James D. Watson). How does it compare to the description in this chapter of a scientific investigation as an iterative process? Can you trace how the confrontations of hypotheses and data led to new hypotheses?

PART I

Comparing Two Treatments

Experiments are often performed to compare two treatments, for example, two different fertilizers, machines, methods, processes, or materials. The object is to determine whether there is any real difference between them, to estimate the difference, and to measure the precision of the estimate.

In this part of the book we discuss the design and analysis of these simple comparative experiments. We describe pitfalls that await the unwary, for example, how violation of the assumption of statistical independence of successive errors can seriously invalidate standard procedures. By understanding the nature of this difficulty and others one is led to appreciate the crucial importance of randomization, replication, and blocking.

In this context significance tests and confidence intervals are introduced, first for comparison of means, and later for comparison of proportions and frequencies. The ideas introduced are of much more general application and are used in the design and analysis of more complicated experiments in later parts of the book.

CHAPTER 2

Use of External
Reference Distribution
to Compare Two Means

In this chapter we consider some data obtained from an experiment to find out whether a modified method gave better results than the standard method. The discussion leads to the introduction and explanation of some basic concepts, such as reference distribution, probability distribution, random sample, significance test, normal distribution, and t distribution, and to basic statistical terms, such as *average*, *mean*, *variance*, *sample*, and *population*.

2.1. RELEVANT REFERENCE SETS AND DISTRIBUTIONS

In the course of his scientific career (see Section 1.4) Peter Minerex had to move himself and his family more than once from one part of the country to another. He found that the prices of suitable houses varied confusingly in different locations and were also quite variable at any given location. He and his wife soon determined the following strategy for choosing a new home. Having found a temporary accommodation, they would spend considerable time merely looking at houses that were available. In this way they built up in their minds a "reference set" or "reference distribution" of available values. Once this distribution was established, it was possible to judge whether a house being considered for purchase was averagely priced, exceptionally (significantly) expensive, or exceptionally (significantly) inexpensive.

The method of statistical inference called *significance testing* (or, equivalently, *hypotheses testing*) parallels this process. The investigator is considering a particular result apparently produced by making some experimental modification of a system. He needs to know whether the result is easily explained by mere chance variation or whether it is exceptional, pointing to the effectiveness of the modification. To make this decision he must in some way produce a relevant reference set that represents a characteristic set of outcomes which could occur if the modification *was entirely without effect*. The *actual* outcome may then be compared with this reference set. If it is found to be exceptional, the result is declared statistically significant.

The analogy given above points to a very important consideration: Minerex must be sure that the reference set of house prices that he and his wife are tacitly using is *relevant* to their present situation. They should not, for instance, use the reference set appropriate to their small, country hometown, if they are looking for a house in a metropolis. We will see that the experimenter must be equally discriminating.

TABLE 2.1. Yield data from an industrial experiment (plant trial)

time order	method	yield
1	A	89.7
2	A	81.4
3	A	84.5
4	A	84.8
5	A	87.3
6	A	79.7
7	A	85.1
8	A	81.7
9	A	83.7
10	A	84.5
11	B	84.7
12	B	86.1
13	B	83.2
14	B	91.9
15	B	86.3
16	B	79.3
17	B	82.6
18	B	89.1
19	B	83.7
20	B	88.5

$$\bar{y}_A = 84.24, \qquad \bar{y}_B = 85.54$$

$$\bar{y}_B - \bar{y}_A = 1.30$$

The following example concerns the assessment of a modification in a manufacturing plant. The principles involved are of course not limited to this application, and you may wish to relate them to an appropriate example in your own field.

An Industrial Experiment: Is the Modified Method Better than the Standard One?

An experiment was performed on a manufacturing plant by making in sequence 10 batches of a chemical using the standard production method (*A*), followed by 10 batches using a modified method (*B*). The results from this plant trial are given in Table 2.1. What evidence do the data provide that method *B* is better than method *A*?

To answer this question the experimenter began properly by plotting the data as shown in Figure 2.1 and calculating the average yields obtained for methods *A* and *B*. He found that

$$\bar{y}_A = 84.24, \qquad \bar{y}_B = 85.54$$

where the bar above the symbol *y* is used to denote an arithmetic average. The modified method thus gave an average that was 1.30 units higher than the one for the standard method. However, because of the considerable variability in the individual test results, the experimenter wondered whether

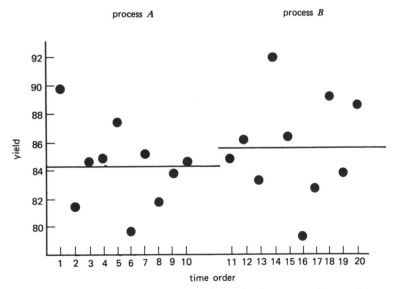

FIGURE 2.1. Yield values plotted in time order for comparative experiment.

the evidence was strong enough to justify making the claim that B would really be better than A in the long run. Might not the difference have arisen from pure chance? Maybe if the experiment were repeated the results would be reversed, with method A this time giving the higher average yield. To consider the issues raised by these questions and to pose the problem in quantitative terms, it is necessary to develop some elementary statistical theory.

2.2. THEORY: PROBABILITY DISTRIBUTIONS, PARAMETERS, AND STATISTICS

Experimental Error

When an operation or experiment is repeated under what are, as nearly as possible, the same conditions, the observed results are never quite identical. The fluctuation that occurs from one repetition to another is called noise, experimental variation, *experimental error*, or merely *error*. In a statistical context the word *error* is used in a technical and emotionally neutral sense. It refers to variation that is often unavoidable. It is not associated with blame.

Many sources contribute to experimental error in addition to errors of measurement, analysis, and sampling. For example, variables such as ambient temperature, skill or alertness of personnel, age and purity of reagents, and efficiency or condition of equipment can all contribute. Moreover, experimental error must be distinguished from careless mistakes such as misplacing a decimal point when recording an observation or using the wrong chemical reagent when doing an experiment.

It has been a considerable handicap to many experimenters that their formal scientific training has left them unequipped to deal with the common situation in which experimental error cannot be safely ignored. Not only is awareness of the possible effects of experimental error essential in the analysis of data, but also its influence is a paramount consideration in *planning* the generation of data, that is, in the *design* of experiments. Therefore, to have a sound base on which to build practical techniques for the design and analysis of experiments, some elementary understanding of experimental error and of associated probability theory is essential.

Experimental Run

We shall say that an experimental *run* has been performed when an apparatus has been set up and allowed to function under a specific set of experimental conditions. For example, in a chemical experiment a run might be made by bringing together in a reactor specific amounts of the chemical reactants, adjusting temperature and pressure to the desired levels, and allowing the reaction to proceed for a particular time. In a psychological experiment a run might consist of subjecting a human subject to controlled stress.

Experimental Data or Results

An experimental result or datum is usually a numerical measurement that describes the outcome of the experimental run. For example, in the chemical experiment the datum of interest might be the yield of product. This would typically be defined as the amount of desired product formed, expressed as a percentage of the maximum theoretically attainable. Thus 10 successive runs made at what were intended to be identical conditions might give the following *data* in the form of percentage yields:

$$66.7 \quad 64.3 \quad 67.1 \quad 66.1 \quad 65.5 \quad 69.1 \quad 67.2 \quad 68.1 \quad 65.7 \quad 66.4$$

Equally in the psychological experiment these data could be the times taken by 10 stressed subjects to perform a specific task.

The Dot Diagram

A dot diagram showing the scatter of these values is presented in Figure 2.2. The dot diagram is a valuable device for displaying the distribution of a small body of data (say up to 20 observations). In particular it shows:

1. The general *location* of the observations (in this example we can see that the yields are clustered near the value 67 rather than, say, 85 or 35).
2. The *spread* of the observations (in the example they extend over about 5 units).

The Frequency Distribution

When a larger number of results is available, the dots become hard to distinguish from each other, and we are better able to appreciate the data by constructing a *frequency distribution*, also called a *frequency diagram* or

FIGURE 2.2.   Dot diagram for a sample of 10 yield observations.

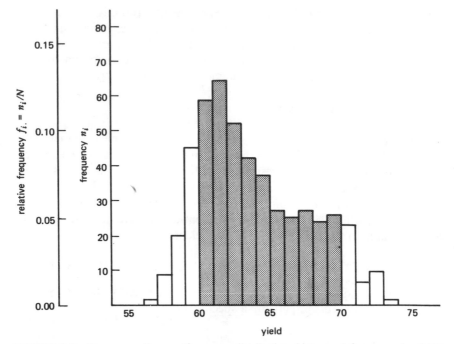

FIGURE 2.3. Frequency diagram (frequency distribution, histogram) for a sample of 500 observations.

histogram. This is accomplished by dividing the horizontal axis into intervals of appropriate size and constructing a rectangle over the ith interval with *area* proportional to n_i, the number (frequency) of observations in that interval. Figure 2.3 shows a frequency distribution for $N = 500$ observations of yield from a production process. Each observation was recorded to one decimal place. Since the smallest of the 500 observations lies between 56 and 57 and the largest between 73 and 74, it is convenient to classify the observations into 18 intervals, each covering a range of one unit. Thus all observations greater than or equal to 56.0 but less than 57.0 are plotted in the first interval, 56 to 57. There are two such observations; hence $n_1 = 2$. Since in this example the intervals are all of length one, the frequency of observations n_i for the ith interval, $i = 1, 2, \ldots, 18$, is directly proportional to the ordinate on the vertical axis.

Most often, as in this example, frequency diagrams have intervals of equal length. Notice, however, that frequency diagrams can be constructed for data in which for some reason the grouping intervals are of different lengths. The important thing to remember is that the *area* of the rectangle constructed on each interval must be proportional to the frequency of observations within that interval.

Figure 2.3 gives a vivid impression of the 500 observations of yield. In particular, it shows their locations and spread. Other characteristics are also brought to our attention. We notice, for instance, that about $\frac{4}{5}$ of the observations lie between 60 and 70. This fraction, more precisely $\frac{382}{500}$, is represented by the shaded area under the frequency diagram between the values 60 and 70.

Exercise 2.1.   Construct a dot diagram for these mileage data, which are given in units of miles per gallon for five test cars: 17.8, 14.3, 15.8, 18.0, 20.2.

Exercise 2.2.   Construct a histogram for these air pollution data, which are given in units of parts per 100 million of ozone: 6.5, 2.1, 4.4, 4.7, 5.3, 2.6, 4.7, 3.0, 4.9, 4.7, 8.6, 5.0, 4.9, 4.0, 3.4, 5.6, 4.7, 2.7, 2.4, 2.7, 2.2, 5.2, 5.3, 4.7, 6.8, 4.1, 5.3, 7.6, 2.4, 2.1, 4.6, 4.3, 3.0, 4.1, 6.1, 4.2.

Hypothetical Population of Results Represented by a Distribution

The total aggregate of observations that conceptually *might* occur as the result of performing a particular operation in a particular way is referred to as the *population* of observations. It is sometimes convenient to think of this population as infinite, but in this book we shall often think of it as finite and having size N, where N is large. The (usually few) observations that have actually occurred are thought of as some kind of *sample* from this population.

For such a large population, bumps in the frequency diagram due to sampling variation would tend to disappear, and we might obtain a histogram having the regular appearance of that in Figure 2.4. (Until further notice the reader should concentrate on this histogram, ignoring for the moment the

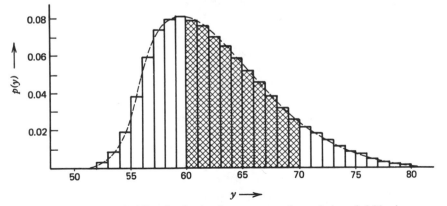

FIGURE 2.4.   Probability distribution for a conceptual population of yield values.

dashed smooth curve.) If we make the area of the rectangle erected on the ith interval of this histogram equal to the *relative frequency* n_i/N of values occurring in that interval, this is equivalent to choosing the vertical scale so that the area under the whole histogram is equal to unity.

Randomness and Probability

A *random* drawing is one where the conditions are such that each member of the population has an equal chance of being chosen. Suppose that we make a random drawing from the population of yields and obtain a value y, which we record as belonging only to a certain group interval. Then:

1. The probability $\Pr(y < y_0)$ that y is less than some value y_0 will be equal to the area under the histogram to the left of y_0 (for illustration, with y_0 equal to 60 in the diagram, $\Pr(y < 60) = 0.361$).
2. The probability $\Pr(y > y_0)$ that y is greater than y_0 will be equal to the area under the histogram to the right of y_0.
3. The probability $\Pr(y_0 < y < y_1)$ that y is greater than y_0, but less than y_1, will be equal to the area under the histogram between y_0 and y_1. The shaded area in Figure 2.4, for example, is equal to 0.545, so that $\Pr(60 < y < 70) = 0.545$.

To the accuracy of the grouping interval, this diagram of relative frequency for the whole population tells us everything we can know about the probability of a randomly chosen member of the population falling within any given range. It is therefore called a *probability distribution*. The vertical ordinate is denoted by $p(y)$ and is called the *probability density*.

We need to discuss the concept of probability density further. For this example the grouping interval happens to be one unit of yield, but more generally suppose that it is h units of yield. Suppose furthermore that for a particular interval the height of the constructed rectangle is $p(y)$ and its area is P. (Recall that this area $P = n/N$ is the probability of the interval containing a randomly chosen y.) Therefore $P = p(y) \times h$ and $p(y) = P/h$. The probability density is thus obtained by dividing the probability associated with a given interval by the length of that interval. Notice that it is always an *area* under a probability distribution that represents probability. The ordinate $p(y)$ of the distribution is not itself a probability and becomes one only when it is multiplied by an appropriate interval.

Probability density has the same status as physical density. Even if we know the density of a metal, we do not know whether a given piece of the metal will be heavy or light. To find this out we must multiply density by volume to obtain its mass. Thus probability = probability density × interval size, just as mass = density × volume.

Representing a Probability Distribution by a Continuous Curve

Since the population in question is supposed to be very large, we can construct a histogram with a very fine grouping interval, and conceptually, at least, we can imagine this grouping interval to be made smaller without limit. Although we can record only to a given accuracy, the physical quantity itself exists, in principle, to any degree of precision we can imagine. Now, if we imagine the interval h taken very small, the probability P associated with any interval becomes proportionately small also. But no matter how far we carry this process, the ratio $p(y) = P/h$ can still remain finite. In the limit, as the interval becomes infinitesimally short, we can conceive of the population being represented by a *continuous* probability distribution like the dashed line in Figure 2.4.

When a mathematical *function* is used to represent this continuous probability distribution, it is sometimes called the (*probability*) *density function*. (One such theoretical function is the normal distribution, which is discussed later in this chapter.) Just as for the discontinuous distribution mentioned earlier, the total area under the continuous probability distribution is unity, and the area between any two values y_0 and y_1 is equal to $\Pr(y_0 < y < y_1)$.

A question sometimes asked is: Given that the population is represented by a *continuous* curve, what is the probability of getting a particular value of, say, $y = 66.3$? If the question refers to the probability of the value y being *exactly* 66.30000000. . . corresponding to a *point* on the horizontal scale in Figure 2.4, the answer must be zero, because $P = p(y) \times h$, and h is zero. However, what is usually meant is: Given that observations are made, say, to the nearest 0.1, what is the probability of getting the value $y = 66.3$? Obviously the answer to this question is not zero. If we suppose that all values between $y = 66.25$ and $y = 66.35$ are recorded as $y = 66.3$, then we require $\Pr(66.25 < y < 66.35)$. The required probability is given by the area under the curve between these two limits and is adequately approximated by $p(66.3) \times 0.1$.

Sample Average and Population Mean

One important feature of a sample is its *average* value, denoted by $\bar{y}$ (read as "y bar"). For the sample of 10 observations plotted in Figure 2.2

$$\bar{y} = \frac{66.7 + 64.3 + \cdots + 66.4}{10} = 66.62 \qquad (2.1)$$

In general we can write, for a sample of n observations,

$$\bar{y} = \frac{y_1 + y_2 + \cdots + y_n}{n} = \frac{\sum y}{n} \qquad (2.2)$$

Since $\bar{y}$ tells us where the scatter of points is centered, it is called a *measure of location* for the sample.

The symbol $\sum$ is the *summation sign*. Thus $\sum y$ means, "Add up all the y's." In some instances where we want to indicate that a *particular* group of y's is to be added up, we have to use a somewhat more elaborate notation. Suppose that $y_1, y_2, y_3, \ldots$ refer to the first, second, third observations and so on, so that y_j means the jth observation. Then, for example, the sum of the observations beginning at the third, y_3, and ending with the eighth, y_8, is written as $\sum_{j=3}^{8} y_j$ and means $y_3 + y_4 + y_5 + y_6 + y_7 + y_8$.

If we imagine a hypothetical population as containing some very large number N of observations, we can denote the corresponding measure of location of the population by the Greek letter η (eta), so that

$$\eta = \frac{\sum y}{N} \tag{2.3}$$

Statistics and Parameters

To distinguish the sample and population quantities we call η the population *mean*, and $\bar{y}$ the sample *average*. In general, a *parameter* like the mean η is a quantity directly associated with the *population*, and a *statistic* like the average $\bar{y}$ is a quantity calculated from a set of data often thought of as some kind of *sample* from a population. Parameters are usually designated by Greek letters; statistics, by Roman letters. In summary we have:

population	sample
a very large set of N observations from which the sample can be imagined to come	a small group of observations actually available

parameter	statistic
population mean: $\eta = \sum y/N$	sample average: $\bar{y} = \sum y/n$

The mean of the population is also called the *expected value* of y, or the *mathematical expectation* of y, and is then denoted by $E(y)$. Thus $\eta = E(y)$.

The Hypothesis of Random Sampling

Since the hypothetical population conceptually contains all values that can occur from a given operation, any set of observations we may collect is *some* kind of sample from the population. An important statistical idea is that in certain circumstances a set of observations may be regarded as a *random sample* from the population, that is, one in which each member of the sample is a random drawing from the whole population. It will be recalled that a *random* drawing is one where each member of the population has an equal chance of being chosen.

Unfortunately this *hypothesis* of random sampling will often *not* apply to actual data. For example consider daily weather data. Positive deviations from the mean are usually followed by other positive deviations (warm days, for instance, tend to follow one another). The generation of such data is not directly representable by *random* drawings from any population. (We see in Chapter 18 how it can be *indirectly* represented by random drawings.) In both the analysis and the design of scientific experiments much hangs on the applicability of this hypothesis of random sampling. The importance of randomness can be appreciated, for example, by considering public opinion polls. A poll conducted on election night at the headquarters of one political party might give an entirely false picture of the standing of its candidate in the voting population.

It is curious that the hypothesis of random sampling is treated in much statistical writing as if it were a law of nature. In fact, for real data it is a property that can never be relied on, although special precautions in the design of an experiment can make the assumption relevant. For this reason we deliberately introduced at the beginning of this chapter industrial data to compare two manufacturing methods for which the random sampling hypothesis is *not* appropriate. In the next section we see what can be done in analyzing the data *without* adopting the random sampling hypothesis. This will lead to a better understanding of subsequent developments.

2.3. THE INDUSTRIAL EXPERIMENT: EXTERNAL REFERENCE DISTRIBUTION

We return to an analysis of the data from the plant trial given in Table 2.1. Recall that, as a start in analyzing the data, the investigator plotted the observations (Figure 2.1) and calculated the averages $\bar{y}_A = 84.24$ and $\bar{y}_B = 85.54$, and their difference $\bar{y}_B - \bar{y}_A = 1.30$. In light of the theory described above, there will be a conceptual population of observations on yield obtained from method B and a corresponding population obtained from method A. The investigator wants to know whether the population mean η_B is greater than η_A. If this is the case, it will be better in the long run to switch to the modified process.

TABLE 2.2. Production record of 210 consecutive batch yield values ("obs." = observation)*

obs.	average 10 obs.	obs.	average 10 obs.	obs.	average 10 obs.	obs.	average 10 obs.	obs.	average 10 obs.	obs.	average 10 obs.
85.5		84.5	84.42	80.5	84.53	79.5	83.72	84.8	84.36	81.1	83.68
81.7		82.4	84.70	86.1	84.09	86.7	83.89	86.6	84.54	85.6	83.91
80.6		86.7	84.79	82.6	84.28	80.5	83.90	83.5	84.58	86.6	83.53
84.7		83.0	85.30	85.4	84.51	91.7	84.50	78.1	84.33	80.0	83.43
88.2		81.8	84.51	84.7	84.33	81.6	83.99	88.8	84.47	86.6	84.06
84.9		89.3	84.90	82.8	83.61	83.9	83.71	81.9	83.98	83.3	84.41
81.8		79.3	84.20	81.9	84.05	85.6	84.04	83.3	83.96	83.1	83.82
84.9		82.7	84.40	83.6	83.94	84.8	83.88	80.0	83.34	82.3	83.68
85.2		88.0	84.82	86.8	84.16	78.4	83.47	87.2	83.65	86.7	84.26
81.9	83.94	79.6	83.73	84.0	83.84	89.9	84.26	83.3	83.75	80.2	83.55
89.4	84.33	87.8	84.06	84.2	84.21	85.0	84.81	86.6	83.93		
79.0	84.06	83.6	84.18	82.8	83.88	86.2	84.76	79.5	83.22		
81.4	84.14	79.5	83.46	83.0	83.92	83.0	85.01	84.1	83.28		
84.8	84.15	83.3	83.49	82.0	83.58	85.4	84.38	82.2	83.69		
85.9	83.92	88.4	84.15	84.7	83.58	84.4	84.66	90.8	83.89		
88.0	84.23	86.6	83.88	84.4	83.74	84.5	84.72	86.5	84.35		
80.3	84.08	84.6	84.41	88.9	84.44	86.2	84.78	79.7	83.99		
82.6	83.85	79.7	84.11	82.4	84.32	85.6	84.86	81.0	84.09		
83.5	83.68	86.0	83.91	83.0	83.94	83.2	85.34	87.2	84.09		
80.2	83.51	84.2	84.37	85.0	84.04	85.7	84.92	81.6	83.92		

32

85.2	83.09	83.0	83.89	82.2	83.84	83.5	84.77	84.4	83.70
87.2	83.91	84.8	84.01	81.6	83.72	80.1	84.16	84.4	84.19
83.5	84.12	83.6	84.42	86.2	84.04	82.2	84.08	82.2	84.00
84.3	84.07	81.8	84.27	85.4	84.38	88.6	84.40	88.9	84.67
82.9	83.77	85.9	84.02	82.1	84.12	82.0	84.16	80.9	83.68
84.7	83.44	88.2	84.18	81.4	83.82	85.0	84.21	85.1	83.54
82.9	83.70	83.5	84.07	85.0	83.43	85.2	84.11	87.1	84.28
81.5	83.59	87.2	84.82	85.8	83.77	85.3	84.08	84.0	84.58
83.4	83.58	83.7	84.59	84.2	83.89	84.3	84.19	76.5	83.51
87.7	84.33	87.3	84.90	83.5	83.74	82.3	83.85	82.7	83.62
81.8	83.99	83.0	84.90	86.5	84.17	89.7	84.47	85.1	83.69
79.6	83.23	90.5	85.47	85.0	84.51	84.8	84.94	83.3	83.58
85.8	83.46	80.7	85.18	80.4	83.93	83.1	85.03	90.4	84.40
77.9	82.82	83.1	85.31	85.7	83.96	80.6	84.23	81.0	83.61
89.7	83.50	86.5	85.37	86.7	84.42	87.4	84.77	80.3	83.55
85.4	83.57	90.0	85.55	86.7	84.95	86.8	84.95	79.8	83.02
86.3	83.91	77.5	84.95	82.3	84.68	83.5	84.78	89.0	83.21
80.7	83.83	84.7	84.70	86.4	84.74	86.2	84.87	83.7	83.18
83.8	83.87	84.6	84.79	82.5	84.57	84.1	84.85	80.9	83.62
90.5	84.15	87.2	84.78	82.0	84.42	82.3	84.85	87.3	84.08

* Read downwards across both pages.

33

The Null Hypothesis

To answer the investigator's question, we act as the devil's advocate and introduce the *null hypothesis*. This is the hypothesis that the modification has made no difference and, in particular, that $\eta_B = \eta_A$. If the null hypothesis is discredited, we say that a *statistically significant* difference has occurred. To proceed we need additional data or additional assumptions.

A Significance Test Using an External Reference Distribution

Luckily in this instance additional data were available in the form of past plant records. Table 2.2 shows yield data for 210 consecutive batches produced in this same plant immediately before the trial discussed above. These observations are plotted in Figure 2.5. All these batches were made using the standard process A, no modification having been made over that period of time. Now the investigator has just performed a plant experiment with 20 consecutive batches and has obtained a difference between the averages of the second 10 (modified process) and the first 10 (standard process) of $85.54 - 84.24 = 1.30$. The question is whether these experimental data are consistent with the null hypothesis that the modification is without effect.

To answer this question it is relevant for the investigator to consider the 210 past data values, all generated using the unmodified process, and to ask, How often have differences occurred between averages of *successive groups* of 10 observations that were at least as great as 1.30? If the answer is "frequently," the investigator will conclude that the observed difference can readily be explained by the ordinary chance variations in the process. But if the answer is "rarely," a more tenable explanation is that the modification has indeed produced a real increase in the process mean, that is,

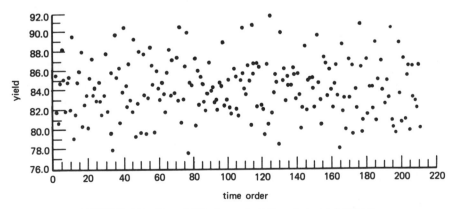

FIGURE 2.5. Plot of 210 observations from industrial process.

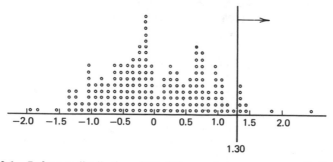

FIGURE 2.6. Reference distribution of 191 differences between averages of adjacent sets of 10 observations.

$\eta_B > \eta_A$. He will then say that a *statistically significant* difference has been established. With the process records for the 210 previous batches given in Table 2.2 are also listed the associated 201 averages of 10 consecutive observations each. For example, the first average listed (83.94) is determined from observations 1 through 10, the second average (84.33) from observations 2 through 11, and so on.

The 191 differences between averages of two adjacent groups of 10 successive batches,* shown in Table 2.3, are plotted in Figure 2.6. They provide a *relevant reference* set with which the observed difference of 1.30, found when the modification was used, may be compared.

It is seen that rather rarely, specifically in only nine instances, do the differences in the reference set exceed +1.30, the difference observed in the trial. The null hypothesis, which would make the observed difference a member of the reference set, is thus somewhat discredited. In statistical parlance the investigator could say that, in relation to this reference set, the observed difference was statistically significant *at the* $\frac{9}{191} = 0.047$ *level of probability.*

This external reference distribution therefore shows that the null hypothesis that the observed difference is due simply to chance is not very plausible. Less than 5 times in 100 would a difference this large be found in the relevant reference set ($0.047 = \frac{47}{1000} < \frac{5}{100}$). Thus it appears likely that a genuine difference does exist and, more specifically, that the modified process is better than the standard process.

---

* These 191 differences were computed as follows. The first, -0.43, was obtained by subtracting the average, 83.94, of batches 1 to 10 *from* the average, 83.51, of batches 11 to 20. This calculation was repeated, using the averages calculated from batches 2 to 11 and 12 to 21, and so on through the table of data.

TABLE 2.3. Reference set of differences between averages of two adjacent sets of 10 successive batches (differences that exceed +1.30 are circled)

	−0.36	−0.32	1.09	−0.43
	−0.52	−0.21	0.87	−1.32
	−1.33	−0.36	1.11	−1.30
	−1.81	−0.93	−0.12	−0.64
	−0.36	−0.75	0.67	−0.58
	−1.02	0.13	1.01	0.37
	0.21	0.39	0.74	0.03
	−0.29	0.38	0.98	0.75
	−0.91	−0.22	(1.87)	0.44
−0.43	0.64	0.20	0.66	0.17
−1.24	−0.17	−0.37	−0.04	−0.23
−0.15	−0.17	−0.16	−0.60	0.97
−0.02	0.96	0.12	−0.93	0.72
−0.08	0.78	0.80	0.02	0.98
−0.15	−0.13	0.54	−0.50	−0.21
−0.79	0.30	0.08	−0.51	−0.81
−0.38	−0.34	−1.01	−0.67	0.29
−0.26	0.71	−0.55	−0.78	0.49
−0.10	0.68	−0.05	−1.15	−0.58
0.82	0.53	−0.30	−1.07	−0.30
0.90	1.01	0.33	−0.30	−0.01
−0.68	(1.46)	0.79	0.78	−0.61
−0.66	0.76	−0.11	0.95	0.40
−1.25	1.04	−0.42	−0.17	−1.06
−0.27	(1.35)	0.30	0.61	−0.13
0.13	(1.37)	1.13	0.74	−0.52
0.21	0.88	1.25	0.67	−1.07
0.24	−0.12	0.97	0.79	−1.40
0.29	0.20	0.68	0.66	0.11
−0.18	−0.12	0.68	1.00	0.46
0.43	−0.37	−0.45	−0.11	−0.01
(1.47)	−1.38	−0.62	−0.40	0.33
(1.33)	−0.90	−0.03	−0.45	−0.87
(2.48)	−0.80	0.54	0.10	−0.18
1.01	−1.04	−0.43	−0.30	0.51
(1.33)	−1.94	−1.24	−0.97	(1.39)
0.29	−0.90	−0.64	−0.82	0.61
0.57	−0.76	−0.86	−1.53	0.50
0.95	−0.63	−1.10	−1.20	0.64
−0.42	−0.94	−0.16	−1.10	−0.53

The Essential Nature of a Significance Test

Our purpose in introducing this example is to make clear the *essential* nature of a significance test.

1. We compute from the data some relevant *criterion* (also called a statistic) appropriate to test a particular *hypothesis* of interest against some *alternative hypothesis*. In this example the difference $\bar{y}_B - \bar{y}_A$ was a criterion appropriate for testing the null hypothesis that the true mean difference $\eta_B - \eta_A$ was zero against the alternative hypothesis that it was greater than zero.
2. We refer the criterion to an *appropriate reference distribution*, which shows how the criterion would be distributed if the tested hypothesis were true.
3. By so doing we make it possible to calculate the probability that a discrepancy at least as large as the one observed would occur by chance if the tested hypothesis were true.
4. This probability is called the *significance level*; if it is sufficiently small, the tested hypothesis is discredited, and we assert that a statistically significant difference has been obtained.

Statistical Independence Not Assumed

Consider the sample of 10 successive observations $y_1, y_2, \ldots, y_{10}$ for method A, listed in Table 2.1. Notice that it has *not* been necessary to assume that the taking of this sample is realistically simulated by 10 *random* choices from some hypothetical population. Such a hypothesis of random sampling would mean that each member of the population had an equal chance of being chosen at any given stage. This would imply, in particular, that the deviation from its mean of, say, the eighth observation y_8 was in no way related to the corresponding deviations of y_7 and y_6, and in fact that one *sequence* of the 10 such deviations was just as likely to occur as any other. As we shall see later, many standard statistical procedures require this hypothesis of random sampling for their validity. Unfortunately these standard techniques are often incorrectly applied to industrial data like those presently being studied and to sequences of meteorological, economic, physiological, psychological, and other data that unfold in some specific time order and are equally unlikely to be representable by random sampling.

For such serial data the probability of getting a high value for any y_8 may very well be greater if y_7 and y_6 are high. In general, when the probability distribution of one observation is affected by the level of another, the observations are said to be *statistically dependent*. By contrast, the hypothesis that the data may be simulated by random sampling from some population implies *statistical independence*.

The test we have employed for the industrial data *does not* involve a hypothesis of random sampling, statistical independence, or equal likelihood for all orderings.

Advantages and Disadvantages of the External Reference Set

From a practical point of view the reference set employed above has appeal because its use supposes only that whatever mechanism gave rise to the (possibly statistically dependent) observations in the past was still operating during the plant trial. To use it, however, we must have a fairly extensive set of past data that can reasonably be regarded as typifying standard process operation. To see how we can tackle the problem of assessing the differences between two averages in other ways we must develop the theory a little further.

Exercise 2.3. Suppose that the industrial situation was the same as above except that only 10 experimental observations were obtained, the first 5 with method A (89.7, 81.4, 84.5, 84.8, 87.3) and the next 5 with B (80.7, 86.1, 82.7, 84.7, 85.5). Using the data in Table 2.2, construct an appropriate reference distribution, and determine the significance level associated with $\bar{y}_B - \bar{y}_A$ for this test.   *Answer*: 0.082.

Exercise 2.4. Consider a trial designed to compare cell growth rate, using a standard way of preparing a biological specimen (method A) and new way (method B). Suppose that four trials done on successive days in the order A, A, B, B gave the results 23, 28, 37, 33, and also that, immediately before this trial, a series of preparations on successive days with method A gave the results 25, 23, 27, 31, 32, 35, 40, 38, 38, 33, 27, 21, 19, 24, 17, 15, 14, 19, 23, 22. Using the external reference distribution, compute the significance level for the null hypothesis $\eta_B = \eta_A$ when the alternative is $\eta_B > \eta_A$. Is there evidence that B gives higher rates than A?

Answer: The observed significance level $= \frac{0}{17} = 0.00$, and evidence is that B gives higher growth rates.

2.4. THEORY: NORMAL AND t DISTRIBUTIONS

We referred earlier to the concept of a population of observations. This population was described as a very large aggregate of observations y that conceptually might occur as the result of repeatedly performing a particular operation, and we showed how it might be described by a probability distribution $p(y)$. The number of observations in the hypothetical population is often regarded as infinite, but we suppose it to be finite and equal to N, where N is some very large number.

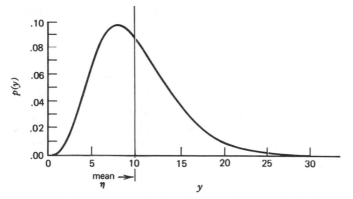

FIGURE 2.7. The mean $\eta = E(y)$ as the center of gravity of a distribution.

Descriptive Measures of the Population: Mean, Variance, and Standard Deviation

We have seen that an important characteristic of the population is its *mean value* $\eta = \sum y/N$. This mean value η is also called the *mathematical expectation* of y and is denoted by $E(y)$. Thus $E(y) = \eta$. It is the first moment or center of gravity of the distribution of y, and it defines the line of balance, as illustrated in Figure 2.7. It thus supplies a measure of the *location* of the distribution. It tells us, for instance, whether the distribution is balanced at 0.6, at 60, or at 60,000. Knowledge of location is useful but incomplete information about a population. To tell a visitor from another world that adult males in the United States have a mean height of approximately 70 inches would be helpful, but he could still believe that some members of this population were 1 inch and others 1000 inches high.

Some measure of the *spread* of the distribution would help him get a better perspective. There are many different measures of spread that might be employed. The most useful for our purpose is the *variance* of the population, denoted by σ^2 (sigma squared).

A measure of how far any particular observation y is from the mean η is the deviation $y - \eta$. The variance is the mean value of the square of such deviations taken over the whole population,

$$\sigma^2 = E(y - \eta)^2 = \frac{\sum(y - \eta)^2}{N} \tag{2.4}$$

Just as we sometimes use a special symbol $E(y)$ to denote the mean value (or expectation), so the special symbol $V(y)$ is sometimes used to denote the variance.

A measure of spread (which has the same units as the original observations) is σ, the positive square root of the variance. This is called the *standard deviation*,

$$\sigma = \sqrt{V(y)} = \sqrt{E(y - \eta)^2} = \sqrt{\sum (y - \eta)^2 / N} \qquad (2.5)$$

Occasionally when the standard deviations of a number of different quantities are under discussion, a subscript will accompany σ. Thus the symbol σ_y (sigma sub y) leaves no doubt that we are talking about the standard deviation of the population of observations y and not, for example, about some other population of observations z.

Descriptive Measures of the Sample: Average, Variance, and Standard Deviation

The data usually available to the investigator are merely a small *sample* from the hypothetical larger set represented by the population. The sample average $\bar{y} = \sum y/n$ supplies a measure of location of the sample. Similarly the sample variance supplies a measure of spread of the sample. The sample variance is usually calculated as

$$s^2 = \frac{\sum (y - \bar{y})^2}{n - 1} \qquad (2.6)$$

The positive square root of the sample variance gives the sample standard deviation,

$$s = \sqrt{\sum (y - \bar{y})^2 / (n - 1)} \qquad (2.7)$$

which has the same units as the observations.

Once again, as was true for the mean η and the average $\bar{y}$, a Greek letter is used for the parameter associated with the population, and a Roman letter for the corresponding statistic associated with the sample; σ^2 and σ denote the population variance and the standard deviation, and s^2 and s denote the sample variance and the standard deviation. A summary is given in Table 2.4.

Degrees of Freedom

The deviations of n observations from their sample average *must* sum to zero. This requirement, that $\sum (y - \bar{y}) = 0$, constitutes a *linear constraint* on the deviations or *residuals* $y_1 - \bar{y}, y_2 - \bar{y}, \ldots, y_n - \bar{y}$ used in calculating $s^2 = \sum (y - \bar{y})^2 / (n - 1)$. It implies that any $n - 1$ of them completely determine the other. The n residuals $y - \bar{y}$ [and hence their sum of squares $\sum (y - \bar{y})^2$ and the sample variance $\sum (y - \bar{y})^2 / (n - 1)$] are therefore said

TABLE 2.4. Population and sample quantities

	population	sample
definition	a hypothetical set of N observations from which the sample of observations actually obtained can be imagined to come (typically N is very large)	a set of n observations actually obtained (typically n is relatively small)
measure of location	population mean $\eta = \sum y/N$	sample average $\bar{y} = \sum y/n$
measure of spread	population variance $$\sigma^2 = \sum (y - \eta)^2/N$$ population standard deviation $$\sigma = \sqrt{\sum (y - \eta)^2/N}$$	sample variance $$s^2 = \sum (y - \bar{y})^2/(n - 1)$$ sample standard deviation $$s = \sqrt{\sum (y - \bar{y})^2/(n - 1)}$$

to have $n - 1$ *degrees of freedom*. In this book the number of degrees of freedom is denoted by the Greek letter v (nu). In the example worked below, in which the yield data for process A are used, the sample variance and sample standard deviation have $v = n - 1 = 10 - 1 = 9$ degrees of freedom. The loss of one degree of freedom is associated with the need to replace the unknown population mean η by $\bar{y}$, the sample average derived from the data.

In later applications we shall encounter examples where, because of the need to calculate several sample quantities to replace unknown population parameters, several constraints are induced on the residuals. When there are p independent linear constraints, we shall say that the n residuals, their sum of squares, and the resulting sample variance all have $v = n - p$ degrees of freedom. It will generally be true that the sample variance will be obtained by dividing a sum of squares of residuals by the appropriate number of degrees of freedom.

The yield data for the standard process A, taken from Table 2.1, are shown in Figure 2.8, together with the sample average $\bar{y} = 84.24$ and the sample standard deviation (Equation 2.7),

$$s = \left[\frac{(89.7 - 84.24)^2 + (81.4 - 84.24)^2 + \cdots + (84.5 - 84.24)^2}{9} \right]^{1/2} \quad (2.8)$$

$$= 2.90$$

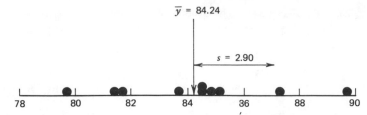

FIGURE 2.8. Yield data for standard process: dot diagram showing sample average and sample standard deviation.

Alternative ways of calculating the sample standard deviation s are given in Appendix 2A.

Exercise 2.5. Calculate the average and standard deviation for the following data on hog–corn ratios (units are price of hogs per 100 pounds/price of corn per bushel): 16.8, 13.3, 11.8, 15.0, 13.2. Confirm that the sum of the residuals $y - \bar{y}$ is zero. Illustrate how you would use this fact to calculate the fifth residual knowing the values of only four of them.   *Answer:* $\bar{y} = 14.02$, $s = 1.924$, with $v = 4$ degrees of freedom.

Exercise 2.6. A psychologist measured (in seconds) the following times required for 10 experimental rats to complete a maze: 24, 37, 38, 43, 33, 35, 48, 29, 30, 38. Find the average, sample variance, and sample standard deviation for these data.
        Answer: $\bar{y} = 35.5$, $s^2 = 48.72$, $s = 6.98$, with $v = 9$ degrees of freedom.

Exercise 2.7. The following observations on the lift of an airfoil (in kilograms) were obtained on successive trials in a wind tunnel: 9072, 9148, 9103, 9084, 9077, 9111, 9096. Calculate the average, sample variance, and sample standard deviation of these observations.
        Answer: $\bar{y} = 9098.71$, $s^2 = 667.50$, $s = 25.84$, with $v = 6$ degrees of freedom.

Exercise 2.8. Given the following number of liters of N normal reagents required to titrate Q grams of a substance: 0.00173, 0.00158, 0.00164, 0.00169, 0.00157, 0.00180, calculate the average, sample variance, and sample standard deviations of these observations.
Answer: $\bar{y} = 0.00167$, $s^2 = 0.798 \times 10^{-8}$, $s = 0.893 \times 10^{-4}$, with $v = 5$ degrees of freedom.

Sample Variance if Population Mean Is Known

If the population mean η is known, the sample variance is calculated as the average of the squared deviations from this known mean,

$$\hat{s}^2 = \frac{\sum (y - \eta)^2}{n} \tag{2.9}$$

with the divisor equal to n, not $n - 1$. This statistic is designated by a dot to distinguish it from s^2. The sum of squares $\sum (y - \eta)^2$ and the associated quantity $\dot{s}^2$ are each said to have n degrees of freedom because all n quantities $y - \eta$ are free to vary; there is no constraint as is the case for s^2. In particular, knowing $n - 1$ of the deviations does not determine the nth.

The Normal Distribution

Repeated observations that differ because of experimental error often vary about some central value in a roughly symmetric distribution in which small deviations occur much more frequently than large ones.

A continuous distribution, which is valuable for representing this situation and occupies an important position in the theory of statistics, is the *Gaussian* or *normal distribution*. The appearance of this distribution and its mathematical formula are shown in Figure 2.9. It is a symmetric curve with its highest ordinate at its center, tailing off to zero in both directions in a way intuitively expected of experimental error. [It has the property that the logarithm of its probability density is a quadratic function of the standardized error $(y - \eta)/\sigma$.]

Reasons for the Importance of the Normal Distribution

Two facts explain the importance of the normal distribution:

1. The central limit effect, which produces a tendency for real error distributions to be "normal-like."
2. The robustness or insensitivity of many commonly used statistical procedures to deviations from theoretical normality.

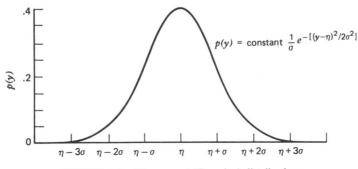

$$p(y) = \text{constant } \frac{1}{\sigma} e^{-[(y-\eta)^2/2\sigma^2]}$$

FIGURE 2.9. The normal (Gaussian) distribution.

The Central Limit Effect Producing Normal-like Error Distributions

Distributions of errors often tend to be approximately normal. That such a tendency is to be expected follows from a theorem in mathematical statistics called the *central limit theorem*.

It is usually true that an "overall" error $y - \eta = \varepsilon$ is an aggregate of a number of component errors (ε is the Greek letter epsilon). Typically, for example, a measurement of yield obtained for a particular experimental run will be subject to analytical error, chemical sampling error, and process error. Process error can be produced by errors in the settings of the experimental conditions, errors due to variation in raw materials, and so on.

Thus the overall error ε will be some function of *many* component errors $\varepsilon_1, \varepsilon_2, \ldots, \varepsilon_n$. If each individual percentage error is fairly small, it is possible to approximate the overall error at a specific set of conditions as a *linear* function of independently distributed component errors

$$\varepsilon = a_1\varepsilon_1 + a_2\varepsilon_2 + \cdots + a_n\varepsilon_n \tag{2.10}$$

where the a's are constants.

The central limit theorem says that, under certain conditions usually met in the real world of experimentation, the distribution of such a linear function of errors will tend to normality as the number of components becomes large, almost irrespective of the individual distributions of the components. An important proviso is that several of the sources of error must make important contributions to the overall error and, in particular, that no single source of error dominates all the rest.

Illustration: Central Limit Tendency for Averages

Figure 2.10a shows the distribution for throws of a single true, six-sided die. The probabilities are all equal for scores of 1, 2, 3, 4, 5, or 6. The mean score is obviously $\eta = 3.5$. Figure 2.10b shows the distribution of the *average* score from two dice, and Figures 2.10c, d, and e show the distribution of average scores calculated from 3, 5, and 10 dice. Now suppose $\varepsilon_1, \varepsilon_2, \ldots, \varepsilon_n$ denote deviations of the scores of the individual dice from the mean value of $\eta = 3.5$. Also suppose ε is the corresponding deviation of the average, that is $\varepsilon = \sum \varepsilon_i/n$. Then ε will satisfy Equation 2.10 if we set $a_1 = a_2 = \cdots = a_n = 1/n$. We note that the original or *parent* distribution of individual observations (scores from single throws) is far from the normal shape. Nevertheless the ordinates of the distribution for averages are approximated by the ordinates of the normal distribution, even for n as small as 5. (The dice throw distributions are, however, necessarily discrete, while the approximating normal distribution is continuous.)

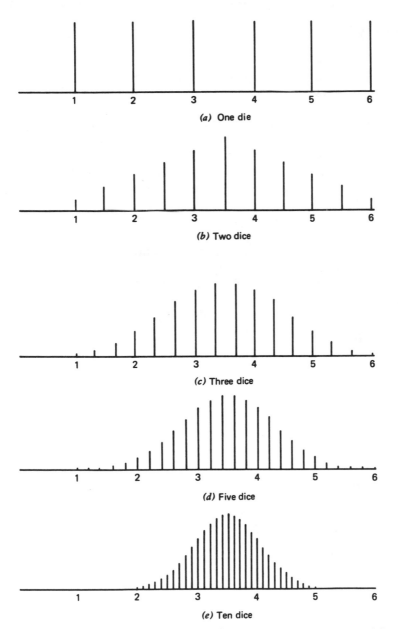

FIGURE 2.10. Distribution of average scores from throwing various numbers of dice.

This example illustrates two important facts:

1. The distribution of an error $\varepsilon = \bar{y} - \eta = (1/n)\varepsilon_1 + (1/n)\varepsilon_2 + \cdots + (1/n)\varepsilon_n$, which is a linear aggregate of a number of component errors, tends to the normal form, even though the distribution of the components is markedly nonnormal.
2. The sample average tends to be normally distributed, even though the individual observations on which it is based are not. Thus statistical methods that depend, not directly on the distribution of individual observations, but on the distribution of one or more averages of observations tends to be insensitive or *robust* to nonnormality.

Robustness of Procedures Derived on the Assumption of Normality

We shall introduce a number of statistical techniques that are derived on the assumption of normality of the original observations. In many cases approximate rather than exact or nearly exact normality is all that is required for these methods to be useful. In this regard they are said to be *robust* to *nonnormality*. Thus, unless specifically warned, the reader should not be unduly worried about exact normality.

Characterizing the Normal Distribution

Once the mean η and the variance σ^2 of a normal distribution are given, the entire distribution is characterized. The terse notation $N(\eta, \sigma^2)$ is often used to indicate a normal distribution having mean η and variance σ^2. Thus the

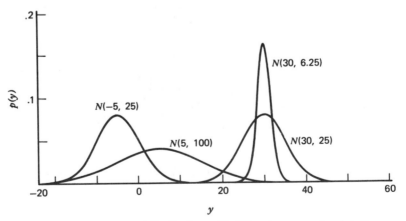

FIGURE 2.11.   Normal distributions with different means and variances.

expression $N(30, 25)$ identifies a normal distribution with a mean of $\eta = 30$ and a variance of $\sigma^2 = 25$. The following collection of normal distributions is shown in Figure 2.11: (a) $N(-5, 25)$; (b) $N(5, 100)$; (c) $N(30, 25)$; (d) $N(30, 6.25)$. These distributions are scaled so that all their areas are equal to unity.

For a normal distribution the standard deviation σ measures the distance from its mean η to the point of inflection of the curve. The point of inflection (see Figure 2.12) is the point at which the slope stops increasing and starts to decrease (or vice versa).

The following statements (see Figure 2.12) should help the reader gain a fuller appreciation of the normal distribution:

1. The probability that a positive deviation from the mean will exceed one standard deviation is 0.1587 (roughly $\frac{1}{6}$). This is the percentage of the total area under the curve taken up by the shaded "tail" area.
2. Because of symmetry, of course, this probability is exactly equal to the chance that a negative deviation from the mean will exceed one standard deviation.
3. From these two statements it is clear that the probability that a deviation *in either direction* will exceed one standard deviation is $2 \times 0.1587 = 0.3174$ (roughly $\frac{1}{3}$), and consequently the probability of such a deviation less than one standard deviation is 0.6826 (roughly $\frac{2}{3}$).
4. The chance that a positive deviation from the mean will exceed two standard deviations is 0.02275 (roughly $\frac{1}{40}$), which is represented by the heavily shaded tail area in Figure 2.12.
5. Again, this is exactly equal to the chance that a negative deviation from the mean will exceed two standard deviations.
6. From these two statements it is clear that the chance that a deviation in either direction will exceed two standard deviations is $2 \times 0.02275 = 0.0455$ (roughly $\frac{1}{20}$ or 0.05).

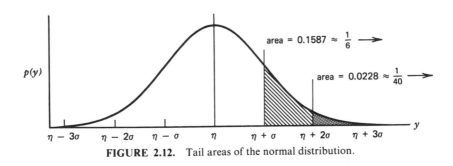

FIGURE 2.12. Tail areas of the normal distribution.

The Standard Normal Distribution

A probability statement concerning some normally distributed quantity y is often best expressed in terms of a *standardized normal deviate* or *unit normal deviate*,

$$z = \frac{y - \eta}{\sigma} \tag{2.11}$$

The distribution of z is $N(0, 1)$, that is, a normal distribution with a mean $\eta = 0$ and a variance $\sigma^2 = 1$. We can therefore rewrite the above statements as follows:

1. $\Pr(y > \eta + \sigma) = \Pr\{(y - \eta) > \sigma\} = \Pr\left\{\left(\frac{y - \eta}{\sigma}\right) > 1\right\}$

$$= \Pr(z > 1) = 0.1587.$$

2. $\Pr(z < -1)$ $= 0.1587.$

3. $\Pr(|z| > 1)$ $= 0.3174.$

4. $\Pr(z > 2)$ $= 0.0228.$

5. $\Pr(z < -2)$ $= 0.0228.$

6. $\Pr(|z| > 2)$ $= 0.0455.$

Using Tables of the Normal Distribution

In general, to determine the probability of y exceeding some value y_0 we compute the normal deviate $z_0 = (y_0 - \eta)/\sigma$ and obtain $\Pr(z > z_0) = \Pr(y > y_0)$ from Table A at the end of this book. It gives $\Pr(z > z_0)$ for various values of z_0.

For illustration suppose that the daily level of an impurity in a reactor feed is known to be approximately normally distributed with a mean of 4.0 and a standard deviation of 0.3. What is the probability that the impurity level on a *randomly chosen* day will exceed 4.4? Here $y_0 = 4.4$, $\eta = 4.0$, and $\sigma = 0.3$, so that

$$z_0 = \frac{y_0 - \eta}{\sigma} = \frac{4.4 - 4.0}{0.3} = 1.33 \tag{2.12}$$

From Table A we find that $\Pr(z > 1.33) = 0.0918$. There is thus a 9% chance that the impurity level on a randomly chosen day will exceed 4.4.

Exercise 2.9. For a random variable having a normal distribution with mean $\eta = 6$ and standard deviation $\sigma = 2$, what is the probability that an individual observation selected at random will be between 5 and 7? *Answer*: 0.38.

Exercise 2.10. For observations with a normal distribution with mean 123.4 and variance 25, what is the chance that a single randomly chosen observation will be less than 120.0? *Answer*: 0.25.

Exercise 2.11. For observations with a normal distribution $N(0.24, 16)$, what is the probability that an individual randomly selected observation will be greater than zero? *Answer*: 0.52.

Student's t Distribution

In the above example we supposed that the standard deviation σ was *known* and was equal to 0.3. In practice it is usually unknown, and we must substitute for it a value s obtained from a small sample of data. Suppose then, somewhat more realistically, that although σ is not exactly known, we have a *sample* standard deviation $s = 0.3$ calculated from seven data values. If the mean impurity level is $\eta = 4.0$, and $s = 0.3$ is an estimate of σ, what can we now say about the random occurrence of an impurity level of $y_0 = 4.4$? Since σ is unknown, we cannot now calculate $z_0 = (y_0 - \eta)/\sigma$ and refer the result to a table of the standard normal distribution. Instead, substituting s for σ, we may calculate

$$t_0 = \frac{y_0 - \eta}{s} = \frac{4.4 - 4.0}{0.3} = 1.33 \tag{2.13}$$

On the basis of certain assumptions, which are discussed shortly, the quantity $t = (y - \eta)/s$ has a known distribution. This was first deduced in 1908 by the chemist W. S. Gosset, who worked for the Guinness brewery in Dublin and, as mentioned in Section 1.5, wrote under the pseudonym "Student." This important distribution is called the t distribution or Student's distribution. The probability points of the t distribution are given in Table B1 at the back of the book. The table is entered with the number of degrees of freedom for the sample standard deviation s. Thus in the present example s has six degrees of freedom, and graphical interpolation yeilds

$$\Pr(t > 1.33) = 0.12 \tag{2.14}$$

To perform this graphical interpolation you need a piece of graph paper and a pencil. Using Table B1 for the row $v = 6$, plot the value of t of 0.718 against the probability 0.25. Similarly plot 1.440 versus 0.1 and 1.943 versus 0.05. Now draw a smooth curve through the points by eye, and then read off the appropriate value (0.12) for the approximate probability against the value

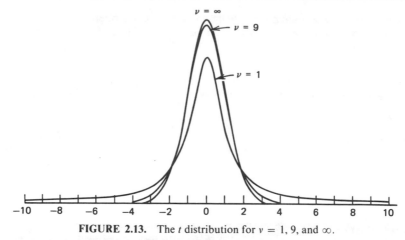

FIGURE 2.13.   The t distribution for $v = 1, 9$, and ∞.

$t = 1.33$. Appropriately, 0.12 is somewhat larger than the value 0.09 obtained on the supposition that the standard deviation was known exactly.

As might be expected, the precise form of the t distribution depends on the degree of uncertainty in s^2, which is measured by the number of degrees of freedom v on which the statistic s^2 is based. Figure 2.13 shows the t distribution for $v = 1$, $v = 9$, and $v = \infty$ degrees of freedom. When $v = \infty$, that is, when the size of the sample is infinite, there is no uncertainty in the estimate s^2, and the t distribution becomes the standard normal distribution of z. When the number of degrees of freedom is small, however, the possibility of variation in s^2 results in a greater probability of extreme deviations and hence in a "heavier" tailed distribution. The following values for $\Pr(t > 2)$ illustrate the point:

$$v = \infty \text{ (normal distribution)} \quad \Pr(t > 2) = 0.023$$
$$v = 9 \qquad\qquad\qquad\qquad\quad \Pr(t > 2) = 0.038$$
$$v = 1 \qquad\qquad\qquad\qquad\quad \Pr(t > 2) = 0.148$$

Except in the extreme tails of the distribution, the normal distribution provides a fair approximation of the t distribution when v is greater than about 15.

Assumptions Needed for the Validity of the t *Distribution*

The quantity $t = (y - \eta)/s$ has a t distribution with v degrees of freedom whose probability points are given in Table B1 if

1. y is normally distributed about η with variance σ^2,
2. s is distributed independently of y,

3. the quantity s^2, which has v degrees of freedom, is calculated from normally and independently distributed observations having variance σ^2.

We discuss these assumptions more fully in the next chapter.

Exercise 2.12. The following data represent the measured performance y, in miles per gallon, of a sample of cars all of the same make and age: 27.8, 24.3, 22.8, 26.0, 24.2.
(a)   For the variance of y calculate (1) the statistic $\hat{s}^2$, assuming that the mean performance for this type of car is known to be $\eta = 25.0$, and (2) the statistic s^2, assuming that the mean performance is unknown.
(b)   The measured performance for a *single* car of a different manufacturer but similar in size to the others is 28.2. Use the t statistic to test the null hypothesis that this difference in manufacturer is without effect and, in particular, that the mean for cars of this size is $\eta = 25.0$, against the alternative hypothesis that it is greater than 25.0, (1) using $\hat{s}^2$ and (2) using s^2.   *Answer*: $\hat{s}^2 = 2.96$, $s^2 = 3.70$, $\hat{t} = 1.86$ ($v = 5$), $t = 1.67$ ($v = 4$).

2.5.   THE INDUSTRIAL EXPERIMENT: AN EXTERNAL REFERENCE DISTRIBUTION BASED ON THE t DISTRIBUTION

Consider again the group of 210 past consecutive yields and averages of successive sets of 10 given in Table 2.2. From these values we can obtain the 10 differences in averages from comparable *nonoverlapping* sequences of observations shown in Table 2.5 and plotted in Figure 2.14. The first entry, 83.94, in column $\bar{y}_1$ is the average of observations 1 through 10, while the first entry, 83.51, under $\bar{y}_2$ is the average of observations 11 through 20. The second entry, 83.99, in column $\bar{y}_1$ is the average of observations 22 through 31, and the second entry, 84.42, under $\bar{y}_2$ is the average of observations 32 through 41. The next pair of entries are computed from observations 43 through 62, and so on. It will be observed that the opportunity has been taken to leave a small gap between the groups of 20 observations to reduce dependence between them. In each group of 20 the average of the first 10 is subtracted from the average of the second 10. The 10 differences between the averages in this case will be nearly *normally* distributed because of the central limit effect. Furthermore, even though in the production record of Table 2.2 successive *individual* batch yields are almost certainly statistically

FIGURE 2.14.   Dot diagram for 10 nearly independent differences, and observed difference.

TABLE 2.5.   Ten nearly independent differences

observed results	$\bar{y}_1$	$\bar{y}_2$	$\bar{y}_2 - \bar{y}_1$
from past records	83.94	83.51	−0.43
	83.99	84.42	0.43
	84.18	84.01	−0.17
	85.18	84.28	−0.90
	83.58	84.38	0.80
	84.42	83.99	−0.43
	84.72	84.21	−0.51
	84.78	83.96	−0.82
	84.09	84.58	0.49
	83.62	84.26	0.64
	$\bar{y}_A$	$\bar{y}_B$	$\bar{y}_B - \bar{y}_A$
from plant trial	84.24	85.54	1.30
variance of differences		$\dot{s}^2 = 0.36$	
standard deviation of differences		$\dot{s} = 0.60$	

dependent, the differences between averages will be distributed approximately independently.

Since the standard method of operation A was used for all 210 individual observations calculated from data in Table 2.2, there is no reason to believe that the mean value of the $\bar{y}_1$'s is any different from that of the $\bar{y}_2$'s. Hence the individual differences $\bar{y}_2 - \bar{y}_1$ may be assumed to have a *known* population mean of zero. Regarding these 10 observed past differences as our basic data, we can compute their sample variance from Equation 2.9:

$$\dot{s}^2 = \frac{(-0.43 - 0)^2 + (0.43 - 0)^2 + \cdots + (0.64 - 0)^2}{10} = 0.36 \quad (2.15)$$

Thus the sample standard deviation of the differences is $\dot{s} = 0.60$, and it has 10 degrees of freedom.

Now let us focus attention on the result of the trial with the unmodified and modified processes A and B when a difference of 1.30 was found. For the null hypothesis $\eta_B - \eta_A = 0$,

$$t_0 = \frac{1.30 - 0}{0.60} = 2.17 \quad (2.16)$$

Graphical interpolation in Table B1 shows that the significance level $\Pr(t > 2.17)$ is about 0.028, which is comparable to the value 0.047 obtained previously with the external reference distribution composed of the 191 differences. This leads, as before, to the null hypothesis being discredited.

It will be seen that the test procedure is equivalent to referring the observed difference 1.30 to a t distribution with 10 degrees of freedom scaled by the standard deviation 0.6. This is illustrated in Figure 3.5b in the next chapter. By comparing Figures 3.5a and 3.5b, one can see that the two procedures we have described give similar results.

Conclusion

In this chapter we have considered two inferential procedures, using a set of external reference data. Both of these procedures have the advantage that they do not require assumptions which tax our credulity (the first requires fewer than the second), but they have the disadvantage that previous data are needed to supply a reference distribution with which the observed difference in averages may be compared. In the next chapter we consider a procedure that does not require external data but does require assumptions which need to be justified.

Exercise 2.13. For the data in Exercise 2.4, construct a figure corresponding to Figure 2.14. What significance level is obtained using "nearly independent" differences? (Use the data in four sets of four, leaving a gap of one between sets.)

Answer: 0.047.

Exercise 2.14. Six temperature readings (°F) were taken on a patient at 5-minute intervals, three before and three after the taking of a drug. The results, recorded as $10(T - 98.0)$, were as follows: before the drug 4, 3, 7; after the drug 10, 6, 8. The same patient when not on the drug gave the following successive results at 5-minute intervals: 5, 5, 9, 7, 3, 4, 5, 8, 9, 12, 14, 8, 9, 11, 14, 9, 10, 10, 6, 5, 4, 2, 3, 3, 3, 8, 2, 3, 4, 6, 5, 3, 2, 4, 6, 4. (a) Construct an external reference distribution for the difference between adjacent averages of three.
(b) Using "nearly independent" differences, construct a reference distribution. (Use the data in five sets of six, leaving a gap of one between sets.) Based on these two reference distributions, what is the significance level for the null hypothesis $\eta_B = \eta_A$ when the alternative is $\eta_B > \eta_A$? *Answer*: (a) $\frac{3}{31} = 0.097$, (b) 0.103.

APPENDIX 2A. CALCULATION OF THE SAMPLE AVERAGE, SAMPLE VARIANCE, AND SAMPLE STANDARD DEVIATION BY CODING DATA

Occasionally, even in this age of computers, we want to make hand calculations. Such calculations are often greatly simplified by coding the data and decoding after the calculations are made. Coding usually consists of subtracting some convenient quantity from all the data values so that they are not unnecessarily large (79 is subtracted from all the data values below). In addition, the data values may be multiplied by 10, 100, or some other power of 10 to eliminate tiresome decimals.

TABLE 2A.1. Calculation of the sample variance and sample standard deviation using coded data

y	$Y = 10(y - 79)$	$Y - \bar{Y} = Y - 52.4$	$(Y - \bar{Y})^2$
89.7	107	54.6	2981.16
81.4	24	-28.4	806.56
84.5	55	2.6	6.76
84.8	58	5.6	31.36
87.3	83	30.6	936.36
79.7	7	-45.4	2061.16
85.1	61	8.6	73.96
81.7	27	-25.4	645.16
83.7	47	-5.4	29.16
84.5	55	2.6	6.76
	524	0.0	7578.40

	coded units	original units
sample average	$\bar{Y} = 52.4$	$\bar{y} = 79 + 5.24 = 84.24$
sample variance	$s_Y^2 = \dfrac{7578.40}{9} = 842.04$	$s_y^2 = 8.4204$
sample standard deviation	$s_Y = 29.02$	$s_y = 2.902$

Addition or subtraction of any constant from the data will add or subtract the same factor from the sample average but will have no effect at all on deviations from the average $y - \bar{y}$ and hence on determining the sample standard deviation or variance. If a multiplicative factor is used in coding, both the average and the standard deviation (which have the same units as the observations) will be multiplied by that factor. Equivalently, the variance will be multiplied by the square of the multiplicative factor.

In Table 2A.1 we illustrate the calculation of the average, the sample variance, and the sample standard deviation for the yield data of Table 2.1 for the standard process (method A), where y denotes the original data, and Y the coded values. Notice that the sum of the deviations from the average is zero, as must always be the case whether or not the data differences are coded.

An Alternative Method for Calculating the Sum of Squares

An alternative method for calculating $\sum (Y - \bar{Y})^2$, the quantity needed to compute the sample variance and hence the sample standard deviation, is to use the mathematical identity

$$\sum (Y - \bar{Y})^2 = \sum Y^2 - n\bar{Y}^2$$

TABLE 2A.2.   **Calculation of corrected sum of squares**

Y	Y^2		
107	11,449		
24	576		
55	3,025		
58	3,364	crude sum of squares	35,036.0
83	6,889	correction for the average	27,457.6
7	49		
61	3,721	corrected sum of squares	7,578.4
27	729		
47	2,209		
55	3,025		
sum 524	35,036		
average		$\bar{Y} = 52.4$	

correction for the average $n\bar{Y}^2 = \dfrac{(\sum Y)^2}{n} = 27{,}457.6$

Here $\sum Y^2$ is sometimes called the "crude" sum of squares; $n\bar{Y}^2$, the "correction for the average" (or, more often, the "correction for the mean" or the "correction factor"); and $\sum (Y - \bar{Y})^2$, the "corrected sum of squares." Thus the identity can be read as

corrected sum of squares = crude sum of squares − correction for the average

If, as before, we work with the coded values, we obtain the results shown in Table 2A.2. This second method yields the value $\sum (Y - \bar{Y})^2 = 7578.4$, as before. Although this method of correcting the "crude" sum of squares provides a somewhat quicker form of computation, it is subject to large rounding errors, particularly for uncoded data. Care must be taken to retain a sufficient number of decimal places, particularly in the "correction for the average."

QUESTIONS FOR CHAPTER 2

1. What is a dot diagram, a histogram, and a probability distribution?
2. What is meant by a *population* of observations, a *sample* of observations, a *random drawing* from the population, and a *random sample* of observations? Define the *population mean, population variance, sample average,* and *sample variance*. Which are *parameters*, and which are *statistics*?
3. What are the *essential* features of a significance test?
4. What is an external reference distribution, and how can it be used in analyzing data? Why is it called *external*? To use an external reference

distribution, what assumptions must be made? Do the data have to be independent? What are the advantages and disadvantages of using external reference distributions? Can you find some data on the comparison of two treatments (or imagine some), preferably in your own field, that could be analyzed using an external reference distribution? Can you think of some data that could not?

5. Why is it better to quote the actual significance level for a set of data than to state, for example, whether or not the results are significant at some predetermined level, say 0.05?

6. What is the central limit effect, and why is it important?

7. What does a normal distribution look like, and why is it important?

8. What two parameters characterize a normal distribution?

9. When must a t distribution be used instead of a normal distribution?

10. What is the meaning of statistical independence?

CHAPTER 3

Random Sampling and the Declaration of Independence

How can we avoid reliance on an extensive external data set to generate a reference distribution? In this chapter we show that, if a certain additional assumption is made, an appropriate reference distribution can be generated from the experimental data themselves, for example, from the 20 observations obtained in the industrial experiment described in the preceding chapter. The additional, critical assumption is that observations may be treated as *random* samples from appropriate populations. This assumption is not one that can be made lightly and has extremely important implications in both the design and the analysis of experimental trials. We therefore consider it carefully.

3.1. THEORY: STATISTICAL DEPENDENCE AND INDEPENDENCE AND THE RANDOM SAMPLING MODEL

At the end of the preceding chapter, in discussing the assumptions needed for the validity of the t distribution, we mentioned the necessity for certain quantities to be *independently distributed*. We could equivalently have said that we required them to be *random variables* that were *statistically independent*. What is meant by these expressions?

Randomness and Random Variables

Suppose that we *know* the distribution $p(y)$ of the population of heights y of recruits for the Patagonian army. Recall that we say a recruit is drawn *at random* from this population if every recruit in the population has an equal chance of being chosen.

Now suppose that a recruit *is* selected randomly. Without seeing him, what do we know of his height? Certainly we do not know it exactly. But it would be incorrect to say that we know nothing about it because, knowing the distribution of heights of recruits, we can make statements of the following kind: the probability that the randomly chosen recruit is shorter than y_0 inches is P_0; the probability that he is taller than y_0 inches, but shorter than y_1 inches, is P_1. A quantity such as the randomly drawn recruit's height, which is not known exactly but for which we know the probability distribution, is called a *random variable*.

Statistical Dependence

Suppose that we are considering two characteristics, for example, the height y_1 in inches and the weight y_2 in pounds of a population of recruits for the Patagonian army. We can imagine that there will be a distribution of heights $p(y_1)$ and weights $p(y_2)$ for this population. Both the height and the weight of a randomly chosen recruit have probability distributions and are thus random variables. Now consider the probability distribution of the weights of all recruits who are 80 inches tall. We can write this distribution as $p(y_2|y_1 = 80)$. It is called the *conditional* distribution of weight y_2, *given* that height y_1 is 80 inches (the vertical line | stands for the word *given*). Clearly we should expect that this probability distribution of weight would be quite *different* from, say, $p(y_2|y_1 = 65)$, the conditional distribution of weight y_2, given that height y_1 is 65 inches. In such a case we say that the random variables y_1 and y_2 are *statistically dependent*.

   Suppose now that y_3 was a measure of the IQ of the recruit. It might well be true that the conditional distribution of IQ would be the same for 80-inch as for 65-inch recruits, that is,

$$p(y_3|y_1 = 80) = p(y_3|y_1 = 65) \tag{3.1}$$

If the conditional distribution of IQ was the same *whatever* the height of the recruit, so that

$$p(y_3|y_1) = p(y_3) \tag{3.2}$$

we would say that y_1 and y_3 were statistically independent. Alternatively, we could say that y_1 and y_3 were *distributed independently*.

The Joint Distribution of Two Random Variables

With height y_1 measured to the nearest inch and weight y_2 to the nearest pound, consider now the *joint probability* of obtaining a recruit with height 65 inches *and* weight 140 pounds. This probability is denoted by $\Pr(y_1 = 65, y_2 = 140)$. One way of finding recruits of this kind is (1) to pick out the special class of recruits who weigh 140 pounds, and (2) to select *from this special class* the recruits who are 65 inches tall. The required joint probability is therefore the probability of finding a recruit who weighs 140 pounds, multiplied by the probability of finding a recruit who is 65 inches tall, *given* that he weighs 140 pounds. In symbols we have

$$\Pr(y_1 = 65, y_2 = 140) = \Pr(y_2 = 140)\Pr(y_1 = 65 | y_2 = 140) \quad (3.3)$$

We could have made the selection equally well by first picking the class of recruits with the desired height and then selecting from this class those with the required weight. Thus it is equally true that

$$\Pr(y_1 = 65, y_2 = 140) = \Pr(y_1 = 65)\Pr(y_2 = 140 | y_1 = 65) \quad (3.4)$$

Now it may be shown that parallel relationships apply for probability densities. Thus, if $p(y_1, y_2)$ is the joint density associated with specific values y_1 and y_2 of two random variables, that density can always be factored as follows:

$$p(y_1, y_2) = p(y_1)p(y_2 | y_1) \quad (3.5)$$

A Special Form for the Joint Distribution of Independent Random Variables

If y_1 and y_2 are *statistically independent*, then $p(y_2 | y_1) = p(y_2)$. Substitution of this expression into Equation 3.5 shows that in the very special circumstance of statistical independence the joint probability density may be obtained by *multiplying individual densities*

$$p(y_1, y_2) = p(y_1)p(y_2) \quad (3.6)$$

A corresponding product formula applies to probabilities of events. Thus, if y_3 was a measure of IQ statistically independent of height, the probability of obtaining a recruit with an IQ greater than 130 *and* a height greater than 72 inches would be given by the product of the individual probabilities,

$$\Pr(y_1 > 72, y_3 > 130) = \Pr(y_1 > 72)\Pr(y_3 > 130) \quad (3.7)$$

The product formula does not apply to variables that are statistically dependent. For instance, the probability that a randomly chosen recruit weighs over 200 pounds ($y_2 > 200$) *and* measures over 72 inches in height ($y_1 > 72$) cannot be found by multiplying the individual probabilities. The product formula does not hold in this instance because $p(y_2 | y_1)$ is not equal to $p(y_2)$ but rather depends on y_1.

Equations 3.6 and 3.7 generalize to any number of variables. For example, if $y_1, y_2, \ldots, y_n$ are *independently distributed*, and $y_{10}, y_{20}, \ldots, y_{n0}$ are particular values of these random variables,

$$\Pr(y_1 > y_{10}, y_2 > y_{20}, \ldots, y_n > y_{n0})$$
$$= \Pr(y_1 > y_{10})\Pr(y_2 > y_{20}) \cdots \Pr(y_n > y_{n0}) \quad (3.8)$$

Application to Repeated Scientific Observations

In the examples above, $y_1, y_2, \ldots, y_n$ represented different kinds of variables—height, weight, and IQ. The formulas also apply when the variables are repeated observations of the same measurement. Suppose that $y_1, y_2, \ldots, y_n$ are repeated measurements of specific gravity recorded in the order in which they were made. Consider the first observation y_1. Suppose that it can be treated as a random variable, that is, it can be characterized as a drawing from some population typified by its density function $p(y_1)$. Suppose also that subsequent observations $y_2, y_3, \ldots, y_n$ can be similarly treated, and that they have density functions $p(y_2), p(y_3), \ldots, p(y_n)$, respectively.

Now it might (or might not*) be true (1) that the individual probability densities $p(y_1), p(y_2), \ldots, p(y_n)$ are all *identical* in form, having exactly the same location, spread, and shape and (2) that $y_1, y_2, \ldots, y_n$ can be treated as statistically *independent*.

Independent and Identically Distributed Random Variables and Random Drawings from a Fixed Population

When the component distributions $p(y_1), p(y_2), \ldots, p(y_n)$ are all identical in form and are statistically independent, $y_1, y_2, \ldots, y_n$ are said to be *independently* and *identically* distributed, and we can write

$$p(y_1, y_2, \ldots, y_n) = p(y_1)p(y_2) \cdots p(y_n) \tag{3.9}$$

where all the distributions on the right are supposed to have the same form. *In that case* the sample of observations $y_1, y_2, \ldots, y_n$ is *as if* it had been generated by random drawings from some fixed population typified by a single probability density function $p(y)$.

Covariance and Correlation as Measures of Linear Dependence

A measure of the linear dependence between, for example, height y_1 and weight y_2 is the *covariance* between y_1 and y_2. It will be recalled (Equation 2.4) that the variance is the mean value in the population of the squared deviation

---

* For example, suppose the operation were such that the first observation made in a sequence always tended to be high. This would imply that $p(y_1)$ was different (and, in particular, had a higher mean) than $p(y_2)$, and the distributions $p(y_1)$ and $p(y_2)$ would then not be identical. Again, suppose the unknown causes that produced experimental errors tended to persist over time so that, whenever there was a positive error in the first observation, making it high, a positive error was likely in the second, making it high also. Then y_1 and y_2 could not be treated as statistically independent.

of an observation y from its mean η. Thus the variances for height and weight are, respectively,

$$V(y_1) = E(y_1 - \eta_1)^2 = \sigma_1^2, \qquad V(y_2) = E(y_2 - \eta_2)^2 = \sigma_2^2 \quad (3.10)$$

The covariance (denoted by Cov) is the mean value in the population of the *product* of $y_1 - \eta_1$ with $y_2 - \eta_2$.

$$\text{Cov}(y_1, y_2) = E(y_1 - \eta_1)(y_2 - \eta_2) = \frac{\sum (y_1 - \eta_1)(y_2 - \eta_2)}{N} \quad (3.11)$$

If y_1 and y_2 were independent, $\text{Cov}(y_1, y_2)$ would be zero.*

In practice, recruits that deviated (positively/negatively) from the mean height would tend to deviate (positively/negatively) from the mean weight. Thus (positive/negative) values of $y_1 - \eta_1$ would tend to be accompanied by (positive/negative) values of $y_2 - \eta_2$, and the covariance between height and weight would be positive. Conversely, the covariance between speed of driver reaction and alcohol consumption would probably be negative. A decrease of reaction speed would be associated with an increase in alcohol consumption, and vice versa.

The covariance itself is dependent on the scales chosen. If, for example, height was measured in feet instead of inches, the covariance would be divided by 12. A "scaleless covariance," called the *correlation coefficient* $\rho(y_1, y_2)$ or simply ρ (the Greek letter rho), is obtained by dividing the deviations by σ_1 and σ_2, respectively. Thus

$$\rho = \rho(y_1, y_2) = E\left[\frac{(y_1 - \eta_1)}{\sigma_1} \frac{(y_2 - \eta_2)}{\sigma_2}\right]$$

$$= \frac{\text{Cov}(y_1, y_2)}{\sigma_1 \sigma_2} = \frac{\text{Cov}(y_1, y_2)}{\sqrt{V(y_1)V(y_2)}} \quad (3.12)$$

Equivalently, we can write $\text{Cov}(y_1, y_2) = \rho\sigma_1\sigma_2$.

The *sample correlation coefficient* between y_1 and y_2 is defined as

$$r(y_1, y_2) = \frac{\sum (y_1 - \bar{y}_1)(y_2 - \bar{y}_2)/(n - 1)}{s_1 s_2} \quad (3.13)$$

The numerator of this expression is called the *sample covariance*, and s_1 and s_2 are the sample standard deviations for y_1 and y_2 (see Equation 2.7).

Exercise 3.1. For these data compute the sample correlation coefficient:

y_1 (height in inches)	65	68	67	70	75
y_2 (weight in pounds)	150	130	170	180	220

Answer: 0.85.

---

* The converse is not true. For example, suppose, apart from error, that y_1 was a quadratic function of y_2 that plotted like the letter U. Then, although y_1 and y_2 would be statistically dependent, their covariance could be zero.

Serial Dependence Measured by Autocorrelation

When data are taken in sequence, there is often a tendency for observations made close together in time (or space) to be more alike than those taken farther apart. This can occur because disturbances, such as high impurity level in the feed to a chemical process or high respiration rate in an experimental animal, are likely to persist somewhat. There are other instances where consecutive observations are *less* alike than those taken further apart. Suppose, for example, that the response measured is the monthly *increase* in weight of an animal. An abnormally high total weight recorded at the end of October (perhaps because of water retention) can result in an unusually high increase being attributed to October, followed by an unusually small increase attributed to November.

If sufficient data are available, tendencies of this kind toward serial dependence can be demonstrated by plotting each observation against the immediately preceding one. In Figure 3.1 such a plot is made for the production record of Table 2.2. The yield y_t, observed at time t, is plotted against the previous yield y_{t-1}, made at time $t-1$, for $t = 210, 209, \ldots, 2$. Similar plots can be made for data two units apart (y_t versus y_{t-2}), three units apart,

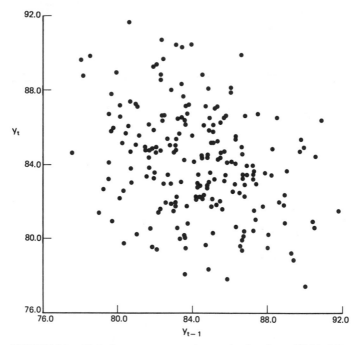

FIGURE 3.1.   Plot of y_t versus y_{t-1}, using production data of Table 2.2.

and so on. The corresponding correlation coefficients are called *auto-correlation coefficients*. The distance between the observations that are so correlated is called the *lag*. The lag k *sample autocorrelation* (coefficient) is defined by*

$$r_k = \frac{\sum (y_t - \bar{y})(y_{t+k} - \bar{y})}{\sum (y_t - \bar{y})^2} \tag{3.14}$$

Exercise 3.2. Compute r_1 for these data: 3, 6, 9, 8, 7, 5, 4.   *Answer*: 0.32.

Note: $0.32 = \frac{9}{28}$, where $\bar{y} = 6$, and the numerator is $(3 - 6)(6 - 6) + (6 - 6)(9 - 6)$ $+ \cdots + (5 - 6)(4 - 6) = 9$ (six terms), and the denominator is $(3 - 6)^2 + (6 - 6)^2$ $+ \cdots + (4 - 6)^2 = 28$ (seven terms).

Exercise 3.3. Compute r_1, r_2, and r_3 for the first 12 observations listed in Table 2.2.
                                                              Answer: $-0.34, -0.16, -0.05$.

Note: Unless the number n of available observations is large, calculated autocorrelation coefficients are unreliable. Usually there is little point in calculating autocorrelations for n less than 50. Exercises 3.2 and 3.3 are included only so that the reader can gain an appreciation of the meaning of the formula for r_k.

If the number of observations increases without limit then under suitable conditions r_k will approach closer and closer to a value ρ_k, called the lag k *theoretical autocorrelation*, or *population autocorrelation*.

For the production data the sample autocorrelation coefficient is $r_1 = -0.29$. For this *large* sample ($n = 210$) the value $r_1 = -0.29$ is highly significantly different from zero. However, for small samples (e.g. $n = 10$), real autocorrelations of this size would be virtually undetectable and yet could seriously invalidate standard tests that assume independence.

Physical Actions Resulting in Independence

It is sometimes possible to induce statistical independence by the physical conduct of an experiment. If we make several throws of a die, or alternatively, simultaneously throw several dice, it can usually safely be assumed that the numbers shown on the individual faces of the dice are distributed independently. For example, imagine a pair of fair dice, one red and one white, thrown many times. Denote the scores shown by the red and the white dice by y_1 and y_2, respectively. We would expect that on a single throw the probability of the white die being a five would be independent

---

* Note that the lag zero ($k = 0$) sample autocorrelation coefficient is always unity, that is, $r_0 = 1$.

of whether or not the red die showed a one, a two, a three, a four, a five, or a six, that is,

$$\Pr(y_2 = 5 | y_1 = 1) = \Pr(y_2 = 5 | y_1 = 2) = \cdots$$
$$= \Pr(y_2 = 5 | y_1 = 6) = \Pr(y_2 = 5) = \tfrac{1}{6} \quad (3.15)$$

Statistical independence is thus achieved in this example by *physically introducing randomness* into the experiment.

Random Sampling, Independence, and Assumptions

Random *sampling* is another way of physically inducing independence. Imagine a very large population of *numbered* red and white tickets in a lottery drum. Suppose that when the red and white tickets were thoroughly mixed and two tickets were drawn at random, one happened to be red, the other white. We would expect the number on the white ticket to be statistically independent of that on the red one; that is, we would expect the probability distribution of the numbers on the white tickets to be independent of that of the numbers on the red tickets.

In the next section we return again to consideration of the data for the industrial experiment introduced in Chapter 2. We saw there how, by using an external reference distribution, significance tests can be constructed that do not rely on the assumption of random sampling. Unfortunately a body of relevant past data, of the kind used there, is usually not available. Inevitably there is pressure to replace by assumption what is unavailable in fact.

Random Samples from Two Populations

A customary assumption is that a set of data, such as the 10 observations from method *A* in the plant trial, may be thought of as a *random* sample from the conceptual *population* of batch yields for method *A*. To visualize what is being assumed, imagine that tens of thousands of batch yields obtained from standard operation with method *A* are written on cards and put into a large lottery drum, which is shown diagramatically at the top left of Figure 3.2. The characteristics of this very large population of cards are represented by the distribution shown in the lower part of the same figure. *On the hypothesis of random sampling* the 10 observations with method *A*, represented by the black dots, would be thought of as 10 *independent random* drawings from lottery drum *A* and hence from distribution *A*. In a similar way the observations with the modified method *B* would be thought of as 10 independent random drawings from lottery drum *B* and hence from population *B*.

A basic null hypothesis that we might wish to test is that the modification is entirely without effect. If this hypothesis is true, the complete set of 20

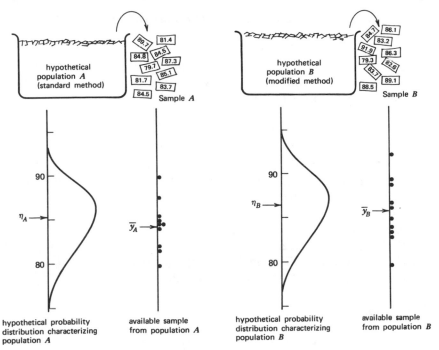

FIGURE 3.2. Random sampling from two hypothetical distributions.

observations can be explained as a random sample from a single *common* population. The alternative hypothesis, which is illustrated in Figure 3.2, is that the distributions from which the two random samples are drawn, although otherwise similar, are located differently and so have different means η_A and η_B.

Why It Would Be Nice If We Could Believe the Random Sampling Model

The assumption that the data can be represented as *random* samples implies that the y's are *independently* distributed about their respective means. Thus the errors $y_1 - \eta_A, y_2 - \eta_A, \ldots, y_{10} - \eta_A, y_{11} - \eta_B, y_{12} - \eta_B, \ldots, y_{20} - \eta_B$ are supposed to vary independently of one another. In particular, statistical independence implies that the *order* in which the runs were made has no influence on the results.

Statistical independence produces certain dramatic simplifications. Most importantly, it endows sample statistics such as $\bar{y}$ with very special distributional properties.

The Mean and Variance of the Average ȳ for Independent, Identically Distributed Observations

If the random sampling model is appropriate, so that the errors are independent and hence uncorrelated, we have the simple rule that $\bar{y}$ varies about the population mean η with variance σ^2/n. Thus

$$E(\bar{y}) = \eta, \qquad V(\bar{y}) = \frac{\sigma^2}{n} \qquad (3.16)$$

When errors in observations varying about a fixed mean are correlated, the expression for the variance of $\bar{y}$ contains a factor C that depends on the degree of correlation, that is, $V(\bar{y}) = C\sigma^2/n$. For a sample of $n = 10$ with only *immediately adjacent* observations autocorrelated, it is shown in Appendix 3A that the factor C can be as large as $(2n - 1)/n = 1.9$ for positively correlated observations or as small as $1/n = 0.1$ for negatively correlated observations. Thus in this example different degrees of lag 1 autocorrelation can change $V(\bar{y})$ by a factor of nineteen!

Sampling Distribution of ȳ

To visualize what is implied by Equations 3.16, suppose that a very large number of white tickets in a white lottery drum represents the population of individual observations y, and that this population has mean η and variance σ^2. Now suppose we (1) randomly draw out a sample of $n = 10$ tickets, (2) calculate the average value $\bar{y}$ and write it down on a blue ticket, (3) put the blue ticket in a second, blue lottery drum, (4) return the white tickets to the white lottery drum and thoroughly mix them, and (5) repeat the whole operation many times until the blue lottery drum is filled with tickets. Then the numbers in the blue drum, which form the population of the averages $\bar{y}$, will have the same mean η as the original distribution but a variance σ^2/n, which is only one-nth as large as that of the original distribution.

The original distribution of the individual observations, represented by the drum full of white tickets, is often called the *parent* distribution. Any distribution derived from the parent distribution by random sampling is called a *sampling* distribution. Thus the distribution of the numbers shown on the tickets in the blue lottery drum is called the *sampling distribution* of $\bar{y}$.

Approach to Normality of the Sampling Distribution of ȳ

To justify the formulae (3.16) for the mean and variance of $\bar{y}$ we need "only" assume that the random sampling model is appropriate. It does not matter

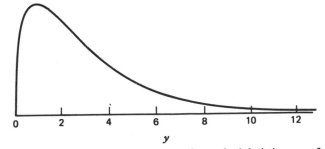

(a) A skewed parent distribution, mean $\eta = 3$, standard deviation $\sigma_y = 2.45$.

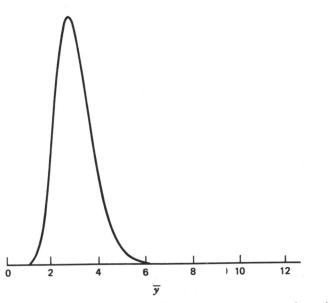

(b) Sampling distribution of the average of 10 random observations, mean $\eta = 3$, standard deviation $\sigma_{\bar{y}} = 2.45/\sqrt{10} = 0.77$.

FIGURE 3.3. Distribution of the average of $n = 10$ observations randomly sampled from a
skewed distribution.

what shape the parent distribution has. In particular, it does not need to be
normal. If it is not normal, then, as illustrated in Figure 3.3 (see also Figure
2.10), on the random sampling model averaging not only reduces the standard
deviation by a factor of $1/\sqrt{n}$ but also simultaneously produces a distribution
for $\bar{y}$ that is more nearly normal. This is due to the central limit effect

discussed in Section 2.4. In summary, then, *on the hypothesis of random sampling*:

	Parent distribution for y	Sampling distribution for $\bar{y}$
mean	η	η
variance	σ^2	σ^2/n
standard deviation	σ	$\sigma/\sqrt{n}$
form of distribution	any*	more nearly normal than parent distribution

Statistics as Estimates of Parameters

The quantity $\bar{y}$ is the average of the particular sample of observations that happens to be available. Without further assumption there is little more we can say about it. If, however, the observations can be regarded as a *random sample* from some population with mean η and variance σ^2, we know the following:

1. $\bar{y}$ has for its mean value the population mean η.
2. $\bar{y}$ varies about η with standard deviation $\sigma/\sqrt{n}$.

Thus, as we imagine taking a larger and larger sample, $\bar{y}$ tends to lie closer and closer to η. In these circumstances it is reasonable to regard $\bar{y}$ as an *estimate* of η. Similarly it may be shown that s^2 has a mean value σ^2 and varies about that value with a standard deviation proportional to $1/\sqrt{n}$. On the statistical assumption of random sampling, therefore, we can regard s^2 as an *estimate* of σ^2.

The problem of choosing estimates in the best possible way is complicated, depending heavily, as the reader might guess, on the definition of best. We will not pursue this subject much further. The important point to remember is that it is the assumption of random sampling that endows the estimates with whatever distributional properties they are supposed to possess.

---

* This statement applies to all parent distributions commonly met in practice. It is not true for certain mathematical toys (e.g., the Cauchy distribution), which need not concern us here.

Sampling Distributions of a Sum and of a Difference

Interest often centers on the distribution of the sum Y of two *independently* distributed random variables y_A and y_B. Let us suppose that:

y_A has a distribution with mean η_A and variance σ_A^2

and

y_B has a distribution with mean η_B and variance σ_B^2

What can we say about the mean and variance of the distribution of $Y = y_A + y_B$? Again we can illustrate this question by considering lottery drums, each with an appropriate population of tickets. Imagine that, after a random drawing from drum A to obtain y_A and another random drawing from drum B to obtain y_B, the sum $Y = y_A + y_B$ is written on a red ticket and put in a third lottery drum. After many such drawings what can we say about the distribution of the sums on the red tickets in the third lottery drum?

It turns out (see Appendix 3A) that the mean value of Y is

$$E(Y) = E(y_A) + E(y_B) = \eta_A + \eta_B \qquad (3.17)$$

Also it can be shown for *independent* drawings that the variance of Y is

$$V(Y) = V(y_A + y_B) = \sigma_A^2 + \sigma_B^2 \qquad (3.18)$$

Correspondingly, for a difference of two random variables, if

$$Y = y_A - y_B \qquad (3.19)$$

then

$$E(Y) = E(y_A) - E(y_B) = \eta_A - \eta_B \qquad (3.20)$$

and for *independent* drawings

$$V(Y) = V(y_A - y_B) = \sigma_A^2 + \sigma_B^2 \qquad (3.21)$$

The result for a difference follows from that for a sum, for if we write $-y_B = y_B'$, then, of course, $V(y_B) = V(y_B')$, and using the expression for the variance of a sum, we have

$$V(y_A - y_B) = V(y_A + y_B') = \sigma_A^2 + \sigma_B^2 \qquad (3.22)$$

These results for the sum and the difference do not depend on the form of the parent distributions; and, because of the central limit effect, the distributions of both sum and difference tend to be closer to the normal than are the original parent distributions.

Special Case When $\sigma_A^2 = \sigma_B^2$

If y_A and y_B are *independent* drawings from the same population with variance σ^2 or, alternatively, from different populations having the same variance σ^2, the variance of the sum and the variance of the difference are identical, that is,

$$V(y_A + y_B) = 2\sigma^2, \qquad V(y_A - y_B) = 2\sigma^2 \qquad (3.23)$$

Random Sampling from a Normal Population

The following important results can be demonstrated on the assumption of *random* sampling from a *normal* parent distribution. These results are illustrated in Figure 3.4 for a sample size of $n = 5$.

If a random sample of n observations is drawn from a normal distribution with mean η and variance σ^2, then:

1. The distribution of $\bar{y}$ is also normal (with mean η and variance σ^2/n).
2. The sample variance s^2 is distributed independently of $\bar{y}$ in a scaled χ^2 (chi-square) distribution. This is a skewed distribution the properties of which are discussed further in Appendix 3B and Section 5.4.
3. The quantity $(\bar{y} - \eta)/(s/\sqrt{n})$ is distributed with $n - 1$ degrees of freedom in the t distribution (discussed in Section 2.4).

Result 3 is particularly remarkable because it allows the deviation $\bar{y} - \eta$ to be judged against an estimate of the standard deviation $s/\sqrt{n}$ of $\bar{y}$, which is obtained *from the internal evidence of the sample itself.* Thus no external data set is needed to obtain the necessary reference distribution, *if* the assumption of random sampling from a normal distribution can be made.

Sufficiency of $\bar{y}$ and s^2

If the assumptions of random sampling from a normal population are exactly satisfied, it can be shown that *all* the information in the sample $y_1, y_2, \ldots, y_n$ about η and σ^2 is contained in the two quantities $\bar{y}$ and s^2. These are then said to be *jointly sufficient* estimates for η and σ^2. The nature of the argument is sketched in Appendix 3C. Now, since η and σ^2 completely define the normal distribution, the standardized residuals

$$\frac{(y_1 - \bar{y})}{s}, \frac{(y_2 - \bar{y})}{s}, \ldots, \frac{(y_n - \bar{y})}{s}$$

or, equivalently, the *relative* magnitudes of the residuals

$$y_1 - \bar{y}, y_2 - \bar{y}, \ldots, y_n - \bar{y}$$

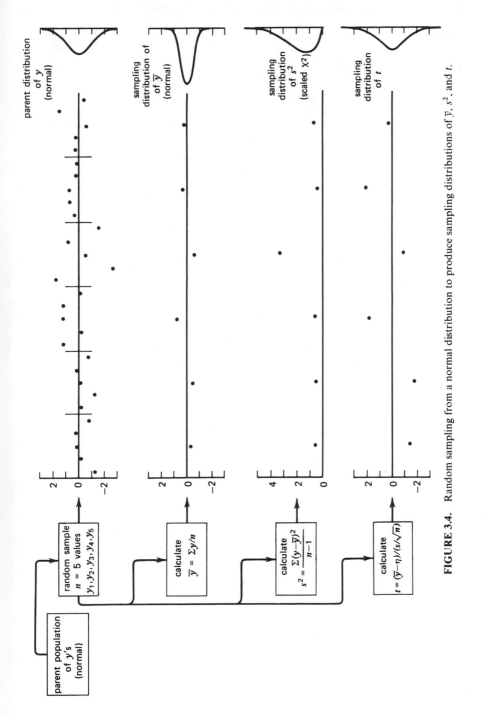

FIGURE 3.4. Random sampling from a normal distribution to produce sampling distributions of $\bar{y}$, s^2, and t.

from such a distribution are informationless. The other side of the coin is that, when *discrepancies* from the hypothesis of random sampling from a normal population occur, these residuals can provide hints about the *nature* of the discrepancies.

Exercise 3.4. The following are 10 measurements of the specific gravity of the same sample of alloy:

date	October 8		October 9		October 10		October 11		October 12	
time of day	A.M.	P.M.	A.M.	P.M.	A.M.	P.M.	A.M.	P.M.	A.M.	P.M.
measured specific gravity y	0.36721	0.36473	0.36680	0.36487	0.36802	0.36396	0.36758	0.36425	0.36719	0.36333

(a) Calculate the average $\bar{y}$ and standard deviation s for this sample of 10 observations.
(b) Plot the residuals $y - \bar{y}$ in time order.
(c) It is proposed to make an analysis based on the assumption that the 10 observations are a random sample from a normal distribution. Do you have any doubts about the validity of this assumption?

A Dubious Application of the t *Distribution: Comparison of an Average and a Hypothetical Mean*

Suppose that the random sampling assumption could be relied on and that the population sampled was approximately normal. Then $(\bar{y} - \eta)/(s/\sqrt{n})$ would be distributed in a t distribution. This would allow us to obtain an appropriate reference distribution for the comparison of an observed average $\bar{y}$ with any hypothetical mean η_0. Consider, for example, the 10 observed data values in the industrial experiment for the *modified* method B. Given the above assumptions, what could we say about the suggestion that the population mean for the modified process is $\eta_0 = 84.12$, against the alternative that it is greater? (The value $\eta_0 = 84.12$ is the mean for the standard process A estimated over a long period of running.)

The relevant calculations are displayed in Table 3.1. Reference of the value $t_0 = 1.22$ to the t table with nine degrees of freedom shows that $\Pr(t > t_0) = 0.13$, which means that in 13 trials out of 100 a discrepancy at least as great as this could be expected from pure chance. Thus, on the basis

TABLE 3.1. Comparison of average with hypothetical mean value

data for modified process (B)	84.7, 86.1, 83.2, 91.9, 86.3, 79.3, 82.6, 89.1, 83.7, 88.5
hypothetical mean value	$\eta_0 = 84.12$
sample average	$\bar{y} = 85.54$
sample variance	$s^2 = \sum (y - 85.54)^2/9 = 13.325$
sample standard deviation	$s = 3.65$
estimated standard error of the average	$s/\sqrt{n} = 3.65/3.16 = 1.16$
t value when mean is η_0	$t_0 = (\bar{y} - \eta_0)/(s/\sqrt{n}) = (85.54 - \eta_0)/1.16$
t value when $\eta_0 = 84.12$	$t_0 = (85.54 - 84.12)/1.16 = 1.22$
significance level for $v = 9$	$\Pr(t \geq t_0) = 0.13$

of this test, there would be no strong evidence to contradict the hypothesis that the modified process had a mean yield equal to that of the standard process, since an event with probability equal to 0.13 is not *that* unusual.

Exercise 3.5. On the model of random sampling from a normal population, compute the significance level for the 10 observations for method A, given the hypothesis that the mean of these values is $\eta_0 = 84.12$ against the alternative, $\eta > \eta_0$. *Answer*: 0.45.

The Weakness of Noncomparative Experiments

The purpose of the illustration in Table 3.1 is to demonstrate how the t distribution, using an internal estimate of error, might be used in a significance test. This application is of dubious value, however, because the design is poor. Apart from the major difficulty arising from the suspected inappropriateness of the random sampling hypothesis, which was discussed earlier, there is a new weakness. The average $\bar{y}$ is compared with an assumed mean yield for the standard process of $\eta_0 = 84.12$. The value $\eta_0 = 84.12$ is supposed to have been established over a long period of running the process. (It is actually the average yield for the 210 results in Table 2.2.) However, things can change over a long period of time. Operating procedures, raw materials, and weather conditions can all change. Since an essential requirement of good design is that like be compared with like, the average for the modified method ought to be compared with the average for the standard process under *current* and *comparable* conditions. It is better, as was done in the actual plant trial, to compare methods A and B using batches manufactured at about the same time.

3.2. THE INDUSTRIAL EXPERIMENT: REFERENCE DISTRIBUTION BASED ON RANDOM SAMPLING MODEL, EXTERNAL VALUE FOR σ

Suppose we believed that the two sets of 10 yields obtained with methods A and B could be treated as *random samples from appropriate populations*. How could we use the results of Section 3.1 to make inferences about the effect of the modification? Let us assume specifically that the standard method A and modified method B give rise to two populations of yields with approximately the same form (and, in particular, the same variance σ^2) but with possibly different means η_A and η_B.

Calculation of the Standard Error of the Difference in Averages

If the data were *random* samples from the populations, with $n_A = 10$ observations from the first population and $n_B = 10$ observations from the second, the variances of the calculated averages $\bar{y}_A$ and $\bar{y}_B$ would be

$$V(\bar{y}_A) = \frac{\sigma^2}{n_A}, \qquad V(\bar{y}_B) = \frac{\sigma^2}{n_B} \tag{3.24}$$

Also, on the random sampling model, $\bar{y}_A$ and $\bar{y}_B$ would be distributed independently, so that

$$V(\bar{y}_B - \bar{y}_A) = \frac{\sigma^2}{n_A} + \frac{\sigma^2}{n_B} = \sigma^2\left(\frac{1}{n_A} + \frac{1}{n_B}\right) \tag{3.25}$$

Thus the standard deviation of the difference in averages, often called the *standard error* of the difference, would be

$$\sqrt{V(\bar{y}_B - \bar{y}_A)} = \sigma\sqrt{\frac{1}{n_A} + \frac{1}{n_B}} \tag{3.26}$$

Even if the distributions of the original observations had been moderately nonnormal, the distribution of the difference $\bar{y}_B - \bar{y}_A$ between sample averages with $n_A = n_B = 10$ would be expected to be nearly normal because of the central limit effect. Therefore, on the assumption of random sampling,

$$z = \frac{(\bar{y}_B - \bar{y}_A) - (\eta_B - \eta_A)}{\sigma\sqrt{1/n_A + 1/n_B}} \tag{3.27}$$

would be approximately a unit normal deviate.

What Shall We Do about σ?

Now σ, the hypothetical population value for the standard deviation, is unknown. We do, however, have the collection of 210 past plant yields (Table 2.2), for which the standard deviation is 2.88. If we use this value for the common standard deviation of the sampled populations, the standard error of the difference $\bar{y}_B - \bar{y}_A = 1.30$ is

$$\sigma \sqrt{\frac{1}{10} + \frac{1}{10}} = \frac{2.88}{\sqrt{5}} = 1.29 \qquad (3.28)$$

If our assumptions are appropriate, the approximate significance level associated with any postulated difference $(\eta_B - \eta_A)_0$ in the population means will then be obtained by referring

$$z_0 = \frac{1.30 - (\eta_B - \eta_A)_0}{1.29} \qquad (3.29)$$

to the table of significance levels of the normal distribution.* In particular, for the null hypothesis $(\eta_B - \eta_A)_0 = 0$, $z_0 = 1.30/1.29 = 1.01$, and $\Pr(z > 1.01) = 0.156$ (or 15.6%). We notice that this significance level of 15.6% differs rather dramatically from the significance levels of 4.7 and 2.8% given by the previous methods, discussed in Sections 2:3 and 2.5, which did not assume random sampling, but employed external reference distributions.

Exercise 3.6. Given the following data and $\sigma = 4$, compute the significance level associated with the hypothesis $(\eta_B - \eta_A)_0 = 0$ for the alternative $\eta_B - \eta_A > 0$.

A	112	116	119	125	
B	113	118	123	126	122

Answer: 0.186.

Exercise 3.7. Given the following data and $\sigma = 4$, compute the significance level associated with the hypothesis $(\eta_B - \eta_A)_0 = 0$ for the alternative $\eta_B - \eta_A > 0$.

A	111	115	118	124	112
B	116	121	126		

Answer: 0.043.

Although the above procedure uses the theory of random sampling to deduce the distribution of $\bar{y}_B - \bar{y}_A$ rather than obtaining it empirically, it

---

* Strictly speaking, the value 2.88 is a sample standard deviation with 209 degrees of freedom. Consequently, it should be designated by s, not σ, and a t table should be used rather than a normal table. But for all practical purposes the t distribution with 209 degrees of freedom is identical to the standardized normal distribution.

still requires an external value for the standard deviation σ. We see in the next section how, on the hypothesis of random sampling, an *internal* estimate of σ may be used.

3.3.   THE INDUSTRIAL EXPERIMENT: REFERENCE DISTRIBUTION BASED ON RANDOM SAMPLING MODEL, INTERNAL ESTIMATE OF σ

Suppose now the only evidence about σ is from the $n_A = 10$ runs made with method A and $n_B = 10$ runs made with method B.

Estimating σ from the Samples

The sample variances are

$$s_A^2 = \frac{\sum (y_A - \bar{y}_A)^2}{n_A - 1} = \frac{75.784}{9} = 8.42,$$

$$s_B^2 = \frac{\sum (y_B - \bar{y}_B)^2}{n_B - 1} = \frac{119.924}{9} = 13.32 \qquad (3.30)$$

On the assumption that the population variances for methods A and B are, to an adequate approximation, equal, these estimates may be combined to provide a *pooled* estimate of s^2 of this common σ^2. This is done by adding the sums of squares in the numerators and dividing by the sum of the degrees of freedom,

$$s^2 = \frac{\sum (y_A - \bar{y}_A)^2 + \sum (y_B - \bar{y}_B)^2}{n_A + n_B - 2} = \frac{75.784 + 119.924}{18} = 10.87 \quad (3.31)$$

On the assumption of random normal sampling, s^2 provides an estimate of σ^2 distributed in standard form with $n_A + n_B - 2 = 18$ degrees of freedom. Substituting the estimate s for the unknown σ on the right-hand side of Equation 3.27, we obtain a quantity

$$t = \frac{(\bar{y}_B - \bar{y}_A) - (\eta_B - \eta_A)}{s\sqrt{1/n_A + 1/n_B}} \qquad (3.32)$$

in which the discrepancy $[(\bar{y}_B - \bar{y}_A) - (\eta_B - \eta_A)]$ is compared with the *estimated* standard error of $\bar{y}_B - \bar{y}_A$. On the assumption of random sampling from normal populations with equal variances, this quantity precisely follows a t distribution with $v = n_A + n_B - 2 = 18$ degrees of freedom.

For the present example $\bar{y}_B - \bar{y}_A = 1.30$ and

$$s\sqrt{1/n_A + 1/n_B} = 1.47.$$

If our assumptions are appropriate, the corresponding significance level associated with any hypothesized difference $(\eta_B - \eta_A)_0$ will be obtained by referring the statistic

$$t_0 = \frac{1.30 - (\eta_B - \eta_A)_0}{1.47} \tag{3.33}$$

to the t table with 18 degrees of freedom. In particular (see Table 3.2), for the null hypothesis $(\eta_B - \eta_A)_0 = 0$, $t_0 = 1.30/1.47 = 0.88$ and

$$\Pr(t > 0.88) = 0.195 \tag{3.34}$$

TABLE 3.2. Calculation of significance level on the basis of random sampling hypothesis

method A	method B	$n_A = 10$	$n_B = 10$
89.7	84.7	sum $= 842.4$	sum $= 855.4$
81.4	86.1		
84.5	83.2	average $= \bar{y}_A = 84.24$	average $= \bar{y}_B = 85.54$
84.8	91.9	difference $= \bar{y}_B - \bar{y}_A = 1.30$	
87.3	86.3		
79.7	79.3	$\sum y_A^2 - \dfrac{(\sum y_A)^2}{n_A} = 75.784*$	$\sum y_B^2 - \dfrac{(\sum y_B)^2}{n_B} = 119.924$
85.1	82.6		
81.7	89.1		
83.7	83.7		
84.5	88.5		

pooled estimate of σ^2

$$s^2 = \frac{75.784 + 119.924}{10 + 10 - 2} = \frac{195.708}{18} = 10.8727$$

with $\nu = 18$ degrees of freedom

estimated variance of $\bar{y}_B - \bar{y}_A$

$$s^2\left(\frac{1}{n_A} + \frac{1}{n_B}\right) = \frac{2s^2}{10} = \frac{s^2}{5}$$

Note: $n_A = n_B = 10$

estimated standard error of $\bar{y}_B - \bar{y}_A$

$$\sqrt{\frac{s^2}{5}} = \sqrt{\frac{10.8727}{5}} = 1.47$$

$$t_0 = \frac{(\bar{y}_B - \bar{y}_A) - (\eta_B - \eta_A)_0}{s\sqrt{1/n_A + 1/n_B}}$$

For $(\eta_B - \eta_A)_0 = 0$, $t_0 = \dfrac{1.30}{1.47} = 0.88$ with $\nu = 18$ degrees of freedom

$$\Pr(t \geq 0.88) = 0.195$$

* $\sum (y_A - \bar{y}_A)^2 = \sum y_A^2 - [(\sum y_A)^2/n_A]$, and similarly for y_B.

Again this significance level (19.5 %) differs dramatically from the significance levels (4.7 and 2.8 %) obtained from the use of the external reference distributions, where no direct appeal to the hypothesis of random sampling was needed. As illustrated in Figure 3.5d, the test is equivalent to referring the observed difference $\bar{y}_B - \bar{y}_A = 1.30$ to a t distribution with a scale factor 1.47 equal to the standard error of the difference $\bar{y}_B - \bar{y}_A$. The fact that this reference distribution uses no external data is an advantage, but it is inappropriate unless the hypothesis of random sampling can be relied on. By contrast, recall that to construct the valid reference sets in Chapter 2 external records were required. Figure 3.5 allows comparison of all the reference distributions considered.

Definition of IIDN $(0, \sigma^2)$

In this book we often refer to the assumption that the errors $\epsilon = y - \eta$ are *independently and identically distributed in a normal distribution with mean zero and variance* σ^2. For convenience we will denote this as the IIDN$(0, \sigma^2)$ assumption.

Exercise 3.8.   Repeat Exercise 3.6, assuming σ unknown.   *Answer*: 0.258.

Exercise 3.9.   Repeat Exercise 3.7, assuming σ unknown.   *Answer*: 0.117.

Exercise 3.10.   Given the following data, compute the significance level associated with $(\eta_B - \eta_A)_0 = 0$:

A	10	8	12	10	10
B	9	8	9	6	

Answer: 0.037.

"Distribution free" Tests

As a means of introducing some elementary principles of statistical analysis and design we have deliberately confronted the reader with a dilemma that is inescapable if we employ the tools so far presented. Even the simplest experimental situation involving the comparison of two means requires either (1) a long sequence of relevant previous records that may not be available or (2) a random sampling assumption that may not be tenable. Experimenters might be forgiven for believing that a solution to these difficulties would be provided by what mathematicians have called "distribution free" tests, also referred to as "nonparametric tests." Unfortunately this is not so.

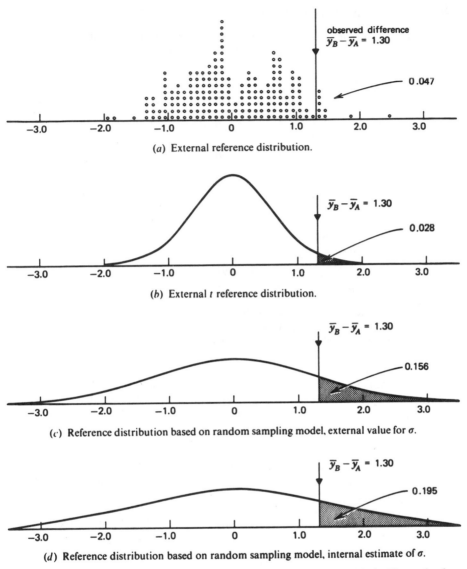

(a) External reference distribution.

(b) External t reference distribution.

(c) Reference distribution based on random sampling model, external value for σ.

(d) Reference distribution based on random sampling model, internal estimate of σ.

FIGURE 3.5. Four reference distributions for industrial experiment, with significance levels shown at right.

Consider once more the industrial experiment. A widely used nonparametric test for the difference in location of two distributions is due to Wilcoxon* and proceeds as follows:

1. Rank the combined samples in order of size from smallest to largest (ties are scored with average rank). Thus for the industrial data (Table 2.1) we obtain the following:

rank	1	2	3	4	5	6	$7\frac{1}{2}$	$7\frac{1}{2}$	$9\frac{1}{2}$	$9\frac{1}{2}$
observation	79.3	79.7	81.4	81.7	82.6	83.2	83.7	83.7	84.5	84.5
method	B	A	A	A	B	B	A	B	A	A

rank	11	12	13	14	15	16	17	18	19	20
observation	84.7	84.8	85.1	86.1	86.3	87.3	88.5	89.1	89.7	91.9
method	B	A	A	B	B	A	B	B	A	B

2. Calculate the sum of the ranks for methods A and B. Check calculations by confirming that the results add to $n_A(n_A + n_B + 1) = 210$.

$$\text{For } A \quad 2 + 3 + 4 + 7\frac{1}{2} + 9\frac{1}{2} + 9\frac{1}{2} + \cdots + 19 = 95.5$$
$$\text{For } B \quad 1 + 5 + 6 + 7\frac{1}{2} + 9\frac{1}{2} + 11 + \cdots + 20 = 114.5$$

3. Refer one of the sums to an appropriate table.
   Alternatively, for equal groups of size n, if S is the larger of the two sums, an approximate significance level is obtained by referring

$$z_0 = \sqrt{12}\,[S - \tfrac{1}{2}n(2n + 1)]/[n(2n + 1)^{1/2}]$$

to tables of the unit normal distribution. For this example $z_0 = 0.718$ and $\Pr(z > 0.718) = 0.236$. This probability of 23.6% is comparable to levels of 15.6 and 19.5% given by the other two approaches that assume random sampling. All these levels differ considerably, however, from the significance levels (4.7 and 2.8%) obtained with tests that do not rely on the random sampling hypothesis.

Wilcoxon's test is not really distribution free. It assumes the random sampling hypothesis† and gives erroneous results when that hypothesis is inappropriate. It is true that

---

* Since we use Wilcoxon's test as an example of what later writers called "distribution free tests," a word should be said about the originator of this procedure. Frank Wilcoxon was a scientist of the first rank who had many years of practical experience working with experimenters on the design and analysis of experiments at the Boyce Thomson Institute and later at the American Cyanamid Laboratories. He was well aware of the essential part that randomization played in the strategy of investigation and in the validation of his procedure. His principal motivation in introducing this test was that it could be *done quickly* at a time when aids to rapid computation were not yet available.

† Somewhat less restrictively, these tests can be justified on the assumption of certain symmetries in the joint error distribution. Thus for $n = 3$ it might be assumed that, for all values of $\epsilon_1, \epsilon_2,$ and $\epsilon_3,$

$$p(\epsilon_1, \epsilon_2, \epsilon_3) = p(\epsilon_1, \epsilon_3, \epsilon_2) = p(\epsilon_2, \epsilon_1, \epsilon_3) = p(\epsilon_2, \epsilon_3, \epsilon_1) = p(\epsilon_3, \epsilon_1, \epsilon_2) = p(\epsilon_3, \epsilon_2, \epsilon_1)$$

It is this kind of assumption that can be rendered absolutely true by the precaution of randomization, which we discuss in the next chapter. But without such a precaution it is as unlikely to be true as the assumption of independence of errors.

TABLE 3.3. **Percentage of 1000 results significant at the 5% level, using the t test (t) and the Wilcoxon test (W)**

a. Without randomization

parent distribution

		rectangular	normal	skew*
	0.0	t 5.6	t 5.4	t 4.7
		W 4.3	W 4.5	W 4.3
autocorrelation between successive observations	−0.4	t 0.5	t 0.3	t 0.1
		W 0.5	W 0.1	W 0.2
	+0.4	t 12.5	t 10.5	t 11.4
		W 11.0	W 9.6	W 10.1

b. With randomization

parent distribution

		rectangular	normal	skew*
	0.0	t 6.0	t 4.3	t 5.9
		W 5.8	W 4.1	W 4.4
autocorrelation between successive observations	−0.4	t 4.8	t 5.5	t 6.3
		W 4.3	W 4.9	W 5.6
	+0.4	t 5.9	t 5.8	t 5.4
		W 4.6	W 5.3	W 4.3

* A chi-square distribution with four degrees of freedom.

the Wilcoxon test and other such tests do not assume that the populations randomly sampled are *normal*. However, the distribution of any test statistic depends on the *joint* distribution $p(\epsilon_1, \epsilon_2, \ldots, \epsilon_n)$ of all the errors considered *simultaneously*. On the *random sampling hypothesis*, this joint distribution has the exceptional property that it may be obtained by multiplying the individual distributions

$$p(\epsilon_1, \epsilon_2, \ldots, \epsilon_n) = p(\epsilon_1)p(\epsilon_2) \cdots p(\epsilon_n) \tag{3.35}$$

If this is so, then the performance of the distribution free test does not depend on the choice of the *individual* distribution function $p(\epsilon)$. It could be a normal distribution, a rectangular distribution, a skew distribution, or any other kind of distribution. However, as we have seen, frequently errors would not be expected to be independently distributed. For example, errors from successive plots in an agricultural field trial or from successive measurements in time on an instrument are likely to be serially correlated. Some idea of the relative effects of different kinds of violations of assumptions on the performance of the t test and the corresponding distribution free test is gained from study of Table 3.3.*

The results in each cell of this table were obtained by taking two samples of 10 observations from identical populations and making a t test (t) and a Wilcoxon test (W) for a change in location. The sampling was repeated 1000 times, and the percentage of results significant at the 5% point were recorded. Since there was no real difference in location, this number should be 5.0. It has a standard deviation of 0.7 because of sampling errors. More accurate results may be obtained by taking larger samples or by using analytical procedures. However, since there is no practical difference between significance levels of, say, 4 and 6%, the present investigation suffices to illustrate the point.

The data were generated from three sets of independently distributed random variables u_i having (a) a rectangular distribution, (b) a normal distribution, and (c) a highly skewed distribution (a χ^2 distribution with four degrees of freedom). In the first two rows of Table 3.3a the observations had independent errors $e_i = u_i$. In the next two rows the observations had errors e_i generated from $e_i = u_i + \theta u_{i-1}$ with θ suitably chosen so that the first lag autocorrelation was $\rho_1 = -0.4$. In the last two rows θ was chosen to give $\rho_1 = +0.4$.

Notice that the significance level of the t test is affected remarkably little by dramatic changes in the parent distribution, for which the nonparametric test provides insurance. However, both the t and Wilcoxon tests *are* seriously impaired by serial dependence. Table 3.3b shows that after randomization (a precaution in the conduct of the experiment, discussed in Chapter 4) all the percentages are reasonably close to 5%. We see therefore that it is the act of randomization, and *not the introduction of a "distribution free" test function*, that plays the major part in making the procedure insensitive to distribution assumptions.

Thus, although nonparametric tests are occasionally useful in the analysis of highly nonnormal data from randomized experiments, their use does not of itself resolve the problem of dependent errors.

3.4. SUMMARY: WHAT HAVE WE LEARNED FROM THE INDUSTRIAL EXPERIMENT EXAMPLE?

Figure 3.5 allows comparison of the first four reference distributions considered. The associated assumptions and disadvantages for the various approaches, labeled (a), (b), (c), (d), and (e), are summarized in Table 3.4.

---

* Results of this kind are of course well known to statisticians but need to be emphasized.

Although we have so far discussed only the problem of the validity of a significance test for comparing two means, it will turn out that the ideas are much farther reaching. Let us summarize the discussion.

In the industrial experiment 10 batches were made under standard conditions and 10 under modified conditions, and it was believed that the 20 test runs were exactly comparable, with customary manufacture in all aspects except in the introduction of the modification. A plant record of 210 previous batch yields using the standard process was available.

1. We wished to assess the possible value of a modification in the manufacturing process.
2. As a first step it was relevant to ask, Can the hypothesis that the modification has had no effect be discounted? Such a hypothesis of no effect is called a *null* hypothesis.
3. To answer the question we needed a *relevant reference* distribution of differences that could be expected to occur if the null hypothesis was true.
4. If, on the one hand, the observed difference was a typical member of the reference distribution, there would be no reason to question the null hypothesis, and we would say that no *statistically significant* difference had been found.
5. If, on the other hand, differences as large as the one observed, or larger, rarely occurred in the reference set, there would be reason to doubt the null hypothesis, and we would say that a *statistically significant* difference had been found. The probability associated with a discrepancy as large as that obtained, or larger, is called the *level of significance*.
6. Obviously a vital question is: What is the relevent reference set for a given experimental situation? If an inappropriate reference set is chosen, the conclusions drawn from the experiment may be quite wrong.
7. For the physical situation described, the external reference set (a) based on the previous plant data is appropriate and yields a significance level of 4.7%. This probability is small, suggesting that the modification has indeed produced a shift in the yield distribution.
8. An alternative reference set (b) employs the fairly innocuous assumption that the 10 *differences in averages* having nonoverlapping sets of observations, computed from the previous 210 data values, are approximately normal and independently distributed. This yields a significance level of 2.8%, which agrees reasonably well with that for external reference set (a).
9. Attractive consequences would follow if the individual observations could be treated as *random samples* from parent populations differing, if at all, only in their mean values. Then $\bar{y}_B - \bar{y}_A$ would be distributed approximately normally with variance $\sigma^2(1/n_A + 1/n_B)$; and if the standard

TABLE 3.4. Assumptions, advantages, and disadvantages associated with various test procedures for the industrial experiment

test procedure	reference distribution for $\bar{y}_B - \bar{y}_A$	assumptions	significance level	advantages	disadvantages
(a)	191 differences of averages of 10 observations from previous 210 production runs with standard process (external)	Past production data can provide relevant reference set for observed difference $\bar{y}_B - \bar{y}_A$.	0.047	No assumption of independence of errors. No need for random sampling hypothesis.	Need relevant, lengthy past records. Construction of reference distribution can be tedious.
(b)	t distribution ($\nu = 10$) with scale factor 0.60 based on 10 approximately independent differences of averages of 10, using 210 production runs with standard process (external)	10 *differences* $\bar{y}_B - \bar{y}_A$ from previous plant data are approximately independent and normal. Past data can provide relevant reference set.	0.028	Rather innocuous additional assumptions supply continuous reference distribution that is less laborious to calculate.	Need relevant, lengthy past records plus (plausible) additional assumptions that *differences* are approximately independent and normally distributed.
(c)	normal distribution with 1.29 for standard deviation of the difference $\bar{y}_B - \bar{y}_A$ (external) $$\left(1.29 = 2.88\sqrt{\frac{1}{10} + \frac{1}{10}}\right)$$	*Individual* observations are as if obtained by *random sampling* from (normal) populations with common standard deviation $\sigma = 2.88$.	0.156	Continuous reference distribution that is easy to calculate.	Need to know σ. Need assumption of *independence* of *individual* errors coming from random sampling hypothesis (almost certainly invalid in the industrial experiment example) and also approximate normality.

(d) *t* distribution ($v = 18$) with scale factor 1.47 (internal)	*Individual* observations are as if obtained by *random sampling* from (normal) populations with unknown common standard deviation σ, estimated by $s = 3.30$ calculated from two sets of 10 observations each.	0.195	No external data needed. Reference distribution calculated from internal data (two sets of 10 observations).	Need assumption of *independence* of *individual* errors coming from random sampling hypothesis (almost certainly invalid in the industrial experiment example) and also approximate normality.
(e) **Wilcoxon**	*Individual* observations are as if obtained by *random sampling* from populations of (almost) any kind.	0.236	Computations are easy. No external data needed. **Populations randomly sampled need not be normal.**	Need assumption of *independence* or symmetry of individual errors arising from random sampling hypothesis (almost certainly invalid in the industrial experiment example).

deviation σ was known (e.g., from previous production data), reference set (c) would be appropriate. Since the random sampling hypothesis is almost certainly invalid for the present example, the value obtained, 15.6%, probably underestimates the significance of the observed difference in averages.

10. If the hypothesis of random sampling from normal populations were true, we would not need external data at all but could use the variation within the groups to provide reference set (d). Again, because of the almost certain violation of the assumption of random sampling, the significance level of 19.5% is inappropriate.

11. Tests are available that have the seductive title "distribution free." Unfortunately they make essentially the same random sampling assumption as the other inappropriate tests discussed. The significance level using such a test is 23.6%.

Exercise 3.11. Construct a figure corresponding to Figure 3.5, and determine the significance levels for the following data, using the five techniques listed in Table 3.4.
Past data on A: 25, 23, 27, 31, 32, 35, 40, 38, 38, 33, 27, 21, 19, 24, 17, 15, 14, 19, 23, 22
Test data: A 23, 28 B 37, 33

                                              Answer: 0.00, 0.05, 0.12, 0.05, 0.17.

Exercise 3.12. Repeat Exercise 3.11 with the following data.
Past data on A: 5, 5, 9, 7, 3, 4, 5, 8, 9, 12, 14, 8, 9, 11, 14, 9, 10, 10, 6, 5, 4, 2, 3, 3, 3, 8, 2, 3, 4, 6, 5, 3, 2, 4, 6, 4.
Test data: A 4, 3, 7 B 10, 6, 8   *Answer*: 0.10, 0.10, 0.12, 0.15, 0.10.

The last three approaches in Table 3.4, (c), (d), (e), all of which are based on assumption of independence of individual errors, give answers for the industrial experiment example that are probably misleading. As a consequence the reader may wonder why we have followed a road that seems to come to a dead end. In traveling this road he does not lack for company; theoretical statisticians frequently make the assumption of independence at the beginning of their writings and rest heavily on it thereafter, making no attempt to justify the assumption even though it might have been thought that "a decent Respect to the Opinions of Mankind requires that they should declare the Causes that impel them" to do so. The mere *declaration* of independence, of course, does not guarantee its existence.

R. A. Fisher, who perhaps advanced the theory of statistics more than any other single person, was himself a lifelong experimenter (experimental animals were to be found not only at his work but in his home as well). He not only saw clearly the dilemma we have raised but also realized near the beginning of his statistical career how to resolve it by randomization. That is the subject of the following chapter.

APPENDIX 3A.   MEAN AND VARIANCE OF A LINEAR COMBINATION OF OBSERVATIONS

Consider the random variables y_1, y_2, y_3 (not necessarily normal) with means η_1, η_2, η_3, variances $\sigma_1^2, \sigma_2^2, \sigma_3^2$, and correlation coefficients $\rho_{12}, \rho_{13}, \rho_{23}$. The mean and variance of the linear combination $Y = a_1 y_1 + a_2 y_2 + a_3 y_3$, where a_1, a_2, a_3 are any positive or negative constants, are, respectively,

$$E(Y) = a_1 \eta_1 + a_2 \eta_2 + a_3 \eta_3 \tag{3A.1}$$

and

$$V(Y) = a_1^2 \sigma_1^2 + a_2^2 \sigma_2^2 + a_3^2 \sigma_3^2 + 2a_1 a_2 \sigma_1 \sigma_2 \rho_{12} + 2a_1 a_3 \sigma_1 \sigma_3 \rho_{13} + 2a_2 a_3 \sigma_2 \sigma_3 \rho_{23} \tag{3A.2}$$

These formulas generalize in an obvious way for a linear combination of n variables. For n variables in the formula for $V(Y)$ there will be n "squared" terms like $a_i^2 \sigma_i^2$ and $n(n-1)/2$ cross-product terms like $2a_i a_j \sigma_i \sigma_j \rho_{ij}$. Specifically,

$$V(Y) = \sum_{i=1}^{n} a_i^2 \sigma_i^2 + 2 \sum_{i=1}^{n} \sum_{j=i+1}^{n} a_i a_j \sigma_i \sigma_j \rho_{ij} \tag{3A.3}$$

Notice that $\sigma_i \sigma_j \rho_{ij}$ is the covariance $\text{Cov}(y_i, y_j)$, so that equation 3A.3 may also be written as

$$V(Y) = \sum_{i=1}^{n} a_i^2 V(y_i) + 2 \sum_{i=1}^{n} \sum_{j=i+1}^{n} a_i a_j \text{Cov}(y_i, y_j) \tag{3A.4}$$

Variance of a Sum and a Difference of Two Random Variables

Since a sum $y_1 + y_2$ can be written as $1y_1 + 1y_2$, and a difference $y_1 - y_2$ can be written as $1y_1 + (-1)y_2$, we have

$$V(y_1 + y_2) = \sigma_1^2 + \sigma_2^2 + 2\rho_{12}\sigma_1\sigma_2 \tag{3A.5}$$

$$V(y_1 - y_2) = \sigma_1^2 + \sigma_2^2 - 2\rho_{12}\sigma_1\sigma_2 \tag{3A.6}$$

We see that, if the correlation between y_1 and y_2 is zero, the variance of the sum is the same as that of a difference. If the correlation is positive, the variance of the sum is greater; if the correlation is negative, the opposite is true.

No Correlation

Consider the statistic Y,

$$Y = a_1 y_1 + a_2 y_2 + \cdots + a_n y_n$$

a linear combination of n random variables $y_1, y_2, \ldots, y_n$, each of which is uncorrelated with any other, then

$$E(Y) = a_1 \eta_1 + a_2 \eta_2 + \cdots + a_n \eta_n \tag{3A.7}$$

which is the same as (the obvious extension of) Equation 3A.1 for correlated variables, but

$$V(Y) = a_1^2\sigma_1^2 + a_2^2\sigma_2^2 + \cdots + a_n^2\sigma_n^2 \tag{3A.8}$$

which is different from Equation 3A.3 for correlated variables.

No Correlation, Equal Variances

If in addition all the variances are equal, say to σ^2, $E(Y)$ remains as before, and

$$V(Y) = (a_1^2 + a_2^2 + \cdots + a_n^2)\sigma^2 \tag{3A.9}$$

Variances of an Average of n Random Variables with All Means Equal to η and All Variances Equal to σ², No Correlation

Since

$$\bar{y} = \frac{y_1 + y_2 + \cdots + y_n}{n} = \frac{1}{n}y_1 + \frac{1}{n}y_2 + \cdots + \frac{1}{n}y_n \tag{3A.10}$$

the average is a linear combination of observations with all the a's equal to $1/n$. Thus, given the assumptions that $E(y) = \eta$, all variances are equal to σ^2, and there is no correlation, then we have for the variance of $\bar{y}$

$$V(\bar{y}) = \left(\frac{1}{n^2} + \frac{1}{n^2} + \cdots + \frac{1}{n^2}\right)\sigma^2 = \frac{n}{n^2}\sigma^2 = \frac{\sigma^2}{n} \tag{3A.11}$$

Variance Unknown

If σ^2 is unknown but an estimate s^2 is available, we can replace σ^2 by s^2 in each of the equations above to obtain an *estimated* value $\hat{V}(Y)$ of the variance of Y. The square root $[\hat{V}(Y)]^{1/2}$ is often called the *standard error* of Y.

Variance of the Average ȳ for Observations That Are Autocorrelated at Lag 1

Suppose that observations $y_1, y_2, \ldots, y_n$ all have the same mean η and the same variance σ^2 and have a lag 1 autocorrelation ρ_1, which is nonzero. Suppose, however, that all other correlations are zero.

Consider the variance of $n\bar{y} = y_1 + y_2 + \cdots + y_n = \sum y$. This is obtained using Equation 3A.3 with

$$\sigma_1^2 = \sigma_2^2 = \cdots = \sigma_n^2 = \sigma^2$$
$$a_1 = a_2 = \cdots = a_n = 1$$
$$\rho_{12} = \rho_{23} = \cdots = \rho_{n-1,n} = \rho_1$$

and with all other correlations zero. Making the necessary substitutions, we have

$$V(n\bar{y}) = \sigma^2[n + 2(n - 1)\rho_1] \tag{3A.12}$$

that is,

$$V(\bar{y}) = \frac{\sigma^2}{n} \left[1 + \frac{2(n-1)}{n} \rho_1 \right] \tag{3A.13}$$

Consider the factor

$$C = 1 + 2\left(\frac{n-1}{n}\right)\rho_1 \tag{3A.14}$$

It may be shown that in the special case being considered, $-\frac{1}{2} < \rho_1 < \frac{1}{2}$. Thus C lies anywhere between $(2n-1)/n$ and $1/n$; for example, for $n = 10$, C lies between 1.9 and 0.1, as mentioned in the chapter. Therefore in this case the presence of first-order autocorrelation can change $V(\bar{y})$ by a factor of 19.

APPENDIX 3B. ROBUSTNESS OF SOME STATISTICAL PROCEDURES

Distribution of s²

As illustrated in Figure 3.4, when observations are randomly drawn from a normal distribution, s^2 has a distribution that is skewed to the right. Suppose that s^2 has v degrees of freedom; then in practice it is most convenient to work with the scaled quantity $\chi^2 = vs^2/\sigma^2$, where χ is the Greek letter chi and χ^2 is called *chi square*. The distribution of chi square has the form

$$p(\chi^2) = \text{constant} \cdot (\chi^2)^{(v/2)-1} e^{-\chi^2/2} \tag{3B.1}$$

This is an important distribution. Its significance points are given in Table C at the back of the book. In addition to providing the distribution of the sample variance, it has many other applications, some of which are discussed in Chapter 5, Section 4.

The mean of the distribution of s^2 is the population variance σ^2. Since the distribution is skewed, the mean does not correspond to the mode, the highest point of the curve. The variance of the distribution of s^2 is $2\sigma^4/v$, and consequently the standard error of s^2 is $\sigma^2\sqrt{2/v}$.

Distribution of ȳ and s² When Parent Distribution Is Not Normal

In analyzing actual data we most often use statistical tables based on the normal distribution. It is important, therefore, to have some idea of what will happen when the underlying distribution is not normal.

On the hypothesis of random sampling, irrespective of the nature of the parent distribution or of the number of observations n, the mean of the distribution of $\bar{y}$ is η and the variance of this distribution is σ^2/n. These are exactly the values found for a normal parent. When the parent distribution is not normal, the distribution of $\bar{y}$ will not be exactly normal. However, because of the central limit effect, the distribution will tend to normality as n is increased. Thus with moderate nonnormality the distribution of $\bar{y}$ will be approximately what it would have been *if its component observations* had a normal distribution. Procedures that rely directly on the distribution of $\bar{y}$ are thus *robust* (insensitive) to nonnormality.

A somewhat different situation exists for the sample variance s^2. No central limit effect like that for the sample average exists for the sample variance. However, irrespective of the parent distribution, the *mean value* of s^2 is σ^2, as it is for the normal parent. The variance of s^2, however, does not in general take the value $2\sigma^4/v$, appropriate when the parent distribution is normal; instead

$$V(s^2) = \frac{2\sigma^4}{n-1}\,c \tag{3B.2}$$

where $V(s^2)$ denotes the variance of s^2, and c is a constant that is equal to unity when the distribution is normal but in general has the value

$$c = 1 + \frac{1}{2}\left(\frac{n-1}{n}\right)\gamma_2 \tag{3B.3}$$

where

$$\gamma_2 = \frac{\mu_4}{\sigma^4} - 3 \tag{3B.4}$$

is a measure of nonnormality called *kurtosis*. For a normal distribution $\gamma_2 = 0$; for a distribution with an excess of density in the tails γ_2 is positive. The quantity μ_4 in Equation 3B.4 is the fourth moment of the population distribution

$$\mu_4 = \frac{\sum(y-\eta)^4}{N} \tag{3B.5}$$

Since many of the distributions met in practice have values of γ_2 that differ somewhat from zero, statistical procedures that depend directly or indirectly on the assumption that this quantity takes its normal theory value must be used with caution.

The situation can be summarized as follows:

	normal parent	nonnormal parent
sample average $\bar{y}$		
(1) mean	η	η
(2) variance	$\dfrac{\sigma^2}{n}$	$\dfrac{\sigma^2}{n}$
sample variance s^2		
(3) mean	σ^2	σ^2
(4) variance	$\dfrac{2\sigma^4}{n-1}$	$\dfrac{2\sigma^4}{n-1}\left(1 + \dfrac{n-1}{2n}\gamma_2\right)$

Robustness to Nonnormality of Certain Statistical Procedures

For the most part the statistical procedures discussed in this book depend for their *approximate* validity on relations (1), (2), and (3) above, or on analogous relations that are the same for normal or nonnormal parent distributions. Rather surprisingly perhaps, this is true of t tests to compare two means, and analysis of variance tests to compare several means that are discussed in later chapters. Occasionally methods are employed that depend on the validity of the normal theory relations (4). Examples are tests to compare several variances, such as Bartlett's test. Such procedures must be treated with much more reserve. When caution is needed, we will point this out.

APPENDIX 3C. FISHER'S CONCEPT OF SUFFICIENCY

In this chapter we mentioned that, for a *random sample* from a *normal* population with mean η and variance σ^2, $\bar{y}$ and s^2 are jointly *sufficient* statistics for η and σ^2. We said this meant that $\bar{y}$ and s^2 contained "all the information" in the sample about η and σ^2. We here go into this matter in a little more detail. For simplicity we suppose in what follows that we know σ^2, and consider what is meant by saying that $\bar{y}$ is a sufficient statistic for η.

Suppose that, instead of $\bar{y}$, we considered an alternative statistic as a measurement of η. For example, if we had an odd number of observations, we might arrange them in order of size and use the middle one (the median m) instead of $\bar{y}$ as an estimate of η. It can be shown that the distribution of m, *given* $\bar{y}$, is not a function of η, that is, $p(m\,|\,\bar{y})$ is independent of η. Furthermore it turns out that this is true *whatever* we choose as an alternative statistic for $\bar{y}$. This means that, once $\bar{y}$ is known, no other statistic can supply any further information about η. In this sense, then, we say that the statistic $\bar{y}$ contains all the information about the parameter η and that it is therefore a sufficient statistic for η. The other side of this coin is that when the assumptions of normality and independence are not true we can expect to find evidence of this in the (otherwise informationless) residuals.

QUESTIONS FOR CHAPTER 3

1. Define statistical dependence, randomness, correlation, covariance, autocorrelation, random sampling, and sampling distribution.
2. Discuss the properties of the sampling distribution of the average $\bar{y}$ from a nonnormal distribution.
3. Discuss the sampling distribution of $\bar{y}$, s^2, and $t = (\bar{y} - \eta)/(s/\sqrt{n})$ from a normal distribution.
4. What is meant by a reference distribution based on the hypothesis of random sampling, and how can it be used in analyzing data?
5. Can you find some data (or imagine some), preferably in your own field,

that might be appropriately analyzed by using a reference distribution based on the hypothesis of random sampling? Can you find (or imagine) some data that almost certainly could *not* be appropriately analyzed in this way?

6. Give formulas for (a) the sum and (b) the differences of correlated random variables. By taking sums and differences of successive pairs of observations, how might these formulas be used to construct a test for lag 1 autocorrelation?

7. Consider the second difference $y_i - 2y_{i-1} + y_{i-2}$. What is its variance if the successive observations y_i, $i = 1, 2, \ldots, n$ are independent? Autocorrelated with $\rho_1 = 0.2$?

CHAPTER 4

Randomization and Blocking with Paired Comparisons

The object of statistical methods is to ensure that iterations occurring in the process of investigation are so planned and the resulting data so analyzed that convergence on a solution is achieved as surely and expeditiously as possible. Because in preceding chapters we have seen how things can go wrong, we can now better appreciate what must be done to ensure that they go right. In this chapter we introduce (1) *randomization* to guarantee inferential validity in the face of unspecified disturbances and (2) *blocking* in pairs to eliminate unwanted sources of variability.

4.1. RANDOMIZATION TO THE RESCUE: TOMATO PLANT EXAMPLE

We have seen that desirable simplifications would follow *if* it could be assumed that data behaved as *random* samples from normal distributions. In particular, given n_A observations with treatment A and n_B with treatment B, the quantity

$$t = \frac{(\bar{y}_B - \bar{y}_A) - (\eta_B - \eta_A)}{s\sqrt{1/n_A + 1/n_B}} \tag{4.1}$$

(where $\bar{y}_A$ and $\bar{y}_B$ are the sample averages, η_A and η_B are the corresponding population means, and s is the pooled sample standard deviation) would be distributed in Student's t distribution with $\nu = n_A + n_B - 2$ degrees of freedom. This distribution would then provide a reference set by which the plausibility of any contemplated value of the difference in means $\eta_B - \eta_A$ could be judged. This procedure has the enormous advantage that the reference set is generated from the samples themselves. Unfortunately the

93

necessary assumption that the data may be treated as *random* samples from normal distributions is often not true.

Fisher pointed out that it is possible to take a precaution in the *actual performance of a comparative experiment* that ensures its validity by guaranteeing an appropriate reference distribution. This precaution is *randomization*.

Example: A Randomized Design Used in the Comparison of Standard and Modified Fertilizer Mixtures for Tomato Plants

The following data were obtained in an experiment conducted by an amateur gardener whose object was to discover whether a change in the fertilizer mixture applied to his tomato plants would result in an improved yield. He had 11 plants set out in a single row; 5 were given the standard fertilizer mixture *A*, and the remaining 6 were fed a supposedly improved mixture *B*. The *A*'s and *B*'s were randomly applied to the positions in the row to give the design shown in Table 4.1. The gardener arrived at this random arrangement by taking 11 playing cards, 5 red corresponding to fertilizer *A* and 6 black corresponding to fertilizer *B*. The cards were thoroughly shuffled and dealt to give the sequence shown in the table. The first card was red, the second was red, the third was black, and so forth.

TABLE 4.1. Results from a randomized experiment (yields of tomatoes in pounds)

position in row	1	2	3	4	5	6	7	8	9	10	11
fertilizer	*A*	*A*	*B*	*B*	*A*	*B*	*B*	*B*	*A*	*A*	*B*
pounds of tomatoes	29.9	11.4	26.6	23.7	25.3	28.5	14.2	17.9	16.5	21.1	24.3

standard fertilizer *A*	modified fertilizer *B*
29.9	26.6
11.4	23.7
25.3	28.5
16.5	14.2
21.1	17.9
	24.3
$n_A = 5$	$n_B = 6$
$\sum y_A = 104.2$	$\sum y_B = 135.2$
$\bar{y}_A = 20.84$	$\bar{y}_B = 22.53$

difference in averages (modified minus standard) $\bar{y}_B - \bar{y}_A = 1.69$

A Test of Significance Based on the Randomization Reference Distribution

Fisher argued that with such a *randomized* experiment it is possible to conduct a significance test without making any assumptions about the distribution (a truly distribution free procedure). The reasoning is as follows. The null hypothesis is that modifying the fertilizer mixture has no effect on the results and therefore, in particular, no effect on the mean. *On this null hypothesis* fertilizers *A* and *B* are mere labels and do not affect the outcome. For example, the first plant would have given 29.9 pounds of tomatoes whether it had been labeled *A* or *B*. The alternative hypothesis is that the modified fertilizer gives a higher mean.

The experimental arrangement is one of the $11!/5!6! = 462$ possible ways* of allocating 5 *A*'s and 6 *B*'s to the 11 trials, any one of which could equally well have been chosen. The randomization distribution appropriate to the hypothesis that the modification is without effect is obtained by calculating the 462 differences in the averages obtained from the 462 rearrangements of the "labels" in the second row of Table 4.1. The histogram in Figure 4.1 shows this randomization distribution. It also shows the difference in averages, $\bar{y}_B - \bar{y}_A = 1.69$, that was actually observed. Specifically, in this example it is found that 154 of the possible 462 arrangements yield differences greater than 1.69, the value actually observed, yielding a significance level of $154/462 = 0.33$ (or 33%). There is therefore no reason to doubt the null hypothesis.

The t Distribution as an Approximation to the Randomization Distribution

On the hypothesis of *random* sampling from *normal* distributions, we would refer the quantity

$$t_0 = \frac{(\bar{y}_B - \bar{y}_A) - (\eta_B - \eta_A)_0}{s\sqrt{1/n_A + 1/n_B}} \tag{4.2}$$

to Student's *t* table with $v = n_A + n_B - 2$ degrees of freedom. Since the mean difference $(\eta_B - \eta_A)_0$ of interest is zero in this example, the quantity

$$t_0 = \frac{1.69 - 0}{3.82} = 0.44 \tag{4.3}$$

is entered in a *t* table with nine degrees of freedom (see calculations in Table 4.2). This would be equivalent to referring the value $\bar{y}_B - \bar{y}_A = 1.69$ itself to a *scaled t* distribution with the standard error of $(\bar{y}_B - \bar{y}_A)$ as the scale

---

* $5! = 5 \cdot 4 \cdot 3 \cdot 2 \cdot 1 = 120$, and. in general. $k! = k(k-1)(k-2) \cdots 3 \cdot 2 \cdot 1$. Note that $0! = 1$.

TABLE 4.2. Calculation of t for tomato experiment

$$\bar{y}_B - \bar{y}_A = 22.53 - 20.84 = 1.69$$

$$s_A^2 = \frac{\sum y_A^2 - (\sum y_A)^2/n_A}{n_A - 1} = \frac{S_A}{\nu_A} = \frac{209.9920}{4} = 52.50$$

$$s_B^2 = \frac{\sum y_B^2 - (\sum y_B)^2/n_B}{n_B - 1} = \frac{S_B}{\nu_B} = \frac{147.5333}{5} = 29.51$$

The pooled variance estimate s^2 is

$$s^2 = \frac{S_A + S_B}{\nu_A + \nu_B} = \frac{\nu_A s_A^2 + \nu_B s_B^2}{\nu_A + \nu_B} = \frac{4(52.50) + 5(29.51)}{4 + 5} = 39.73$$

with $\nu = n_A + n_B - 2 = \nu_A + \nu_B = 9$ degrees of freedom.
The estimated variance of $\bar{y}_B - \bar{y}_A$ is $s^2(1/n_B + 1/n_A) = 39.73(\frac{1}{6} + \frac{1}{5}) = 14.57$

The standard error of $\bar{y}_B - \bar{y}_A$ is $\sqrt{14.57} = 3.82$

$$t_0 = \frac{(22.53 - 20.84) - (\eta_B - \eta_A)_0}{\sqrt{39.73(\frac{1}{6} + \frac{1}{5})}} = \frac{1.69 - (\eta_B - \eta_A)_0}{3.82}$$

where $(\eta_B - \eta_A)_0$ is the hypothesized value of $\eta_B - \eta_A$. If $(\eta_B - \eta_A)_0 = 0$,

$$t_0 = 0.44$$

and

$$\Pr(t \geq t_0) = 0.34$$

factor, that is, 3.82. The appropriately scaled t distribution is superimposed on the randomization distribution in Figure 4.1. The two are in fair agreement; in particular, the significance levels are as follows:

randomization distribution   0.33
t distribution   0.34

It would be tedious to compute the randomization distribution every time a test of significance was made. However, it is possible to show that the randomization reference distribution is usually approximated reasonably well by the appropriately scaled t distribution. Hence, *provided that we randomize*, we can employ t tests (and other procedures to be mentioned later) as *approximations to exact randomization tests*, and we will be free of the random sampling assumption as well as the assumption of exact normality.

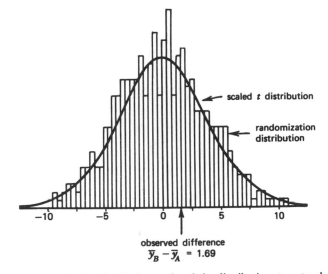

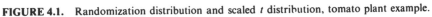

FIGURE 4.1. Randomization distribution and scaled t distribution, tomato plant example.

Exercise 4.1. Given the following data from a randomized experiment, construct the randomization reference distribution and the approximating scaled t distribution. What is the significance level in each case?

$$A \quad B \quad B \quad A \quad B$$
$$3 \quad 5 \quad 5 \quad 1 \quad 8$$

Answer: 0.05, 0.04.

Exercise 4.2. Repeat Exercise 4.1 with these data:

$$B \quad A \quad B \quad A \quad A \quad A \quad B \quad B$$
$$32 \quad 30 \quad 31 \quad 29 \quad 30 \quad 29 \quad 31 \quad 30$$

Answer: 0.02, 0.01.

4.2. RANDOMIZED PAIRED COMPARISON DESIGN: BOYS' SHOES EXAMPLE

Often we can greatly increase precision by making comparisons *within matched pairs* of experimental material. Randomization can again ensure the validity of such experiments. We illustrate with data from a trial on boys' shoes.

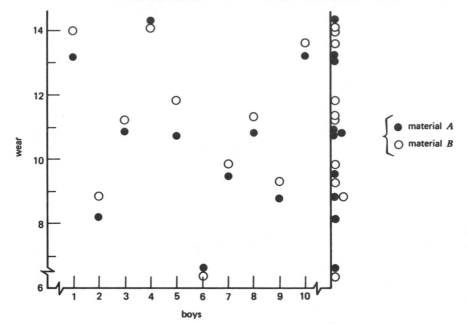

FIGURE 4.2. Data on two different materials *A* and *B*, used for making soles of boys' shoes.

An Experiment on Boys' Shoes

The data in Figure 4.2 are measurements of the amount of wear of the soles of shoes worn by 10 boys. The shoe soles were made of two different synthetic materials, *A* and *B*. As is seen from the marginal dot diagram on the right side of the figure, the results overlap extensively and do not at first sight suggest that one material is better than the other. Calculation with the form of *t* test given by Equation 4.2 confirms this conclusion.

An important fact so far omitted, however, is that the experiments were run *in pairs*. Each boy wore a special pair of shoes, the sole of one shoe having been made with *A* and the sole of the other with *B*. The decision as to whether the left or the right sole was made with *A* or *B* was determined by the flip of a coin. Figure 4.3 shows the *difference* in wear ($B - A$) for each boy. The marginal dot diagram on the right-hand side of this figure now clearly indicates that, for these 10 boys, material *A* usually shows less wear than *B*.

Why Do the Diagrams Seem to Tell Different Stories?

In Figure 4.3 the variability among the boys has been eliminated. During the test no two boys walked, ran, jumped, kicked, and scuffed the same on the

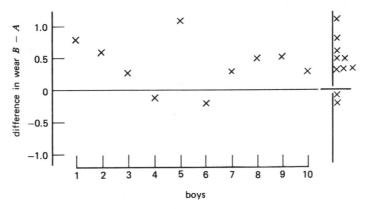

FIGURE 4.3. Differences $B - A$ for data in Figure 4.2, boys' shoes example.

same surfaces or in the same puddles for the same amount of time. The rates at which they wore out the soles of the shoes, therefore, were quite varied. However, pairs of feet go around together, and by working with the 10 differences $B - A$ most of this boy-to-boy variation could be eliminated. An experimental design of this kind is called a *randomized paired comparison design*.

Statistical Analysis of the Data from the Paired Comparison Design

Material A was standard, and B was a cheaper substitute that, it was feared, might result in an increased amount of wear. The immediate purpose of the experiment was to test the hypothesis that no change in wear resulted when switching from A to B against the alternative that there was increased wear with material B. The data for this experiment are given in Table 4.3.

A Test of Significance Based on the Randomization Reference Distribution

Randomization for the paired design was accomplished by tossing a coin. A head meant that material A was used on the right foot. The sequence of tosses actually obtained was as follows:

$$T \quad T \quad H \quad T \quad H \quad T \quad T \quad T \quad H \quad T$$

leading to the allocation of treatments shown in Table 4.3.

Consider the null hypothesis that the amounts of wear associated with materials A and B are the same. On this hypothesis the labeling given to a

TABLE 4.3.   Data on the amount of wear measured with
two different materials A and B, boys' shoes
example*

boy	material A	material B	$B - A$ difference d
1	13.2(L)	14.0(R)	0.8
2	8.2(L)	8.8(R)	0.6
3	10.9(R)	11.2(L)	0.3
4	14.3(L)	14.2(R)	−0.1
5	10.7(R)	11.8(L)	1.1
6	6.6(L)	6.4(R)	−0.2
7	9.5(L)	9.8(R)	0.3
8	10.8(L)	11.3(R)	0.5
9	8.8(R)	9.3(L)	0.5
10	13.3(L)	13.6(R)	0.3
		average difference	= 0.41

* (L) indicates this material was used on the left sole; (R),
that it was used on the right sole.

pair of results is quite unrelated to their individual values and merely affects
the *sign* associated with the differences. The sequence of 10 coin tosses is
one of $2^{10} = 1024$ equiprobable outcomes. To test the null hypothesis,
therefore, the average difference 0.41 actually observed may be compared
with the other 1023 average differences that could have occurred as a result
of different outcomes of coin tosses. The complete set of 1024 possible differences is obtained by averaging the 10 differences with all possible combinations of plus and minus signs, that is,

$$\bar{d} = \frac{\pm 0.8 \pm 0.6 \pm \cdots \pm 0.3}{10} \tag{4.4}$$

Clearly the value $\bar{d} = 0.41$ that actually occurred is quite unusual. Only three
samples produced by the randomization process give values of $\bar{d}$ greater than
0.41. These are samples in which the eight positive differences remain the
same and one or both of the negative differences (boys 4 and 6) have opposite
signs. Four further samples give values of $\bar{d}$ equal to 0.41. If we include half
these ties, we obtain a significance level of $5/1024 = 0.005$ (0.5%). Thus the
conclusion from this randomization test is that a highly statistically significant increase in the amount of wear is associated with the cheaper material B.

A t Distribution as an Approximation to the Reference Distribution

Consider again the differences $0.8, 0.6, \ldots, 0.3$. If we accepted the hypothesis of random sampling of the d's from a normal population with mean δ, we could use the t distribution to compare $\bar{d}$ and δ. (See Figure 3.4 and the discussion at the end of Section 3.1, but notice that the present design is not open to the criticisms advanced there.) If there are n differences, then $(\bar{d} - \delta)/(s_d/\sqrt{n})$ is distributed as t with $n - 1$ degrees of freedom, where

$$s_d^2 = \frac{\sum (d - \bar{d})^2}{n - 1} = \frac{\sum d^2 - (\sum d)^2/n}{n - 1} \tag{4.5}$$

Thus

$$s_d^2 = \frac{3.030 - 1.681}{9} = 0.149 \tag{4.6}$$

$$s_d = \sqrt{0.149} = 0.386, \quad \text{and} \quad \frac{s_d}{\sqrt{n}} = \frac{0.386}{\sqrt{10}} = 0.122$$

On the null hypothesis δ equals zero, the reference distribution against which the observed $\bar{d} = 0.41$ may be viewed is a scaled t distribution with nine degrees of freedom centered at zero with a scale factor of 0.122. The value of t_0 associated with the null hypothesis $\delta = 0$ is

$$\frac{0.41}{0.12} = 3.4 \tag{4.7}$$

and by referring this to the t table with nine degrees of freedom we get

$$\Pr(t \geq 3.4) \simeq 0.004 \tag{4.8}$$

This significance level of 0.4% agrees closely with the result given by the randomization distribution.

Once more, *provided that we take the precaution of randomization* (necessary in any case to ensure the validity of the experiment), we may employ a randomization test to assess statistical significance, a sufficiently close approximation to which is usually obtained by appropriate use of the t distribution. Usually, therefore, we will not go far astray if we analyze randomized paired experiments using the appropriate t statistic.

Exercise 4.3. Given these data from a randomized paired comparison design and using the usual null hypothesis, construct the randomization distribution and the corresponding scaled t distribution, and calculate the significance levels based on each:

B	A	A	B	A	B	B	A	A	B
7	9	3	5	8	12	11	4	4	6

Exercise 4.4.   Repeat Exercise 4.3 with these data:

B	A	B	A	A	B	A	B	B	A	B	A	A	B	B	A
29	29	32	30	23	25	36	37	39	38	31	31	27	26	30	27

4.3.   BLOCKING AND RANDOMIZATION

The device of pairing observations is a special case of *blocking* that has important applications in many kinds of experiments. A block is a portion of the experimental material (the two shoes of one boy in this example) that is expected to be more homogeneous than the aggregate (all the shoes of all the boys). By confining treatment comparisons within such blocks, greater precision can often be obtained.

In the paired design the "block size" is two, and we compare two treatments *A* and *B*. Suppose, however, that we were interested in comparing four treatments applied to horseshoes. Then, making use of the happy circumstance that a horse has four feet, we could run this experiment with a block size of four. Such a trial could employ a number of horses, with each horse having each of the four differently treated horseshoes applied randomly to one of his four feet. We discuss such uses of larger blocks in Chapters 7 and 8.

Pairs (Blocks) in Time and Space

Runs made close together in time or space are often (but not always) more similar than runs further apart and hence can often provide a basis for blocking. For example, suppose that in the comparison of two treatments *A* and *B* two runs could be made each day. If there was reason to think that conditions on the same day would, on the whole, be more alike than those on different days, it could be advantageous to run the trial as a paired test. In this test a block would be an individual day, and the order of running on a particular day would be decided at random, for example, by flipping a coin.

In the comparison of methods of treating leather specimens, pieces of leather close together on the hide would be expected to be more alike than pieces farther apart. Thus, in an experiment to compare treatments *A* and *B* applied to leather, a number of 6-inch squares might be cut from several hides. Each square might then be cut in two, and treatments *A* and *B* randomly applied to the halves.

A Possible Improvement in the Design of the Experiment on Tomato Plants

At the beginning of this chapter we considered a randomized experiment carried out by a gardener to compare the effects of two different fertilizer mixtures on his tomato plants. The fully randomized arrangement he used was a valid one. However, an equally valid arrangement in randomized pairs would probably have been more sensitive, in the sense that the error in the estimated difference would be smaller and hence the chance of detecting small differences would be greater. Plots close together might be expected to be more alike and could be used as the basis for pairing. With six plots allocated to each fertilizer, the arrangement might have looked like that shown below, where parentheses indicate a pair (or block) of adjacent plots:

$$(B\ A)\quad (B\ A)\quad (A\ B)\quad (B\ A)\quad (A\ B)\quad (B\ A)$$

The relevant error would now arise only from differences between *adjacent* plots, and this could be considerably less than the error of the unpaired design, which measures differences among all the plots.

The paired design, however, is not always more sensitive. In the industrial data (Table 2.1), for example, adjacent observations are *negatively* correlated, so that comparisons within pairs are *less* alike than other comparisons. Furthermore in an experiment using, for example, 12 plots for an unpaired design the reference t distribution would have 10 degrees of freedom. For the paired design it would have only 5 degrees of freedom. Thus we would gain from the paired design only if the reduction in variance from pairing outweighed the effect of the decrease in the number of degrees of freedom of the t distribution.

Block What You Can and Randomize What You Cannot

Fisher saw the lack of independence in experimental material both as an opportunity and as a challenge. Consider the tomato plant example. On the one hand, positive correlation between adjacent plot yields may be exploited by blocking to obtain greater precision. On the other hand, randomization can approximately validate the t test derived on the assumption of *random* sampling from normal populations.

Although blocking and randomization are valuable devices for dealing with *unavoidable* sources of variability, they should not be used as excuses to avoid hard thinking about sources of variability that are *avoidable*. A good experimenter will identify important extraneous factors ahead of time and eliminate them insofar as is practical from comparisons between treatments *within* a block (pair). Representative variation *between* blocks should however be encouraged. For instance, whereas care should be taken that the two shoes in each pair are as alike as possible, except in the property under consideration, it would be advantageous to introduce pairs of shoes of different styles and boys of different habits reasonably representative of the real world.

Randomization and "Distribution free" Tests

The randomization tests mentioned in this chapter and introduced by Fisher in 1935 and 1936 were the first examples of what were later to be called, in spite of Fisher's protest, "nonparametric" or "distribution free" tests. It seems a pity that the name *randomization test* was ever replaced, for it is, as the reader will have appreciated, the physical act of randomization applied to the experimental material that renders the procedure distribution free. Without randomization, as we have seen, "distribution free" tests are just as seriously affected by lack of independence in the parent distribution as are other tests (see Table 3.3). It is not suggested, of course, that mathematical theorists are unaware of this limitation, but it is our experience that users of the tests are frequently misled by the inappropriate name. In particular many users do not know that (1) *unless* randomization has been performed, the "distribution free" tests do not possess the properties claimed for them, and (2) *if* randomization is performed, standard parametric tests such as the *t* test usually supply adequate approximations.

4.4. NOISE STRUCTURE, MODELS, AND RANDOMIZATION

The industrial experiment described earlier illustrated how discrepancies from assumption, undetectable in available data, can invalidate inferential procedures. Fisher has shown how we can extricate ourselves from this seemingly insurmountable difficulty. By introducing randomization as part of the physical conduct of the experiment itself, we can validate our inferential procedures, whatever the form of the unknown disturbances. Autocorrelation between errors such as occurred in the industrial data is only one of an infinity of possible discrepancies from ideal assumptions.

A statistical model is a statement of the kind

$$y = \text{structural model} + \text{error}$$

or

$$y = f(\mathbf{X}_1) + \epsilon \qquad (4.9)$$

In this equation $\mathbf{X}_1$ stands for one or more variables whose values we know in each experimental run and whose effect we wish to study (in a comparative experiment $\mathbf{X}_1$ could indicate whether a particular plot was treated with A or with B). Now it is convenient mathematically to endow the errors ϵ with properties that make problems easy to solve. For example, it is often assumed that they can be regarded as *random* variables with a very specific form of distribution (independent, identical, and normal with mean zero and constant variance). More realistically, however, the model should be written as

$$y = f(\mathbf{X}_1) + \epsilon(\mathbf{X}_2) \qquad (4.10)$$

where X_2 stands for all the additional variables that might affect the result, but about which we know little or nothing.

Now, in fact, all kinds of odd things go into the variables X_2. Consider the gardener and his tomato plants. It could be (unknown to him) that the fifth and ninth plots were unusually fertile because they were close to decaying organic matter, that the second and sixth plots had an unusually large number of stones hidden beneath the surface, making root growth very difficult, or that because of a shadow cast by a nearby tree the west end of the strip of land tended to get more sunshine than the east end. All the factors just mentioned are *fixed* (but unknown) disturbances, and it is questionable whether their joint effect could be approximately represented in the formal way described above unless some appropriate element of chance was specifically introduced into the experiment. That element of chance is randomization. Randomization ensures that the probability that any particular plot receives any particular treatment is a matter of chance, and this, in turn, ensures the validity of the randomization test and hence approximately of the appropriate t test.

4.5. SUMMARY: COMPARISON, REPLICATION, RANDOMIZATION, AND BLOCKING IN SIMPLE COMPARATIVE EXPERIMENTS

What have we learned about the conduct of experiments to assess the possible difference between two treatments* A and B?

1. Experiments should be comparative. For example, if we are testing a modification, the modified and unmodified procedure should be run side by side in the same experiment.
2. There should be genuine replication. Both A and B runs should usually be carried out several times. Furthermore, this should be done in such a way that variation among replicates can provide an accurate measure of errors that affect comparisons made between an A run and a B run.
3. Whenever appropriate, blocking (pairing) should be used to reduce error. Similarity of basic conditions for pairs of runs provides a basis for blocking, for example, those made on the same day, from the same blend of raw materials, with animals from the same litter, or on shoes from the same boy.

---

* *Treatment* is a generic term borrowed from agricultural research that stands for fertilizers, machines, methods, processes, materials, or whatever the two (possibly more) things are that are being compared. It is widely used in the literature of statistical experimental design.

4. Having eliminated "known" sources of discrepancy, either by holding them constant during the experiment or by blocking, unknown discrepancies should be forced by randomization to contribute homogeneously to both *A* and *B* runs. This will generate an estimate of error that is appropriate to the comparisons made and will also approximately validate standard tests.

REFERENCES AND FURTHER READING

Fisher, R. A. (1966), *Design of Experiments*, 8th ed., Hafner (Macmillan).
Cox, D. R. (1958). *Planning of Experiments*, John Wiley.

QUESTIONS FOR CHAPTER 4

1. What is randomization, and what is its value?
2. How can experiments be randomized in practice? Be specific.
3. What is the value of pairing and blocking? In practice, how can a design be blocked?
4. Can you find (or imagine) an example of an experiment, preferably in your own field, in which both randomization and blocking were (or could have been) used?
5. Can you describe an actual situation, preferably in your own field, in which trouble was encountered because the data were not obtained from a randomized experiment? If you cannot describe an actual situation, can you imagine one?
6. What is a randomization distribution?

CHAPTER 5

Significance Tests and Confidence Intervals for Means, Variances, Proportions, and Frequencies

In order to introduce elementary ideas in the design and analysis of experiments it has been convenient so far to talk exclusively in terms of significance tests. The reader at this stage will be in some danger of over-emphasizing the importance of this particular mode of statistical inference. Often the investigator already knows that there will *be* an effect; what he wants to do is to *estimate* its magnitude and calculate an *interval* within which the true value almost certainly lies. Such an interval is called a *confidence interval*. Suppose that η, the amount of lead in a sample of water, is measured in milligrams per cubic centimeter and that a water chemist has made a number of tests that produce an estimate of η equal to 0.0054. A confidence interval statement for η would be of the form "my tests show that the interval extending from 0.0047 to 0.0061 includes the true value η with 99 % probability."

In this chapter we consider how such intervals may be calculated. The results are first applied to the comparison of means. Then significance tests and associated confidence intervals are considered for variances, pro-portions, and frequencies. The chapter ends with a discussion of the contin-gency table, an important tool for the analysis of categorical data.

5.1. A MORE DETAILED DISCUSSION OF SIGNIFICANCE TESTS

Consider again the experiment on boys' shoes of Section 4.2, where it was feared that the use of a cheaper material B instead of A might increase the amount of wear of the shoe sole. Ten wear differences d were observed in a randomized paired design. Their average was $\bar{d} = 0.41$, and their estimated standard deviation was $s_d = 0.39$, yielding an estimated standard error for the average difference of $s_{\bar{d}} = 0.39/\sqrt{10} = 0.12$.

That the mean difference δ had some hypothesized value δ_0 could be tested by referring

$$t_0 = \frac{\bar{d} - \delta_0}{s_{\bar{d}}} = \frac{0.41 - \delta_0}{0.12} \tag{5.1}$$

to the t table with nine degrees of freedom. In particular, for $\delta_0 = 0$, $t_0 = 0.41/0.12 = 3.4$, and

$$\Pr(t > t_0) = \Pr(t > 3.4) \simeq 0.004 \tag{5.2}$$

In other words, if the true mean difference δ were zero, a deviation as large as that experienced, or larger, would occur by chance *in the expected positive direction* only about 4 times in 1000.

Somewhat more formally we could say that

H_0, the *hypothesis to be tested*, is the "null" hypothesis that the true mean difference δ is zero

and

H_1, the *alternative hypothesis*, is that δ is greater than zero

or, more briefly,

H_0 is that $\delta = \delta_0 = 0$, H_1 is that $\delta > \delta_0 = 0$

One- and Two-Sided Significance Tests

If the modification under test could have affected the wear equally well in either direction, we might wish to test the hypothesis that the true difference was zero against the alternative that it could be *greater or less* than zero, that is,

H_0 is that $\delta = \delta_0 = 0$, H_1 is that $\delta \neq \delta_0 = 0$

We would now ask how often t would exceed 3.4 *or fall short of* -3.4. Because the t distribution is symmetric, the required probability is obtained by doubling the previously obtained probability, that is,

$$\Pr(|t| > |t_0|) = \Pr(|t| > 3.4) = 2 \times \Pr(t > 3.4) \simeq 0.008 \tag{5.3}$$

If the true difference were zero, a deviation *in either direction* as large as that experienced, or larger, would occur by chance only about 8 times in 1000.

Exercise 5.1. Given the data below from a randomized paired design, calculate the t statistic for testing the hypothesis $\delta = 0$ and the probability associated with the two-sided significance test (interpolate graphically in the t table):

A	B	A	B	B	A	A	B	B	A
3	5	8	12	11	4	2	10	9	6

Answer: 0.014.

Conventional Significance Levels

A series of conventional "critical" significance levels is in common use. These levels correspond to probabilities representing varying degrees of skepticism. When the probability that a discrepancy as large as that observed, or larger, might occur is smaller than one of these critical probabilities, the discrepancy between observation and hypothesis is said to be "significant"* at that level. As a guide, it could be said that, when one's attitude is a priori "neutral" to a particular type of discrepancy, one begins to be slightly suspicious of a discrepancy at the 0.20 level, somewhat convinced of its reality at the 0.05 level, and fairly confident of it at the 0.01 level. In practice, an experimenter's *prior* belief in the possibility of a particular type of discrepancy must affect his attitude. If the alternative hypothesis was plausible a priori, the experimenter would feel much more confident of a result significant at the 0.05 level than if it seemed to contradict all previous experience.

Although conventional significance levels provide a yardstick, it is always best to state the probability itself, rather than to say that a result is either significant or not significant at some conventional level. The statement that a particular deviation is "not significant at the 0.05 level" is sometimes found to mean, on closer examination, that the actual probability is 0.06. The difference in mental attitude to be associated with a probability of 0.05 and one of 0.06 is negligible, of course.

Significance testing in general has been a greatly overworked procedure, and in many cases where significance statements have been made it would have been better to provide an interval within which the value of the parameter would be expected to lie. Interval statements of this kind are now discussed.

---

* The choice of the word *significant* in this context is perhaps unfortunate. It refers not to the *importance* of a hypothesized effect but to its *plausibility* in the light of the data. In other words, a particular result may be statistically significant but scientifically unimportant.

5.2. CONFIDENCE INTERVALS FOR A DIFFERENCE IN MEANS: PAIRED COMPARISON DESIGN

The hypothesis H_0 of interest is not always the null hypothesis of "no difference." Suppose that in the boys' shoes example increased wear of the cheaper material could be tolerated so long as it was not greater than 0.10. It might then be asked: Do the data contradict the hypothesis H_0 that $\delta = \delta_0 = 0.10$ (the alternative H_1 being that $\delta > 0.10$)?

From Equation 5.1 the appropriate test statistic is

$$t_0 = \frac{0.41 - 0.10}{0.12} = 2.6 \tag{5.4}$$

and since for nine degrees of freedom

$$\Pr(t > 2.6) \simeq 0.014 \tag{5.5}$$

the hypothesis that $\delta = 0.10$ is discredited at the 0.014 level of significance.

If the alternative had been that δ could be greater or less than 0.10, the appropriate significance level would be $\Pr(|t| > 2.6) \simeq 0.029$. We have then the following results:

tested hypothesis H_0	significance level (two-sided test)
$\delta_0 = 0.00$	0.008
$\delta_0 = 0.10$	0.029

Obviously we could substitute a whole series of values for δ_0 and calculate the corresponding levels of significance.

A $1 - \alpha$ confidence interval for the true difference is such that, using a two-sided significance test, all values of δ within the confidence interval do not produce a significant discrepancy with the data at the chosen level of probability α, whereas all the values of δ outside the interval do show such a discrepancy. The quantity $1 - \alpha$ is sometimes called the *confidence coefficient*.

For nine degrees of freedom $\Pr(|t| > 2.262) = 0.05$. Thus all values of δ for which

$$\left| \frac{0.41 - \delta}{0.12} \right| < 2.262 \tag{5.6}$$

would not be discredited by a two-sided significance test made at the 0.05 level. Consequently they would define a 0.95 or 95% confidence interval. The 95% confidence *limits* for δ are thus

$$0.41 \pm 2.262 \times 0.12 \tag{5.7}$$

or

$$0.41 \pm 0.27 \tag{5.8}$$

This interval extends from 0.14 to 0.68. Figure 5.1 shows how the two values $\delta_- = 0.14$ and $\delta_+ = 0.68$, called confidence *limits*, are the values with which the observed difference $\bar{d} = 0.41$ just achieves a significant discrepancy.

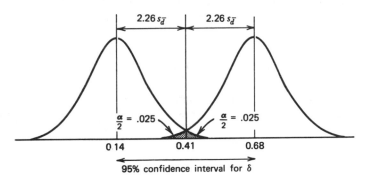

FIGURE 5.1. Generation of 95% confidence interval for δ, boys' shoes example.

Exercise 5.2. For the data given in Exercise 5.1, compute 80% confidence limits for the mean difference δ.   *Answer*: 3.0, 6.6.

In general the $1 - \alpha$ confidence limits for δ are

$$\delta_- = \bar{d} - t_{\alpha/2} s_{\bar{d}} \quad \text{and} \quad \delta_+ = \bar{d} + t_{\alpha/2} s_{\bar{d}}$$

that is,

$$\bar{d} \pm t_{\alpha/2} s_{\bar{d}} \tag{5.9}$$

where, with n differences, $d_1, d_2, \ldots, d_n$,

$$s_{\bar{d}}^2 = \frac{s_d^2}{n} = \frac{\sum_{u=1}^{n} (d_u - \bar{d})^2}{(n-1)n} \tag{5.10}$$

The value $t_{\alpha/2}$ denotes the t deviate in Table B1 at the end of this book, corresponding to the single-tail area $\alpha/2$.

The interval can be derived in the following way:

$$\Pr\left(\left|\frac{\bar{d} - \delta}{s_{\bar{d}}}\right| < t_{\alpha/2}\right) = 1 - \alpha \tag{5.11}$$

$$\Pr(|\bar{d} - \delta| < t_{\alpha/2}\, s_{\bar{d}}) = 1 - \alpha \tag{5.12}$$

$$\Pr(\bar{d} - t_{\alpha/2}\, s_{\bar{d}} < \delta < \bar{d} + t_{\alpha/2}\, s_{\bar{d}}) = 1 - \alpha \tag{5.13}$$

The meaning of Equation 5.13 is as follows. If a *series* of random sets of n differences is sampled from the appropriate normal distribution of differences with mean δ and with any fixed σ, and a $1 - \alpha$ confidence interval $\bar{d} \pm t_{\alpha/2}\, s_{\bar{d}}$ is constructed from each set, a proportion $1 - \alpha$ of these intervals will include the value δ and a proportion α will not. For example, for $\alpha = 0.05$, 95% of such intervals will include the population mean value δ and 5% will not.

Confirmation Using Randomization Theory

Randomization theory can be used to verify the values $\delta_{-} = 0.14$ and $\delta_{+} = 0.68$, which just produce a significant discrepancy with the data and so delineate the confidence interval. The necessary data for making the calculations are given in Table 5.1. The original differences are taken from Table 4.3.

TABLE 5.1. Calculations for confirmation via randomization theory of confidence interval (original differences from Table 4.1), boys' shoes example

(1) original difference	(2) difference -0.68	(3) difference -0.14
0.80	0.12	0.66
0.60	-0.08	0.46
0.30	-0.38	0.16
-0.10	-0.78	-0.24
1.10	0.42	0.96
-0.20	-0.88	-0.34
0.30	-0.38	0.16
0.50	-0.18	0.36
0.50	-0.18	0.36
0.30	-0.38	0.16
average 0.41	-0.27	0.27

Consider the hypothesis that the parent populations, otherwise identical, have a difference in mean equal to the upper confidence limit $\delta_+ = 0.68$. Following the paired randomization test argument, the hypothesis implies that after subtracting 0.68 from each of the differences the signs (plus or minus) carried by the resulting quantities is a pure matter of random labeling. These quantities, which are given in column (2) of Table 5.1, have an average of $0.41 - 0.68 = -0.27$. It is easily shown that by sign switching just 25 out of the 1024 possible averages (i.e., 2.4%) have equal or smaller values. Similarly, after subtracting $\delta_- = 0.14$ to obtain the third column of the table, whose average is $0.41 - 0.14 = 0.27$, sign switching produces just 19 out of 1024 possible averages (i.e., 1.9%) with equal or larger values. These values approximate* reasonably closely the two probabilities of 2.5% provided by the t statistic.

We shall proceed from now on to calculate confidence intervals using normal random sampling theory (and, in particular, the usual procedures based on the t distribution) in the analysis of randomized designs on the basis that these procedures could usually be verified to an adequate approximation with randomization theory.

Exercise 5.3.   Using the data in Exercise 5.1, determine the approximate 80% confidence interval, using the randomization distribution. *Answer*: From 2.8 to 6.5 approximately.

Sets of Confidence Intervals

A better understanding of the uncertainty associated with an estimate is provided by a *set* of confidence intervals. For instance, using Equation 5.9 and Table B1, one can compute the set of confidence intervals given in Table 5.2. These intervals are shown diagramatically in Figure 5.2.

TABLE 5.2.   Confidence intervals, boys' shoes example

significance level	confidence coefficient	confidence interval	
		δ_-	δ_+
0.001	0.999	−0.16	0.98
0.01	0.99	0.01	0.81
0.05	0.95	0.14	0.68
0.10	0.90	0.19	0.63
0.20	0.80	0.24	0.58

* An adjustment for continuity due to Yates (see R. A. Fisher, *Design of Experiments*, 8th ed., 47) which allows for the fact that the randomization distribution is discrete produces even closer agreement. This refinement, however, will not concern us here.

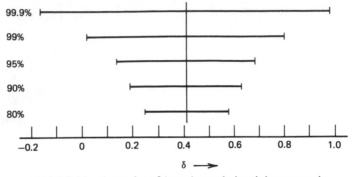

FIGURE 5.2.   A set of confidence intervals, boys' shoes example.

A Confidence Distribution

A comprehensive summary of all confidence interval statements is supplied by a *confidence distribution*. This is a curve having the property that the area under the curve between any two limits is equal to the confidence coefficient for that pair of confidence limits. Such a curve is shown for the shoe data in Figure 5.3. For example, the area under the confidence distribution between the 90% limits 0.19 and 0.63 is exactly 90% of the total area under the curve. The curve is a scaled t distribution centered at $\bar{d} = 0.41$ and having scale factor $s_{\bar{d}} = 0.12$.

Note that, on the argument we have given, this confidence distribution merely provides a convenient way of summarizing all possible confidence interval statements. It

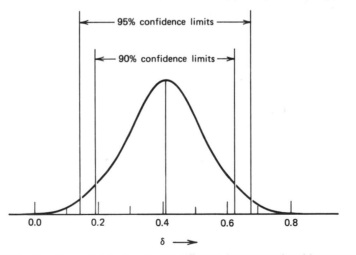

FIGURE 5.3.   Confidence distribution for the difference in wear produced by two materials, boys' shoes example.

is *not* the probability distribution of δ (which, according to the particular theory of statistical inference here discussed, is a fixed constant and does *not* have a distribution). It is interesting, however, that, according to an alternative statistical theory called *Bayesian inference*, it could be interpreted as a probability distribution. We do not pursue this tempting topic here.

Confidence Intervals Are More Useful than Single Significance Tests

The information given by a set of confidence intervals subsumes that supplied by a significance test, and provides more besides. For example, consider the statement that the 95% interval for δ extends from $\delta_- = 0.14$ to $\delta_+ = 0.68$. That the observed difference $\bar{d} = 0.41$ is significant for $\delta = 0$ at the 0.05 level is included in the confidence interval statement, since the postulated value $\delta = 0$ lies outside the interval. The interval statement, however, provides the experimenter with such additional facts as these:

1. At the 5% level the data contradict any assertion that the modification causes a change in the amount of wear *greater* than 0.68. Such a statement could be very important in this investigation, where the possible extent of a deleterious effect was under study.
2. Contemplation of the whole 95% interval (0.14, 0.68) in relation to the average wear (which is about 11) makes it clear that, although we have demonstrated a "highly significant" difference between the amounts of wear obtained with A and B materials, the *percentage change* is quite small. In fact, it may be as small as $(0.14/11) \times 100 = 1\%$ and is likely to be no larger than $(0.68/11) \times 100 = 6\%$.
3. The width of the interval, $0.68 - 0.14 = 0.54$, is large compared with the size of the average difference 0.41. If this difference had to be estimated more precisely, the information about the standard deviation of the difference would be useful for deciding roughly *how many* further experiments would be needed to reduce the confidence interval to any desired extent. For example, a further test with 30 more pairs of shoes would be expected to about halve the interval. (The length of the interval is inversely proportional to the *square root* of the total number of pairs.)

5.3. CONFIDENCE INTERVALS FOR A DIFFERENCE IN MEANS: UNPAIRED DESIGN

Consider now a fully randomized (unpaired) design to compare the means of two populations, using for illustration the trial on tomato plants (Section 4.1), in which standard (A) and modified (B) fertilizers were compared. A confidence interval for the mean difference $\eta_B - \eta_A$ may be obtained by an argument similar to that used above. Recall that the difference in the averages is $\bar{y}_B - \bar{y}_A = 22.53 - 20.84 = 1.69$. The hypothesis that the mean difference $\delta = \eta_B - \eta_A$ has some value δ_0 can be tested by referring

$$t_0 = \frac{(\bar{y}_B - \bar{y}_A) - \delta_0}{s\sqrt{1/n_A + 1/n_B}} = \frac{1.69 - \delta_0}{3.82} \tag{5.14}$$

to a table of the t distribution with $(5 - 1) + (6 - 1) = 9$ degrees of freedom. For a two-tailed test $\Pr(|t| > 2.262) = 0.05$.
    Thus all values of $\delta = \eta_B - \eta_A$ for which

$$\left|\frac{1.69 - \delta}{3.82}\right| < 2.262 \tag{5.15}$$

are not discredited at the 5 % level by a two-sided significance test. The 95 % confidence limits are therefore

$$1.69 \pm 3.82 \times 2.262 \tag{5.16}$$

that is,

$$1.69 \pm 8.64 \tag{5.17}$$

and the 95 % interval for δ extends from -6.95 to 10.33. These limits may be verified approximately by a randomization test as before.
    In general, the $1 - \alpha$ limits for $\delta = \eta_B - \eta_A$ are

$$\bar{y}_B - \bar{y}_A \pm t_{\alpha/2}\, s\sqrt{\frac{1}{n_A} + \frac{1}{n_B}} \tag{5.18}$$

where

$$s^2 = \frac{(n_A - 1)s_A^2 + (n_B - 1)s_B^2}{(n_A - 1) + (n_B - 1)} \tag{5.19}$$

Exercise 5.4.   Given the following data from a randomized experiment, compute the probability associated with the two-sided significance test, using the t table:

A   B   B   A   B

3   5   5   1   8

Answer: 0.08.

Exercise 5.5.   Repeat Exercise 5.4 with these data:

B   A   B   A   A   A   B   B

32   30   31   29   30   29   31   30

Answer: 0.02.

Exercise 5.6. Using the data in Exercise 5.4, compute 95, 90, and 80% confidence intervals.   *Answer*: $(-0.7, 8.7)$, $(0.5, 7.5)$, $(1.6, 6.4)$.

Exercise 5.7. Using the data in Exercise 5.5, compute 99, 95, and 90% confidence intervals. Draw a confidence distribution for the difference in means δ.
Answer: $(-0.4, 3.4)$, $(0.3, 2.7)$, $(0.5, 2.5)$.

Exercise 5.8. Following the approach described in this chapter, derive the formula for a $1 - \alpha$ confidence interval for the mean η, given a random sample $y_1, y_2, \ldots, y_n$ from a normal population.
Answer: $\bar{y} \pm t_{n-1, \alpha/2} s/\sqrt{n}$, where $s^2 = \sum (y - \bar{y})^2/(n - 1)$.

Exercise 5.9. Using your answer to Exercise 5.8, compute the 90% confidence interval for η, given the following mileage readings: 20.4, 19.3, 22.0, 17.5, 14.3 miles per gallon. List all your assumptions.   *Answer*: 18.7 ± 2.8.

In this chapter we have obtained formulas for $1 - \alpha$ confidence intervals of differences in means for paired and unpaired experiments, Equations 5.9 and 5.18. Notice that both these important formulas are of the form

$$\text{statistic} \pm t_{\alpha/2} \times (\text{standard error of the statistic}) \qquad (5.20)$$

where the standard error is the square root of the estimated variance of the statistic. For regression statistics, which are discussed later, as indeed for any statistic that is a linear function of the approximately normally distributed data, similar intervals can be constructed.

5.4. INFERENCES ABOUT VARIANCES OF NORMALLY DISTRIBUTED DATA

Most often we are interested in possible differences in the mean level of response produced by different methods or treatments. Sometimes, however, it is the degree of *variation* of the data that is of interest. In manufacturing industries it is important, for example, that the dye uptake of nylon thread, the potency of an antityphus vaccine, and the speed of color film vary as little as possible about appropriate specified values. Process modifications that reduce variance, even though they leave the mean unchanged, can be of great importance. Again it may be of interest to compare the *variation* of two or more analytical methods. We now consider, therefore, how tests of significance and confidence intervals may be obtained for variances.

The Chi-Square Distribution

Suppose that $z_1, z_2, \ldots, z_v$ are a set of v (nu) independently distributed unit normal deviates, each having mean zero and variance unity. Then their sum of squares $\sum_{u=1}^{v} z_u^2 = z_1^2 + z_2^2 + \cdots + z_v^2$ has a distribution of special importance, called a χ^2 (*chi-square*) *distribution*. The number of *independent* squared normal variables determines an important parameter of the distribution called the number of degrees of freedom v. We write

$$\sum_{u=1}^{v} z_u^2 \sim \chi_v^2$$

to mean that $\sum z_u^2$ has a chi-square distribution (is distributed as χ^2) with v *degrees of freedom. Thus the sum of squares of v independent unit normal deviates has a chi-square distribution with v degrees of freedom.*

The χ_v^2 distribution has mean v and variance $2v$. It is skewed to the right, but the skewness becomes less as v increases, and, for v greater than 50, χ^2 is approximately normal. The percentage points of the χ^2 distribution are given in Table C at the end of this book. (See also Appendix 3B; the mathematical form of the chi-square distribution is given in Equation 3B.1. The curve in Figure 3.3a is a χ^2 distribution having $v = 4$ degrees of freedom.)

Distribution of Sample Variances Calculated from Normally Distributed Data

In what follows we suppose that $y_1, y_2, \ldots, y_n$ are independent, normally distributed random variables having mean η and variance σ^2. Since $z_u = (y_u - \eta)/\sigma$ is normally distributed with mean zero and variance unity, the standardized sum of squares $\sum z_u^2$ of deviations from the population mean has a chi-square distribution with $v = n$ degrees of freedom, that is,

$$\frac{\sum (y_u - \eta)^2}{\sigma^2} \sim \chi_n^2$$

Now the variance estimate appropriate when the population mean is known is $\hat{s}^2 = \sum (y_u - \eta)^2/n$. It follows, therefore, that

$$\frac{n\hat{s}^2}{\sigma^2} \sim \chi_n^2 \quad \text{or, equivalently,} \quad \hat{s}^2 \sim \frac{\sigma^2}{n} \chi_n^2 \tag{5.21}$$

The more common circumstance is that in which η is unknown and the sample average $\bar{y}$ must be used instead of the population mean. It may be shown that the standardized sum of squares of deviations from $\bar{y}$ has a chi-square distribution with $v = n - 1$ degrees of freedom. Thus

$$\frac{\sum (y_u - \bar{y})^2}{\sigma^2} \sim \chi_{n-1}^2 \tag{5.22}$$

Since the variance estimate appropriate when η is unknown is

$$s^2 = \frac{\sum (y_u - \bar{y})^2}{(n-1)},$$

it follows that

$$\frac{(n-1)s^2}{\sigma^2} \sim \chi^2_{n-1} \qquad \text{or, equivalently,} \qquad s^2 \sim \frac{\sigma^2}{n-1} \chi^2_{n-1} \qquad (5.23)$$

The distribution of s^2 is thus a scaled chi-square distribution with scale factor $\sigma^2/(n-1)$.

Significance of a Hypothetical Variance σ^2: Normally Distributed Data

Given a sample value s^2, one may use the chi-square distribution to check any hypothesis concerning the population variance σ^2. This is done by referring vs^2/σ^2 to the appropriate entry in the χ^2 table.

Example

Suppose that $s^2 = 13$ has been calculated from what is believed to be a random sample of $n = 6$ items from a normal population, and we wish to test the hypothesis that $\sigma^2 = 10$ against the alternative that $\sigma^2 > 10$. Now

$$\frac{vs^2}{\sigma^2} = \frac{(n-1)s^2}{\sigma^2} = \frac{5 \times 13}{10} = 6.5$$

Entering the χ^2 table with $v = 5$ degrees of freedom, we see that this value is exceeded by chance about 25 % of the time. There is little reason, therefore, to question the hypothesis that $\sigma^2 = 10$.

Confidence Limits for σ^2 for Normally Distributed Data

Suppose that A and B are values obtained from Table C of the χ^2_{n-1} distribution for probabilities $(1 - \frac{1}{2}\alpha)$ and $\frac{1}{2}\alpha$, respectively. Given the assumptions stated, if the population variance is σ^2, then we have

$$\Pr\left\{ A < \frac{(n-1)s^2}{\sigma^2} < B \right\} = 1 - \alpha \qquad (5.24)$$

Rearranging this statement, we have equivalently

$$\Pr\left\{ \frac{(n-1)s^2}{B} < \sigma^2 < \frac{(n-1)s^2}{A} \right\} = 1 - \alpha \qquad (5.25)$$

The lower and upper $1 - \alpha$ confidence limits for σ^2 are thus given by

$$\frac{(n-1)s^2}{B} \quad \text{and} \quad \frac{(n-1)s^2}{A} \tag{5.26}$$

As always, the confidence limits are the values of the parameter that will just make the sample value significant at the stated level of probability, and from repeated samples a proportion $1 - \alpha$ of such intervals will include the parameter σ^2.

Example

Suppose that we wish to obtain a 95% confidence interval for σ^2 in the preceding example, where we have $s^2 = 13$ with five degrees of freedom. The χ^2 table shows that the 0.975 and 0.025 probability points of χ^2 with five degrees of freedom are

$$A = 0.831 \quad \text{and} \quad B = 12.83$$

Thus the required confidence limits are

$$\frac{5 \times 13}{12.83} = 5.97 \quad \text{and} \quad \frac{5 \times 13}{0.831} = 78.22$$

The reader should notice that, unless the number of degrees of freedom is high, the variance and its square root, the standard deviation, cannot be estimated very precisely. As a rough guide the standard deviation of an estimate s expressed as a percentage of σ is $100/\sqrt{2v}$. Thus, for example, if we wanted an estimate of σ having no more than a 5% standard deviation we would need a sample of about 200 observations!

Exercise 5.10.   Using the data of Exercise 5.9, obtain a 90% confidence interval for the variance of the mileage readings. Carefully state your assumptions.

Ratio of Two Variances

Suppose that a sample of n_1 observations is randomly drawn from a normal distribution having variance σ_1^2, a second sample of n_2 observations is randomly drawn from a second normal distribution having variance σ_2^2, and estimates s_1^2 and s_2^2 of the two population variances are calculated, having v_1 and v_2 degrees of freedom. Then s_1^2/σ_1^2 is distributed as $\chi_{v_1}^2/v_1$, and s_2^2/σ_2^2 is distributed as $\chi_{v_2}^2/v_2$.

Now the ratio $(\chi^2_{v_1}/v_1)/(\chi^2_{v_2}/v_2)$ has an F distribution having v_1 and v_2 degrees of freedom, whose probability points are given in Table D at the end of the book. Thus

$$\frac{s_1^2/\sigma_1^2}{s_2^2/\sigma_2^2} \sim F_{v_1, v_2} \qquad \text{or, equivalently,} \qquad \frac{s_1^2}{s_2^2} \sim \frac{\sigma_1^2}{\sigma_2^2} F_{v_1, v_2} \qquad (5.27)$$

The F distribution may thus be used to check hypotheses concerning σ_1^2/σ_2^2, the ratio of population variances of two normal populations. In particular, the null hypothesis that the variances are equal ($\sigma_1^2 = \sigma_2^2$) may be tested by referring the ratio of the sample variances s_1^2/s_2^2 directly to the F table. The table is arranged on the assumption that the larger of the two sample variances is in the numerator.

Example

The sample variances for replicated analyses performed by an inexperienced chemist 1 and an experienced chemist 2 are $s_1^2 = 0.183$ ($v_1 = 12$) and $s_2^2 = 0.062$ ($v_2 = 9$). Assuming that the chemists' results can be treated as normally and independently distributed random variables having variances σ_1^2 and σ_2^2, let us test the null hypothesis that $\sigma_1^2 = \sigma_2^2$ against the alternative that $\sigma_1^2 > \sigma_2^2$. According to the null hypothesis $\sigma_1^2 = \sigma_2^2$, the ratio s_1^2/s_2^2 is distributed as $F_{12,9}$. Referring the ratio $0.183/0.062 = 2.95$ to the table of F with $v_1 = 12$ and $v_2 = 9$, we find that the 10 and 5% points are, respectively, 2.38 and 3.08. The null hypothesis is somewhat discredited, therefore, and there is a suggestion that the inexperienced chemist has a larger variance.

Confidence Limits for σ_1^2/σ_2^2 for Normally Distributed Data

Arguing exactly as before, if A and B are the $1 - \frac{1}{2}\alpha$ and $\frac{1}{2}\alpha$ probability points of the F distribution with v_1 and v_2 degrees of freedom, the lower and upper $1 - \alpha$ confidence limits for σ_1^2/σ_2^2 are

$$\frac{s_1^2/s_2^2}{B} \qquad \text{and} \qquad \frac{s_1^2/s_2^2}{A}$$

The $1 - \frac{1}{2}\alpha$ probability point of the F distribution may be obtained from the tabulated $\frac{1}{2}\alpha$ points. This is done by interchanging v_1 and v_2 and taking the reciprocal of the tabled value.

Suppose, for instance, we need the 90% limits for the variance ratio σ_1^2/σ_2^2 for the analyst example above, where a value $s_1^2/s_2^2 = 2.95$ was obtained from variance estimates having, respectively, 12 and 9 degrees of freedom.

Entering the table of the 0.05 probability point with $v_1 = 12$ and $v_2 = 9$, we find $B = 3.07$. To obtain the 0.95 probability point, we enter the table with

$v_1 = 9$ and $v_2 = 12$ and take the reciprocal of the value 2.80 to obtain $A = 1/2.80 = 0.357$. Thus the required 90% confidence limits are

$$\frac{2.95}{3.07} = 0.96 \quad \text{and} \quad \frac{2.95}{0.357} = 8.26$$

Lack of Robustness of Tests on Variances

In discussing tests on variances we have especially emphasized the assumption of normality of the underlying distribution of the observations. Tests to compare means (the t test and the analysis of variance tests to be described later) are rather insensitive to the normality assumption. But this is not true for direct tests on variances such as those given above. It is sometimes possible to avoid this difficulty by converting a test on variances to a test on means. The following example illustrates this idea.

Example

Each week two analysts each performed five tests on identical repeated samples. The source of the identical samples was changed each week, and the samples were included at random in the usual routine analysis and were not identifiable. The variances calculated for the results were as follows.

week	s_1^2	$\log 100s_1^2$	s_2^2	$\log 100s_2^2$	$\log 100s_1^2 - \log 100s_2^2$
1	0.142	1.15	0.043	0.63	0.52
2	0.091	0.96	0.079	0.90	0.06
3	0.214	1.33	0.107	1.03	0.30
4	0.113	1.05	0.037	0.43	0.62
5	0.082	0.91	0.045	0.65	0.26

$$\bar{d} = 0.352$$

$$s_d = 0.226$$

$$s_{\bar{d}} = \frac{s_d}{\sqrt{n}} = 0.101$$

$$t = \frac{\bar{d} - 0}{s_{\bar{d}}} = \frac{\bar{d} - 0}{s/\sqrt{n}} = \frac{0.352}{0.101} = 3.49$$

The logarithm of the sample variance s^2 is much more nearly normally distributed than is s^2 itself. Also the variance of $\log s^2$ is independent of the population variance σ^2. We can carry out an approximate test that is insensitive to nonnormality of the original data by taking logarithms of the sample variances and performing a t test on these "observations." To allow for changes from week to week in the nature of samples analyzed, we have in this instance performed a *paired t* test. The value of $t = 3.49$ with four degrees of freedom is significant at the 5% level. Hence analyst 2 is probably more accurate.

In this example there is a natural division of the data into groups. When this is not so, the observations can be divided randomly into small groups and the same device used.

5.5. INFERENCES ABOUT PROPORTIONS: THE BINOMIAL DISTRIBUTION

Data sometimes occur as the *proportion of times* a certain event happens. We now consider how such data can be treated. We first introduce an important discrete distribution that on certain specific assumptions describes the way in which proportions vary.

The Binomial Distribution

Major Denis Bloodnok makes money by betting with a pocketful of biased pennies that he knows come up heads, on the average, 8 times out of 10. We shall say that for each of his pennies the *probability p* of a head is 0.8 and the *probability $q = 1 - p$* of a tail is 0.2. Suppose he decides to throw $n = 5$ of these biased pennies and to bet on *the number y* that come down heads. For expository purposes we suppose that the pennies may be distinguished in some way (e.g., by their dates), and we call them penny A, penny B, and so on. To make his bets appropriately Bloodnok needs to know the probability, with five pennies tossed, of each of the following *events*: no heads, one head, two heads, and so on, that is, he needs the $n + 1 = 6$ values

$$\Pr(y = 0), \Pr(y = 1), \ldots, \Pr(y = 5)$$

We will call the tossing of the five pennies a *trial* and will denote the *outcome* of this trial by listing heads (H) and tails (T) in order. Thus, if pennies A, B, and C fall tails and pennies D and E fall heads, we write the outcome of this trial as (T T T H H).

$\Pr(y = 0)$
The event $y = 0$ can happen only as a result of the outcome (T T T T T). If, as is reasonable, the result of tossing any particular penny is independent of the result of tossing any other, we have

$$\Pr(y = 0) = \Pr(T\,T\,T\,T\,T) = q \times q \times q \times q \times q = q^5 = 0.2^5 = 0.00032$$

$Pr(y = 1)$

Five separate outcomes can yield the event $y = 1$. The single head can occur with penny A, B, C, D, or E. Thus

$$Pr(y = 1) = Pr(H\,T\,T\,T\,T) + Pr(T\,H\,T\,T\,T) + Pr(T\,T\,H\,T\,T)$$
$$+ Pr(T\,T\,T\,H\,T) + Pr(T\,T\,T\,T\,H)$$

Now the probability of *each one* of these five outcomes is $pq^4 = 0.8 \times 0.2^4 = 0.00128$. Thus

$$Pr(y = 1) = 5pq^4 = 5 \times 0.8 \times 0.2^4 = 5 \times 0.00128 = 0.00640$$

$Pr(y = 2)$

The outcome "two heads and three tails in some *specific* order," say (H H T T T), has probability p^2q^3. But there are 10 different possible orders in which two heads and three tails can occur:

(H H T T T) (H T H T T) (H T T H T) (H T T T H) (T H H T T)
(T H T H T) (T H T T H) (T T H H T) (T T H T H) (T T T H H)

Thus

$$Pr(y = 2) = 10p^2q^3 = 10 \times 0.8^2 \times 0.2^3 = 0.05120$$

General Expression for $Pr(y)$

Evidently to obtain $Pr(y)$ in general we can:

1. Calculate the probability $p^y q^{n-y}$ of y heads and $n - y$ tails *occurring in some specific order*. Note that $q = 1 - p$.
2. Compute the number of different orders in which y heads and $n - y$ tails can occur: this is called the *binomial coefficient* and is denoted by $\binom{n}{y}$.
3. Multiply the results together to obtain $Pr(y) = \binom{n}{y} p^y q^{n-y}$.

Binomial Coefficient Formula

A general formula for the binomial coefficient is

$$\binom{n}{y} = \frac{n!}{y!(n-y)!}$$

In this formula $n!$ is read as n *factorial* and is equal to

$$n \times (n-1) \times (n-2) \times \cdots \times 2 \times 1$$

For example, the number of orderings for two heads and three tails is

$$\binom{5}{2} = \frac{5 \times 4 \times 3 \times 2 \times 1}{(2 \times 1)(3 \times 2 \times 1)} = \frac{5 \times 4}{2} = 10$$

which is the number we just obtained by direct enumeration.

Pascal's Triangle

An alternative procedure for obtaining binomial coefficients, when n is not too large, uses Pascal's triangle:

n	binomial coefficients
0	1
1	1 1
2	1 2 1
3	1 3 3 1
4	1 4 6 4 1
5	1 5 10 10 5 1
6	1 6 15 20 15 6 1
7	1 7 21 35 35 21 7 1

To construct the triangle write 1 in the center of the space opposite $n = 0$, and suppose that imaginary slots to the left and right of 1 are filled with zeros. Each entry in successive rows is obtained by adding in pairs the entries above it. Thus $0 + 1 = 1$ and $1 + 0 = 1$, yielding the numbers 1, 1 in the row $n = 1$. For the row $n = 2$, we have $0 + 1 = 1$, $1 + 1 = 2$, $1 + 0 = 1$, and so on. Proceeding in this way, we can readily construct the table given above. For $n = 5$ we obtain 1, 5, 10, 10, 5, 1, the binomial coefficients needed for Bloodnok's problem, some of which have been calculated already. Notice the symmetry of the binomial coefficients about the center value(s).

Exercise 5.11. Obtain the binomial coefficients

$$\binom{8}{8}, \ \binom{8}{7}, \ \binom{8}{6}, \ \binom{8}{5}, \ \binom{8}{4}, \ \binom{8}{3}, \ \binom{8}{2}, \ \binom{8}{1}, \ \binom{8}{0}$$

(a) from the formula for $\binom{n}{y}$ and (b) by extending Pascal's triangle.

Answer: 1, 8, 28, 56, 70, 56, 28, 8, 1.

We have seen that the probability of obtaining y heads when p is the probability of a head, q is the probability of a tail, and n is the number of throws is

$$\Pr(y) = \binom{n}{y}p^y q^{n-y} = \frac{n!}{y!(n-y)!}p^y q^{n-y} \qquad (5.28)$$

TABLE 5.3. Binomial distribution for y heads in n trials when $p = 0.8, n = 5$

number of heads y	binomial coefficient $\binom{5}{y} = 5!/y!(5 - y)!$	$p^y q^{5-y}$	Pr(y)
0	1	0.00032	0.000
1	5	0.00128	0.006
2	10	0.00512	0.051
3	10	0.02048	0.205
4	5	0.08192	0.410
5	1	0.32768	0.328
			1.000

The $n + 1$ probabilities Pr(0), Pr(1), Pr(2), ..., Pr(n) yield the *binomial distribution* of y. For Bloodnok's problem the binomial distribution of the number of heads is calculated in Table 5.3 and plotted in Figure 5.4a.

Figure 5.4b shows the distribution of heads for a trial with five fair pennies ($p = 0.5$). Although for five fair pennies the probability of *at least* four heads [given by Pr(4) + Pr(5)] is only 0.19, for Bloodnok's biased pennies it is 0.74. Thus a wager at even money that the major can throw at least four heads appears unfavorable to him, assuming he is using fair pennies, but using his biased pennies, he can make an average of 48 cents for every 1 dollar bet! To see this, suppose that he bets a single dollar on this basis 100 times. In about 74 cases he will make a dollar, and in about 26 he will lose a dollar. His net gain is thus $74 - 26 = 48$ dollars per 100 bet.

Exercise 5.12.   Obtain Pr(y) for $y = 0, 1, 2, ..., 5$ for five fair pennies ($p = 0.5$), and confirm the result Pr(4) + Pr(5) = 0.19. What would Bloodnok's average profit or loss be if we accepted his wager but made him use fair pennies?

Answer: A loss of 62 cents for every 1 dollar bet.

Figures 5.4c and d show binomial distributions for throws of 20 biased pennies and 20 fair pennies.

General Properties of the Binomial Distribution

The binomial distribution might be applied to describe the proportion of animals surviving a certain dosage of drug in a toxicity trial or the proportion of manufactured components passing a destructive test in routine quality

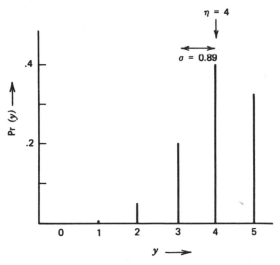

(a) Binomial distribution with $p = 0.8$ and $n = 5$.

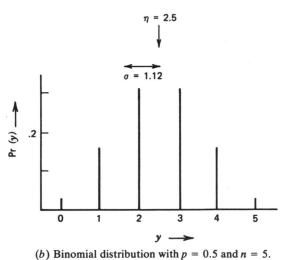

(b) Binomial distribution with $p = 0.5$ and $n = 5$.

FIGURE 5.4. Binomial distributions for various choices of p and n.
(continued on following page)

127

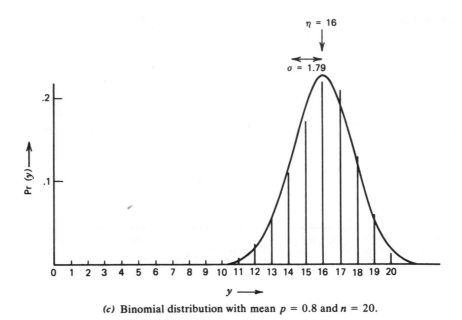

(c) Binomial distribution with mean $p = 0.8$ and $n = 20$.

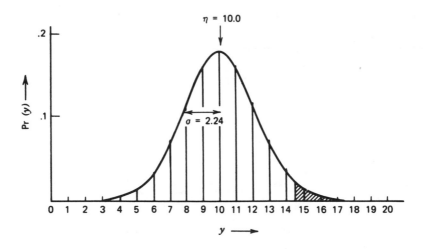
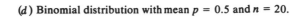

(d) Binomial distribution with mean $p = 0.5$ and $n = 20$.

FIGURE 5.4. (continued)

128

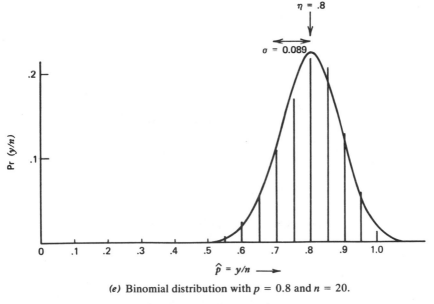

(e) Binomial distribution with $p = 0.8$ and $n = 20$.

FIGURE 5.4. (continued)

inspection. In all such trials there are two outcomes: (head or tail), (survived or died), (passed or failed test). In general, it is convenient to call one outcome a *success* and the other a *failure*. Thus we can say the binomial distribution (Equation 5.28) gives the probability of y successes out of a possible total of n in a trial where the fixed probability of an individual success is p and of a failure is $q = 1 - p$.

Mean and Variance of the Binomial Distribution

The mean and variance of the binomial distribution are

$$\eta = np, \qquad \sigma^2 = npq \tag{5.29}$$

Thus the standard deviation is $\sigma = \sqrt{npq}$. For example, for the distribution of the gambler's five biased pennies ($p = 0.8$, $n = 5$),

$$\eta = 5 \times 0.8 = 4.0, \qquad \sigma = \sqrt{5 \times 0.8 \times 0.2} = 0.89$$

as illustrated in Figure 5.4a.

Distribution of the Observed Proportion of Successes

In some cases we may wish to consider not the number of successes but the *proportion of successes* y/n. The probability of obtaining some value y/n is the same as the probability of obtaining y, so

$$\Pr\left(\frac{y}{n}\right) = \binom{n}{y}p^y q^{n-y} \tag{5.30}$$

The distribution of y/n is obtained simply by rescaling the horizontal axis of the binomial distribution $\Pr(y)$ (compare Figures 5.4c and e). The *observed* proportion of heads y/n is an *estimate*, $\hat{p}$, of the probability p.* The distribution of y/n shown in Figure 5.4e thus shows how reliable such an estimate of p would be for a sample size of $n = 20$.

The mean and standard deviation of the distribution of $\hat{p} = y/n$ are

$$\eta_{\hat{p}} = p, \qquad \sigma_{\hat{p}} = \sqrt{pq/n} \tag{5.31}$$

Exercise 5.13. Verify the mean and standard deviations for the distributions shown in Figures 5.4b, c, d, and e.

Approach to Normality

In Figures 5.4c, d, and e the continuous lines show normal distributions having the same means and standard deviations as the respective binomial distributions. In general, the normal approximation steadily improves as n is increased and is better when p is not near the extremes of zero or unity. A rough rule is that for $n > 5$ the normal approximation will be adequate if the absolute value of $(1/\sqrt{n})(\sqrt{q/p} - \sqrt{p/q})$ is less than 0.3.

A Test of Significance

Suppose that we were suspicious of Bloodnok and managed to get hold of one of his pennies. Suppose that we tossed it 20 times, and it came down heads 15 times. Would this confirm our suspicions?

Giving him the benefit of the doubt, we might entertain the null hypothesis that his penny was fair ($p = 0.5$) against the alternative that it was biased in his favor ($p > 0.5$). We would then ask: If the null hypothesis were true, how often would an event as extreme as that observed ($y = 15$), or more extreme, occur? Applying the binomial formula, we find, given that $p = 0.5, n = 20$,

$$\Pr(y \geq 15) = \Pr(y = 15) + \Pr(y = 16) + \Pr(y = 17) + \Pr(y = 18) + \Pr(y = 19) + \Pr(y = 20)$$
$$= 0.015 \quad\quad + 0.005 \quad\quad + 0.001 \quad\quad + 0.000 \quad\quad + 0.000 \quad\quad + 0.000$$
$$= 0.021$$

---

* A caret ˆ over a symbol, or group of symbols, means "an estimate of," thus $\hat{p}$ is read, "an estimate of" p.

The observed event is significant at the 2.1 % level. We might therefore regard our suspicions as justified.

Normal Approximation

It might be supposed that the normal approximation to $\Pr(y > y_0)$ would be obtained by referring $(y_0 - \eta)/\sigma = (y_0 - np)/\sqrt{npq}$ to standard normal tables. In fact, a much closer approximation is obtained by applying a correction due to Yates, in which $y_0 - \frac{1}{2}$ is substituted for y_0 in the above formula. Figure 5.4d shows why. An approximation to $\Pr(y = 15)$ is supplied by the area under the normal curve from $y = 14.5$ to $y = 15.5$. A similar approximation to $\Pr(y = 16)$ is the area under the normal curve between $y = 15.5$ and $y = 16.5$. Thus $\Pr(y \geq 15)$ is approximated by the area under the normal curve *to the right of* $y = 14.5$. In general, if y has a binomial distribution with parameters p and n, then $\Pr(y > y_0)$ is approximated by $\Pr(z > z_0)$, where z is a unit normal variable and $z_0 = (y_0 - \frac{1}{2} - np)/\sqrt{npq}$.

If Yates's adjustment is not applied, significance is overestimated by the normal approximation. The computed significance probability is too small. For the present example, using the adjustment,

$$z_0 = \frac{14.5 - 10.0}{\sqrt{5}} = 2.012$$

From the normal tables, therefore, we have

$$\Pr(y \geq 15) \doteq \Pr(z \geq 2.012) = 0.022 = 2.2\%$$

which agrees closely with the exact value of 2.1 %.

A Confidence Interval for p

Suppose that our null hypothesis had been that $p = 0.8$ (which happens to be true). Then, after obtaining $y = 15$ heads, we could again evaluate $\Pr(y \geq 15)$ by adding binomial probabilities with $p = 0.8$. In this way we would obtain for $p = 0.8$, $n = 20$,

$$\Pr(y \geq 15) = 0.175 + 0.218 + 0.205 + 0.137 + 0.058 + 0.012 = 0.805$$

At this stage, therefore, we could say that, whereas 0.5 is an implausible value for p, 0.8 is not. Now imagine a series of values of p confronting the data $y = 15$, with the appropriate significance computed for each value. As p was increased from zero, there would be some value p_- less than y/n that *just* produced significance at, say, the 2.5 % level for the null hypothesis $p = p_-$, tested against the alternative $p > p_-$. Similarly, there would be some other value p_+ greater than y/n that *just* produced significance at the 2.5% level for the null hypothesis $p = p_+$, tested against the alternative $p < p_+$. In fact, the values that do this are $p_- = 0.51$ and $p_+ = 0.94$. They are the limits of a 95% confidence interval for p. In repeated sampling, 95% of the intervals calculated in this way will include the true value of p. Confidence limits for p based on the estimated probability $\hat{p} = y/n$ may be conveniently read from the charts given in Table F at the end of this book.

Some Uses of the Binomial Distribution

Not all readers will wish to pursue the risky career of gambling with biased coins. The following exercises illustrate other uses of the binomial distribution.

Exercise 5.14. An IQ test is standardized so that 50% of male adults in the general population obtain scores over 100. Among 43 male applicants tested at an army recruiting office only 10 obtained scores higher than 100. Is there significant evidence that the applicants are not 'a random sample from the population used to standardize the test? *Answer*: Yes. For $\hat{p} = \frac{10}{43}$ the significance level is 0.0003 on the hypothesis that $p = 0.5$ against the alternative $p < 0.5$. The normal approximation gives 0.0004.

Exercise 5.15. Using a table of random numbers, police in a certain state randomly stopped 100 cars passing along a highway and found that 37 of the drivers were wearing seat belts. Obtain confidence limits for the probability of wearing seat belts on the highway in question. Explain your assumptions.
Partial answer: 95% confidence limits are 0.275 and 0.475.

Exercise 5.16. In a plant that makes ball bearings, samples are routinely subjected to a stringent crushing test. A sampling inspection scheme is required such that if out of a random sample of n bearings more than y fail, the batch is rejected. Otherwise it is accepted. Denote by p the proportion of bearings in a large batch that will fail the test. It is required that there be a 95% chance of accepting a batch for which p is as low as 0.3 and a 95% chance of rejecting a batch for which p is as high as 0.5. Using the normal approximation, find values for n and p that will be satisfactory.
Hint: Show that to satisfy the requirements it is necessary that

$$\frac{(y_0 - \frac{1}{2}) - 0.3n}{\sqrt{n \times 0.3 \times 0.7}} = \frac{0.5n - (y_0 - \frac{1}{2})}{\sqrt{n \times 0.5 \times 0.5}} = 1.645$$

Answer: $n = 62$, $y = 25$.

Variance Stabilizing Transformation for the Binomial Distribution: Testing Mothproofing Agents

Mothproofing agents for wool are sometimes compared by placing 20 moth larvae in contact with a treated wool sample and noting the number that die

TABLE 5.4.   Comparison of two methods of application of mothproofing agent in seven laboratories, tests with twenty moth larvae

method of application	different ways of measuring results	laboratory						
		1	2	3	4	5	6	7
A	number dead	8	7	1	16	10	19	9
	percentage dead	40	35	5	80	50	95	45
	score	43	40	14	71	50	86	47
B	number dead	12	6	3	19	15	20	11
	percentage dead	60	30	15	95	75	100	55
	score	57	37	25	86	67	100	53
difference in percentages		20	−5	10	15	25	5	10
difference in scores		14	−3	11	15	17	14	6

	average difference	standard error of average difference	t value	significance level (%) (two-sided test)
percentage	11.43	3.73	3.06	2.2
score	10.57	2.63	4.03	0.7

in a given period of time. It is known that tests done in different laboratories typically give widely different results. A cooperative experiment involving seven different laboratories resulted in the data shown in Table 5.4. The objective was to compare two different methods A and B of applying the agent to the wool sample. Randomization was employed in selecting the particular larvae and wool samples for each trial.

If the number of larvae dying y (or, equivalently, the percentage dying $100y/n$) could be assumed to be approximately normally distributed with constant variance, a paired t test could be used to assess the null hypothesis that, irrespective of which application method was used, the mean proportion dying was the same in each laboratory. Such a test yields the value $t = 3.06$ with six degrees of freedom. This is significant at the 2.2% level, suggesting that method B causes greater mortality of the larvae.

Now the proportion dying varies greatly from laboratory to laboratory. Indeed, it looks from the data as if p might easily vary from 0.05 to 0.95. If this were so, the variances npq could differ by a factor of 5. Fisher suggested that, before treating binomial proportions $y/n = \hat{p}$ by standard normal theory techniques using t statistics (and the analysis of variance and regression methods discussed later), a transformation should be made to a different

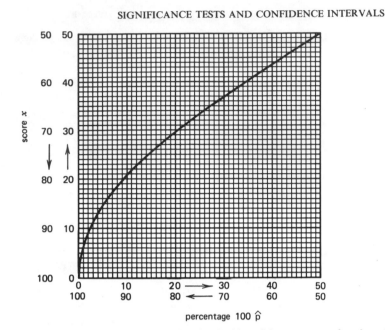

FIGURE 5.5. Variance stabilizing transformation for the binomial: score x as a function of the percentage $100\hat{p}$.

"metric" x, given by

$$\sin x = \sqrt{\hat{p}} \qquad\qquad (5.32)$$

The quantity x may simply be thought of as a score derived from p but having more convenient properties. [Technically, as it is used here, it is an *angle* measured in grads (100 grads $= 90°$).] A graph of this score x versus the proportion $\hat{p}$ (measured as a percentage) is shown in Fig. 5.5. The purpose of this transformation is to stabilize the variance. Thus, whereas the variance of p is very different for different values of p, the variance of x is approximately constant. It will be seen from the graph that this stabilization is achieved by stretching out the scale at the ends of the range, at the expense of the center. Not only does the transformation achieve a standardized variance, but also it turns out that x is more nearly normally distributed over the whole range than is $\hat{p}$. The rationale for choice of transformations is given in Section 7.8.

To carry out the modified test, one first transforms the percentages, using the graph in Figure 5.5. For example, the entry 40% in the first column of the table transforms to a score of 43. A paired t test applied to the scores now gives a value $t = 4.03$, which is even more significant than that obtained from the raw percentages. Because the assumptions on which the t test is derived are more nearly met by the transformed scores, their use will, on the average, produce significant tests of greater sensitivity, as happens in this example. The reason the transformation produces a marked difference in the significance level is that for this example the observed values y/n cover a wide range. If this were not so, the transformation would have had little effect on the significance level.

The variance of the empirical percentage $100\hat{p} = 100y/n$, where y has a binomial distribution, is $10,000pq/n$, which, of course, depends on p. The variance of the transformed score x is very nearly independent of p and has a theoretical value of $1013/n$. Even though this theoretical variance is available, it is usually better to employ a variance calculated from the data, as was done in the t test used above. More explicitly, in the example above we could have argued that each transformed value has a theoretical variance of $1013/20 = 50.7$, and we could have based our test on this value. However, if we use that route, we must make a direct assumption of the exact applicability of standard binomial sampling. It is always best to avoid assumptions we do not *have* to make. In this instance we have done so by employing the standard paired t test, which, since it employs a variance estimate computed directly from variation in the *actual data*, is more robust to departures from assumption. We now consider some of the ways in which departure from standard binomial sampling can occur.

What Happens When the Probability Varies?

If Major Bloodnok throws 20 pennies for *each* of which $p = 0.8$, the distribution of the number of heads y is the binomial shown in Figure 5.4c. For this distribution the mean and variance are $\eta = 20 \times 0.8 = 16$ and $\sigma^2 = 20 \times 0.8 \times 0.2 = 3.2$. Now suppose instead that the probabilities differed among the 20 biased pennies. We can imagine, for example, that the probabilities were distributed among the 20 pennies as follows:

number of pennies	2	4	6	8		
probability of penny falling heads	0.6	0.7	0.8	0.9	$\bar{p} = 0.8,$	$\sigma_p^2 = 0.01$

Thus, of the 20 pennies, 2 would have a probability of 0.6 of falling heads, 4 a probability of 0.7, and so on. The reader may confirm that for this particular distribution the mean probability is $\bar{p} = 0.8$ and the variance of this distribution is $\sigma_p^2 = 0.01$ ($\sigma_p = 0.1$).

Probability Varying within Each Trial

If the Major makes a series of trials in each of which all 20 of the variously biased pennies are thrown, the distribution of the number of heads y will no longer follow the binomial distribution exactly but will have a somewhat smaller spread. In general, it may be shown that

$$\eta_y = n\bar{p}, \qquad \sigma_y^2 = n\bar{p}\bar{q} - n\sigma_p^2 \tag{5.33}$$

For example, for the distribution of probabilities used above for illustration ($\bar{p} = 0.8$ and $\sigma_p = 0.1$),

$$\eta_y = 16, \qquad \sigma_y^2 = 3.20 - 0.20 = 3.00$$

Probability Varying between Trials

Suppose instead that Bloodnok (1) takes a penny at random from his pocket (which contains the 20 variously biased coins), (2) conducts a trial by tossing that particular penny 20 times and recording the number of heads y, and (3) returns the penny to his pocket and repeats the whole operation with another penny drawn randomly from his pocket. The values of y from a long series of such trials would have a distribution that would again differ from the binomial. However, in this case the variance would be greater than that for a binomial with $p = 0.8$ and $n = 20$. In general, the mean and variance of y would be

$$\eta_y = n\bar{p}, \qquad \sigma_y^2 = n\bar{p}\bar{q} + n(n-1)\sigma_p^2 \tag{5.34}$$

Therefore, for the distribution of probabilities with $\bar{p} = 0.8$ and $\sigma_p = 0.1$,

$$\eta_y = 16, \qquad \sigma_y^2 = 3.20 + 3.80 = 7.00$$

Variation of Probability within and between Trials

In general, denoting the variance within a trial by σ_{1p}^2 and the variance between trials by σ_{2p}^2, we have

$$\sigma_y^2 = n\bar{p}\bar{q} - n\sigma_{1p}^2 + n(n-1)\sigma_{2p}^2 \tag{5.35}$$

Equivalently, the variance of $\hat{p} = y/n$ is

$$\sigma_{\hat{p}}^2 = \frac{1}{n}\left[(\bar{p}\bar{q} - \sigma_{1p}^2) + (n-1)\sigma_{2p}^2\right] \tag{5.36}$$

Typically, variation of p within a trial produces a minor reduction in variance, but variation of p between trials can result in a larger increase in variance (about n times as great for the same σ_p^2).

Application: Sampling Inspection of Ball Bearings

Each day 20 ball bearings randomly sampled from routine production are subjected to a crushing test. The numbers of bearings failing the test on 10 successive days were as follows:

day	1	2	3	4	5	6	7	8	9	10	
number y failing out of 20	8	4	10	6	2	3	7	12	5	7	$\bar{y} = 6.4$

Is there evidence of day-to-day variation in the probability p of failure?

Consider the null hypothesis that p (the probability that a bearing fails) does not vary from day to day. On this hypothesis the distribution of $(y - \eta)/\sigma = (y - np)/\sqrt{npq}$ may be approximated by a standard normal distribution. We do not know p, but an estimate is provided by $\hat{p} = \bar{y}/20 = 6.4/20 = 0.32$.

Now, for a sample of k observations y from a fixed binomial population, $\sum_{j=1}^{k} (y_j - n\hat{p})^2/n\hat{p}\hat{q}$ is approximately distributed as a χ^2 distribution with $k - 1$ degrees of freedom. Thus, to test the null hypothesis, the value

$$\frac{\sum (y_j - 6.4)^2}{20 \times 0.32 \times 0.68} = \frac{86.40}{4.352} = 19.85$$

may be referred to a χ^2 distribution with $k - 1 = 9$ degrees of freedom. This value yields significance at the 2.5 % level. There is evidence, therefore, of real variation from day to day in failure rates as measured by the crushing test.

Now the variance of y_j is estimated by $\sum_{j=1}^{k} (y_j - 6.4)^2/9 = 9.6$. Thus, substituting estimates in Equation 5.34

$$\sigma_y^2 = n\bar{p}\bar{q} + n(n - 1)\sigma_p^2$$

we have

$$9.600 = 4.352 + 380\sigma_p^2$$

A (very rough) estimate of σ_p^2 is, therefore, $(9.600 - 4.352)/380 = 0.014$, yielding an estimated standard deviation for p of $\sqrt{0.014} = 0.12$. Now p is estimated as 0.32, so that the estimated standard deviation of p is relatively quite large (about one third of $\hat{p}$).

Notice that, although the investigation allows the existence of variation in p to be demonstrated and makes it possible to estimate σ_p very approximately, it does not tell us why p is varying. This variation could arise, for example, from inadequacies in the test procedure or the sampling method, as well as from inadequate process control. Further iterations in the investigation should take up these questions.

5.6. INFERENCES ABOUT FREQUENCIES: THE POISSON DISTRIBUTION

When p is small but n is large, the binomial distribution approaches an important limiting distribution, the Poisson distribution, which we now discuss.

How Many Driving Accidents Does Minnie Bannister Have in a Year?

When Minnie Bannister is driving her car, the probability that she will have an accident *in any given minute* is small. Suppose that it is equal to $p = 0.00021$. Suppose also that in the course of 1 year she is at risk for $n = 10,000$ minutes.* Then, in spite of the small probability of an accident in a given minute, her *mean frequency of accidents per year* is not negligible. In fact, it is $\eta = np = 2.1$. Thus, if the appropriate conditions apply, her frequency of accidents per year will be distributed in the binomial distribution having $p = 0.00021$ and $n = 10,000$, shown in the second row (*b*) of Table 5.5.

We see that in 12.2% of years (about 1 year in 7) Minnie will have no accidents. In 25.7% of years (about 1 year in 4) she will have one accident, and so on. Occasionally, specifically in $(4.2 + 1.5 + 0.4 + 0.1)\% = 6.2\%$ of years, she will have five accidents or more. We must remember of course that this will be a chance phenomenon and will occur not because she is less careful in these years.

Now on the same basis consider the record of Henry Crun, who is not such a good driver as Minnie. Henry's probability p' of having an accident in any given minute is $p' = 0.00042$ (twice that of Minnie), but since he drives for only $n' = 5000$ minutes in a year (half as much as Minnie), his mean number of accidents per year $\eta = n'p' = 2.1$ is the same as hers. The remarkable fact is, however, that not only Henry's mean but also Henry's *distribution* of accidents is virtually the same as Minnie's. This distribution, calculated from the binomial formula with $n' = 5000$ and $p' = 0.00042$, is shown in the first row (*a*) of Table 5.5.

The Poisson Distribution as a Limit to the Binomial Distribution

The reason for the close agreement is this. If, in the expression

$$\Pr(y) = \binom{n}{y} p^y (1 - p)^{n-y}$$

for a binomial distribution, p is made smaller and smaller *at the same time* that n is made larger and larger, so that the mean $\eta = np$ remains finite and fixed, it may be shown that $\Pr(y)$ tends closer and closer to the limit

$$\Pr(y) = \frac{e^{-\eta}\eta^y}{y!} \tag{5.37}$$

This limiting distribution is called the *Poisson distribution*, after its discoverer. It has the property that the probability of a particular value y of the frequency

---

* To do this she will need to drive her car, on the average, for a little less than half an hour a day.

TABLE 5.5. Distribution of accidents per-year, assuming binomial distribution with (a) $p = 0.00042$, $n = 5000$, (b) $p = 0.00021$, $n = 10,000$, (c) limiting Poisson distribution with $\eta = 2.1$

	probability that yearly frequency of accidents is								
	0	1	2	3	4	5	6	7	8
(a) Henry Crun (exact binomial $p = 0.00042$, $n = 5000$) $\eta = 2.1$, $\sigma^2 = 2.0991$	0.12240	0.25715	0.27007	0.18905	0.09924	0.04166	0.01457	0.00437	0.00115
(b) Minnie Bannister (exact binomial $p = 0.00021$, $n = 10,000$) $\eta = 2.1$, $\sigma^2 = 2.0996$	0.12243	0.25716	0.27004	0.18903	0.09923	0.04167	0.01458	0.00437	0.00115
(c) limiting Poisson distribution $p \to 0$, $n \to \infty$ $\eta = 2.1$, $\sigma^2 = 2.1$	0.12246	0.25716	0.27002	0.18901	0.09923	0.04168	0.01459	0.00438	0.00115

139

depends *only* on the mean frequency η. Thus η is the *only* parameter of this distribution.

For both Minnie and Henry the mean frequency of accidents per year is $\eta = 2.1$. For this mean frequency probabilities calculated from the Poisson distribution are

$$\Pr(0) = e^{-2.1} = 0.12246, \qquad \Pr(1) = e^{-2.1} \times 2.1 = 0.25716$$

$$\Pr(2) = e^{-2.1} \times \frac{2.1^2}{2} = 0.27002, \qquad \text{etc.}$$

These values are given in the third row (c) of Table 5.5 and are seen to very closely approximate the values calculated from the binomial distributions. For each frequency the three calculated results diverge, in fact, only in the fifth figure. Usually probabilities are required to much less accuracy. An important fact is that the binomial distribution tends to the Poisson *quite rapidly*. In practice n and p do not need to be very extreme before the limiting Poisson distribution supplies a good approximation. This limiting distribution is of importance because there are many everyday situations of the kind where the probability p of an event at a particular time or place is small but the number of opportunities n for it to happen is large.

Example: How Many Raisins in a Spoonful?

A manufacturer of raisin bran cereal wants to ensure that on the average 90% of spoonfuls will each contain *at least one* raisin. On the assumption that raisins are randomly distributed in the cereal, to what value should the mean η of raisins per spoonful be adjusted to ensure this?

If one imagines a spoonful divided up into raisin sized cells, the number n of cells will be large, but the probability of any given cell being occupied by a raisin will be small. Hence the Poisson distribution should supply a good approximation.

The required probability is (1 − probability of getting no raisins). Thus we require that $1 - e^{-\eta} = 0.90$, that is, $\eta = \log_e 10 = 2.3$. Hence the manufacturer should ensure that on the average there are 2.3 raisins per spoonful.

Exercise 5.17.   A random spoonful of raisin bran is found to contain six raisins. Does this contradict a null hypothesis that on the average there are 2.3 raisins per spoonful?

The Variance of a Poisson Distribution Is Equal to Its Mean: $\sigma^2 = \eta$

As p approaches zero, $q = 1 - p$ approaches 1, so that if $\eta = np$ remains finite, $\sigma^2 = npq$ tends to $\sigma^2 = \eta = np$. Thus a frequency y distributed as a

Poisson variable with mean η also has variance equal to η. For example, the approximating Poisson distribution for Minnie's and Henry's accidents has mean $\eta = 2.1000$ and so has variance $\sigma^2 = 2.1000$. The variances for the exact binomial distributions are as follows:

Henry: $n'p'q' = 5,000 \times 0.00042 \times 0.99958 = 2.0991$
Minnie: $npq = 10,000 \times 0.00021 \times 0.99979 = 2.0996$

The agreement is very close.

Approach to Normality of the Poisson Distribution When η Is Not Too Small

As the mean frequency increases, the Poisson distribution approaches normality quite quickly. Thus Figure 5.6a shows the Poisson distribution with mean $\eta = \sigma^2 = 2.1$ (representing Minnie's and Henry's accident distributions), while Figure 5.6b shows a Poisson distribution with $\eta = \sigma^2 = 10$. The former is quite strongly skewed, but the latter could be fairly well approximated by a normal distribution. When employing the normal approximation to compute a tail area, Yates's adjustment should be used, as with the binomial.

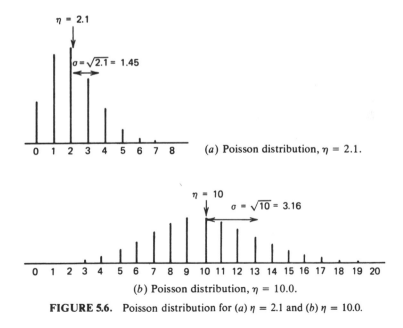

(a) Poisson distribution, $\eta = 2.1$.

(b) Poisson distribution, $\eta = 10.0$.

FIGURE 5.6. Poisson distribution for (a) $\eta = 2.1$ and (b) $\eta = 10.0$.

Confidence Limits for a Poisson Frequency

Suppose that we have an observed value y_0 for a Poisson distributed frequency. What are the $1 - \alpha$ confidence limits η_- and η_+ for the mean η frequency? As before, η_- is some value less than y_0 that just produces a result significant at the $\frac{1}{2}\alpha$ level against the alternative $\eta > \eta_-$. Correspondingly η_+ is some value greater than y_0 that just produces significance at the $\frac{1}{2}\alpha$ level against the alternative $\eta < \eta_+$. It is tedious to compute these values directly from the distribution. Table G at the back of the book, however, gives 95 and 99 % limits for $y_0 = 1, 2, \ldots, 50$.

Exercise 5.18. Obtain expressions giving approximate confidence limits for a Poisson frequency, using the normal approximation with Yates's adjustment. Check the 95 % limits for $y_0 = 10$ and $y_0 = 20$ obtained using your formula against the values in Table G.

Additive Property of the Poisson Distribution

The Poisson distribution has an important additive property. If $y_1, y_2, \ldots, y_k$ are observations from independent Poisson distributions with means $\eta_1, \eta_2, \ldots, \eta_k$, then $Y = y_1 + y_2 + \cdots + y_k$ has a Poisson distribution with mean $\eta_1 + \eta_2 + \cdots + \eta_k$.

Sum of Normalized Squared Deviations Approximately Distributed as Chi Square

With η of moderate size (say greater than 5) the Poisson distribution can be very roughly approximated by a normal distribution with mean η and variance η. Thus $(y - \eta)/\sqrt{\eta}$ is approximately a standard normal deviate, and its square $(y - \eta)^2/\eta$ is distributed approximately as χ^2 with one degree of freedom. It follows that, if $y_1, y_2, \ldots, y_k$ are k observations from independent Poisson distributions with means $\eta_1, \eta_2, \ldots, \eta_k$, then

$$\sum_{j=1}^{k} \frac{(y_j - \eta_j)^2}{\eta_j} \sim \chi_k^2 \tag{5.38}$$

In words: the sum of the *observed minus the expected frequency squared divided by the expected frequency* is approximately distributed as chi square with k degrees of freedom.

Effect of Inhomogeneity

We saw for the binomial that, if the probability varies within samples, the variance is slightly decreased, but if the probability varies between trials, it is markedly increased. For corresponding limiting Poisson distributions, on realistic assumptions, only variation between trials affects the variance. Specifically, if η_j, the mean frequency in the jth trial, varies from trial to trial with variance σ_η^2 about a mean η, the variance σ_y^2 of the frequency y from trial to trial is not η but

$$\sigma_y^2 = \eta + \sigma_\eta^2$$

We may, therefore, test for the existence of variation in the mean η_j using the chi-square approximation above.

Example

The supervisor of the raisin bran production unit suspects that the raisins are not being added to the cereal in accordance with the process specification. In a test the quality control inspector randomly selects one box per hour over a 12-hour period, thoroughly shakes the box, and counts the raisins in one standard-size scoopful. The specification requires a mean of 36 raisins per scoopful. Random sampling variation alone would thus produce Poisson distributed frequencies with mean $\eta = 36$ and standard deviation $\sigma = \sqrt{36} = 6$.

The inspector's data are as follows:

hour	1	2	3	4	5	6	7	8	9	10	11	12	total
number of raisins in a scoopful	43	46	50	40	38	29	31	35	41	52	48	37	490

average frequency = 40.83

yielding

$$\frac{\sum (y_j - 36)^2}{36} = \frac{866}{36} = 24.06$$

Entering the chi-square table with 12 degrees of freedom, we find this value to be significant at about the 2% point.

Two possible reasons for the discrepancy are as follows:

1. The mean frequency of raisins η is not 36.0.
2. The variance from box to box is greater than can be accounted for by random sampling variation.

To test the first possibility we may use the additive property of the Poisson distribution. The sum of the frequencies (490) is compared with the value to

be expected ($12 \times 36 = 432$) if the mean frequency per scoopful were really
36. We enter the value

$$\frac{(490 - 432)^2}{432} = 7.79$$

in the chi-square table with one degree of freedom, to find that it is significant
at about 0.5% level. It appears, therefore, that the mean frequency is greater
than 36, the value to be expected if the process were operating correctly.

In testing the second possibility we are confronted by a difficulty. Although
we have discredited the possibility that $\eta = 36.0$, we do not know the true
mean frequency at which the process is working. However, it may be shown
that for samples drawn from a Poisson distribution of unknown mean the
quantity $(y_j - \bar{y})^2/\bar{y}$, in which $\bar{y}$ is substituted for η, will be approximately
distributed as χ^2 with $k - 1$ degrees of freedom. For our example,
$\bar{y} = 490/12 = 40.83$, and the value

$$\frac{\sum_{j=1}^{12} (y_j - 40.83)^2}{40.83} = 14.34$$

is entered in the chi-square table with $k - 1 = 12 - 1 = 11$ degrees of
freedom. The significance level is then about 20%.

This analysis indicates, therefore, that there are statistically significant
discrepancies from the specification frequencies, and that these are likely to
be due to an excessively high mean level rather than to unexplained variation
from hour to hour. In our enthusiasm to illustrate some pretty applications
of the chi-square distribution the reader will have noticed that we have
omitted what should be an essential first step in any real analysis: *we have
not plotted the data.* You are invited to remedy this negligence and so to
suggest what additional enquiries the supervisor might make in furtherance
of this iterative investigation.

Variance Stabilizing Transformation for the Poisson Distribution

Often it is desired to analyze frequency data using standard normal theory
procedures involving t statistics and the analysis of variance and regression
techniques discussed later. But because the variance of a Poisson variable
is equal to its mean, contrary to the assumptions of these normal theory
procedures, the variance alters as the mean changes. If the frequencies cover
wide ranges, therefore, it is best to work with scores obtained by appropriate
transformation. The appropriate variance stabilizing transformation in this
instance is the square root. If y is distributed as a Poisson variable, $\sqrt{y}$ has
an approximately constant variance equal to 0.25.

Exercise 5.19. Reanalyze the raisin bran data, using the square root transformation.

The square root transformation not only stabilizes variance but also improves the normal approximation to the distribution.

5.7. CONTINGENCY TABLES AND TESTS OF ASSOCIATION

The data in Table 5.6 show the results from five hospitals of a surgical procedure designed to improve the functioning of certain joints impaired by disease. In this study the meanings to be attached to "no improvement," "partial functional restoration," and "complete functional restoration" were carefully defined in terms of measurable phenomena.

TABLE 5.6. **Results (percentages) from a certain surgical procedure in five hospitals**

	hospital					overall percentage
	A	B	C	D	E	
no improvement	27.7	16.1	10.1	16.4	52.4	24.5
partial functional restoration	38.3	32.3	45.6	43.8	35.4	40.6
complete functional restoration	34.0	51.6	44.3	39.8	12.2	34.9

The figures, expressed as percentages, certainly suggest differences in success at the five hospitals. However, percentages can be very misleading. For example, 28.6% can, on closer examination, turn out to represent either 2/7 or 286/1000. To make an analysis, therefore, we must consider the *original data* on which these percentages are based. These are the frequencies shown as large numerals in Table 5.7.

An entry, say y_h, in the hth cell of the table, is a frequency of occurrence and might be expected to be distributed about its mean η_h in a Poisson distribution. A null hypothesis that allowed expected values $\eta_1, \eta_2, \ldots, \eta_{15}$ to be calculated for each cell *independently of the the data* could therefore be tested approximately by referring $\sum (y_h - \eta_h)^2/\eta_h$ to a chi-square table with 15 degrees of freedom. However, the hypothesis of interest in this case is not of

TABLE 5.7. Results of surgical procedure: original frequencies shown in large type, expected frequencies in top right corner, contribution to chi-square in top left corner of each cell

				hospital			
	A	B	C	D	E	total
no improvement	0.19 11.53 **13**	0.89 7.60 **5**	6.67 19.37 **8**	3.44 31.39 **21**	26.05 20.11 **43**	**90**
partial functional restoration	0.06 19.08 **18**	0.53 12.59 **10**	0.48 32.07 **36**	0.31 51.97 **56**	0.55 33.29 **29**	**149**
complete functional restoration	0.01 16.39 **16**	2.49 10.81 **16**	2.02 27.55 **35**	0.91 44.64 **51**	12.10 28.60 **10**	**128**
total	**47**	**31**	**79**	**128**	**82**	**367**

χ^2 value 56.7, degrees of freedom 8, significance level $<0.001\%$

146

that kind. Instead, we require to know whether the data are explicable on the hypothesis that the mean frequencies in the various categories are distributed *in the same proportions* from hospital to hospital. The only evidence of what the expected frequencies should be on this hypothesis comes from the marginal totals of the data. For instance, from the right-hand margin, 90 out of 367 or 24.523 % of all patients showed no improvement. Since there was a total of 47 patients from hospital A, 24.523 % of 47 or 11.53 would be the expected frequency of patients showing no improvement in hospital A if the null hypothesis were true. In a similar way the other empirical expected frequencies can be computed from the marginal totals. In general, the expected frequency $\hat{\eta}_{ij}$ to be entered in the ith row and jth column is $F_i F_j / F$, where F_i is the marginal frequency in the ith row, F_j is the marginal frequency in the jth column, and F is the total frequency. These values are shown in the top right-hand corner of each cell. The value of chi-square contributed by the cell in the ith row and jth column, $(y_{ij} - \hat{\eta}_{ij})^2 / \hat{\eta}_{ij}$, is shown in the top left-hand corner of each cell. For the first entry, for example, the contribution is $(13 - 11.53)^2 / 11.53 = 0.19$. Adding together the 15 contributions, we have a total chi-square value of 56.7.

R. A. Fisher showed that, when empirical expected frequencies were calculated from the marginal totals in this way, the chi-square statistic could still properly be referred to the standard chi-square table, provided that the number of degrees of freedom with which the table was entered was suitably chosen. The correct number of degrees of freedom for a table with r rows and c columns is $(r - 1)(c - 1)$. It corresponds to the number of cells in the table less the number of known relationships that the method of calculation has imposed on the expected values. Thus the calculated value of 56.7 should be referred to a chi-square table with $2 \times 4 = 8$ degrees of freedom. The value is highly significant, that is, the corresponding small probability is far beyond the range of the table.

Hospital E is different from the other four hospitals in that it is a referral hospital. It seems relevant, therefore, to ask two questions:

1. How much of the discrepancy is associated with differences between the referral hospital, on the one hand, and the nonreferral hospitals *taken together*, on the other?
2. How much of the discrepancy is associated with differences among the four nonreferral hospitals?

Table 5.8, which has the same format as Table 5.7, permits the appropriate comparisons to be made. It is evident that the differences are mainly between the referral hospital E and the nonreferral hospitals. Differences among nonreferral hospitals are readily explained by sampling variation.

TABLE 5.8. Comparisons among hospitals

a. Hospital E (referral) compared to hospitals A, B, C, D (nonreferral)

restoration	hospital		total
	$A + B + C + D$	E	
none	17.50 69.89 47	26.06 20.11 43	90
partial	0.16 115.71 120	0.55 33.29 29	149
complete	3.48 99.40 118	12.10 28.60 10	128
total	285	82	367

χ^2 value   49.8
degrees of freedom   2
significance level   <1%

b. Comparison of hospitals A, B, C, D (nonreferral)

restoration	hospital				total
	A	B	C	D	
none	3.55 7.75 13	0.00 5.11 5	1.94 13.03 8	0.00 21.11 21	47
partial	0.16 19.79 18	0.71 13.05 10	0.23 33.26 36	0.08 53.89 56	120
complete	0.61 19.46 16	0.78 12.84 16	0.16 32.71 35	0.18 53.00 51	118
total	47	31	79	128	285

χ^2 value   8.3
degrees of freedom   6
significance level   about 20%

We notice that the former discrepancies arise because the results from the referral hospital are *not as good* as those from hospitals *A*, *B*, *C*, and *D*. This might bc expected because typically the more difficult cases are sent to the referral hospital.

The 2 × 2 Table: Comparison of Repair Records for Television Sets

A special case of some importance occurs when there are only two rows and two columns, and consequently the resulting chi-square has only one degree of freedom. For instance, the data in Table 5.9 were obtained by a consumer organization from 412 buyers of new color television sets made by manufacturers *A* and *B*. The frequencies in the column headed "required service" denote the number of consumers whose sets needed service at least once during a 2-year period.

TABLE 5.9. Comparison of two brands of color television sets

brand	required service		did not require service		total
A	2.74	129.87	2.79	143.13	
		111		162	273
B	5.38	66.13	4.89	72.87	
		85		54	139
total		196		216	412

A null hypothesis of interest is that the proportions requiring service are the same for both manufacturers, observed discrepancies being due to sampling variation. To test the hypothesis expected frequencies may be calculated from the marginal totals as before. These are shown in the top right-hand corner of the table. By adding the contributions from the four cells, we obtain a value of $\chi^2 = 15.51$, which is referred to the chi-square table with one degree of freedom. This gives a significance level of less than 0.01 %. Thus, if these samples can be regarded as *random* samples from the two manufacturers, there is little doubt that sets by maker *A* have a better repair record than sets by maker *B*.

Shortcut Calculation of the 2×2 *Table*

If we label the four frequencies in a 2×2 table as a, b, c, d, it may be shown that the value of χ^2 is given by

$$\chi^2 = \frac{(ad - bc)^2(a + b + c + d)}{(a + b)(c + d)(a + c)(b + d)} \tag{5.39}$$

For the above example, where $a = 111, b = 162, c = 85$, and $d = 54$, we have, as before,

$$\chi^2 = \frac{(7776)^2 \times 412}{196 \times 216 \times 273 \times 139} = 15.51$$

Yates's Adjustment

The frequencies that occupy the cells of any contingency table must be integers. For this application, therefore, the *continuous* chi-square distribution that is tabulated provides only an approximation to the true and essentially discrete distribution of the chi-square statistic. The approximation is good, provided that the number of cells is not too small and the expected frequencies in the cells are not too small. For the 2×2 contingency table, where the first proviso is most strained, the test as described above tends to exaggerate significance. A considerably improved approximation is obtained by applying Yates's adjustment (already discussed in approximating the binomial distribution by a normal curve). In the present context this adjustment consists of changing the observed frequencies by half a unit to give *smaller* deviations from the expected values. The test is then performed with these less extreme frequencies. For the color television data, for example, we should change the observed frequencies from 111, 162, 85, 54 to 111.5, 161.5, 84.5, 54.5, which yield a slightly smaller value for χ^2 of 14.70 with a significance level of about 0.01 %.

The shortcut formula with Yates's adjustment reduces to

$$\chi^2 = \frac{[|ad - bc| - \frac{1}{2}(a + b + c + d)]^2(a + b + c + d)}{(a + b)(c + d)(a + c)(b + d)} \tag{5.40}$$

where $|ad - bc|$ is the absolute value of $ad - bc$.

Exercise 5.20. Using the shortcut formula with Yates's adjustment, confirm that the television data give a χ^2 value of 14.70.

QUESTIONS FOR CHAPTER 5

1. Why is a set of confidence intervals more useful than a significance test? Illustrate with examples.

2. What is the formula for the confidence interval of the difference between two means for a paired comparison design? For an unpaired (fully randomized) design? How are the formulas similar? How are they different?

3. Suppose that you are given data from a simple comparative experiment in which n_A observations have been made with treatment A and n_B observations with treatment B. To use Equation 5.9 is it necessary that $n_A = n_B$? If $n_A = n_B$, does Equation 5.9 always give a shorter 90% confidence interval than Equation 5.18? Explain how would you decide which equation to use? How would you verify each interval using a randomization distribution?

4. Think of a specific problem of comparing two treatments (e.g., methods or brands), preferably from your own field. How would you organize an experiment to find out which one is better? What are some problems that might arise? How would you analyze the results? How could past data be used to assist in choosing the design and the number of runs to be made?

5. Let y be the number of sixes recorded in the throw of three dice. Find $\Pr(y = 0), \Pr(y = 1), \Pr(y = 2), \Pr(y = 3), \Pr(y > 1), \Pr(y < 3)$. Suppose that, in 16 throws of a single die, a six is thrown on eight occasions; does this discredit the hypothesis that the die is fair? Use these data to obtain a 95% confidence interval for p.

6. Possibly meaningful signals have been received from outer space. The data take the form of the number of pulses y received in each of a sequence of 127 minutes. A skeptic suggests that the variation in the frequencies observed $y_1, y_2, \ldots, y_{127}$ might be ascribed to chance causes alone. Describe any test you might make of the skeptic's hypothesis.

7. You are assigned to a research program to assess the effectiveness of anticancer drugs in animals and human beings. Write a report for the physician in charge of the program (who has no statistical knowledge) describing how methods using (a) the binomial distribution, (b) the Poisson distribution, and (c) contingency tables might be of value in assessing the results from the program. Illustrate your discussion with hypothetical experiments and data, and give appropriate analyses.

8. Illustrate how the binomial and Poisson distributions may be used to design sampling inspection schemes.

Problems for Part I

Whether or not specifically asked, the reader should always (1) plot the data in any potentially useful way, (2) state the assumptions made, (3) comment on the appropriateness of these assumptions, and (4) consider alternative analyses.

1. For the following observations: 5, 4, 8, 6, 7, 8, calculate:
   (a) The sample average.
   (b) The sample variance.

2. The following are recorded shear strengths of spot welds (in pounds):

12,560	12,900	12,850	12,710

   Calculate the average, the sample variance, and the sample standard deviation.

3. Four ball bearings were taken at random from a production line, and their diameters were measured. The results (in centimeters) were as follows:

1.0250	1.0252	1.0249	1.0249

   Calculate the sample standard deviation. Calculate the standard error of the mean.

4. (a) The following are results of fuel economy tests (miles per gallon) obtained from a sample of 50 automobiles in 1965:

17.74	12.17	12.22	13.89	16.47
15.88	16.10	16.74	17.54	17.43
14.57	12.90	12.81	14.95	16.25
17.13	14.46	14.20	16.90	11.34
12.57	13.15	16.53	13.60	13.34
13.67	14.23	15.81	16.63	11.40
14.94	13.66	9.79	13.08	14.57
14.93	14.01	14.43	16.35	15.65
11.52	17.46	14.67	15.92	16.02
13.46	13.70	14.98	14.57	15.72

   Plot a histogram for these data, and calculate the sample mean and the sample standard deviation.

(b) The following are 50 results obtained in 1975:

24.57	24.79	22.21	25.84	25.35
22.19	24.37	21.32	22.74	23.38
25.10	28.03	29.09	29.34	24.41
25.12	25.27	27.46	27.65	27.95
21.67	22.15	24.36	26.32	24.05
28.27	26.57	26.10	24.35	30.04
25.18	27.42	24.50	23.21	25.10
23.59	26.98	22.64	25.27	25.84
27.18	24.69	26.35	23.05	23.37
25.46	28.84	22.14	25.42	21.76

Plot a histogram for these data, and calculate the sample mean and the sample standard deviation. Considering both data sets, comment on any interesting aspects you find. Make a list of questions you would like answered that would allow you to investigate further.

5. If a random variable has a normal distribution with mean 80 and standard deviation 20, what is the probability that it assumes a value:
   (a) Less than 77.4?
   (b) Between 61.4 and 72.9?
   (c) Greater than 90.0?
   (d) Less than 67.6 or greater than 88.8?
   (e) Between 92.1 and 95.4?
   (f) Between 75.0 and 84.2?

6. Suppose that daily mill yields at a certain plant are normally distributed with a mean of 10 tons and a standard deviation of 1 ton. What is the probability that tomorrow the value will be greater than 9.5? To obtain your answer is it necessary to assume that the observations behave as statistically independent random variables?

7. Four samples are taken every hour from a production line, and the average hourly measurement of a particular impurity is recorded. Approximately one out of six of these *averages* exceeds 1.5% when the mean value is approximately 1.4%. State assumptions that would enable you to determine the proportion of the *individual* readings exceeding 1.6%. Make the assumptions, and do the calculations. Are these assumptions likely to be true? If not, how might they be violated?

8. Suppose that 95% of the bags of a certain kind of cement mix weigh between 49 and 53 pounds. *Averages of three* successive bags were plotted, and 47.5% of these were observed to lie between 51 and x pounds.

Estimate the value of x. State assumptions you make, and say whether these assumptions are likely to be true for this example.

9. For each sample of 16 boxes of cereal an *average box weight* is recorded. Past records indicate that, on the average, 1 out of 40 of these *average box weights* exceeds 8.4 ounces, and 1 out of 40 is less than 8.1 ounces. What is the probability, then, that an *individual box* of cereal, selected at random, will weigh less than 8.0 ounces? State any assumptions you make.

10. The lengths of bolts produced in a factory may be taken to be normally distributed. The bolts are checked on two "Go–NoGo" gauges and those shorter than 2.9 or longer than 3.1 inches are rejected.

    (a) A random sample of 397 bolts is checked on the gauges. If the (true but unknown) mean length of the bolts produced at that time was 3.06 inches and the (true but unknown) standard deviation was 0.03 inches, what values would you expect for n_1, the number of bolts found to be too short, and n_2, the number of bolts found to be too long?

    (b) A random sample of 50 bolts from another factory is also checked. If, for these, $n_1 = 12$ and $n_2 = 12$, estimate the mean and the standard deviation for these bolts. State your assumptions.

11. The mean weight of individual items coming from a production line is 83 pounds. On the average, 19 out of 20 of these individual weights are between 81 and 85 pounds. If the *average weight of six* randomly selected items is plotted, what proportion of these points will fall between the limits 82 and 84 pounds? What is the precise nature of the random sampling hypothesis that you need adopt to answer this question? Consider how it might be violated, and consider what the effect of such violation might be on your answer.

12. After examination of some thousands of observations it was determined that the observed thrusts for a certain aircraft engine were approximately normally distributed with mean 1000 pounds and standard deviation 10 pounds. What percentage of such measured thrusts would you expect to be:

    (a) Less than 990 pounds?

    (b) Greater than 1010 pounds?

    (c) Between 990 and 1020 pounds?

13. A civil engineer tested two different types (A and B) of a special reinforced concrete beam. He made nine test beams (five A's and four B's) and measured the strength of each. From the following data he wants to decide whether there is any real difference between the two types. What

assumptions does he need to draw conclusions? What might he conclude? Give reasons for your answer.

Strength (coded units)

A	B
67	45
80	71
106	87
83	53
89	

14. Every 90 minutes, routine readings are made of the level of asbestos fiber present in the air at an industrial plant. A salesman claimed that spraying with a chemical, S-424, could be beneficial. Arrangements were made for a comparative trial in the plant itself. Four consecutive readings, the first two without S-424 and the second two with S-424, were as follows:

$$8 \quad 6 \quad 3 \quad 4$$

In the light of past data collected without the chemical spray, given below, do you think this additive works? Explain your answer.

asbestos levels (112 consecutive readings)

Dec. 6	9 10 9 8 9 8 8 8 7 6 9 10 11 9 10 11
Dec. 7	11 11 11 10 11 12 13 12 13 12 14 15 14 12 13 13
Dec. 8	12 13 13 13 13 13 10 8 9 8 6 7 7 6 5 6
Dec. 9	5 6 4 5 4 4 2 4 5 4 5 6 5 5 6 5
Dec. 10	6 7 8 8 8 7 9 10 9 10 9 8 9 8 7 7
Dec. 11	8 7 7 7 8 8 8 8 7 6 5 6 5 6 7 6
Dec. 12	6 5 6 6 5 4 3 4 5 5 6 5 6 7 6 5

15. Nine samples were taken from two streams, four from one and five from the other, and the following data obtained:

pollution level in stream 1 (ppm)	pollution level in stream 2 (ppm)
16	9
12	10
14	8
11	6
	5

It is claimed that the data prove that stream 2 is cleaner than stream 1. A statistician asked the following questions. When were the data taken? All in one day? On different days? Were data taken during the same time period for the two streams? Were the temperatures of the two streams the same? Where in the streams were the data taken? Why were these points chosen? Are they representative? Are they comparable?

Why do you think he asked these questions? Are there other questions he should have asked? Is there any set of answers to these questions (and to others you invent) that would justify the use of a t test to draw conclusions? What conclusions?

16. Fifteen judges rated two randomly allocated brands of beer A and B according to taste (scale: 1 to 10) as follows:

brand A	2	4	2	1	9	9	2	2
brand B	8	3	5	3	7	7	4	

Stating assumptions, test the null hypothesis that $\eta_A = \eta_B$ against the alternative that $\eta_A \neq \eta_B$. Can you think of a better design for this experiment? Write *precise* instructions for the conduct of your preferred experiment.

17. These data were obtained in a comparison of two different methods for determining dissolved oxygen concentration (in milligrams per liter):

sample	1	2	3	4	5	6
method A (amperometric)	2.62	2.65	2.79	2.83	2.91	3.57
method B (visual)	2.73	2.80	2.87	2.95	2.99	3.67

Estimate the difference between the methods. Give a confidence interval for the mean difference. What assumptions have you made?

Do you think that this experiment provides an adequate basis for choosing one method over the other? If not, what further experiments and data analysis would you recommend? Imagine different outcomes that could result from your analysis, and say what each might show.

18. The following are smoke readings taken on two different camshafts:

	smoke reading	
engine	camshaft A	camshaft B
1	2.7	2.6
2	2.9	2.6
3	3.2	2.9
4	3.5	3.3

Suppose that your supervisor brings you these data and says that one group in your company claims there is essentially no difference in the camshafts and the other says there really is a difference. What questions would you ask about this experiment? What answers would make further analysis useful? What further analysis? Assuming any answers to your questions that you like, write a report on the experiment and the data. Somewhere in your report explain the difference between "statistically significant" and "technically important."

19. Two different configurations for a solar energy collector were tested with the following results. The measured quantity was power (in watts).

configuration A	$1.8^{(1)}$	$1.9^{(2)}$	$1.1^{(5)}$	$1.4^{(7)}$
configuration B	$1.9^{(3)}$	$2.1^{(4)}$	$1.5^{(6)}$	$1.5^{(8)}$

The data were collected at eight different, comparable time periods. The random order of the tests is indicated by the superscripts in parentheses. Is there evidence that a statistically significant difference exists between the mean values for power attainable from these two configurations? Do you need to assume normality to make a test?

20. In a cathode interference resistance experiment, with the filament voltage at 6.9, the plate current was measured on seven twin triodes. Analyze the data given below. Is there any evidence of a systematic difference between the plate current readings of compartments A and B?

tube	plate current readings	
	A	B
1	1.176	1.279
2	1.230	1.000
3	1.146	1.146
4	1.672	1.176
5	0.954	0.699
6	1.079	1.114
7	1.204	1.114

Explain your assumptions.

21. An agricultural engineer obtained the following data on two methods of drying corn and asked for a statistical analysis:

drying rate

with preheater	without preheater
16	20
12	10
22	21
14	10
19	12

(a) What questions would you ask him?

(b) Under what circumstances would you be justified in analyzing the data, using (i) a paired t test, (ii) an unpaired t test, (iii) something else?

(c) Analyze the data for options (i) and (ii).

22. The following are data on the effect of two different methods for administering a bronchodilating aerosol. Method A is by hand, and method B is by an automatic inhalation device. The quantity measured was the specific airways resistance 30 minutes after the administration of the aerosol. What questions would you ask on how the experiment was conducted? What analysis might be appropriate?

A	17.00	22.80	21.60	20.40	11.20	14.00	52.25	7.50	12.20	18.85	6.05	4.05
B	11.60	11.60	13.65	17.22	8.25	6.20	41.50	6.96	8.40	9.00	5.18	3.00

[*Source*: These data were selected from a larger set reported by F. J. McIlneath and B. M. Cohen, *J. Med.*, **1**, 229 (1970).]

23. An agricultural engineer is trying to develop a special piece of equipment for processing a crop immediately after harvesting. Two configurations of this equipment are to be compared. Twenty runs are to be performed during a period of 5 days. Each run involves setting up the equipment in either configuration I or configuration II, harvesting a given amount of crop, processing it on the equipment, obtaining a quantitative measure of performance, and cleaning the equipment. Since there is only one piece of equipment, the tests must be done one at a time. The engineer has asked you to consult on this problem.

(a) What questions would you ask him?

(b) What advice might you give him about planning the experiment?

(c) The engineer believes a most accurate experiment requires that an equal number of runs be conducted with each configuration. What assumptions would make it possible to demonstrate in a quantitative way that this is true?

24. Assuming a standard deviation $\sigma = 0.4$, calculate a 90% confidence interval for mean reaction time from the following data (in seconds):

    1.4   1.2   1.2   1.3   1.5   1.0   2.1   1.4   1.1

    Carefully state and criticize any assumptions you make. Repeat this problem, assuming that σ is unknown.

25. Five student groups in a surveying class obtained the following measurements of the distance (in meters) between two points:

    420.6   421.0   421.0   420.7   420.8

    Stating precisely your assumptions, find an approximate 95% confidence interval for the mean measured distance.

26. Given the following data on egg production from 12 hens randomly allocated to two different diets, estimate the mean difference produced by the diets and obtain a 95% confidence interval for this mean difference. Explain your answer and the assumptions you make.

Diet A	166	174	150	166	165	178
Diet B	158	159	142	163	161	157

27. Two species A and B of trees were planted on a total of 20 randomly selected plots, 10 for A and 10 for B. The average height per plot was measured after 6 years. The results (in meters) were as follows:

A	3.2	2.7	3.0	2.7	1.7	3.3	2.7	2.6	2.9	3.3
B	2.8	2.7	2.0	3.0	2.1	4.0	1.5	2.2	2.7	2.5

    Obtain a 95% confidence interval for the difference in means.

28. A chemical reaction was studied by making 10 runs with a new, supposedly improved method (method B) and 10 runs with the standard method (method A). The following yield results were obtained:

method A	method B
$54.6^{(16)}$	$74.9^{(12)}$
$45.8^{(10)}$	$78.3^{(19)}$
$57.4^{(11)}$	$80.4^{(14)}$
$40.1^{(2)}$	$58.7^{(6)}$
$56.3^{(20)}$	$68.1^{(8)}$
$51.5^{(18)}$	$64.7^{(3)}$
$50.7^{(9)}$	$66.5^{(7)}$
$64.5^{(15)}$	$73.5^{(4)}$
$52.6^{(1)}$	$81.0^{(17)}$
$48.6^{(5)}$	$73.7^{(13)}$

The superscripts in parentheses denote the time order in which the runs were performed. Comment on the data, and analyze them.

29. [These data are taken from U.S. Patent 3,505,079 (April 7, 1970) for a process for preparing a dry, free-flowing baking mix.] Five recipes were used for making a number of cakes, starting from two different types A and B of premix. The difference between the two premixes was that A was aerated and B was not. The volumes of the cakes were measured with the following results:

	Volume	
recipe	A	B
1	83	65
2	90	82
3	96	90
4	83	65
5	90	82

The five recipes differed somewhat in amount of water added, beating time, baking temperature, and baking time. The patent claims that significantly greater volume was obtained with A. Do these data support this claim? Make any relevant comments you think are appropriate. Include in your answer a calculated 95% confidence interval for the true difference between the volumes obtained with A and B. State any assumptions you make.

30. The following are results from a larger study on the pharmacological effects of nalbuphine. The measured response obtained from 11 subjects was change in pupil diameter (in millimeters) after 28 doses of nalbuphine (B) or morphine (A).

A	+2.4	+0.08	+0.8	+2.0	+1.9	+1.0
B	+0.4	+0.2	−0.3	+0.8	0.0	

Assume that the subjects were randomly allocated to the drugs. What is the 95% confidence interval for $\eta_B - \eta_A$? What is the 90% confidence interval? State your assumptions.
[Source: These data are from H. W. Elliott, G. Navarro, and N. Nomof, J. Med., 1, 77 (1970).]

31. In a set of six very expensive dice the pips on each face are diamonds. It is suggested that the weight of the stones will cause the "five" or "six" faces to fall downwards, and hence the "one" and "two" faces to fall upwards, more frequently than they would with fair dice. To test this

conjecture a trial is conducted as follows: the throwing of a one or a two is called a success, the whole set of six dice is then thrown 64 times, and the frequencies of throws with 0, 1, 2, ..., 6 successes are as follows:

Total

Successes out of six (number of dice showing a one or a two)	0	1	2	3	4	5	6	
frequency	0	4	19	15	17	7	2	64

(a) What would be the theoretical probability of success and the mean and variance of the above frequency distribution if all the dice were fair?

(b) What is the empirical probability of success calculated from the data, and what is the sample average and variance?

(c) Test the hypotheses that the mean and the variance have their theoretical values.

(d) Calculate the expected frequencies in the seven "cells" of the table on the assumption that the probability of a success is exactly $\frac{1}{3}$.

(e) Make a chi-square test of agreement of the observed with the expected frequencies (combine the frequencies for zero and one successes and also for five and six successes, so that expected frequencies are not less than five in any cell). The number of degrees of freedom for the test is the number of frequencies compared *less one*, since the total of the expected frequencies has been chosen to match the actual total number of trials.

(f) Calculate the expected frequencies based on the empirical estimate of probability, and make a chi-square test of agreement with the observed frequencies. The number of degrees of freedom is now the number of cell frequencies compared *less two*, since expected frequencies have been chosen so that the total frequency and the mean both match the data.

(g) Can you think of a better design for this trial?

32. The level of radiation in the control room of a nuclear reactor is to be automatically monitored by a Geiger counter. The monitoring device works as follows. Every tenth minute the number (frequency) of "clicks" occurring in t seconds is counted automatically. A scheme is required such that, if this frequency exceeds some number c, an alarm will sound. The scheme should have the following properties: if the number of *clicks per second* is less than four there should be only about 1 chance in 500 that the alarm will sound, but if the number reaches 16 there should only be about 1 chance in 500 that the alarm will not sound. What values should be chosen for t and c?

Hint: Recall that the square root of a Poisson frequency is roughly normally distributed with standard deviation 0.5. [*Answer*: $t \simeq 2.25$, $c \simeq 20$ (closest integer to 20.25).]

33. Check the approximate answer given for problem 32 by actually evaluating the Poisson frequencies.

34. Consider the following data on recidivism, as measured by re-arrest of persons who have committed at least one previous offense. There are a number of points on which you would wish to be reassured before drawing conclusions on such data. What are they? Assuming that your questions received satisfactory answers, what conclusions could you draw from the following data?

	treatment received		
	no psychotherapy	group therapy	individual therapy
back in prison within 1 year	24	10	41
back in prison in 1 to 10 years	25	13	32
never back in prison	12	20	9
total	61	43	82

PART II

Comparing More
Than Two Treatments

Sometimes experimenters want to compare more than two treatment means. The "treatments" may be four diets, six brands of lawnmowers, or five fertilizers. The investigator may want to assess the evidence that any real differences exist and to estimate these differences.

In this part of the book we explain some basic ideas about the design and analysis of comparative experiments of this kind. Designs that permit direct comparison of k treatments are discussed in Chapter 6, and designs that allow for the elimination of block differences in Chapters 7 and 8. The technique of analysis of variance is introduced. We show how simple plots of residuals can reveal model inadequacy and how reference distributions can summarize major conclusions.

CHAPTER 6

Experiments to Compare k Treatment Means

So far we have discussed the comparison of only two means. Frequently, however, it is important to compare more than two. In this chapter we discuss designs selected to do this in which k treatments have been randomly allocated to the experimental material.

6.1. BLOOD COAGULATION TIMES WITH FOUR DIFFERENT DIETS

Table 6.1 gives coagulation times for samples of blood drawn from 24 animals receiving four different diets A, B, C, and D. (To help the reader concentrate on essentials we have adjusted the data so that the averages come out to be whole numbers.) These data are plotted in Figure 6.1. The diets were randomly allocated to the animals, and the blood samples were taken and tested in random order. The methods discussed below can be precisely justified on the supposition that the data can be treated as random samples from four normal populations having the same variance σ^2 and differing, if at all, only in their means. Alternatively the analysis resulting from these assumptions may be approximately justified by randomization theory in a manner similar to that employed in the comparison of two means. This justification requires, of course, that randomization has actually been carried out.

Consider this question: Is there evidence to indicate any real differences among the mean values associated with the different treatments (diets)? The null hypothesis to be tested, then, is that the treatment means η_A, η_B, η_C, and η_D are all the same, and the alternative hypothesis is that they are not. The calculations to explore these hypotheses are best set out in an analysis of variance table, a valuable device due to Fisher. Essentially this analysis

TABLE 6.1. Coagulation time (seconds) for blood drawn from 24 animals randomly allocated to four different diets

	diet (treatment)			
	A	B	C	D
	$62^{(20)}$	$63^{(12)}$	$68^{(16)}$	$56^{(23)}$
	$60^{(2)}$	$67^{(9)}$	$66^{(7)}$	$62^{(3)}$
	$63^{(11)}$	$71^{(15)}$	$71^{(1)}$	$60^{(6)}$
	$59^{(10)}$	$64^{(14)}$	$67^{(17)}$	$61^{(18)}$
		$65^{(4)}$	$68^{(13)}$	$63^{(22)}$
		$66^{(8)}$	$68^{(21)}$	$64^{(19)}$
				$63^{(5)}$
				$59^{(24)}$
treatment average	61	66	68	61
grand average	64			

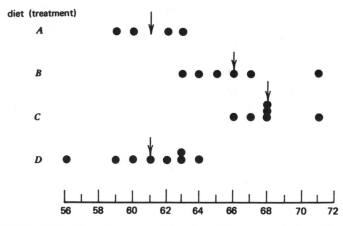

FIGURE 6.1. Plot of data, blood coagulation times. Diet averages indicated by ar

determines whether the discrepancies *between* the treatment averages are greater than could reasonably be expected from the variation that occurs *within* the treatment classifications.

The superscripts in parentheses in Table 6.1 show the random order in which the observations were obtained. The grand average is defined as the sum of all the observations divided by the total number of observations, in this case $1536/24 = 64$.

6.2. ESTIMATING THE AMOUNT OF VARIATION WITHIN AND BETWEEN TREATMENTS

Variation within Treatments

An estimate of the internal consistency of the data may be obtained by pooling the individual sample variances obtained from within each of the data sets corresponding to the different treatments. For example, for the first set of $n_1 = 4$ observations with average $\bar{y}_1 = 61$, the sum of squares S_1 (of deviations from this average) is

$$S_1 = (62 - 61)^2 + (60 - 61)^2 + (63 - 61)^2 + (59 - 61)^2$$
$$= 1^2 + (-1)^2 + 2^2 + (-2)^2 = 10 \tag{6.1}$$

There are three degrees of freedom ($v_1 = n_1 - 1 = 4 - 1 = 3$), so that the sample variance determined from the first treatment is

$$s_1^2 = \tfrac{10}{3} = 3.3 \tag{6.2}$$

Similarly,

$$s_2^2 = \tfrac{40}{5} = 8.0, \qquad s_3^2 = \tfrac{14}{5} = 2.8, \qquad s_4^2 = \tfrac{48}{7} = 6.8 \tag{6.3}$$

In general, k sample variances (all believed to provide estimates of the same unknown true variance σ^2) may be combined to give the pooled estimate

$$s_R^2 = \frac{v_1 s_1^2 + v_2 s_2^2 + \cdots + v_k s_k^2}{v_1 + v_2 + \cdots + v_k}$$

$$= \frac{S_1 + S_2 + \cdots + S_k}{(n_1 - 1) + (n_2 - 1) + \cdots + (n_k - 1)} = \frac{S_R}{N - k} = \frac{S_R}{v_R} \tag{6.4}$$

where n_t is the number of observations within the tth treatment.*

---

* The subscript R is used to designate "residuals." In this example the individual residuals are the quantities $y_{ti} - \bar{y}_t$; for instance, 1, -1, 2, and -2 in Equation 6.1 are residuals.

In this example, therefore, *the within-treatment sum of squares* is

$$S_R = 10 + 40 + 14 + 48 = 112 \tag{6.5}$$

and the number of within-treatment degrees of freedom is

$$v_R = 3 + 5 + 5 + 7 = 20 \tag{6.6}$$

Hence the pooled estimate of the error variance, sometimes called the *within-treatment mean square*, is

$$s_R^2 = \frac{S_R}{v_R} = \frac{112}{20} = 5.6 \tag{6.7}$$

(Note that the upper case S stands for the sum of squares, and the lower case s^2 for the mean square. This convention is followed throughout this book.) On the random sampling assumption the mean square $s_R^2 = 5.6$ supplies an estimate of the intrinsic within-treatment error variance σ^2 based on 20 degrees of freedom.

Algebraically, if we use y_{ti} to denote the ith observation in the tth treatment, the sum of squares within the tth treatment is

$$S_t = \sum_{i=1}^{n_t} (y_{ti} - \bar{y}_t)^2 \tag{6.8}$$

The *within-treatment sum of squares* for all treatments is therefore

$$S_R = S_1 + S_2 + \cdots + S_k = \sum_{t=1}^{k} \sum_{i=1}^{n_t} (y_{ti} - \bar{y}_t)^2 \tag{6.9}$$

and the *within-treatment mean square* is

$$s_R^2 = \frac{S_R}{N - k} = \sum_{t=1}^{k} \sum_{i=1}^{n_t} \frac{(y_{ti} - \bar{y}_t)^2}{N - k} \tag{6.10}$$

where k is the number of treatments.

Exercise 6.1. Five different treatments were administered to 17 human subjects. The allocation of treatments to subjects was random. Their performance on a specific task was then measured. Scores, in coded units, are given below. Calculate s_R^2 for these data.

A	B	C	D	E
1	9	6	3	14
3	5	6	3	10
5	5	3	0	18
	5		6	

Answer: 6.33.

Variation between Treatments

The grand average $\bar{y}$ is the sum of all the observations divided by the total number of observations (in this case, $1536/24 = 64$). *If* there were no real differences between the treatment means, a second estimate of σ^2 could be obtained from the variation of the treatment averages about this grand average.

Specifically, the estimate is provided by the *between-treatment mean square*

$$s_T^2 = \frac{\sum_{t=1}^k n_t(\bar{y}_t - \bar{y})^2}{k-1} \tag{6.11}$$

which is the ratio of the *between-treatment sum of squares*

$$S_T = \sum_{t=1}^k n_t(\bar{y}_t - \bar{y})^2 \tag{6.12}$$

and the between-treatment degrees of freedom $v_T = k - 1$. It will be seen that S_T is a *weighted* sum of squares of deviations of treatment averages from the grand average with the numbers $n_1, n_2, \ldots, n_k$ in the treatments acting as weights.

The reader may better understand Equation 6.11 by considering the situation where all the treatment groups are of the same size ($n_t = n$). Then the variance of a treatment average obtained from a random sample is σ^2/n. Thus $\sum_{t=1}^k (\bar{y}_t - \bar{y})^2/(k-1)$ would provide an estimate of σ^2/n; hence $\sum_{t=1}^k n(\bar{y}_t - \bar{y})^2/(k-1)$ would provide an estimate of σ^2 if there were no real differences in the treatment means.

In the present example the calculation of the between-treatment mean squares is as follows:

	treatment				
	A	B	C	D	
$\bar{y}_t$	61	66	68	61	
$\bar{y}_t - \bar{y}$	-3	2	4	-3	$\bar{y} = 64$
n_t	4	6	6	8	

so that

$$S_T = 4(-3)^2 + 6(2)^2 + 6(4)^2 + 8(-3)^2 = 228 \tag{6.13}$$

$$v_T = 3 \tag{6.14}$$

$$s_T^2 = \frac{S_T}{v_T} = \frac{228}{3} = 76.0 \tag{6.15}$$

Clearly, if the true means *do* vary from treatment to treatment, this between-treatment estimate s_T^2 of σ^2 will tend to be inflated, that is, it will reflect variations among the treatment means as well as the intrinsic error variance σ^2. By contrast, the within-treatment mean square s_R^2 will be *unaffected* by differences in treatment means.

Exercise 6.2. Calculate s_T^2 for the data in Exercise 6.1.   *Answer*: 64.5.

Comparison of between- and within-Treatment Estimates

On the null hypothesis that there are no differences between the treatment means, we now have two estimates of σ^2: the residual within-treatment estimate $s_R^2 = 5.6$, with $v_R = 20$ degrees of freedom, and the between-treatment estimate $s_T^2 = 76.0$, with $v_T = 3$ degrees of freedom. This between-treatment estimate of σ^2 is many times larger than the within-treatment estimate. Consequently we are led to suspect that the null hypothesis is untenable and that some of this between-treatment variation must be caused by real differences between the treatment means. Before discussing how such a suspicion may be tested more objectively (Section 6.6), we show how the necessary calculations can be conveniently arranged in an analysis of variance table.

6.3. THE ARITHMETIC AND GEOMETRY OF THE ANALYSIS OF VARIANCE TABLE

A measure of the *overall* variation in the data could have been obtained by ignoring the separation into treatments and calculating the sample variance for the whole aggregate of N results. This would be done by calculating

$$S_D = \sum_{t=1}^{k} \sum_{i=1}^{n_t} (y_{ti} - \bar{y})^2 \tag{6.16}$$

the total sum of squares of deviations about the grand average $\bar{y}$ (sometimes referred to as the total "corrected" sum of squares) and dividing by the appropriate number of degrees of freedom $v_D = N - 1$. Thus

$$s_D^2 = \frac{\sum_{t=1}^{k} \sum_{i=1}^{n_t} (y_{ti} - \bar{y})^2}{N - 1} = \frac{S_D}{v_D} \tag{6.17}$$

For this particular set of data

$$S_D = (62 - 64)^2 + (60 - 64)^2 + \cdots + (59 - 64)^2 = 340 \qquad (6.18)$$

$$\nu_D = N - 1 = 23 \qquad (6.19)$$

$$s_D^2 = \frac{340}{23} = 14.8 \qquad (6.20)$$

The various calculated quantities may now be arranged as shown in Table 6.2.

TABLE 6.2.   An analysis of variance table

source of variation	sum of squares	degrees of freedom	mean square
between treatments	$S_T = 228$	$\nu_T = 3$	$s_T^2 = 76.0$
within treatments	$S_R = 112$	$\nu_R = 20$	$s_R^2 = 5.6$
total about the grand average	$S_D = 340$	$\nu_D = 23$	$s_D^2 = 14.8$

Notice that for the sum of squares $228 + 112 = 340$ and for the degrees of freedom $3 + 20 = 23$. Thus this table has a remarkable feature. In both the sum of squares and the degrees of freedom columns the values for *between* and *within* treatments add up to give the corresponding value for the *total* about the grand average. The additive property of the sum of squares arises because of the algebraic identity (true for any set of data whatever)

$$\sum_{t=1}^{k} \sum_{i=1}^{n_t} (y_{ti} - \bar{y})^2 = \sum_{t=1}^{k} n_t(\bar{y}_t - \bar{y})^2 + \sum_{t=1}^{k} \sum_{i=1}^{n_t} (y_{ti} - \bar{y}_t)^2 \qquad (6.21)$$

or

$$S_D \qquad = \qquad S_T \qquad + \qquad S_R \qquad (6.22)$$

In words:

total sum of squares of deviations from the grand average	=	between-treatment sum of squares	+	within-treatment (residual) sum of squares

Exercise 6.3. Construct an analysis of variance table for the data in Exercise 6.1.
Answer: $S_T = 258$, $v_T = 4$, $s_T^2 = 64.5$, $S_R = 76$, $v_R = 12$, $s_R^2 = 6.33$, $S_D = 334$, $v_D = 16$, $s_D^2 = 20.9$.

The Contribution from the Grand Average

The quantity S_D, which appears in the analysis of variance table as the total sum of squares of deviations from the grand average $\bar{y}$, is often calculated using the formula

$$S_D = \sum_{t=1}^{k} \sum_{i=1}^{n_t} y_{ti}^2 - N\bar{y}^2 \tag{6.23}$$

In this expression the term $N\bar{y}^2$ is the sum of squares due to the grand average, which is sometimes called the *correction for the average* (or the *correction factor*). It will be denoted by S_A, that is, $S_A = N\bar{y}^2$. If we write G for the grand total of all the observations,

$$S_A = \frac{G^2}{N} = N\bar{y}^2 \tag{6.24}$$

The term $\sum_{t=1}^{k} \sum_{i=1}^{n_t} y_{ti}^2$, which is called the *total* (or *crude*) *sum of squares*, will be denoted by S. Equation 6.23 can thus be written as

$$S_D = S - S_A \tag{6.25}$$

Therefore

$$S = S_A + S_D \tag{6.26}$$

so that, finally, using Equation 6.22, we can split up the sum of squares of the original N observations into three additive parts:

$$\sum_{t=1}^{k} \sum_{i=1}^{n_t} y_{ti}^2 \quad = \quad N\bar{y}^2 \quad + \sum_{t=1}^{k} n_t(\bar{y}_t - \bar{y})^2 + \sum_{t=1}^{k} \sum_{i=1}^{n_t} (y_{ti} - \bar{y}_t)^2 \tag{6.27}$$

$$S \quad = \quad S_A \quad + \quad S_T \quad + \quad S_R \tag{6.28}$$

In words:

total sum of squares (about the origin zero)	=	sum of squares (or correction) due to the average	+	between-treatment sum of squares	+	within-treatment (residual) sum of squares

The associated degrees of freedom are

$$N \quad = \quad 1 \quad + \quad k-1 \quad + \quad N-k \tag{6.29}$$

For our particular problem we have for the corresponding sums of squares

$$98{,}644 \quad = \quad 98{,}304 \quad + \quad 228 \quad + \quad 112$$

$$(6.30)$$

and for the degrees of freedom

$$24 \quad = \quad 1 \quad + \quad 3 \quad + \quad 20$$

$$(6.31)$$

With the contribution for the grand average specifically included, then, the analysis of variance takes the form shown in Table 6.3. The analysis of variance expressed algebraically takes the form shown in Table 6.4.

TABLE 6.3. **Full analysis of variance table for the data given in Table 6.1**

source of variation	sum of squares	degrees of freedom	mean square
average	98,304	1	98,304
between treatments	228	3	76.0
within treatments	112	20	5.6
total	98,644	24	

TABLE 6.4. **Analysis of variance table showing explicitly the contribution associated with the grand average**

source of variation	sum of squares	degrees of freedom	mean square
average	$S_A = N\bar{y}^2 = \left(\sum\limits_{t}^{k}\sum\limits_{i}^{n_t} y_{ti}\right)^2 / N$	$v_A = 1$	$s_A^2 = S_A/v_A$
between treatments	$S_T = \sum\limits_{t}^{k} n_t(\bar{y}_t - \bar{y})^2$	$v_T = k - 1$	$s_T^2 = S_T/v_T$
within treatments	$S_R = \sum\limits_{t}^{k}\sum\limits_{i}^{n_t} (y_{ti} - \bar{y}_t)^2$	$v_R = N - k$	$s_R^2 = S_R/v_R$
total	$S = \sum\limits_{t}^{k}\sum\limits_{i}^{n_t} y_{ti}^2$	N	

Each summation in this table starts at either $i = 1$ or $t = 1$.

Exercise 6.4.   Calculate S_A and S for the data in Exercise 6.1.   *Answer*: 612, 946.

Often the contribution associated with the grand average $\bar{y}$ is of no practical interest,* and consequently an analysis of variance table of the form of Table 6.2 is reported in preference to one like Table 6.3. Examples occur, however, where this is not so. For instance, observations might be measures of departure from some standard that represented desired performance. To answer the question of whether the performance during the experimental period had differed from this norm one would need to study the sum of squares associated with the grand average itself.

Suppose that $\eta = 63$ represented an established norm for blood coagulation time. It might then be instructive to consider all our results with this established norm as the origin, the "observations" now being deviations of the data values from 63. Their grand average would be 1, the "correction" due to the average would be 24, the total sum of squares about the origin 63 would be 364, and the decomposition of the sum of squares would be as follows:

$$\begin{array}{cccccccc} \text{total} & & \text{average} & & \text{between treatments} & & \text{within treatments} \\ 364 & = & 24 & + & 228 & + & 112 \end{array}$$

The analysis of variance table would therefore be as shown in Table 6.5.

TABLE 6.5.   **Analysis of variance table for observations considered as deviations from 63**

source of variation	sum of squares	degrees of freedom	mean square
average deviation from norm	24	1	24.0
between treatments	228	3	76.0
within treatments	112	20	5.6
total	364	24	

In this example differences are indicated between the treatment means. But suppose that this was not so and the between-treatment mean square was similar in magnitude to the within-treatment mean square. Then a further question would arise: Given that no treatment differences are detectable, what can we say about the possibility that the overall treatment mean (supposed common) is different from the established norm,

---

* This is the reason that it is customarily dismissed somewhat disparagingly as a "correction factor" and that $\sum y^2$ suffers the indignity of being called a "crude" sum of squares, whereas $\sum (y - \bar{y})^2$ enjoys the title of "corrected" sum of squares. It is common to publish the analysis of variance table with only the corrected sum of squares, with its $(n - 1)$ degrees of freedom.

$\eta = 63$? In addressing this question the mean square associated with the average would be compared to that within treatments.

Exercise 6.5. Using $s^2 = 5.6$ as a measure of variance, calculate the t value appropriate for testing the hypothesis that the grand mean for all the data is $\eta = 63$. How is this value of t related to the entries in Table 6.5?   *Answer*: $t_0 = 2.07$, $t_0^2 = s_A^2/s_R^2 = 24.0/5.6$.

6.4. DECOMPOSITION OF THE OBSERVATIONS IMPLIED BY THE ANALYSIS

Consideration of the formulas will show that what we have done is simply to break up the data in the manner illustrated in Table 6.6. The table shows that each individual result is regarded as being made up of the grand average $\bar{y}$, a *between*-treatment deviation $\bar{y}_t - \bar{y}$, and, finally, a *within*-treatment deviation (residual) $y_{ti} - \bar{y}_t$, which represents the part of the observation not already accounted for. Each of the four entries can be regarded as 24 elements of a vector (see Appendix 6B). The vectors are denoted by **Y**, **A**, **T**, and **R**, respectively, and, following the usual rules of vector addition, we can write our decomposition as

$$\mathbf{Y} = \mathbf{A} + \mathbf{T} + \mathbf{R} \qquad (6.32)$$

The sums of squares of the analysis of variance table are seen to be nothing more than the squares of the individual vector elements added up. They represent, therefore, the squared lengths of the vectors **Y**, **A**, **T**, and **R**.

The degrees of freedom reflect the number of ways in which the entries in each part of the table are free to vary. The value of v_A is unity because, whatever the data values may be, there will be only one number, namely, the value of the grand average (64 for this example), that is free to change in this part of the table. The value v_T is 3 because, whatever the data, given that there are four treatments, only three of the four deviations (-3, 2, 4, -3 for this particular set of data) are free to vary because they are the deviations of the treatment averages from the overall average. In other words, since $\sum_{t=1}^{k} n_t \bar{y}_t = \sum_{t=1}^{k} n_t \bar{y}$,

$$\sum_{t=1}^{k} n_t(\bar{y}_t - \bar{y}) = 0 \qquad (6.33)$$

If we are told, for example, that

$$\bar{y}_2 - \bar{y} = 2, \qquad \bar{y}_3 - \bar{y} = 4, \qquad \bar{y}_4 - \bar{y} = -3 \qquad (6.34)$$

then substituting in Equation 6.33, we have

$$4(\bar{y}_1 - \bar{y}) + 6(2) + 6(4) + 8(-3) = 0 \qquad (6.35)$$

TABLE 6.6. Arithmetic breakup of data given in Table 6.1

	observations y_{ii}		grand average $\bar{y}$		treatment deviations $\bar{y}_t - \bar{y}$		residuals (within-treatment deviations) $y_{ii} - \bar{y}_t$

$$
\begin{bmatrix}
62 & 63 & 68 & 56 \\
60 & 67 & 66 & 62 \\
63 & 71 & 71 & 60 \\
59 & 64 & 67 & 61 \\
   & 65 & 68 & 63 \\
   & 66 & 68 & 64 \\
   &   &   & 63 \\
   &   &   & 59
\end{bmatrix}
=
\begin{bmatrix}
64 & 64 & 64 & 64 \\
64 & 64 & 64 & 64 \\
64 & 64 & 64 & 64 \\
64 & 64 & 64 & 64 \\
   & 64 & 64 & 64 \\
   & 64 & 64 & 64 \\
   &   &   & 64 \\
   &   &   & 64
\end{bmatrix}
+
\begin{bmatrix}
-3 & 2 & 4 & -3 \\
-3 & 2 & 4 & -3 \\
-3 & 2 & 4 & -3 \\
-3 & 2 & 4 & -3 \\
   & 2 & 4 & -3 \\
   & 2 & 4 & -3 \\
   &   &   & -3 \\
   &   &   & -3
\end{bmatrix}
+
\begin{bmatrix}
1 & -3 & 0 & -5 \\
-1 & 1 & -2 & 1 \\
2 & 5 & 3 & -1 \\
-2 & -2 & -1 & 0 \\
   & -2 & 0 & 2 \\
   & -1 & 0 & 3 \\
   &   &   & 2 \\
   &   &   & -2
\end{bmatrix}
$$

vector*	**Y**		**A**		**T**		**R**
sum of squares**	98,644	=	98,304	+	228	+	112
degrees of freedom†	24	=	1	+	3	+	20

* In standard notation each vector would be represented by a column of twenty-four elements. A rearrangement of these elements is used above to emphasize the structure of the design.

** Squared length of vector.

† Number of dimensions in which the vector is constrained to lie.

176

so that $\bar{y}_1 - \bar{y}$ will have to be equal to -3 since the deviations must add to zero.

Finally, for the within-treatment (or residual) component of the table the 24 numbers are subject to four constraints since each column of deviations must add to zero. In other words, for each value of t, $\sum_{i=1}^{n_t} y_{ti} = n_t \bar{y}_t$, whence $\sum_{i=1}^{n_t} (y_{ti} - \bar{y}_t) = 0$. The resulting number of degrees of freedom for v_R is thus $24 - 4 = 20$.

Exercise 6.6.   For the data in Exercise 6.1 construct a table corresponding to Table 6.6.

Basic Quantities in Practical Analysis

Table 6.6 shows the essential nature of the analysis of variance. The same quantities, from which all the calculations can be made, are set out more compactly in Table 6.7. In Appendix 6A more expeditious methods for calculating the analysis of variance are described.

TABLE 6.7.   Decomposition of observations for completely randomized design*

	treatment			
	A	B	C	D
within-treatment deviations (residuals)	$1^{(20)}$ $-1^{(2)}$ $2^{(11)}$ $-2^{(10)}$	$-3^{(12)}$ $1^{(9)}$ $5^{(15)}$ $-2^{(14)}$ $-1^{(4)}$ $0^{(8)}$	$0^{(15)}$ $-2^{(7)}$ $3^{(1)}$ $-1^{(17)}$ $0^{(13)}$ $0^{(21)}$	$-5^{(23)}$ $1^{(3)}$ $-1^{(6)}$ $0^{(18)}$ $2^{(22)}$ $3^{(19)}$ $2^{(5)}$ $-2^{(24)}$
deviations of individual treatment averages from grand average	-3	2	4	-3
grand average	$\bar{y} = 64$			

* The superscripts in parentheses recorded with the residuals indicate the time order in which the corresponding observations were made.

In general, performance of an adequate data analysis requires the basic quantities contained in the following:

1. A table like 6.7 showing the decomposition of the observations implied by the model.
2. A table like 6.8 showing estimates $\hat{y}_{ti}$ of expected values of the entries in the table, assuming that the model is correct. In this particular instance the estimated values $\hat{y}_{ti}$ in the tth column are all equal to $\bar{y}_t$, the average appropriate for the tth treatment. They can be regarded as the 24 elements of a vector $\hat{\mathbf{Y}}$ of estimated values.

TABLE 6.8. Table of estimated values $\hat{y}_{ti}$, the 24 elements of the vector $\hat{\mathbf{Y}}$

treatment			
A	B	C	D
61	66	68	61
61	66	68	61
61	66	68	61
61	66	68	61
	66	68	61
	66	68	61
			61
			61

Often nowadays a computer is used for the calculations. It is not enough to ensure that the computer produces analysis of variance tables like Table 6.3; it should also produce tables corresponding to 6.7 and 6.8. Moreover, one must ensure that these tables are carefully studied.

Geometric Interpretation of the Analysis of Variance

Since there are 24 observations, a full geometric representation would require 24 dimensions. Fortunately, however, we can illustrate the geometry for this example in three dimensions, which are tacitly understood to represent the multidimensional space of the actual problem. (For a brief description of the relevant mathematical points, see Appendix 6B.)

In Figure 6.2a the vector $\mathbf{Y}$, which has the 24 observations as its elements [62, 60, ..., 59], is resolved into two components: $\mathbf{A}$, associated with the

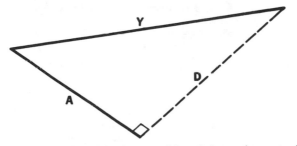

(*a*) Geometric representation of the decomposition of observation vector **Y** in terms of vector **A** (corresponding to the grand average) and vector **D** (deviations from the grand average).

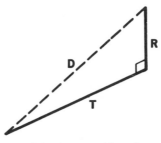

(*b*) Geometric representation of the decomposition of vector **D** in terms of vector **T** (treatment deviations) and vector **R** (residuals).

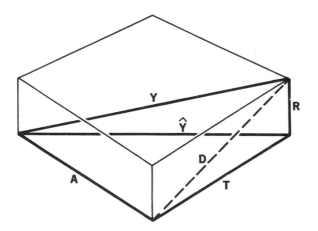

(*c*) Geometric representation of the analysis of variance in terms of an orthogonal decomposition of the vector **Y** in terms of vectors **A**, **T**, and **R**.

FIGURE 6.2. Geometry of the analysis of variance.

grand average, which also has 24 elements [64, 64, ..., 64], and **D**, whose 24 elements are the deviations [(62–64), (60–64), ..., (59–64)]. The vector **D** is orthogonal to **A** since $\sum_{j=1}^{N} \bar{y}(y_j - \bar{y}) = 0$. In Figure 6.2b vector **D** is, in turn, resolved into two components: **T**, associated with the 24 treatment deviations $(-3, -3, -3, -3, 2, 2, ..., -3, -3)$, and **R**, which has for its 24 elements the residuals $(1, -1, ..., 2, -2)$. Finally, in Figure 6.2c we see the observation vector **Y** resolved into three orthogonal components: **A** for the grand average, **T** for the treatment deviations, and **R** for the residuals. That these vectors **A**, **T**, and **R** are orthogonal to one another is readily confirmed by noting that their inner products are equal to zero. This may be checked for this particular example by actually summing the products of the elements of the vectors **A**, **T**, and **R** from Table 6.6. That it is generally true follows from the following algebraic identities:

$$\sum_{t=1}^{k} \sum_{i=1}^{n_t} \bar{y}(\bar{y}_t - \bar{y}) = 0 \qquad \text{indicates that } \mathbf{A} \text{ is perpendicular to } \mathbf{T}$$

$$\sum_{t=1}^{k} \sum_{i=1}^{n_t} \bar{y}(y_{ti} - \bar{y}_t) = 0 \qquad \text{indicates that } \mathbf{A} \text{ is perpendicular to } \mathbf{R}$$

$$\sum_{t=1}^{k} \sum_{i=1}^{n_t} (\bar{y}_t - \bar{y})(y_{ti} - \bar{y}_t) = 0 \quad \text{indicates that } \mathbf{R} \text{ is perpendicular to } \mathbf{T}$$

In the 24-dimensional sample space, for different possible samples with four treatments, the component **A** is constrained to lie in a 1-dimensional space (a line), the component **T** in a 3-dimensional space, and the component **R** in a 20-dimensional space. The degrees of freedom in the analysis of variance are additive because the dimensions of the components of **Y** must sum to 24.

Consider the squared lengths of these vectors. The additive relationship $S = S_A + S_T + S_R$ of the analysis of variance is seen to arise from the 24-dimensional analog of Pythagoras' theorem, which relates the square of the length of the "hypotenuse" **Y** to the sum of the squares of the lengths of the other three sides of the figure, **A**, **T**, and **R**. (See Appendix 6B for further explanation of these ideas in terms of a three-dimensional example.)

Geometrically, the estimated values are the elements of the vector $\hat{\mathbf{Y}}$ in Figure 6.2c. Note that $\hat{\mathbf{Y}} = \mathbf{A} + \mathbf{T}$ indicates how the estimated values are obtained, and $\mathbf{Y} = \hat{\mathbf{Y}} + \mathbf{R}$ indicates the relation between the observations, the estimated values, and the residuals.

Frequently the analysis of variance is written down after the "elimination" of the grand average. Table 6.2 shows an example of such an analysis. The corresponding breakup of the observations is shown in Table 6.9. The component **D** illustrated in Figure 6.2b represents the deviations from the grand average $y_{ti} - \bar{y}$ after the component **A** representing the grand average has

TABLE 6.9. Arithmetic breakup of deviations from the grand average

deviations from grand average

$$y_{ti} - \bar{y}$$

$$\begin{bmatrix} -2 & -1 & 4 & -8 \\ -4 & 3 & 2 & -2 \\ -1 & 7 & 7 & -4 \\ -5 & 0 & 3 & -3 \\ & 1 & 4 & -1 \\ & 2 & 4 & 0 \\ & & & -1 \\ & & & -5 \end{bmatrix}$$

treatment deviations

$$\bar{y}_t - \bar{y}$$

$$= \begin{bmatrix} -3 & 2 & 4 & -3 \\ -3 & 2 & 4 & -3 \\ -3 & 2 & 4 & -3 \\ -3 & 2 & 4 & -3 \\ & 2 & 4 & -3 \\ & 2 & 4 & -3 \\ & & & -3 \\ & & & -3 \end{bmatrix}$$

residuals
(within-treatment deviations)

$$y_{ti} - \bar{y}_t$$

$$+ \begin{bmatrix} 1 & -3 & 0 & -5 \\ -1 & 1 & -2 & 1 \\ 2 & 5 & 3 & -1 \\ -2 & -2 & -1 & 0 \\ & -1 & 0 & 2 \\ & 0 & 0 & 3 \\ & & & 2 \\ & & & -2 \end{bmatrix}$$

	vector	sums of squares*	degrees of freedom†
$[\mathbf{D} = \mathbf{Y} - \mathbf{A}]$		340	23
$= \mathbf{T}$		228	3
$+ \mathbf{R}$		112	20

* Squared length of vector.
† Number of dimensions in which the vector is constrained to lie.

181

been subtracted from the observation vector $\mathbf{Y}$. The analysis of variance table after the elimination of the grand average is concerned with the resolution of the component $\mathbf{D}$ into two orthogonal parts, $\mathbf{T}$ and $\mathbf{R}$, as shown in Figure 6.2b, where once again $\mathbf{T}$ represents the contribution of the individual treatment deviations and $\mathbf{R}$ is the residual component.

6.5. DIAGNOSTIC CHECKING OF THE BASIC MODEL

The resolution of the data discussed above which leads to the analysis of variance table is a purely arithmetic one with a readily appreciated geometric meaning. The relationships that exist between the various vectors and their sums of squares are identities and would be true for *any* set of data. However, as we mentioned in Section 6.1, this particular mode of resolution is relevant in relation to a specific model that links the observations and the various decompositions with the underlying parameters of the sampled populations. In particular, it is relevant if the data are random samples from four *normal* populations having the same variance but possibly different means. In this case the model could be written* as

$$y_{ti} = \eta_t + \epsilon_{ti} \tag{6.36}$$

where η_t is the population mean for the tth treatment, and the errors ϵ_{ti} are IIDN$(0, \sigma^2)$, that is, independently and identically distributed according to a normal distribution with mean zero and unknown but fixed variance σ^2. If this were so, then s_R^2 and $\bar{y}_1, \bar{y}_2, \ldots, \bar{y}_k$ would together be sufficient statistics for σ^2 and $\eta_1, \eta_2, \ldots, \eta_k$, the only parameters in the model.

Thus, if the IIDN$(0, \sigma^2)$ assumption were appropriate, all the relevant information about $\eta_1, \eta_2, \ldots, \eta_k$ and σ^2 would be supplied by the k treatment averages $\bar{y}_1, \bar{y}_2, \ldots, \bar{y}_k$ and s_R^2. *If we could rely on the exactness of this assumption*, we could say that after these statistics had been calculated no further relevant information remained in the original data, and we could, therefore, ignore the residuals and the original observations and concentrate entirely on the interpretation of these statistics.

In practice it would be very unwise to do this without further checks because data may contain valuable information not allowed for in the assumed mathematical model and therefore not revealed by the associated analysis of variance table. For example, suppose that during the taking of the blood samples for Table 6.1 there had been a steady fall in temperature in the laboratory, resulting in a trend in blood coagulation times. This is not specifically

---

* An alternative form of the model is $y_{ti} = \eta + \tau_t + \epsilon_{ti}$, where $\eta_t = \eta + \tau_t$, η is the overall mean, and τ_t is the incremental effect of treatment t. For further discussion see Section 6.6.

allowed for in the model (Equation 6.36), but randomization of the order of taking and testing the blood samples would ensure that errors due to this systematic trend were applied randomly to the treatment groups. This would approximately validate the significance tests to be discussed later, but the additional variation produced by the trend would decrease the sensitivity of the tests. However, plotting the residuals against the time order given in parentheses in Table 6.7 could reveal the existence of such a trend. This could be important because (1) it would point to a source of variation in the analysis that had not previously been allowed for and that could be controlled in the future, and (2) it could lead to a more precise analysis of the differences in coagulation times in which the trend with time was explicitly taken into account, not randomly mixed in with the error.

Analysis of Residuals

Discrepancies of many different kinds between a tentative model and the data can be detected by studying residuals $y_{ti} - \hat{y}_{ti}$, the elements of the vector **R**. These residuals are the quantities remaining after the systematic contributions associated with the assumed model (in this case the contributions for the treatment averages) are removed, and are shown, for example, in the body of Table 6.7. When assumptions concerning the adequacy of the model are true, we expect to find that, apart from restrictions introduced by the analysis itself, the residuals vary randomly. (An example of a restriction imposed by the analysis is that the sum of the residuals within every treatment will equal zero.) If, however, we find that the residuals contain unexplained systematic tendencies, we shall be suspicious of the model. Therefore, as an *automatic preliminary* to further statistical analysis, a table of residuals should always be constructed and studied.

The particular discrepancies described below should be looked for as a matter of routine, but the experimenter should also be on the alert for other abnormalities.

Distribution of Residuals

A general inspection should first be carried out by plotting a dot diagram of the residuals. This is done for the blood coagulation data in Figure 6.3a. If the IIDN(0, σ^2) assumption is true, this plot will have roughly the appearance of a sample from a normal distribution centered at zero. Considerable fluctuation will arise when the number of observations is small, so that any appearance of moderate nonnormality in that case is not necessarily indicative of an underlying cause. When gross abnormalities occur, however, possible explanations must be sought.

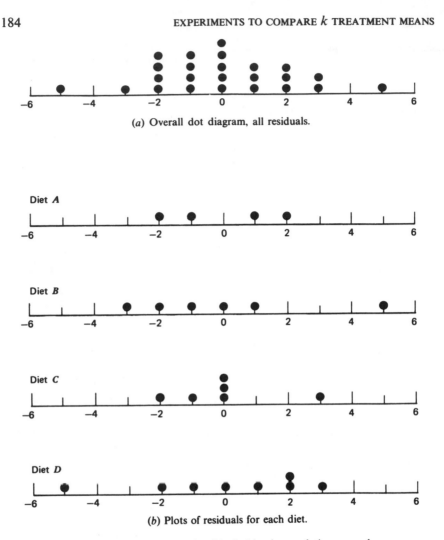

(a) Overall dot diagram, all residuals.

(b) Plots of residuals for each diet.

FIGURE 6.3. Plots of residuals, blood coagulation example.

The most common form of discrepancy revealed by such a plot occurs when one or more of the residuals is very much larger or smaller than any of the others. Since the most likely explanation for such a value is a copying or arithmetic mistake, the whole series of records from the original taking of the observations onward must be painstakingly reviewed. If no mistake of this kind can be found, the circumstances surrounding the carrying out of the

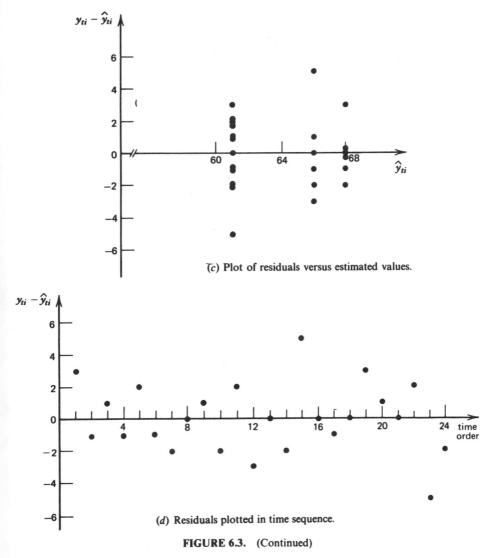

(c) Plot of residuals versus estimated values.

(d) Residuals plotted in time sequence.

FIGURE 6.3. (Continued)

apparently discrepant run should be carefully investigated and considered. What is important is not just seeking an excuse for the possible rejection of the outlying observation, but, rather, investigating the possibility that the discrepant result may have unexpected implications worthy of further investigation. The plot in Figure 6.3a gives no indication of such abnormalities for the blood coagulation times.

Abnormalities Associated with Particular Treatments

Abnormal behavior of residuals may be associated with a particular treatment (diet). To detect possible tendencies of this sort, individual dot diagrams for the four diets are shown in Figure 6.3b. The suggestion, for example, that excessively large variation in blood coagulation time was associated with a particular diet could be an important lead in further investigation. The plots in Figure 6.3b, however, do not suggest anomalous behavior associated with any particular diet.

Relation between the Size of Residuals and the Expected Value of Response

If the mathematical model is appropriate, the residuals should be unrelated to the levels of any known variable. In particular, they should be unrelated to the levels of the response itself. This latter point may be investigated by plotting the residuals $y_{ti} - \hat{y}_{ti}$ against the estimated values $\hat{y}_{ti}$, as is done for the coagulation data in Figure 6.3c.

Sometimes the variance increases as the value of the response increases. If the coagulation times had constant *percentage* error, for example, the absolute values of the residuals would tend to increase as the size of the observations increased, and the plot would have a *funnel-like* appearance. No such tendency, however, is apparent.

Plot of the Residuals in Time Sequence

Sometimes a chemical standard may drift or the skill of the experimenter may improve as experiments proceed. Tendencies of this kind may be uncovered by plotting residuals against their time order as in Figure 6.3d. There seems to be no reason to suspect any such effect for the coagulation data.

Plot of the Residuals versus Variables of Interest

Residuals should be plotted against any variable of possible relevance. Thus it might make sense to plot the coagulation time residuals against the temperature in the laboratory. A positive result (*a*) could contribute to basic knowledge, (*b*) could suggest that closer temperature control should be maintained in later phases of the investigation, or (*c*) could result in the inclusion of temperature as a factor to be studied in further work.

Exercise 6.7. For the data in Exercise 6.1, make plots of residuals.

6.6. USE OF THE ANALYSIS OF VARIANCE TABLE

Once we are satisfied by careful inspection of the residuals that the model $y_{ti} = \eta_t + \epsilon_{ti}$ does not contain obvious inadequacies, we can study the analysis of variance table with more assurance. In particular, the table provides the basis for a formal test of the hypothesis that the treatment means are all equal, that is, $\eta_A = \eta_B = \eta_C = \eta_D$.

A Test of the Hypothesis That the Treatment Means Are Equal

An analysis of variance table for the data in Table 6.1 is set out in Table 6.10.

TABLE 6.10. Analysis of variance table showing calculation of the s_T^2/s_R^2 ratio

model: $y_{ti} = \eta_t + \varepsilon_{ti}$

source of variation	sum of squares	degrees of freedom	mean square	ratio
between treatments	$S_T = 228$	3	$s_T^2 = 76.0$	$s_T^2/s_R^2 = 76.0/5.6 = 13.6$
within treatments	$S_R = 112$	20	$s_R^2 = 5.6$	
total about the grand average	$S_D = 340$	23		

In the present example, if the null hypothesis were true and the IIDN$(0, \sigma^2)$ assumption held for the errors, the ratio s_T^2/s_R^2 would follow an F distribution with 3 and 20 degrees of freedom (see Figure 6.4). The area under this curve

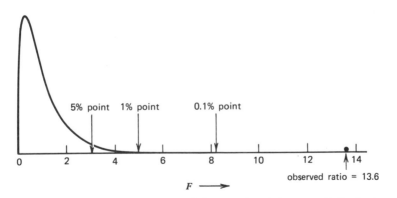

FIGURE 6.4. Observed value of the ratio $s_T^2/s_R^2 = 13.6$ in relation to an F distribution with 3 and 20 degrees of freedom, blood coagulation example.

to the right of the observed ratio 13.6, which is the significance level for this result, is 0.000046. Exact probability values such as this one may be obtained from certain pocket calculators or larger computers. Percentage points adequate for most purposes can be obtained from Table D at the back of this book. By consulting this table, it will be found that the upper 5, 1, and 0.1 % significance points of the F distribution with 3 and 20 degrees of freedom are 3.10, 4.94, and 8.10, respectively. The data, then, discredit the null hypothesis, and we prefer to believe that real differences exist among the treatment means.

A randomization distribution can be generated by calculating F from every possible allocation of the data to the treatment groups. It has been shown that the randomization F distribution can usually be approximated quite well by the tabulated normal theory F distribution. Thus, as before, the normal theory test can be regarded as an approximation for the randomization test, which is its ultimate justification.

The analysis we have been considering is sometimes called a *one-way* or *single-classification* analysis of variance because the data are classified in only one way, namely, by treatment. In later chapters the analysis is extended to designs in which there is more than one basis of classification.

Exercise 6.8.   For the data in Exercise 6.1, test the hypothesis that all treatment means are equal against the alternative that they are not equal. State your assumptions.
    Answer: $s_T^2/s_R^2 = 10.2$; $F_{4, 12}(0.001) = 9.63$, so result is significant at the 0.001 level.

Treatment Effect

It is often useful to think in terms of treatment "effects" or increments For the tth treatment we can define such an effect in terms of the deviation of the true treatment mean η_t from $\eta = \sum n_t \eta_t/N$, the weighted mean of all treatment means. Then the treatment effect for the tth group may be denoted by

$$\tau_t = \eta_t - \eta \tag{6.37}$$

and the model $y_{ti} = \eta_t + \epsilon_{ti}$ becomes

$$y_{ti} = \eta + \quad \tau_t \quad + \quad \epsilon_{ti} \tag{6.38}$$

The corresponding analysis of the data

$$y_{ti} = \bar{y} + (\bar{y}_t - \bar{y}) + (y_{ti} - \bar{y}_t) \tag{6.39}$$

has been used throughout this chapter.

The model and the corresponding breakdown of the data are illustrated for the blood coagulation example in Figures 6.5a and 6.5b. We now describe certain results that can be obtained, assuming this model to apply.

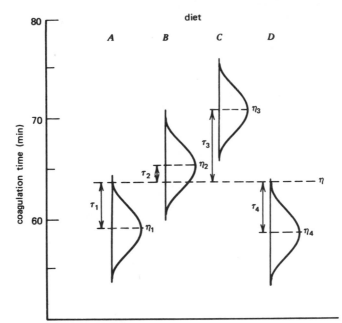

(a) A graphical representation of the conceptual model of Equation 6.38, blood coagulation example.

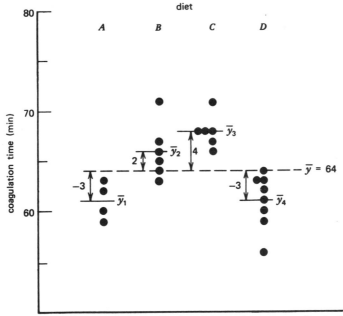

(b) The data regarded as a particular realization of the model corresponding to the analysis of Equation 6.39.

FIGURE 6.5. The model in relation to the data.

Expected Values of Mean Squares

When real differences occur between the treatment means, the expected value of the between-treatment mean square is inflated. The size of this inflation depends on a weighted sum of squares of the τ_t's. In fact the expected value of s_T^2, the treatment mean square, is

$$E(s_T^2) = \frac{E \sum n_t(\bar{y}_t - \bar{y})^2}{k-1} = \sigma^2 + \frac{\sum n_t(\eta_t - \eta)^2}{k-1} = \sigma^2 + \frac{\sum n_t \tau_t^2}{k-1} \qquad (6.40)$$

Algebraic expressions for the various sums of squares occurring in the analysis of variance, together with calculation formulas, are shown in Table 6.11.

TABLE 6.11. Analysis of variance table showing expected values for mean squares

model: $y_{ti} = \eta + \tau_t + \epsilon_{ti}$, where $\tau_t = \eta_t - \eta$

source of variation	sum of squares	degrees of freedom	mean square	expected value of mean square
treatments	$S_T = \sum\limits_t^k n_t(\bar{y}_t - \bar{y})^2$	$\nu_T = k - 1$	s_T^2	$\sigma^2 + \left[\sum\limits_t^k n_t \tau_t^2/(k-1)\right]$
within treatments	$S_R = \sum\limits_t^k \sum\limits_i^{n_t} (y_{ti} - \bar{y}_t)^2$	$\nu_R = N - k$	s_R^2	σ^2
total about the grand average	$S_D = \sum\limits_t^k \sum\limits_i^{n_t} y_{ti}^2 - N\bar{y}^2$	$N - 1$		

6.7. USE OF A REFERENCE DISTRIBUTION TO COMPARE MEANS

We have seen that the hypothesis that the blood coagulation means are equal is discredited by the data. But how do they differ? Is one much higher than the other three? Are all four distinctly different from one another? We now consider a simple graphical device that is often useful for comparing k treatment means.

Suppose that σ were known exactly and that the number of observations n for each treatment had been the same. Then any treatment average $\bar{y}_t$ would be distributed about its mean η_t with standard deviation $\sigma/\sqrt{n}$. If k treatment averages $\bar{y}_1, \bar{y}_2, \ldots, \bar{y}_k$ had the *same* mean η then they would behave as k observations from the same common, nearly normal distribution with standard deviation $\sigma/\sqrt{n}$. Suppose, therefore, that we constructed such a normal distribution and imagined it capable of being slid along the horizontal axis. There should be some position in which it could rest which would make plausible the proposition that $\bar{y}_1, \bar{y}_2, \ldots, \bar{y}_k$ were random samples from it. For the example concerning coagulation times, σ is not known and the sample sizes are not all equal. As a rough but useful approximation for this example in which the n_i's are not dramatically different, we have therefore replaced the normal distribution with a t distribution* with scale factor $\sqrt{s^2/\bar{n}} = \sqrt{5.6/6} = 0.97$, where $\bar{n} = \Sigma n_i/k = 6$ is the average sample size. We refer to this distribution, shown in Figure 6.6, as the approximate *reference distribution* for the averages $\bar{y}_t$.

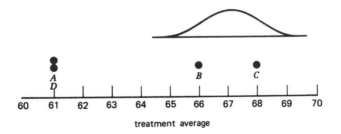

FIGURE 6.6. Sample averages in relation to a reference t distribution with scale factor $\sqrt{s_R^2/\bar{n}} = \sqrt{5.6/6} = 0.97$, blood coagulation example.

It is easy to sketch this reference t distribution using Table B2 given at the end of this book. This table shows the ordinates of the t distribution for various values of t and for various numbers of degrees of freedom v. For the present example $v = 20$ and the scale factor $s/\sqrt{\bar{n}}$ is 0.966. Entering Table B2 with $v = 20$, we obtain

value of t	0	0.5	1.0	1.5	2.0	2.5	3.0
t ordinate	0.394	0.346	0.236	0.129	0.058	0.023	0.008
$t \times 0.966$	0	0.48	0.97	1.45	1.93	2.42	2.90

* When s is substituted for σ, the quantities $\sqrt{n}(\bar{y}_t - \eta_t)/s$ are not independent since they all use the same estimate s. However, this does not seriously invalidate the suggested procedure unless the number of degrees of freedom associated with s is small (say less than 10).

To draw the reference distribution, an arbitrary origin η was first chosen in the neighborhood of the averages to be compared (in the case illustrated we took $\eta = 67.05$). The ordinates were then plotted at η, $\eta \pm 0.48$, $\eta \pm 0.97$, and so on, and a smooth curve drawn through the points. If we required the area under the distribution curve to be unity it would be necessary to multiply the t ordinates by $(0.966)^{-1}$. However, for our purpose only the relative values of the ordinates are of importance, and this step is unnecessary since we can always imagine the vertical scale to be suitably adjusted.

Consider the sample averages shown in relation to the approximate reference distribution in Figure 6.6. One should imagine that the reference distribution can be slid along the horizontal axis. This allows contemplation of different hypotheses. Note that there is no place to center the distribution so that *all four* treatment averages appear to be typical observations randomly selected from it. This is the graphical equivalent of what has already been formally demonstrated by the F test. But in addition, the reference distribution makes it clear that, although η_B and η_C are probably greater than η_A and η_D, the data do not permit one to distinguish between η_B and η_C, or between η_A and η_D.

Effect of Selection

Contemplation of a plot of the sample averages against a background of the reference distribution is a good way for the investigator to perceive and get in touch with what his experiment has told him. One result of showing the observed distribution of sample averages in the light of their approximate reference distribution is that visual allowance is made for selection. For example, in Figure 6.7a, even with the reference distribution slid into its most plausible position, it is clearly unlikely that the two averages plotted there arose from a single population. In Figure 6.7b 16 averages are being compared, and the figure has been drawn so that the largest and smallest averages occupy the same positions as the two averages in Figure 6.7a. Notice however that there is now no particular reason to regard these two as not coming from the same population. The 16 averages regarded as a complete sample are of the kind that one would expect to arise from the given reference distribution. For further illustration of the use of reference distributions, see Section 7.6. The reference distribution technique is a rough method for making what are called *multiple comparisons*. More formal statistical techniques for multiple comparisons are described in Appendix 6C.

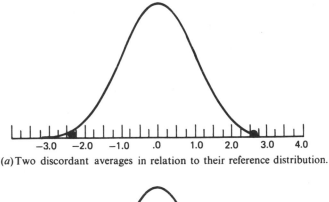

(a) Two discordant averages in relation to their reference distribution.

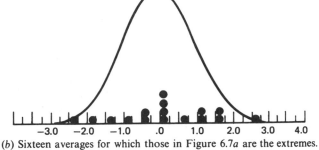

(b) Sixteen averages for which those in Figure 6.7a are the extremes.

FIGURE 6.7. Reference distributions and the problem of selection.

Exercise 6.9. For the data in Exercise 6.1, construct an appropriate reference distribution for the treatment averages.

Partial answer: Scale factor $= \sqrt{6.33/3.4} = 1.36$. Treatment E appears to have a mean larger than the means of the remaining treatments; the latter values do not seem to differ appreciably from one another.

6.8. SUMMARY

The essential steps in analyzing data from a randomized design to compare k treatments are as follows:

1. Plot the original data.
2. Construct analysis tables corresponding to Tables 6.7 and 6.8 (and 6.6, if desired).

3. Perform diagnostic checks on the residuals.
4. If the checks appear satisfactory, construct the analysis of variance table corresponding to Table 6.10, and make an appropriate overall test of significance.
5. Plot the individual treatment averages, and contemplate them against a background of the appropriate reference distribution.

The simple additive model (Equation 6.38)

$$y_{ti} = \eta + \tau_t + \epsilon_{ti}$$

which underlies the usual analysis of variance procedure, may have inadequacies that, once revealed, can point the way to a better model. It is essential, therefore, to carry out careful diagnostic checks. In particular, on a routine basis *residuals should be plotted* in all the ways described in Section 6.5 and any others that seem appropriate.

The *reference distribution* (Section 6.7) is a pictorial representation of the data that not only shows the plausibility of the overall null hypothesis but *also helps the experimenter form and evaluate other hypotheses*. It provides a readily understood summary of the main conclusions of an investigation.

Today it is possible to have most, if not all, of the calculating and plotting done by a computer. The reader should not be content with computer programs that omit important items mentioned above such as the calculation and plotting of residuals. He must also remember that even the tables and displays produced by a perfect program are useless unless accompanied by careful *study* and *thought*.

APPENDIX 6A. SHORTCUT METHOD FOR CONSTRUCTING THE ANALYSIS OF VARIANCE TABLE

The one-way classification model* is

$$y_{ti} = \eta + \tau_t + \epsilon_{ti}$$

where $t = 1, 2, \ldots, k$, and $i = 1, 2, \ldots, n_t$.

* The model discussed in this chapter is called a *one-way classification* model because the data are classified in only a single way, namely, according to a treatment. In the next chapter we consider a two-way classification model; the two classifications are treatments and blocks.

The data from a single classification experiment may be displayed as shown in the following table (sometimes called a *one-way layout*):

	treatment classification				
	1	2	3	$\cdots$	k
observations within a treatment	y_{11} y_{12} y_{13} $\vdots$ y_{1n_1}	y_{21} y_{22} y_{23} $\vdots$ y_{2n_2}	y_{31} y_{32} y_{33} $\vdots$ y_{3n_3}	$\cdots$ $\cdots$ $\cdots$ $\cdots$	y_{k1} y_{k2} y_{k3} $\vdots$ y_{kn_k}
treatment totals	T_1	T_2	T_3	$\cdots$	T_k $\quad G = \sum_{t=1}^{k} T_t = $ grand total
number of observations within treatment	n_1	n_2	n_3	$\cdots$	n_k $\quad N = \sum_{t=1}^{k} n_t = $ total number of observations
treatment averages	$\bar{y}_1$	$\bar{y}_2$	$\bar{y}_3$	$\cdots$	$\bar{y}_k$ $\quad \bar{y} = \dfrac{G}{N} = $ grand average
treatment deviations* $(\bar{y}_t - \bar{y})$	$\bar{y}_1 - \bar{y}$	$\bar{y}_2 - \bar{y}$	$\bar{y}_3 - \bar{y}$	$\cdots$	$\bar{y}_k - \bar{y}$

* This line is not required for the calculation of the analysis of variance table.

The analysis of variance table for these data is obtained by computing:

(1) $\sum y^2 = \sum_{t=1}^{k} \sum_{i=1}^{n_t} y_{ti}^2$ The (crude) sum of squares of all the observations.

(2) $S_A = \dfrac{G^2}{N} = \dfrac{(\sum y)^2}{N}$ The grand total squared divided by the total number of observations. S_A is often called the *correction factor*.

(3) $S_D = \sum y^2 - S_A$ The quantity S_D is the total sum of squares corrected for the average, or, more simply, the total corrected sum of squares. (Note that $\sum y^2$ will indicate the crude sum of squares where S_D is the corrected sum of squares.) The quantities $\sum y^2$ and S_A usually do not appear in an analysis of variance table.

(4) $S_T = \sum\limits_{t=1}^{k} \dfrac{T_t^2}{n_t} - S_A$

The total for each treatment is squared and then divided by the number of observations comprising that total. These quantities are then summed, and the correction factor is subtracted to give S_T, the between-treatment sum of squares. If all $n_t = n$, then $S_T = (1/n) \sum_i^k T_i^2 - S_A$.

(5) $S_R = \sum y^2 - S_A - S_T$

The within-treatment sum of squares is obtained by subtraction.

(6) $s_T^2 = \dfrac{S_T}{k-1}$

$s_R^2 = \dfrac{S_R}{N-k}$

The between-treatment and within-treatment mean squares are obtained by dividing the sums of squares by their corresponding degrees of freedom.

Thus the analysis of variance table is as follows:

source of variation	sum of squares	degrees of freedom	mean square
between treatments	$S_T = \sum T_t^2/n_t - S_A$	$k - 1$	s_T^2
within treatments	$S_R = \sum y^2 - S_A - S_T$	$N - k$	s_R^2
total corrected sum of squares	$S_D = \sum y^2 - S_A$	$N - 1$	

Using the data given in Table 6.1, we have

	treatment				
	A	B	C	D	
	62	63	68	56	$\sum y^2 = 98{,}644$
	60	67	66	62	
	63	71	71	60	$S_A = \dfrac{G^2}{N} = \dfrac{(1536)^2}{24} = 98{,}304$
	59	64	67	61	
		65	68	63	$S_D = \sum y^2 - S_A = 340$
		66	68	64	
				63	
				59	
total T_t	244	396	408	488	$G = 1536$
n_t	4	6	6	8	$N = 24$

$$S_T = \frac{(244)^2}{4} + \frac{(396)^2}{6} + \frac{(408)^2}{6} + \frac{(488)^2}{8} - 98,304 = 98,532 - 98,304 = 228$$

$$S_R = S_D - S_T = 340 - 228 = 112$$

APPENDIX 6B.   VECTORS AND GEOMETRY ASSOCIATED WITH THE ANALYSIS OF A SAMPLE

In this chapter we considered the analysis of a data vector in terms of certain relevant orthogonal components. Since there were 24 observations in the example, the decomposition took place in $N = 24$ dimensions. For those unfamiliar with vectors and coordinate geometry, this appendix is an introduction to the basic ideas. These are illustrated for only three dimensions but can be extended to any number of dimensions.

Three-Dimensional Vectors and Their Geometric Representation

1. A set of $n = 3$ numbers $[y_1, y_2, y_3]$ arranged in a row or column that might, for example, be three data values, can be regarded as the elements of a *vector*. A vector may be represented by a single letter in bold-faced type; thus $y = [y_1, y_2, y_3]$.
2. A vector has *magnitude* and *direction* and can be represented geometrically by a line in space pointing from the origin to the point with coordinates (y_1, y_2, y_3). Thus, if the coordinates are labeled as in Figure 6B.1 and we set $y_1 = 3$, $y_2 = 5$, and $y_3 = 4$, the line y represents the vector $y = [3, 5, 4]$. However, a vector holds no particular position in space and need not originate at the origin. Any other line having the same length and pointing in the same direction represents the same vector. The vector $-y = [-3, -5, -4]$.
3. The *length* of a vector is the square root of the sum of squares of its elements. For example, the length of the data vector $y = [3, 5, 4]$ in Figure 6B.1 is $\sqrt{9 + 25 + 16} = \sqrt{50} = 7.07$.
4. Two vectors are *summed* by summing corresponding elements. For example,

$$y + x = y + x$$
$$\begin{pmatrix} 3 \\ 5 \\ 4 \end{pmatrix} + \begin{pmatrix} 1 \\ -2 \\ -3 \end{pmatrix} = \begin{pmatrix} 4 \\ 3 \\ 1 \end{pmatrix}$$

The geometrical representation of this addition is shown in Figure 6B.1. The origin of the vector x is placed at the end of the vector y; the vector from the origin of y to the end of x, signed as shown and completing the triangle, now represents y + x.
5. The *inner product* of two vectors y and x is the sum of products $\sum yx$ of their corresponding elements. Thus the inner product of the vectors $y = [3, 5, 4]$, $x = [1, -2, -3]$ is

$$\{3 \times 1\} + \{5 \times (-2)\} + \{4 \times (-3)\} = -19$$

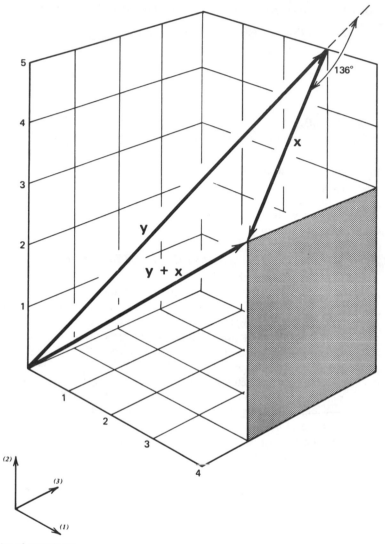

order of coordinates

FIGURE 6B.1. Geometric representation of the vectors $y = [3, 5, 4]$, $x = [1, -2, -3]$, and their sum $y + x = [4, 3, 1]$.

198

The normalized inner product $\sum yx/(\sum y^2 \sum x^2)^{1/2} = -19/(50 \times 14)^{1/2} = -0.72$ is the *cosine* of the angle between the two vectors. Thus in Figure 6B.1 the angle between $\mathbf{y}$ and $\mathbf{x}$ is the angle $136°$, whose cosine is -0.72. Notice in particular that when $\sum yx = 0$ the cosine of the angle between the vectors is zero, and consequently they are at right angles. The vectors are then said to be *orthogonal*.

Extensions to n Dimensions

Although it is not easy to imagine a space of more than three dimensions, all the properties listed above can be shown to have meaning for vectors in n-dimensional space.

As was first demonstrated by R. A. Fisher, vector representation can provide considerable insight into the nature of statistical procedures such as the t test and analysis of variance. We illustrate below, again limiting the demonstration to three dimensions.

A t Test and the Corresponding Analysis of Variance

Suppose that a random sample of three manufactured items is measured for conformance with a certain standard, and the data [3, 5, 4] are recorded as deviations from the target value. To test the hypothesis that their mean η has not deviated from zero we might calculate the t statistic:

$$t_0 = \frac{\bar{y} - \eta_0}{s/\sqrt{n}} = \frac{4 - 0}{1/\sqrt{3}} = 4\sqrt{3} = 6.9$$

where $s^2 = \sum (y - \bar{y})^2/(n - 1) = 1$. Referring to the t table with $v = n - 1 = 2$ degrees of freedom, we find that $Pr(t > 6.9) \simeq 0.01$, which discredits the null hypothesis that $\eta = 0$ against the alternative $\eta > 0$. If the mean η could equally well deviate in either direction, the appropriate two-sided significance level would be $Pr(|t| > 6.9) \simeq 0.02$.

Alternatively, we could test the hypothesis using an analysis of variance table as shown in Table 6B.1. The significance level $Pr(F > 48.0) \simeq 0.02$ is the same as that for the two-sided t test. This must be so since, when the F ratio has a single degree of freedom in the numerator, $t^2 = F$.

TABLE 6B.1. Analysis of variance for conformance data

	sum of squares		degrees of freedom	mean square	
average	$S_{\bar{y}} = n\bar{y}^2$	$= 48$	1	48	$F = 48.0$
residuals	$S_{y-\bar{y}} = \sum (y - \bar{y})^2$	$= 2$	$n - 1 = 2$	1	
total	$S_y = \sum y^2$	$= 50$	$n = 3$		

Specifically,

$$t^2 = \frac{n\bar{y}^2}{s^2} = \frac{n\bar{y}^2}{\sum(y - \bar{y})^2/(n-1)} = F$$

Thus

$$\Pr(F > 48.0) = \Pr(t^2 > 6.9^2) = \Pr(|t| > 6.9)$$

Orthogonal Vectors and Addition of Sums of Squares

Table 6B.2 shows the relevant analysis of the observation vector into a average vector and a residual vector. This decomposition parallels that of Table 6.6. Figure 6B.2a shows the same analysis geometrically. Since $\sum \bar{y}(y - \bar{y}) = 0$, the vectors $\bar{y}$ and $y - \bar{y}$ are orthogonal, as is indicated in the diagram. The three vectors y, $\bar{y}$, and $y - \bar{y}$ thus form a right triangle.

TABLE 6B.2. Data for illustration

	observation vector	=	average vector	+	residual vector
	$\mathbf{y}$		$\bar{\mathbf{y}}$		$\mathbf{y} - \bar{\mathbf{y}}$
	$\begin{bmatrix} 3 \\ 5 \\ 4 \end{bmatrix}$	$=$	$\begin{bmatrix} 4 \\ 4 \\ 4 \end{bmatrix}$	$+$	$\begin{bmatrix} -1 \\ +1 \\ 0 \end{bmatrix}$
sum of squares	50	=	48	+	2
degrees of freedom	3	=	1	+	2

Now the *length* of the vector $\bar{y}$, indicated by $|\bar{y}|$, is

$$|\bar{y}| = (\bar{y}^2 + \bar{y}^2 + \bar{y}^2)^{1/2} = \sqrt{3}\,\bar{y} = \sqrt{n}\,\bar{y}$$

and the length of the vector $y - \bar{y}$ is

$$|y - \bar{y}| = [(y_1 - \bar{y})^2 + (y_2 - \bar{y})^2 + (y_3 - \bar{y})^2]^{1/2} = [\sum(y - \bar{y})^2]^{1/2}$$

The algebraic identity

$$\sum y^2 = n\bar{y}^2 + \sum(y - \bar{y})^2$$

or (see Table 6B.1)

$$S_y = S_{\bar{y}} + S_{y-\bar{y}}$$

may be written as

$$|\mathbf{y}|^2 = |\bar{\mathbf{y}}|^2 + |\mathbf{y} - \bar{\mathbf{y}}|^2$$

From Figure 6B.2a it is seen that this is merely a statement of Pythagoras' theorem that the square on the hypotenuse of a right triangle is equal to the sum of the squares of the other two sides.

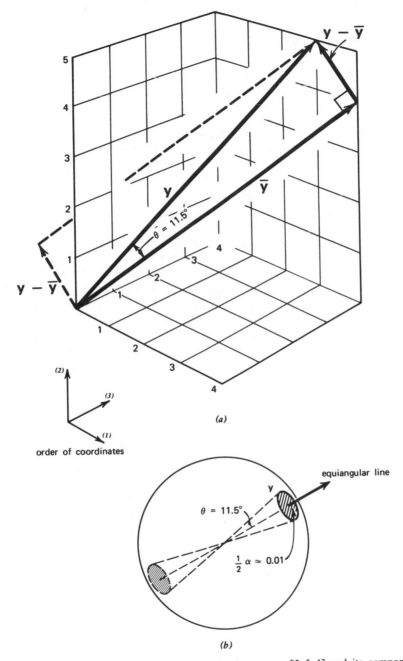

order of coordinates

(a)

equiangular line

$\theta = 11.5°$

$\frac{1}{2} \alpha \simeq 0.01$

y

(b)

FIGURE 6B.2 Geometric representation of the data vector $[3, 5, 4]$ and its components $\bar{y} = [4, 4, 4]$ and $y - \bar{y} = [-1, 1, 0]$ in relation to the t test and associated analysis of variance.

Geometry of the t *Test*

The expression for t_0 can be written as

$$t_0 = \sqrt{n-1} \left\{ \frac{n\bar{y}^2}{\sum (y - \bar{y})^2} \right\}^{1/2} = \sqrt{2} \left\{ \frac{3\bar{y}^2}{\sum (y - \bar{y})^2} \right\}^{1/2} = \frac{\sqrt{2}|\bar{y}|}{|y - \bar{y}|}$$

Now (see Figure 6B.2a), the length of the vector $\bar{y}$ divided by the length of the vector $y - \bar{y}$ is the cotangent of the angle θ (greek letter *theta*) that y makes with $\bar{y}$. Thus

$$t_0 = \sqrt{2} \cot \theta \qquad \text{or, in general,} \qquad t_0 = \sqrt{n-1} \cot \theta$$

Now $\cot \theta$ gets larger as θ gets smaller, and we see that the quantity t_0, which we routinely refer to the t table, is a measure of how *small* the angle is between y and $\bar{y}$. Notice that, no matter what data we analyze, the vector $\bar{y} = [\bar{y}, \bar{y}, \bar{y}]$ will lie on the equiangular line— the line that makes an equal angle with each of the coordinate axes. In fact $\bar{y}$ is obtained by dropping a perpendicular from the data vector y onto the equiangular line.

The data vector $y = [3, 5, 4]$ strongly suggests the existence of a consistent nonzero component because this vector y makes only a small angle $[(\cot^{-1}(t_0/\sqrt{n-1}) = \cot^{-1}(6.9/\sqrt{2}) = \cot^{-1} 4.88 = 11.5°]$ with the vector $\bar{y} = [4, 4, 4]$ on the equiangular line. In this case the angle θ is small, t_0 is large, and the significance level is small.

On the other hand, an inconsistent data vector like $y = [-6, 14, 4]$, yielding the same average and hence the same vector $\bar{y} = [4, 4, 4]$, produces a much larger angle ($\theta = 63.9°$), a much smaller value of t_0, and a much larger significance probability.

Exercise 6B.1. Obtain t_0, θ, and the significance level for the data sets $[3.9, 4.1, 4.0]$ and $[-6, 14, 4]$. *Answer*: $(69.3, 1.2°, < 0.0005)$: $(0.69, 63.9°, 0.28)$.

Significance Tests

From the geometry of t it is seen that a test of the hypothesis $\eta = 0$ against $\eta > 0$ amounts to asking what the probability is that a vector y would, by chance, make so small an angle as θ with the equiangular line. Now (see Figure 6B.2b), if $\eta = 0$ the elements of the data vector y are a set of independent normally distributed errors. For such errors the vector y has the property that its tip is equally likely to lie anywhere on a sphere drawn with center at the origin and radius $|y|$.

Now consider the shaded cap shown in the figure on the surface of the sphere and included in a cone obtained by rotating y about the equiangular line. The required probability is evidently equal to the surface area of this cap, expressed as a fraction of the total surface area of the sphere. For a two-sided test we must include the surface area of a second complementary cap, shown in the diagram.

Analysis of Variance

We have seen that the sums of squares of the various elements in Table 6B.1 indicating the squared length of corresponding vectors are additive because the components $\bar{y}$

and $\mathbf{y} - \bar{\mathbf{y}}$ are orthogonal and consequently Pythagoras' theory applies. The degrees of freedom indicate the number of dimensions in which the vectors are free to move. Thus before the data are collected the vector $\mathbf{y}$ is unconstrained and has $n = 3$ degrees of freedom; the vector $\bar{\mathbf{y}}$, which has elements $(\bar{y}, \bar{y}, \bar{y})$ and is constrained to lie on the equiangular line, has only 1 degree of freedom; the vector $\mathbf{y} - \bar{\mathbf{y}}$, which is constrained to lie on a plane perpendicular to $\mathbf{y}$, has $n - 1 = 2$ degrees of freedom. The analysis of variance of Table 6B.2 conveniently summarizes these facts.

In general, each statistical model discussed in this book determines a certain line, plane or space on which *if there were no error* the data *would have* to lie. For the example of this section, for instance, the model is $y = \eta + \epsilon$. Thus, without the errors ϵ, the data would *have to* lie on the equiangular line at some point $[\eta, \eta, \eta]$. The t and F criteria measure the angle that the actual data vector, which is subject to error, makes with the appropriate line, plane and space dictated by the model. The corresponding tables indicate probabilities that angles as small or smaller will occur by chance. These probabilities are dependent on the dimensions of the model and of the data through the degrees of freedom in the table.

Generalization

The vector breakdown of Table 6.6 for the general one-way analysis of variance is a direct extension of that of Table 6B.2. The analysis of variance of Table 6.3 is a direct extension of that of Table 6B.1. The geometry and resulting distribution theory for the general case is essentially an elaboration of that given above.

APPENDIX 6C. MULTIPLE COMPARISONS

Formal procedures for allowing for the effect of selection in making comparisons have been the subject of considerable research (see, e.g., O'Neill and Wetherill, 1971, and Miller, 1977, also the references listed therein).

Confidence Interval for a Particular Difference in Means

A confidence interval for the true difference between the means of, say, the pth and qth treatments may be obtained as follows. The observed difference $\bar{y}_p - \bar{y}_q$ has variance $\sigma^2(1/n_p + 1/n_q)$, and σ^2 is estimated by the within-treatment mean square s^2. Thus the estimated variance of $\bar{y}_p - \bar{y}_q$ is $s^2(1/n_p + 1/n_q)$, and a confidence interval for this single *preselected* difference is provided by

$$(\bar{y}_p - \bar{y}_q) \pm t_{v,\,\alpha/2}\, s\, \sqrt{\frac{1}{n_p} + \frac{1}{n_q}} \tag{6.C1}$$

where $v = v_R$, the degrees of freedom associated with s^2.

For the example discussed in this chapter, a confidence interval for the true difference between the means of treatments A and B can be established as follows. We have

$\bar{y}_B - \bar{y}_A = 66 - 61 = 5$, $s_R^2 = 5.6$ with $v = 20$ degrees of freedom, $n_B = 6$ and $n_A = 4$, and the estimated variance for $\bar{y}_B - \bar{y}_A$ is $5.6\,(\frac{1}{4} + \frac{1}{6}) = 2.33$. Thus the 95% confidence limits for the mean difference $\eta_B - \eta_A$ are $5 \pm 2.09\sqrt{2.33}$, that is, 5 ± 3.2, where 2.09 is the value of t appropriate for 20 degrees of freedom, which is exceeded, positively or negatively, a total of 5% of the time.

The $1 - \alpha$ confidence limits calculated in this way will be valid for any *single* chosen difference; the chance that the specific interval given above includes the true difference $\eta_B - \eta_A$ on the stated assumptions will be equal to $1 - \alpha$. For k treatments, however, there are $k(k - 1)/2$ treatment pairs, and the differences between each one of these pairs can be used to construct a confidence interval. Whereas for each interval individually the chance of including the true value is exactly equal to $1 - \alpha$, the chance that all the intervals will *simultaneously* include their true values is less than $1 - \alpha$.

Tukey's Paired Comparison Procedure

In comparing k averages, suppose that we wish to state the confidence interval for $\eta_i - \eta_j$, taking account of the fact that all possible comparisons may be made. It has been shown by Tukey (1949) that the confidence limits for $\eta_i - \eta_j$ are then given by

$$(\bar{y}_i - \bar{y}_j) \pm \frac{q_{k,v,\alpha/2}}{\sqrt{2}} s \sqrt{\frac{1}{n_i} + \frac{1}{n_j}} \qquad (6\text{C}.2)$$

where $q_{k,v}$ is the appropriate upper significance level of the *studentized range* for k means, and v the number of degrees of freedom in the estimate s^2 of variance σ^2. This formula is exact if the numbers of observations in all the averages are equal, and approximate if the averages are based on unequal numbers of observations.

The size of the confidence interval for any given level of probability is larger when the range statistic $q_{k,v}$ is used rather than the t statistic, since the range statistic allows for the possibility that any one of the $k(k - 1)/2$ possible pairs of averages might have been selected for the test. Critical values of $q_{k,v}/\sqrt{2}$ have been tabulated; see, for instance, Pearson and Hartley (1966), Table 29. As an example, in an experimental program on the bursting strengths of diaphragms the treatments consisted of $k = 7$ different types of rubber, and $n = 4$ observations were run with each type. The data were as follows:

treatment t	A	B	C	D	E	F	G
average $\bar{y}_t$	63	62	67	65	65	70	60
estimates of variance s_t^2	9.2	8.7	8.8	9.8	10.2	8.3	8.0

For this example, $k = 7$, $s^2 = 9.0$, $v = 21$, $\alpha = 0.05$, and $q_{k,v,\alpha/2}/\sqrt{2} = 3.26$; these values give for the 95% limits

$$\pm \frac{q_{k,v,\alpha/2}}{\sqrt{2}} \sqrt{\left(\frac{1}{n_i} + \frac{1}{n_j}\right)s^2} = \pm 3.26\sqrt{(\frac{1}{4} + \frac{1}{4})9.0} = \pm 6.91 \qquad (6\text{C}.3)$$

Thus any observed difference greater in absolute value than 6.91 could be considered statistically significant; hence we could say that the corresponding true difference is not likely to be zero. The $7 \times 6/2 = 21$ differences are listed in the following table. Those that are statistically significant are circled. The *total* error rate is $\alpha = 0.05$.

treatment	A	B	C	D	E	F	G
average $\bar{y}_t$	63	62	67	65	65	70	60
difference $\bar{y}_i - \bar{y}_j$	*	1	−4	−2	−2	⊘−7	3
		*	−5	−3	−3	⊘−8	2
			*	2	2	−3	⊘7
				*	0	−5	5
					*	−5	5
						*	⊘10
							*

Dunnett's Procedure for Multiple Comparisons with a Standard

Experimenters often use a control or standard treatment as a benchmark against which to compare the specific treatments. The question then arises whether any of the treatment means may be considered to be different from the mean of the control. In the above example suppose that A was the control. The statistics of interest now are the $k - 1$ differences $\bar{y}_t - \bar{y}_A$, where $\bar{y}_A$ is the observed average response for the control treatment. The $1 - \alpha$ confidence intervals for all $k - 1$ differences from the control are as given by Equation 6C.2, except that the value of $q_{k, v, \alpha/2}/\sqrt{2}$ is replaced with Dunnett's t. For tabulated values of this quantity, $t_{k, v, \alpha/2}$, see Dunnett (1964). Thus in the above example we have $t_{k, v, \alpha/2} = 2.80$, giving for the 95% limits

$$\pm t_{k, v, \alpha/2} s \sqrt{\frac{1}{n_A} + \frac{1}{n_t}} = \pm 2.80 \times 3.00\sqrt{\tfrac{1}{4} + \tfrac{1}{4}} = \pm 5.94 \tag{6C.4}$$

Therefore any observed difference from the control greater than 5.94 in absolute value can be considered statistically significant. The $k - 1 = 6$ differences are as follows:

treatment	A (control)	B	C	D	E	F	G
average	63	62	67	65	65	70	60
difference	*	1	−4	−2	−2	⊘−7	3

Only the difference $\bar{y}_F - \bar{y}_A$ is indicative of a real difference between the means of six treatments and the control treatment.

For the special case of comparisons against a standard or a control it is good practice to allot more observations n_A to the control treatment than to each of the other treatments n_t. The ratio n_A/n_t should be approximately equal to the square root of the number of treatments, that is, $n_A/n_t = \sqrt{k}$.

Other Procedures

Other techniques are also available for making multiple comparisons between treatment averages. One method, to be used only if the F test has shown evidence of statistically significant differences, is the Newman–Keuls (Newman, 1939, and Keuls, 1952). An alternative has been suggested by Duncan (1955). A method for constructing an interval statement appropriate for *all possible comparisons* among the k treatments, not merely their differences, has been proposed by Scheffé (1953). The Scheffé method is the most conservative, that is, it produces the widest interval statements.

Use of Formal Tests for Multiple Comparisons

In practice it is questionable how far we should go with such formal tests. The difficulties are as follows:

1. How exact should we be about uncertainty? We may ask, for example, "How much difference does it make to know whether a particular probability is exactly 0.04, exactly 0.06, or about 0.05?"
2. Significance levels and confidence coefficients are arbitrarily chosen.
3. In addition to the procedures we have mentioned, others employ still other bases for making multiple comparisons. The subtleties involved are not easy to understand, and the experimenter may find himself provided with an exact measure of the uncertainty of a proposition he does not fully comprehend.

For many practical situations a satisfactory alternative is careful inspection of the treatment averages in relation to a sliding reference distribution, as described in this chapter. The procedure is admittedly approximate, but, we believe, not misleadingly so.

REFERENCES AND FURTHER READINGS

An authoritative text on analysis of variance is:

Scheffé, H. (1953). *Analysis of Variance*, Wiley.

For further information on multiple comparisons, see these articles and the references listed therein:

O'Neill, R., and G. B. Wetherill. (1971). The present state of multiple comparison methods, *J. Roy. Stat. Soc., Ser. B*, **33**, 218.
Miller, R. G., Jr., (1977). Developments in multiple comparisons, 1966–1976, *J. Am. Stat. Assoc.*, **72**, 779.

The following are the references mentioned in Appendix 6C on multiple comparisons:

Tukey, J. W. (1949). Comparing individual means in the analysis of variance, *Biometrics*, **5**, 99.

Pearson, E. S., and H. O. Hartley. (1966). *Biometrika Tables for Statisticians*, Vol. 1, 3rd ed., Cambridge University Press.

Dunnett, C. W. (1964). New tables for multiple comparisons with a control, *Biometrics*, **20**, 482.

Newman, D. (1939). The distribution of the range in samples from a normal population expressed in terms of an independent estimate of the standard deviation, *Biometrika*, **31**, 20.

Keuls, M. (1952). The use of the Studentized range in connection with an analysis of variance, *Euphytica*, **1**, 112.

Duncan, D. B. (1955). Multiple range and multiple *F* tests, *Biometrics*, **11**, 1.

Scheffé, H. (1953). A method for judging all contrasts in the analysis of variance, *Biometrika*, **40**, 87.

QUESTIONS FOR CHAPTER 6

1. What are the basic ideas of the analysis of variance?
2. Invent some data for three treatments with four replications each. How can the data vector be decomposed into three separate parts? What are these parts? Construct an analysis of variance table.
3. What is the usual model for a one-way analysis of variance? What are its possible shortcomings?
4. Why is the assumption of normality made in analysis of variance? If the experiment is properly randomized, is this assumption necessary?
5. How is Pythagoras' theorem related to the analysis of variance?
6. What are residuals? How can they be calculated? How can they be plotted? Why should they be plotted?
7. How can a reference distribution diagram be constructed for the comparison of k means? What can one tell from such a diagram but not from an analysis of variance table?

CHAPTER 7

Randomized Blocks and Two-Way Factorial Designs

In this chapter and in the following one, the principle of paired comparisons first considered in Chapter 4 is extended to the comparison of more than two treatments, using randomized designs with larger block sizes. The organization of the relevant chapters is as follows:

	unblocked arrangements	blocked arrangements
comparison of 2 treatments	Chapters 2–5	Sections 4.2 and 5.2
comparison of k treatments	Chapter 6	Chapters 7 and 8

In blocked designs two kinds of effects are contemplated: those of the treatments, which are of major interest to the experimenter, and those of the blocks, whose contribution it is desired to eliminate. In practice, blocks might be, for example, different litters of animals (a direct extension of the twin idea), blends of chemical material, strips of land, or contiguous periods of time.

The chapter ends with discussion of a different kind of two-way arrangement, a replicated (factorial) design in which the *main effects* of two factors and their *interaction* are all of equal interest. An example illustrates the advantages sometimes obtainable by suitably transforming the data. Methods for transforming data to achieve more nearly constant variance and additivity are presented.

7.1. EXAMPLE: COMPARISON OF FOUR VARIANTS OF A PENICILLIN PRODUCTION PROCESS

An example of a randomized block experiment is shown in Table 7.1. In this example a process for the manufacture of penicillin was being investigated, and yield was the response of primary interest. There were $k = 4$ variants of the basic process to be studied, denoted as treatments A, B, C, and D.

It was known that an important raw material, corn steep liquor, was quite variable. Fortunately blends sufficient for four runs could be made, thus supplying the opportunity to run all $k = 4$ treatments within each of $n = 5$ blocks (blends of corn steep liquor). The experiment was protected from extraneous unknown sources of bias by running the treatments in random order within each block.

A randomized block design of this kind has the following advantages:

1. It provides the opportunity to eliminate blend-to-blend (block) variation from comparison of the treatments.
2. It provides a wider inductive basis than an experiment run with uniform raw material—the treatments are tested, not with just one, but with five *different* blends of corn steep liquor.

In addition to the observations themselves, three other quantities useful for the subsequent analysis are recorded in Table 7.1. These are the block

TABLE 7.1. Results from randomized block design on penicillin manufacture*

block (blend of corn steep liquor)	treatment				block average
	A	B	C	D	
blend 1	$89^{(1)}$	$88^{(3)}$	$97^{(2)}$	$94^{(4)}$	92
blend 2	$84^{(4)}$	$77^{(2)}$	$92^{(3)}$	$79^{(1)}$	83
blend 3	$81^{(2)}$	$87^{(1)}$	$87^{(4)}$	$85^{(3)}$	85
blend 4	$87^{(1)}$	$92^{(3)}$	$89^{(2)}$	$84^{(4)}$	88
blend 5	$79^{(3)}$	$81^{(4)}$	$80^{(1)}$	$88^{(2)}$	82
treatment average	84	85	89	86	$86 = \dfrac{\text{grand}}{\text{average}}$

* The superscripts in parentheses associated with the observations indicate the random order in which the experiments were run within each blend.

averages, the treatment averages, and the grand average. (For expository purposes the yield values have been altered somewhat to give whole number averages.)

7.2. A MODEL WITH CORRESPONDING DECOMPOSITION OF OBSERVATIONS

An observation made with the tth treatment applied to the ith block will be denoted by y_{ti}, the ith block average by $\bar{y}_i$, the tth treatment average by $\bar{y}_t$, and the grand average by $\bar{y}$. Thus, algebraically, the results of a randomized block design can be represented as in Table 7.2.

A Mathematical Model

Now consider a mathematical model that might describe such a set of data. The simplest such model supposes that an observation y_{ti} can be represented as the sum of a general mean η, a block effect β_i, a treatment effect τ_t, and an error ϵ_{ti}; that is,

$$y_{ti} = \eta + \beta_i + \tau_t + \epsilon_{ti} \qquad (7.1)$$

Associated with such a model is the decomposition of the observations:

$$y_{ti} = \bar{y} + (\bar{y}_i - \bar{y}) + (\bar{y}_t - \bar{y}) + (y_{ti} - \bar{y}_i - \bar{y}_t + \bar{y}) \qquad (7.2)$$

The last quantity $y_{ti} - \bar{y}_i - \bar{y}_t + \bar{y}$ is called the *residual* because it represents what is left after the grand average, block differences, and treatment differences

TABLE 7.2. **Results from randomized block design, general case**

		treatment					block average
		1	2 $\cdots$	t	$\cdots$	k	
block	1	y_{11}	y_{21} $\cdots$	y_{t1}	$\cdots$	y_{k1}	
	2	y_{12}	y_{22} $\cdots$	y_{t2}	$\cdots$	y_{k2}	
	$\vdots$	$\vdots$	$\vdots$	$\vdots$		$\vdots$	$\vdots$
	i	y_{1i}	y_{2i} $\cdots$	y_{ti}	$\cdots$	y_{ki}	$\bar{y}_i$
	$\vdots$	$\vdots$	$\vdots$	$\vdots$		$\vdots$	$\vdots$
	n	y_{1n}	y_{2n} $\cdots$	y_{tn}	$\cdots$	y_{kn}	
treatment average			$\cdots$ $\bar{y}_t$ $\cdots$				$\bar{y} =$ grand average

have all been allowed for. This decomposition for the penicillin data set is given in Table 7.3, which the reader should now study carefully.

In vector notation the decomposition can be written as

$$\mathbf{Y} = \mathbf{A} + \mathbf{B} + \mathbf{T} + \mathbf{R} \tag{7.3}$$

In this equation each of the symbols $\mathbf{Y}, \mathbf{A}, \mathbf{B}, \mathbf{T}$, and $\mathbf{R}$ represents a vector containing the $N = nk$ elements of the corresponding two-way table. Each vector is obtained by setting end to end the four columns of the appropriate two-way table. For the penicillin data, $N = nk = 5 \times 4 = 20$, and, for example, $\mathbf{Y} = [89, 84, 81, 87, 79, 88, 77, \ldots, 88]$.

Exercise 7.1. In data set (a) the treatments are three different methods for determining a pollutant level, the blocks are two technicians, and the observations are measured levels of pollutant (in parts per million) in six identical samples. In data set (b) the treatments are three different brands of a product, the blocks are four laboratories, and the observations are performance readings (the higher the better). For each set of data construct the table corresponding to Table 7.3.

		data set (a) treatment				*data set (b)* treatment		
		A	B	C		A	B	C
block	1	7	36	2	1	6.5	7.4	7.4
	2	13	44	18	2	6.8	7.3	6.9
					block 3	6.4	7.2	8.0
					4	6.7	6.9	6.5

Partial answer: (a) $\bar{y} = 20$, residuals $= (2, -2, 1, -1, -3, 3)$; (b) $\bar{y} = 7.0$, residuals $= (-0.2, 0.2, -0.4, 0.4, 0.1, 0.1, -0.2, 0, 0.1, -0.3, 0.6, -0.4)$.

Sums of Squares and Degrees of Freedom

The "sums of squares" are sums of squares of the elements occurring in each part of Table 7.3 and represent the squared lengths of the vectors $\mathbf{Y}, \mathbf{A}, \mathbf{B}, \mathbf{T}$, and $\mathbf{R}$. For example,

$$S = 89^2 + 84^2 + 81^2 + \cdots + 84^2 + 88^2 = 148,480 \tag{7.4}$$

$$S_A = 86^2 + 86^2 + 86^2 + \cdots + 86^2 + 86^2 = 147,920 \tag{7.5}$$

As before, the number of degrees of freedom is the number of dimensions in which each vector is free to move and is equal to the number of elements that can be selected arbitrarily. For example, in the part of the table headed "block deviations" consider the $n = 5$ deviations of the block averages from the grand average. Only $n - 1 = 4$ of these could be selected arbitrarily

TABLE 7.3. Decomposition of observations for randomized block experiment, penicillin example

		grand average $\bar{y}$		block deviations $(\bar{y}_i - \bar{y})$		treatment deviations $(\bar{y}_i - \bar{y})$		residuals $(y_{ti} - \bar{y}_i - \bar{y}_i + \bar{y})$

analysis of observations

$$
\begin{bmatrix}
89 & 88 & 97 & 94 \\
84 & 77 & 92 & 79 \\
81 & 87 & 87 & 85 \\
87 & 92 & 89 & 84 \\
79 & 81 & 80 & 88
\end{bmatrix}
=
\begin{bmatrix}
86 & 86 & 86 & 86 \\
86 & 86 & 86 & 86 \\
86 & 86 & 86 & 86 \\
86 & 86 & 86 & 86 \\
86 & 86 & 86 & 86
\end{bmatrix}
+
\begin{bmatrix}
6 & 6 & 6 & 6 \\
-3 & -3 & -3 & -3 \\
-1 & -1 & -1 & -1 \\
2 & 2 & 2 & 2 \\
-4 & -4 & -4 & -4
\end{bmatrix}
+
\begin{bmatrix}
-2 & -1 & 3 & 0 \\
-2 & -1 & 3 & 0 \\
-2 & -1 & 3 & 0 \\
-2 & -1 & 3 & 0 \\
-2 & -1 & 3 & 0
\end{bmatrix}
+
\begin{bmatrix}
-1 & -3 & 2 & 2 \\
3 & -5 & 6 & -4 \\
-2 & 3 & -1 & 0 \\
1 & 5 & -2 & -4 \\
-1 & 0 & -5 & 6
\end{bmatrix}
$$

	y_{ti}	grand average	block deviations	treatment deviations	residuals
vector notation	$\mathbf{Y}$	$\mathbf{A}$	$\mathbf{B}$	$\mathbf{T}$	$\mathbf{R}$
	=	=	+	+	+
sums of squares (squared length of vector)	S 148,480	S_A 147,920	S_B 264	S_T 70	S_R 226
	=	=	+	+	+
degrees of freedom (number of dimensions in which vector is constrained to lie)	20	1	4	3	12

212

TABLE 7.4. **Analysis of variance table, penicillin example**

source of variation	sum of squares	degrees of freedom
average	$S_A = 147{,}920$	1
blends (blocks)	$S_B = 264$	4
treatments	$S_T = 70$	3
residuals	$S_R = 226$	12
total	$S = 148{,}480$	20

while maintaining the essential property that their sum be zero. Similarly, only $k - 1 = 3$ of the $k = 4$ treatment deviations could be selected arbitrarily. Finally, bearing in mind that each column and row of the table of the residuals must sum to zero, only $(n - 1)(k - 1) = 12$ of these entries could be selected arbitrarily.

Additivity of Sums of Squares and Degrees of Freedom

The sums of squares and associated degrees of freedom of Table 7.3 have been inserted in the analysis of variance table, Table 7.4. The general formulas for these quantities are given in Table 7.5. Note the additive property of the sum of squares, on the one hand, and the degrees of freedom, on the other. For the sum of squares we have

$$S = S_A + S_B + S_T + S_R \qquad (7.6)$$

Exercise 7.2. For the data sets in Exercise 7.1, find the appropriate sums of squares and degrees of freedom.
Partial answer: (a) $S_A = 2400$, $S_B = 150$, $S_T = 1200$, $S = 3778$, $v_B = 1$, $v_T = 2$, $v_R = 2$, $v = 6$; (b) $S_A = 588$, $S_B = 0.42$, $S_T = 0.96$, $S = 590.46$, $v_B = 3$, $v_T = 2$, $v_R = 6$, $v = 12$.

The Analysis of Variance and Pythagoras' Theorem

The vectors **A**, **B**, **T**, and **R** are all at right angles to one another, as is easily verified. For example, for **B** and **R** we have the inner product

$$(6)(-1) + (-3)(3) + (-1)(-2) + \cdots + (-4)(6) = 0 \qquad (7.7)$$

The additive property of the sums of squares (Equation 7.6) is an immediate consequence of a multidimensional version of Pythagoras' theorem, as was

TABLE 7.5. **Algebraic decomposition of sums of squares for the randomized block design, general formulas***

source of variation	sum of squares	degrees of freedom
average (correction factor)	$S = nk\bar{y}^2$	1
between blocks	$S_B = k \sum_{i}^{n} (\bar{y}_i - \bar{y})^2$	$n - 1$
between treatments	$S_T = n \sum_{t}^{k} (\bar{y}_t - \bar{y})^2$	$k - 1$
residuals	$S_R = \sum_{t}^{k} \sum_{i}^{n} (y_{ti} - \bar{y}_i - \bar{y}_t + \bar{y})^2$	$(n - 1)(k - 1)$
total	$S = \sum_{t}^{k} \sum_{i}^{n} y_{ti}^2$	$N = nk$

* Abbreviated calculation methods for the construction of this table are given in Appendix 7A.

true for the completely randomized design discussed in Chapter 6. Figure 7.1 illustrates the relation

$$D = B + T + R \qquad (7.8)$$

where $D = Y - A$ is a vector of deviations of the data from the grand average. Since the vectors B, T, and R are mutually orthogonal, we have

$$S_D = S_B + S_T + S_R \qquad (7.9)$$

In words, the sum of squares of deviations from the grand average is equal to the sum of squares for blocks plus the sum of squares for treatments plus the sum of squares of residuals.

Exercise 7.3. By computing appropriate inner products (see Equation 7.7), show that all the vectors A, B, T, and R in Table 7.3 are orthogonal to each other, but Y is not orthogonal to any of them.

Exercise 7.4. Repeat Exercise 7.3 for the data sets in Exercise 7.1.

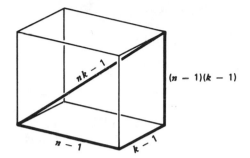

(*a*) Dimensionality (degrees of freedom).

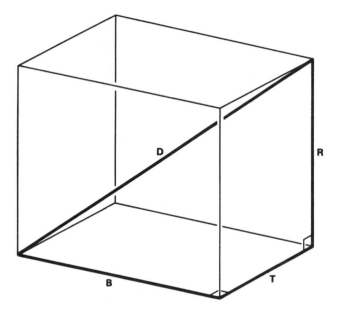

(*b*) Vectors.

FIGURE 7.1. Analysis of variance for a randomized block design: Vector decomposition of the data according to

$$
\begin{array}{ccccccc}
\mathbf{D} & = & \mathbf{B} & + & \mathbf{T} & + & \mathbf{R} \\
y_{ti} - \bar{y} & = & (\bar{y}_i - \bar{y}) & + & (\bar{y}_t - \bar{y}) & + & (y_{ti} - \bar{y}_i - \bar{y}_t + \bar{y})
\end{array}
$$

Summary Tables

Table 7.6 shows the essential quantities involved in the decomposition already given in more extended form in Table 7.3. The corresponding decomposition of a general randomized block design with n blocks and k treatments is expressed algebraically in Table 7.7. Given in the margins are the grand average $\bar{y}$, the deviations of the individual block averages from the grand average $(\bar{y}_i - \bar{y})$, and the deviations of the treatment averages from the grand average $(\bar{y}_t - \bar{y})$. In the body of the table are shown the residuals $(y_{ti} - \bar{y}_i - \bar{y}_t + \bar{y})$ remaining after the deviations due to blocks and treatments have been allowed for.

Exercise 7.5.   For the data sets in Exercise 7.1 construct the tables corresponding to Table 7.6.
Partial answer: (a) Block 1 residuals: 2, 1, -3; Block 2 residuals: -2, -1, 3; treatment deviations: -10, 20, -10; block deviations: -5, 5, (b) Block 3 residuals: -0.4, -0.2, 0.6; treatment deviations: -0.4, 0.2, 0.2; block deviations: 0.1, 0; 0.2, -0.3.

The reader should now be able to understand the relationship between:

1. The numbers in Tables 7.3 and 7.6.
2. Their algebraic representation in Table 7.7.
3. The formulas for their sums of squares in Table 7.5.
4. Their geometric meaning in Figure 7.1.

TABLE 7.6.   Summary table: residuals and treatment and block deviations, penicillin example*

| block | residual for treatment: | | | | deviation of block averages from grand average |
	A	B	C	D	
blend 1	$-1^{(1)}$	$-3^{(3)}$	$2^{(2)}$	$2^{(4)}$	6
blend 2	$3^{(4)}$	$-5^{(2)}$	$6^{(3)}$	$-4^{(1)}$	-3
blend 3	$-2^{(2)}$	$3^{(1)}$	$-1^{(4)}$	$0^{(3)}$	-1
blend 4	$1^{(1)}$	$5^{(3)}$	$-2^{(2)}$	$-4^{(4)}$	2
blend 5	$-1^{(3)}$	$0^{(4)}$	$-5^{(1)}$	$6^{(2)}$	-4
deviation of treatment averages from grand average	-2	-1	3	0	$86 = \dfrac{\text{grand}}{\text{average}}$

* The superscripts in parentheses associated with each residual indicate the time order in which the corresponding observations were run within each blend.

TABLE 7.7.   Summary table: residuals and treatment and block deviations for a general randomized block design

		|	1	2 ...	treatment t	...	k	deviation of block averages from grand average
	1	|		...		...		
	2	|		...		...		
	⋮	|			⋮			⋮
block	i	|		...	$(y_{ti} - \bar{y}_i - \bar{y}_t + \bar{y})$	...		$(\bar{y}_i - \bar{y})$
	⋮	|			⋮			⋮
	n	|		...		...		

deviation of treatment averages from grand average   ... $(\bar{y}_t - \bar{y})$ ... $\bar{y}$ = grand average

Before going on with the book, the reader should feel entirely comfortable with all the relationships and meanings that these tables and diagrams imply. If still hesitant about vectors and their geometry, reread Appendix 6B.

Estimated Values

An important adjunct to Table 7.6 is Table 7.8, which gives the estimated values $\hat{y}_{ti}$ obtained by adding the observed block deviations and treatment

TABLE 7.8.   Table of estimated values $\hat{y}_{ti}$, penicillin example

		treatment			
		A	B	C	D
	1	90	91	95	92
	2	81	82	86	83
block	3	83	84	88	85
	4	86	87	91	88
	5	80	81	85	82

deviations to the grand average,

$$\hat{y}_{ti} = \bar{y} + (\bar{y}_i - \bar{y}) + (\bar{y}_t - \bar{y}) \tag{7.10}$$

These are the values obtained when the portion due to residuals is omitted; that is, denoting the vector of estimated values $\hat{y}_{ti}$ by $\hat{\mathbf{Y}}$, we have

$$\hat{\mathbf{Y}} = \mathbf{A} + \mathbf{B} + \mathbf{T} \tag{7.11}$$

If the model of Equation 7.1 is adequate, the estimated values will provide the best available estimates of the response for each block treatment combination.

Exercise 7.6.   For each data set in Exercise 7.1 construct the table corresponding to Table 7.8.

Partial answer: Values for first row: (a) 5, 35, 5; (b) 6.7, 7.3, 7.3.

7.3. IMPLICATIONS OF THE ADDITIVE MODEL

The decomposition of the observations shown in Table 7.3, which leads to the analysis of variance table, Table 7.4, is a purely algebraic process motivated by a model of the form

$$y_{ti} = \eta + \beta_i + \tau_t + \epsilon_{ti} \tag{7.12}$$

Thus the underlying expected response

$$\eta_{ti} = \eta + \beta_i + \tau_t \tag{7.13}$$

is supposed to follow the pattern illustrated in Figure 7.2. The model is called *additive* because, for example, if increment τ_3 provided an increase of six units in the response and if the influence of block β_4 increased the response by four units, the increase of both together would be assumed to be $6 + 4 = 10$ units in the response. Although this simple additive model would often provide an adequate approximation, there are circumstances where it would not.

If the block effect and treatment effect were not additive, an *interaction* would be said to occur between blocks and treatments. Consider, for instance, the comparison of four catalysts A, B, C, and D with five blends of raw material representing blocks. It could happen that a particular impurity occurring in blend 3 poisoned catalyst B and made it ineffective, even though the impurity did not affect the other catalysts. This would lead to a low response for observation y_{23}, where these two influences came together, and would constitute an interaction between blend and catalyst.

Another way in which interactions can occur is when an additive model does apply, but not in the metric (scale, transformation) in which the data are originally measured. Suppose, for example, that in the original metric

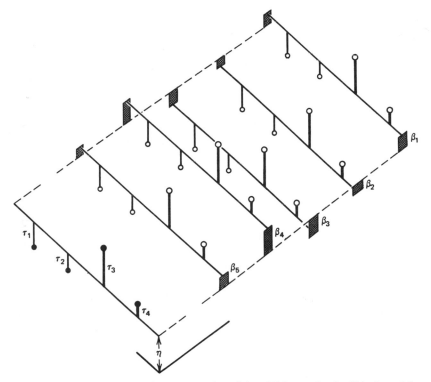

FIGURE 7.2. Diagrammatic representation of the additive randomized block model.

the response relationship was multiplicative so that

$$\eta_{ti} = \eta \beta_i \tau_t$$

Then, if the response covered a wide range, nonadditivity (interaction) between block effects β_i and treatment effects τ_t could seriously invalidate any linear model we attempted to fit. However, denoting logged values by primes, we can rewrite the above model as

$$\eta'_{ti} = \eta' + \beta'_i + \tau'_t$$

Interaction effects would disappear and the linear model would be applicable if, instead of y, $\log y$ were now analyzed. Interactions can thus be conveniently thought of as belonging to one of two categories: *transformable interactions* which may be eliminated by analyzing, for example, the log, square root, or reciprocal of the original data; *nontransformable interactions*, such as the blend–catalyst interaction discussed above, which cannot be eliminated in this way.

Ideally, blocks represent influences that one would expect only to raise or lower the general level of the response and that are to be eliminated rather than studied for their inherent interest. However, unexpected effects can occur that can be important and informative. Therefore for each individual data set careful examination of the results—and particularly of the residuals—must be undertaken, with departures from additivity as one possibility to be looked for.

A two-way arrangement in which both the row and column factors are of equal interest, as well as the interactive influence that one factor may have on the effect of the other, is *not* a randomized block design but a two-way factorial. A design of this kind, which usually should be replicated, is discussed in Section 7.7.

7.4. DIAGNOSTIC CHECKING OF THE MODEL

Analysis conducted without careful study of the original data and associated analysis of residuals requires blind faith that (1) the experiment will supply information only on the questions in mind at the start and (2) the assumptions implied by the analysis will be adequately met. While blind faith in a particular model is foolhardy, refusal to associate data with *any* model is to eschew a powerful tool. As implied earlier, a middle course may be followed. On the one hand, inadequacies in proposed models should be looked for; on the other, if a model appears reasonably appropriate, advantage should be taken of the greater simplicity and clarity of interpretation that it provides.

Critical examination of crude data and residuals for a randomized block experiment closely follows that discussed for the one-way classification in Chapter 6 with some additions.

For the penicillin data, examination of Figures 7.3a, b, and c, which are self-explanatory, reveals nothing of special interest. Concerning Figure 7.3c we have already noted that, if the model is appropriate, there should be no association between the size of the residuals and the estimated values $\hat{y}_{ti}$, and it will be remembered that one discrepancy to look for in this plot is a funnel shape suggesting a relationship between mean and variance. Now that we are dealing with a two-way (blocks and treatments) design, this plot takes on further potential interest.

Curvilinear Plot Suggesting Transformable Nonadditivity

The plot of the residuals versus the predicted values sometimes shows a curvilinear relationship. Thus the residuals may tend to be positive for low values of $\hat{y}$, become negative for intermediate values, and be positive again for high values. This appearance suggests *nonadditivity* between the block and

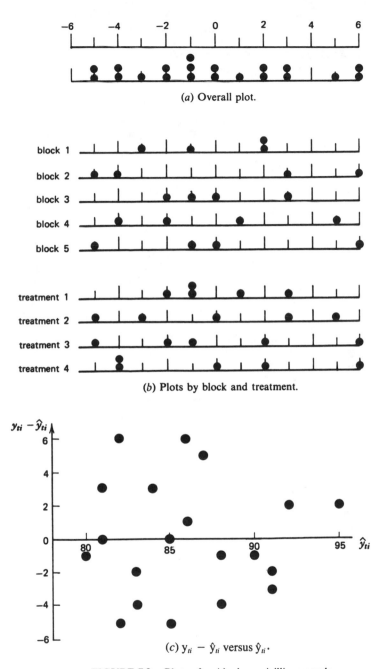

(a) Overall plot.

(b) Plots by block and treatment.

(c) $y_{ti} - \hat{y}_{ti}$ versus $\hat{y}_{ti}$.

FIGURE 7.3. Plots of residuals, penicillin example.

221

treatment effects such as might be eliminated by suitable transformation of the response. The reason is as follows.

Suppose that the model (Equation 7.12) is appropriate, not for the response y itself, but for some nonlinearly transformed response Y such as $\log y$ or $\sqrt{y}$. Then Y will plot against y as a curve so that the mean value of Y_{ti} will be approximately linearly related to $\eta_{ti} - \beta\eta_{ti}^2$, where β is some unknown constant measuring curvature of the relation, and, approximately, the expected response (of Equation 7.13) is therefore

$$\eta_{ti} = \eta' + \beta_i' + \tau_t' \;\; (+ \beta\eta_{ti}^2)$$

Now $\hat{y}_{ti}$ approximates η_{ti}, so that, when β is nonzero, residuals $y_{ti} - \hat{y}_{ti}$, obtained after fitting Equation 7.12, should show a quadratic tendency when plotted against $\hat{y}_{ti}$.

For data seriously in need of transformation, funnel and curvilinear tendencies can occur together producing a cornucopia-shaped plot.

One Degree of Freedom for Transformable Nonadditivity

A formal test for transformable nonadditivity based on the detection of a curvilinear relationship between $y - \hat{y}$ and $\hat{y}$ is due to J. W. Tukey. We illustrate with the penicillin data. The residuals $y_{ti} - \hat{y}_{ti}$ are given in Table 7.6, and estimated values in Table 7.8. Now treat the values $q_{ti} = \hat{y}_{ti}^2$ as data,* fit the model to the q_{ti} exactly as before, and then calculate residuals $q_{ti} - \hat{q}_{ti}$.

The quantities $y_{ti} - \hat{y}_{ti}$ and $q_{ti} - \hat{q}_{ti}$ are shown in Table 7.9. The test seeks a correlation between these two sets of residuals.

If P is the sum of products of the two sets of residuals and Q is the sum of squares of the residuals $q_{ti} - \hat{q}_{ti}$, then a least squares estimate (see Chapter 14) of β is $\hat{\beta} = P/Q = 0.023$. Corresponding to such a relationship, an associated sum of squares S_{na} for transformable nonadditivity having one degree of freedom is given by $P^2/Q = 2.00$. The component S_{na} is part of the residual sum of squares $S_R = 226$. Thus, we have the following analysis:

source of variation	sum of squares	degrees of freedom	mean square	
transformable nonadditivity	$S_{na} = $ 2	1	2	$\Big\}\, F = 0.10$
remainder	$S_R - S_{na} = 224$	11	20.4	
residual	$S_R = 226$	12		

* Exactly the same final result is obtained by setting $q_{ti} = (\hat{y}_{ti} - c)^2$, where c may be $\bar{y}$ or any other convenient constant.

TABLE 7.9. Values of $\begin{cases} y_{ti} - \hat{y}_{ti} \\ q_{ti} - \hat{q}_{ti} \end{cases}$ when $q_{ti} = \hat{y}_{ti}^2$, penicillin example

		treatment			
		A	B	C	D
block	1	-1 -24	-3 -12	2 36	2 0
	2	3 12	-5 6	6 -18	-4 0
	3	-2 4	3 2	-1 -6	0 0
	4	1 -8	5 -4	-2 12	-4 0
	5	-1 16	0 8	-5 -24	6 0

$$P = \sum (y - \hat{y})(q - \hat{q}) = 86$$

$$Q = \sum (q - \hat{q})^2 \qquad = 3696$$

$$S_{na} = \frac{P^2}{Q} \qquad = 2.00$$

It can be shown that on the standard normal theory assumptions the ratio of mean squares is exactly distributed as F when the appropriate null hypothesis is true. Obviously in this example there is no evidence of transformable interaction.

A significant result may indicate the need for transformation of the data. One way to choose a suitable transformation is to make the analyses for various transformations and then select a suitable one that shows no evidence of the existence of transformable interaction. With the fast computers now available this can be done quickly and inexpensively.

Special formulas are available from which Tukey's test may be computed for particular designs. The advantage of the general approach given here is that it may be used with any analysis of variance from any design or from any regression (least squares) analysis.

7.5. USE OF THE ANALYSIS OF VARIANCE TABLE

In the present example the analysis of residuals fails to reveal anything of interest. We will therefore complete the analysis on the assumption that theoretical requirements are approximately met. The sums of squares and

TABLE 7.10. Analysis of variance table for randomized block design, penicillin example

$$\text{model:} \quad y_{ti} = \eta + \beta_i + \tau_t + \epsilon_{ti}$$

$$\epsilon_{ti} \sim \text{IIDN}(0, \sigma^2)$$

hypotheses tested: (i) all β_i are zero, (ii) all τ_t are zero

source of variation	sum of squares	degrees of freedom	mean square	expected value of mean squares	ratio of mean squares
between blocks (blends)	$S_B = 264$	$n - 1 = 4$	$s_B^2 = 66.0$	$\sigma^2 + k \sum_i \beta_i^2/(n-1)$	$s_B^2/s_R^2 = 3.51$
between treatments	$S_T = 70$	$k - 1 = 3$	$s_T^2 = 23.3$	$\sigma^2 + n \sum_t \tau_t^2/(k-1)$	$s_T^2/s_R^2 = 1.24$
residuals	$S_R = 226$	$(n - 1)(k - 1) = 12$	$s_R^2 = 18.8$	σ^2	
total (corrected)	$S = 560$	$N - 1 = 19$			

associated degrees of freedom were given earlier in Table 7.4. The complete analysis of variance is shown in Table 7.10.

If we assume the additive model of Equation 7.12, with the nk quantities ϵ_{ti} as a sample of independently distributed random variables, each having mean zero and variance σ^2, the expected value of the residual mean square s_R^2 is equal to σ^2 whatever the magnitudes of the block and treatment effects. If the null hypothesis that all the treatment means are equal was true, the expected value of the treatment mean square s_T^2 would also equal σ^2. On the other hand, if the treatment means differed, this mean square would be inflated by the term $n\Sigma\tau_t^2/(k-1)$. Now, (1) on the further assumption that the distribution of the errors is normal* or (2) as an approximation to randomization theory†, we may test the null hypothesis that all the treatment means are equal by referring the ratio s_T^2/s_R^2 to the F distribution with $k-1=3$ and $(n-1)(k-1)=12$ degrees of freedom. In this example the ratio is $23.3/18.8 = 1.24$. Reference to the F tables with 3 and 12 degrees of freedom shows that this value of F would not be unusual if the null hypothesis was true, and hence the data do not contradict the hypothesis that the means of the four treatments are equal (see Figure 7.4a). In fact, given this hypothesis, we find $\Pr(s_T^2/s_R^2 > 1.24) = 0.33$. Thus the four different treatments (variants for the manufacture of penicillin) have not been demonstrated to give different yields. The variability among the treatment averages can be reasonably attributed to experimental error.

Since randomization has been applied to treatments *within* blocks, a similar test applied to compare block means cannot appeal to the justification of randomization. For further illustration, however, we make the assumption that the errors are IIDN($0, \sigma^2$) and proceed to test the hypothesis that all the block means are equal. The ratio of s_B^2 to s_R^2 is $66.0/18.8 = 3.51$, and from the F tables $\Pr(s_B^2/s_R^2 \geq 3.51) = 0.04$ (see also Figure 7.4b). Not unexpectedly, then, the null hypothesis of no blend-to-blend variation is discredited by the experiment.

Increase in Efficiency Due to Elimination of Block Differences

The analysis of variance (Table 7.10) shows the advantage of using the randomized block arrangement. Of the total sum of squares not associated with treatments or with the mean, more than half (264) is accounted for by block-to-block variation. If the experiment had been arranged on a completely random basis with no blocks, the error variance could have been much larger. A completely random arrangement would have been equally *valid*,

---

* We now have the IIDN($0, \sigma^2$) assumption.
† This requires only random allocation of treatments within the blocks.

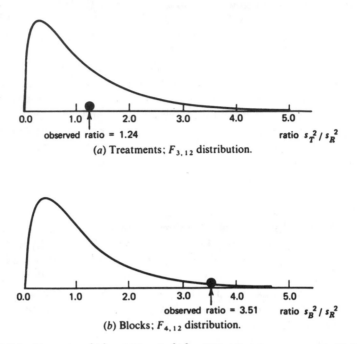

(a) Treatments; $F_{3, 12}$ distribution.

(b) Blocks; $F_{4, 12}$ distribution.

FIGURE 7.4. The ratios $s_T^2/s_R^2 = 1.24$ and $s_B^2/s_R^2 = 3.51$ referred to appropriate F distributions, penicillin example.

but the randomized block arrangement is more *sensitive* for distinguishing real differences from random error.

Exercise 7.7. For the data sets in Exercise 7.1, construct the tables corresponding to Table 7.10
Partial answer: For sums of squares and degrees of freedom, see answers to Exercise 7.2.
(a) $s_B^2 = 150$, $s_T^2 = 600$, $s_R^2 \doteq 14$; (b) $s_B^2 = 0.14$, $s_T^2 = 0.48$, $s_R^2 = 0.18$.

7.6. THE USE OF REFERENCE DISTRIBUTIONS TO COMPARE INDIVIDUAL MEANS

As in Chapter 6, more detailed understanding of what may be concluded about the relative values of treatment means can be obtained by plotting the treatment averages in relation to an appropriate reference distribution or, more formally, by using numerical multiple comparison tests such as those mentioned in Appendix 6C. For comparison of the treatments in the penicillin

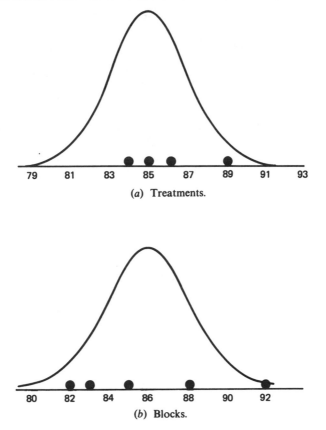

FIGURE 7.5. Reference distribution diagrams, penicilllin example.

example the reference distribution is $t_\nu\sqrt{s^2/n}$, that is, a t distribution with $\nu = (n - 1)(k - 1) = 12$ degrees of freedom, scaled by the factor $\sqrt{s^2/n} = \sqrt{18.8/5} = 1.94$. The four treatment averages are shown in relation to this reference distribution in Figure 7.5a. For the example the figure merely confirms the verdict of the overall F test. Figure 7.5b shows the corresponding picture for the *block* averages, where now the reference distribution is a t distribution with 12 degrees of freedom and scale factor $\sqrt{18.8/4} = 2.17$. This distribution cannot be located so as to make the block averages seem to be a typical sample. If we position the reference distribution so as to make the lowest average, 82, correspond to a reasonably large probability density, then the highest average, 92, is necessarily associated with a low probability density. The evidence for differences among the block means is strong but not overwhelming, and this is precisely the conclusion reached by the overall F test, which shows significance at the 0.04 level.

The overall F test is often misunderstood. An experimenter who found a value significant at the 4 % level might easily jump to the conclusion that there was much more evidence for separation among the means than really exists. Inspection of Figure 7.5b, however, would force to his attention the truth that, although it is likely that differences of some kind among the means exist (e.g., blends 1 and 5 probably have different means), they are not well estimated. In particular, if differences of, say, 5 units were of economic importance, a much larger experiment would be needed to estimate them.

We now consider a different kind of two-way arrangement by means of which two factors and their interaction can be studied.

7.7. A TWO-WAY (FACTORIAL) DESIGN

Look at the data of Table 7.11, which appear in a paper by Box and Cox (1964). These are survival times of groups of four animals randomly allocated to three poisons and four treatments. The experiment was part of an investigation to combat the effects of certain toxic agents. This arrangement is called

TABLE 7.11. Survival times (unit, 10 hours) of animals in a 3 × 4 replicated factorial experiment, toxic agents example

poison	treatment			
	A	B	C	D
I	0.31	0.82	0.43	0.45
	0.45	1.10	0.45	0.71
	0.46	0.88	0.63	0.66
	0.43	0.72	0.76	0.62
II	0.36	0.92	0.44	0.56
	0.29	0.61	0.35	1.02
	0.40	0.49	0.31	0.71
	0.23	1.24	0.40	0.38
III	0.22	0.30	0.23	0.30
	0.21	0.37	0.25	0.36
	0.18	0.38	0.24	0.31
	0.23	0.29	0.22	0.33

a 3×4 factorial design and is replicated four times. There is no blocking, and both factors, poisons and treatments, are of equal interest, as is the possibility that these factors interact. If they interact, they will not behave in the additive manner illustrated in Figure 7.2. Instead, the mean difference in survival times between some treatments will be different for different poisons.

The analysis of the data is best considered in two stages. First we ignore poison and treatment labels and regard the experiment as an example of the one-way arrangement of Chapter 6 with 12 groups of four animals each. The appropriate analysis of variance is then as follows:

source of variation	sum of squares × 1000	degrees of freedom	mean square × 1000
between groups	2205.5	11	200.5
within groups (error)	800.7	36	22.2
total (corrected)	3006.2	47	

With y_{tij} the survival time of the jth animal given the ith poison and the tth treatment, this analysis is associated with a model and a corresponding breakdown of the data as follows:

$$y_{tij} = \eta_{ti} + \epsilon_{tij} \tag{7.14}$$

$$y_{tij} = \bar{y}_{ti} + (y_{tij} - \bar{y}_{ti}) \tag{7.15}$$

Now, if poisons and treatments act additively, $\eta_{ti} = \eta + \tau_t + \beta_i$, where τ_t is the mean increment in survival time associated with the tth treatment, and β_i is the corresponding increment associated with the ith poison. If, however, there is interaction, an additional increment $\omega_{ti} = \eta_{ti} - \eta - \tau_t - \beta_i$ is needed to make the equation balance.* This part of the model and the related data breakdown are therefore as follows:

$$\eta_{ti} = \eta + \quad \tau_t \quad + \quad \beta_i \quad + \quad \omega_{ti} \tag{7.16}$$

$$\bar{y}_{ti} = \bar{y} + (\bar{y}_t - \bar{y}) + (\bar{y}_i - \bar{y}) + (\bar{y}_{ti} - \bar{y}_t - \bar{y}_i + \bar{y}) \tag{7.17}$$

In this decomposition τ_t and β_i are called the *main effects* of treatments and poisons, and the ω_{ti} the *interaction effects*.

The arithmetic of the further analysis of the group averages now parallels that for the randomized block design (see Equations 7.1 and 7.2). Group averages $\bar{y}_{ti}$ replace basic data, and an interaction sum of squares replaces the residual sum of squares of the randomized block analysis. Also, since the $\bar{y}_{ti}$

* ω is the Greek letter *omega*.

TABLE 7.12. Analysis of variance table for replicated two-way factorial arrangement, toxic agents example

source of variation	sum of squares $\times 1000$	degrees of freedom	mean square $\times 1000$	expected value of mean square	ratio of mean squares
poisons	$S_p = 1033.0$	$n - 1 = 2$	$s_p^2 = 516.5$	$\sigma^2 + mk \sum \beta_i^2/(n-1)$	$s_p^2/s_e^2 = 23.2$
treatments	$S_T = 922.4$	$k - 1 = 3$	$s_T^2 = 307.5$	$\sigma^2 + mn \sum \tau_t^2/(k-1)$	$s_T^2/s_e^2 = 13.8$
interaction	$S_I = 250.1$	$(n-1)(k-1) = 6$	$s_I^2 = 41.7$	$\sigma^2 + m \sum \sum \omega_{ti}^2/(n-1)(k-1)$	$s_I^2/s_e^2 = 1.9$
error	$S_e = 800.7$	$nk(m-1) = 36$	$s_e^2 = 22.2$		
total (corrected)	$S = 3006.2$	$nkm - 1 = 47$			

are averages of four observations, the sums of squares have 4 as an additional multiplier. The between-group sum of squares is thus further analyzed as follows:

source of variation	sum of squares × 1000	degrees of freedom	mean square × 1000
poisons	1033.0	2	516.5
treatments	922.4	3	307.5
interaction (residual)	250.1	6	41.7
between groups	2205.5	11	

Combining the two analyses, we obtain Table 7.12, where, in general, it is supposed that there are n levels of some factor P ($n = 3$ poisons), k levels of some factor T ($k = 4$ treatments), and m replications ($m = 4$ animals per group). Then the corresponding sums of squares, S_P for factor P, S_T factor T, S_I for the interaction between P and T, S_e for error, and S for total, are given by the formulas

$$S_P = mk \sum_i (\bar{y}_i - \bar{y})^2, \qquad S_T = mn \sum_t (\bar{y}_t - \bar{y})^2,$$

$$S_I = m \sum_t \sum_i (\bar{y}_{ti} - \bar{y}_t - \bar{y}_i + \bar{y})^2$$

$$S_e = \sum_t \sum_i \sum_j (y_{tij} - \bar{y}_{ti})^2, \qquad S = \sum_t \sum_i \sum_j (y_{tij} - \bar{y})^2$$

Exercise 7.8. Make a decomposition of the observations for the poisons–treatments example that parallels the one in Table 7.3. There should be six tables of elements associated with the original observations, the grand average, the poison main effects, the treatment main effects, the interaction, and the residuals. Confirm to your own satisfaction (a) that the various components are orthogonal, (b) that the sums of squares of the individual elements in the tables give correctly the sums of squares of the analysis of variance table, and (c) that the allocated numbers of degrees of freedom are logical.

7.8. SIMPLIFICATION AND INCREASED SENSITIVITY FROM TRANSFORMATION

If we could assume that the model was adequate and, in particular, that the errors ϵ_{tij} in Equation 7.14 were normally and independently distributed with constant variance, then, using the F table with Table 7.11, the effects of poisons

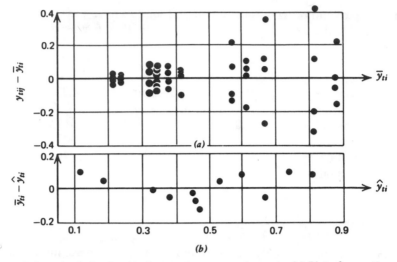

FIGURE 7.6. Analysis of residuals, experiment on toxic agents. (a) Plot of $y_{tij} - \bar{y}_{ti}$ versus $\bar{y}_{ti}$, showing funnel shape. (b) Plot of $\bar{y}_{ti} - \hat{y}_{ti}$ versus $\hat{y}_{ti}$, showing curvilinear tendency.

and treatments would be judged highly significant. This was expected. The principal object of the experiment was to estimate these effects.

Now the poisons were variants of the same substance, and the treatments were also similar to one another. It would have been convenient and not very surprising, therefore, if these two factors had behaved additively. The analysis of variance table shows, however, that there is some suggestion of interaction between poisons and treatments.

As soon as an analysis of residuals is carried out for these data, however, it is immediately obvious that the model considered above is dramatically inadequate. Figure 7.6a shows a plot of $y_{tij} - \bar{y}_{ti}$ versus $\bar{y}_{ti}$. The funnel-shaped plot strongly suggests that, contrary to assumption, the standard deviation σ increases as the mean value η increases. Furthermore, if it is supposed that interaction terms might be dispensed with, the estimated values for the cell means would be $\hat{y}_{ti} = \bar{y}_t + \bar{y}_i - \bar{y}$. The quantities $\bar{y}_{ti} - \hat{y}_{ti}$ obtained by subtracting these estimated values from the cell averages are plotted against $\hat{y}_{ti}$ in Figure 7.6b. The tendency to a curvilinear relation suggests the existence of transformable nonadditivity.

Variance Stabilizing Transformations

The residual analysis suggests that, contrary to assumption, σ_y is a function of η. If this is so, it may be possible to find a convenient data transformation $Y = f(y)$ that has constant variance.

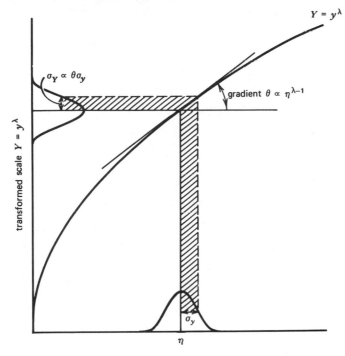

FIGURE 7.7.   Transmission of error from y to $Y = y^\lambda$.

Suppose that the standard deviation σ_y of y is proportional to some *power* of the mean η of y,*

$$\sigma_y \propto \eta^\alpha$$

and that we make a power transformation of the data,

$$Y = y^\lambda$$

Then (see Figure 7.7)

$$\sigma_Y = \theta\sigma_y \propto \theta\eta^\alpha$$

and θ, the gradient of the graph of Y plotted against y, depends on the mean value η of y and so might more appropriately be denoted by θ_η.

Now it can be shown that, if $Y = y^\lambda$, the gradient θ_η is proportional to $\eta^{\lambda-1}$. Thus

$$\sigma_Y \propto \eta^{\lambda-1}\eta^\alpha = \eta^{\lambda+\alpha-1}$$

* α and λ are the Greek letters *alpha* and *lambda*, and $\propto$ stands for "is proportional to."

TABLE 7.13. **Variance stabilizing transformations when** $\sigma_y \propto \eta^\alpha$

Dependence of σ_y on η	α	$\lambda = 1 - \alpha$	variance stabilizing transformation	example
$\sigma \propto \eta^2$	2	-1	reciprocal	
$\sigma \propto \eta^{3/2}$	$1\frac{1}{2}$	$-\frac{1}{2}$	reciprocal square root	
$\sigma \propto \eta$	1	0	log	sample variance
$\sigma \propto \eta^{1/2}$	$\frac{1}{2}$	$\frac{1}{2}$	square root	Poisson frequency
$\sigma \propto$ constant	0	1	no transformation	

Thus Y is chosen so that σ_Y does not depend on η, if

$$\lambda = 1 - \alpha$$

By taking the mathematical limit appropriately, one can also show that a value $\lambda = 0$ implies the appropriateness of the log transformation. Some values of α with appropriate transformations are summarized in Table 7.13.

In Chapter 5 the log transformation was used for a sample variance s^2 and the square root transformation for a Poisson frequency. These transformations make direct use of the results set out in Table 7.13.

The extent of the curvature that a transformation produces over a given range may be regarded as determining its "strength." Of the transformations specifically considered above, the reciprocal is the strongest and the square root the mildest. For a binomial proportion y/n the mean value is p, and the standard deviation is proportional to $\sqrt{pq} = \sqrt{p - p^2}$. The standard deviation is thus related to the mean, but not by a power law. By an extension of the argument given above, however, it may be shown that the appropriate variance stabilizing transformation is such that

$$\sin x = \sqrt{\hat{p}} = \sqrt{y/n}$$

so that

$$x = \sin^{-1}\sqrt{\hat{p}}$$

We have used this transformation in Chapter 5.

Empirical Determination of α

For replicated data a value for α can sometimes be found empirically. Suppose that for the jth set of experimental conditions $\sigma_j \propto \eta_j^\alpha$; then $\log \sigma_j = \text{const} + \alpha \log \eta_j$. Thus $\log \sigma_j$ would yield a straight line plot against $\log \eta_j$ with slope α. In practice we do not know σ_j and η_j, but estimates s_j and $\bar{y}_j$ may be sub-

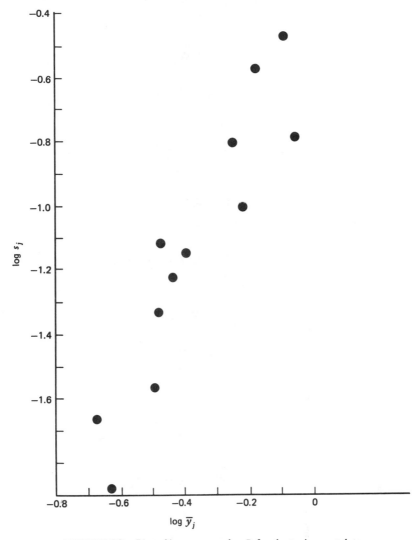

FIGURE 7.8. Plot of log s_j versus log $\bar{y}_j$ for the toxic agent data.

stituted. Illustrating with the animal data, we show in Figure 7.8 the plot of log s_j versus log $\bar{y}_j$ for the 12 cells in Table 7.11. The slope α of a line drawn through the points is evidently close to 2. From Table 7.13 a value $\alpha = 2$ implies the need for reciprocal transformation. We shall see shortly that an analysis carried out for reciprocal survival time (rate of dying) has great advantages.

TABLE 7.14. Calculation of nonadditivity sum of squares, toxic agents example

	I				II				III			
	A	B	C	D	A	B	C	D	A	B	C	D
$\bar{y}_{ti}$	0.4125	0.8800	0.5675	0.6100	0.3200	0.8150	0.3750	0.6675	0.2100	0.3350	0.2350	0.3250
$\hat{y}_{ti}$	0.4523	0.8148	0.5306	0.6723	0.3792	0.7417	0.4575	0.5992	0.1111	0.4736	0.1894	0.3311
$q_{ti} = \hat{y}_{ti}^2$	0.2046	0.6639	0.2816	0.4520	0.1438	0.5501	0.2093	0.3590	0.0123	0.2243	0.0359	0.1096
$\bar{y}_{ti} - \hat{y}_{ti}$	−0.0398	0.0652	0.0369	−0.0623	−0.0592	0.0733	−0.0825	0.0683	0.0989	−0.1386	0.0456	−0.0061
$q_{ti} - \hat{q}_{ti}$	−0.0456	0.0545	−0.0240	0.0151	−0.0215	0.0256	−0.0113	0.0071	0.0671	−0.0802	0.0353	−0.0223

$P = 4\sum(\bar{y}_{ti} - \hat{y}_{ti})(q_{ti} - \hat{q}_{ti}) = 0.1104,\ Q = 4\sum(q_{ti} - \hat{q}_{ti})^2 = 0.0793$

$S_{na} = P^2/Q = 0.1537$

Nonadditivity

Tukey's test may be performed for the cell averages. The required quantities are set out in Table 7.14 for the untransformed data.

The interaction sum of squares is then split into components to give the following analysis:

source	sum of squares × 1000	degrees of freedom	mean square × 1000
interaction $\begin{cases} \text{transformable} \\ \text{remainder} \end{cases}$	$250.1 \begin{cases} 153.7 \\ 96.4 \end{cases}$	$6 \begin{cases} 1 \\ 5 \end{cases}$	$41.7 \begin{cases} 153.7 \\ 19.3 \end{cases} F = 6.9$
error	800.7	36	22.2

It is seen that there is strong evidence of transformable interaction in the data. A similar analysis carried out on reciprocals of the data, however, shows no indication of transformable interaction.

Advantages of the Transformation

The mean squares and degrees of freedom for the analyses of variance for untransformed and reciprocally transformed data are shown in Table 7.15. The fact that the data have been used to choose the transformation is approximately allowed for (see Box and Cox, 1964) by reducing the number of degrees of freedom in the within-group mean square by unity (from 36 to 35).

TABLE 7.15. Analyses of variance for untransformed and transformed data, toxic agents example

	untransformed		transformed $Y = y^{-1}$	
	degrees of freedom	mean square × 1000	degrees of freedom	mean square × 1000
poisons (P)	2	516.5	2	1743.9
treatments (T)	3	307.5	3	680.5
$P \times T$ interaction	$6 \begin{cases} 1 \\ 5 \end{cases}$	$41.7 \begin{cases} 153.7 \\ 19.3 \end{cases}$	6	$26.2 \begin{cases} 45.4 \\ 22.3 \end{cases}$
within groups (error)	36	22.2	35	24.7

For this particular example the effects of transformation are very striking. An analysis is obtained in the reciprocal scale that is more sensitive and easier to interpret. Specifically:

1. The mean squares associated with poisons and treatments are now much larger relative to the within-group (error) mean square. The sensitivity of the experiment has been increased almost threefold. This is equivalent to increasing the size of the experiment by a factor of nearly three.
2. The poison × treatment interaction mean square that previously gave some indication of significance is now much closer in size to the error mean square.

Thus the effects of poison and treatments on *rates of dying* ($Y = 1/y$) are roughly additive. This allows a very simple interpretation of the data in which the effects of poisons and treatments can be entirely described in terms of their main effects.

The greater sensitivity of the analysis in the transformed metric is illustrated in Figure 7.9, which shows reference distribution diagrams for the comparisons of poisons I, II, and III for the data before and after transformation. To facilitate comparison, the scales are chosen so that averages cover the same range on both scales. In the upper diagram the reference distribution is a t distribution with $v = 36$, scaled by $\hat{\sigma}_{\bar{y}_i} = \sqrt{0.0222/16} = 0.037$. In the lower distribution $v = 35$, and the scale factor is $\hat{\sigma}_{\bar{Y}_i} = \sqrt{0.242/16} = 0.123$.

Exercise 7.9. Carry through the analysis for $Y = y^{-1}$. Plot residuals, and consider whether in the new metric there is evidence of model inadequacy of any kind. Compare *treatment* averages in the two scales as is done for poison averages in Figure 7.9.

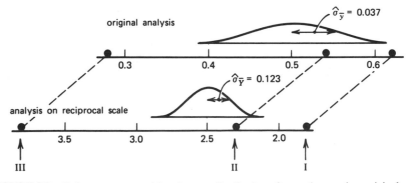

FIGURE 7.9. Poison averages with reference distributions for analyses using original and reciprocal scales, toxic agents example.

7.9. LIKELIHOOD ESTIMATION OF THE TRANSFORMATION

Most of the methods of analysis we discuss are totally appropriate and efficient when the models (*a*) are *structurally adequate*, and the (supposedly independent) errors (*b*) have *constant variance* and (*c*) are *normally distributed*. The possibility of transforming the response *y* makes these three desiderata less burdensome and greatly broadens the class of problems to which the methods may be applied, for we then need ask only that the requirements be approximately satisfied *for some transformation* of *y* and not necessarily for the measurement *y* itself.

In the toxic agents example we considered how analysis of residuals could show (*a*) structural inadequacy (Tukey's analysis revealed transformable nonadditivity) and (*b*) transformable variance inequality. Each of these separately could throw some light on a suitable choice of transformation. Furthermore, had the data been more numerous, an analysis of residuals might also have revealed transformable nonnormality, which could have further facilitated the choice of transformation.

It is natural to seek a transformation that, so far as possible, both satisfies and combines information from the three desiderata simultaneously. Box and Cox (1964) showed how this could be done. The method is applicable for almost any kind of statistical model and any kind of transformation, but for illustration consider again the choice of a power transformation y^λ for the toxicity data. It is tentatively assumed that an *additive* model with *normally* distributed errors having *constant* variance is appropriate for *some* y^λ. The usual parameters of the model (such as row and column means) can then be estimated *simultaneously* with λ by the method of maximum likelihood. It turns out that this can be done as follows. For various values of λ perform a standard analysis on

$$y^{(\lambda)} = \frac{y^\lambda - 1}{\lambda \dot{y}^{\lambda-1}}, \qquad (y^{(0)} = \dot{y} \ln y)$$

The value $\lambda = 0$ corresponds to the log transformation, and it may be shown that $y^{(0)} = \dot{y} \ln y$. In these expressions the quantity $\dot{y}$ is the geometric mean* of all the data. The maximum likelihood value for λ is that for which the residual sum of squares (S_λ, say) from the fitted model is minimized.

For the toxicity data the sum of squares S_λ, after fitting rows and columns only, has 42 degrees of freedom. Values of S_λ for various values of λ are as follows:

λ	-2.5	-2.0	-1.6	-1.4	-1.2	-1.0	-0.8	-0.6	-0.4	-0.2	0.0	0.5	1.00
S_λ	1.333	0.664	0.463	0.401	0.359	0.333	0.323	0.326	0.343	0.375	0.424	0.635	1.051

Values close to the minimum are plotted in Figure 7.10, yielding the estimate $\hat{\lambda} = -0.75$. An approximate 95% confidence interval may also be obtained from the graph by calculating a critical sum of squares S from

$$S = S_\lambda \left[1 + \frac{t_\nu^2(0.025)}{\nu} \right]$$

In this expression $S = 0.322(1 + 4.08/42) = 0.353$. The values $\lambda_- = -1.18$ and $\lambda_+ = -0.32$ read off from the curve now provide the required confidence limits. Notice

* The geometric mean is obtained by averaging log *y* and taking the antilog of the result.

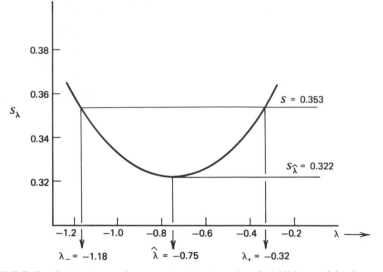

FIGURE 7.10. Residual sum of squares S_λ plotted against λ. Additive model using as data $y^{(\lambda)} = (y^\lambda - 1)/\lambda \dot{y}^{\lambda - 1}$, toxic agents example.

that these limits include the value $\lambda = -1$, and in practice the easily understood reciprocal was actually used.

In the analysis we have presented, information on additivity, variance stability, and normality is appropriately combined to estimate λ. Box and Cox (1964) discuss in their paper how each of these attributes may be separately studied and related to the simpler methods we have discussed here.

General Comments on Transformations

The possibility of transformation should always be kept in mind. Often there is nothing in particular to recommend the original metric in which the measurements happen to be taken. A research worker studying athletics may measure the time t in seconds that a subject takes to run 1000 meters, but he could equally well have considered $1000/t$, which is the athlete's speed in meters per second. If the times vary over a sufficiently wide range, the data may be able to indicate that the model assumptions are more nearly met in a particular metric. The *analysis* should be conducted in that metric. After analysis the results may be transformed back and reported in whatever scale is most easily understood.

In the toxic agents example, the transformation had a profound effect. There the data covered a wide range of values and a strong (reciprocal) transformation was found to be appropriate. If the data had covered a narrow range or the appropriate transformation had been milder, a plot of Y versus y would have been much more nearly linear. A linear recoding $Y = (y - c)/k$ (such as is often used to simplify calculations) has no effect on standard analyses. For example, t and F values are unaffected, and confidence intervals for Y transfer back identically to confidence intervals for y. Correspondingly, a *nearly* linear recoding such as is produced by mild transformation over a narrow range has little effect on the analysis. In particular transformations of the kind listed in Table 7.13 are unlikely to produce much effect unless y_{max}/y_{min} is moderately large (say, greater than three).

7.10. SUMMARY

The reader should reread the summary for Chapter 6 (Section 6.8) because most of the points made there are equally applicable to the present chapter. An important new idea is error reduction by random application of treatments within blocks of relatively homogeneous material.

In the two-way replicated factorial design both variables are of equal interest, as is their interaction. As with the randomized block arrangement, it is important to check the model by analysis of residuals—in particular, to consider whether transformation might improve model adequacy and possibly facilitate the interpretation of results.

The need for transformation can be indicated by a relationship between cell variances and cell averages. Transformable interaction (nonadditivity) is indicated by a quadratic tendency in the plot of $y - \hat{y}$ versus $\hat{y}$. When appropriate transformation is made for data covering a wide range of values, a considerable gain in precision is possible.

APPENDIX 7A. CALCULATIONS FOR CONSTRUCTING ANALYSIS OF VARIANCE TABLE FOR RANDOMIZED BLOCK DESIGN

METHOD 1
Using the deviations from Table 7.7 in formulas given in Table 7.5

$$S_A = (5)(4)(86)^2 = 147,920$$
$$S_B = 4[(6)^2 + (-3)^2 + (-1)^2 + (2)^2 + (-4)^2] = 264$$
$$S_T = 5[(-2)^2 + (-1)^2 + (3)^2 + (0)^2] = 70$$
$$S_R = (-1)^2 + (-3)^2 + (2)^2 + \cdots + (-5)^2 + (6)^2 = 226$$
$$S = (89)^2 + (88)^2 + \cdots + (80)^2 + (88)^2 = 148,480$$
$$= S_B + S_T + S_R + S_A$$

METHOD 2

Using row and column averages

$$S_A = nk\bar{y}^2 = 147{,}920$$

$$S_B = k \sum_{i=1}^{n} (\bar{y}_i - \bar{y})^2 = k\left(\sum_{i=1}^{n} \bar{y}_i^2 - n\bar{y}^2\right) = k \sum_{i=1}^{n} \bar{y}_i^2 - nk\bar{y}^2$$

$$= 4[(92)^2 + (83)^2 + (85)^2 + (88)^2 + (82)^2] - 5(4)(86)^2 = 264$$

$$S_T = n \sum_{t=1}^{k} (\bar{y}_t - \bar{y})^2 = n\left(\sum_{t=1}^{k} \bar{y}_t^2 - k\bar{y}^2\right) = n \sum_{t=1}^{k} \bar{y}_t^2 - nk\bar{y}^2$$

$$= 5[(84)^2 + (85)^2 + (89)^2 + (86)^2] - 147{,}920 = 70$$

$$S = \sum_{t=1}^{k} \sum_{i=1}^{n} y_{ti}^2 = 148{,}480$$

$$S_R = S - S_B - S_T - S_A = 226$$

METHOD 3

Using row and column totals

$$S_A = nk\bar{y}^2 = nk\left(\frac{1}{nk} \sum_{t=1}^{k} \sum_{i=1}^{n} y_{ti}\right)^2 = \frac{(\sum_{t=1}^{k} \sum_{i=1}^{n} y_{ti})^2}{nk}$$

$$= \frac{(\text{grand total})^2}{\text{total number of observations}} = \frac{(1{,}720)^2}{20} = 147{,}920.$$

$$S_B = k \sum_{i=1}^{n} (\bar{y}_i - \bar{y})^2 = k \sum_{i=1}^{n} \bar{y}_i^2 - nk\bar{y}^2 = k \sum_{i=1}^{n} \left(\frac{1}{k} \sum_{t=1}^{k} y_{ti}\right)^2 - S_A$$

$$= \frac{1}{k} \sum_{i=1}^{n} \left(\sum_{t=1}^{k} y_{ti}\right)^2 - S_A$$

$$= \frac{1}{k} \sum_{\text{blocks}} (\text{block total})^2 - S_A = \frac{1}{k} \sum_{\text{rows}} (\text{row sum})^2 - S_A$$

$$= \tfrac{1}{4}[(368)^2 + (332)^2 + (340)^2 + (352)^2 + (328)^2] - 147{,}920$$

$$= 264$$

Similarly

$$S_T = \frac{1}{n} \sum_{\text{columns}} (\text{column sum})^2 - S_A$$

$$= \tfrac{1}{5}[(420)^2 + (425)^2 + (445)^2 + (430)^2] - 147{,}490 = 70$$

$$S = (89)^2 + (88)^2 + \cdots + (80)^2 + (88)^2 = 148{,}480$$

$$S_R = S - S_B - S_A - S_T = 226$$

APPENDIX 7B. ALGEBRAIC DEMONSTRATION OF THE ADDITIVITY OF THE SUMS OF SQUARES IN A RANDOMIZED BLOCK

Consider the decomposition

$$y_{ti} = \bar{y} + (\bar{y}_i - \bar{y}) + (\bar{y}_t - \bar{y}) + (y_{ti} - \bar{y}_i - \bar{y}_t + \bar{y}) \tag{7B.1}$$

We make use of this identity:

$$\sum_t^k \sum_i^n y_{ti}^2 = \sum_t^k \sum_i^n (y_{ti} - \bar{y})^2 + nk\bar{y}^2 \tag{7B.2}$$

Subtracting $\bar{y}$ from both sides of Equation 7B.1 and summing the squares $(y_{ti} - \bar{y})^2$ over all t and i, we have

$$\sum_t^k \sum_i^k (y_{ti} - \bar{y})^2 = \sum_t^k \sum_i^n (\bar{y}_i - \bar{y})^2 + \sum_t^k \sum_i^n (\bar{y}_t - \bar{y})^2 + \sum_t^k \sum_i^n (y_{ti} - \bar{y}_i - \bar{y}_t + \bar{y})^2$$

$$+ 2 \sum_t^k \sum_i^n (\bar{y}_i - \bar{y})(\bar{y}_t - \bar{y})$$

$$+ 2 \sum_t^k \sum_i^n (\bar{y}_i - \bar{y})(y_{ti} - \bar{y}_i - \bar{y}_t + \bar{y})$$

$$+ 2 \sum_t^k \sum_i^n (\bar{y}_t - \bar{y})(y_{ti} - \bar{y}_i - \bar{y}_t + \bar{y}) \tag{7B.3}$$

Then

$$\sum_t^k \sum_i^n (y_{ti} - \bar{y})^2 = k \sum_i^n (\bar{y}_i - \bar{y})^2 + n \sum_t^k (\bar{y}_t - \bar{y})^2 + \sum_t^k \sum_i^n (y_{ti} - \bar{y}_i - \bar{y}_t + \bar{y})^2 \tag{7B.4}$$

The right-hand side of Equation 7B.4 comes from the first three terms on the right-hand side of Equation 7B.3. The last three terms (the cross-product terms) vanish because

$$2 \sum_t^k \sum_i^n (\bar{y}_i - \bar{y})(\bar{y}_t - \bar{y}) = 2 \sum_i^n (\bar{y}_i - \bar{y}) \sum_t^k (\bar{y}_t - \bar{y})$$

$$= 2n(\bar{y} - \bar{y})k(\bar{y} - \bar{y}) = 0 \tag{7B.5}$$

$$2 \sum_t^k \sum_i^n (\bar{y}_i - \bar{y})(y_{ti} - \bar{y}_i - \bar{y}_t + \bar{y}) = 2 \sum_i^n (\bar{y}_i - \bar{y}) \sum_t^k (y_{ti} - \bar{y}_i - \bar{y}_t + \bar{y})$$

$$= 2 \sum_i^n (\bar{y}_i - \bar{y})k(\bar{y}_i - \bar{y}_i - \bar{y} + \bar{y}) = 0 \tag{7B.6}$$

and

$$2 \sum_t^k \sum_i^n (\bar{y}_t - \bar{y})(\bar{y}_{ti} - \bar{y}_i - \bar{y}_t + \bar{y}) = 2 \sum_t^k (\bar{y}_t - \bar{y}) \sum_i^n (y_{ti} - \bar{y}_i - \bar{y}_t + \bar{y})$$

$$= 2 \sum_t^k (\bar{y}_t - \bar{y})n(\bar{y}_t - \bar{y}_t - \bar{y} + \bar{y}) = 0 \tag{7B.7}$$

Therefore, finally, substituting Equation 7B.4 into Equation 7B.2, we have

$$\sum_t^k \sum_i^n y_{ti}^2 = nk\bar{y}^2 + k\sum_i^n (\bar{y}_i - \bar{y})^2 + n\sum_t^k (\bar{y}_t - \bar{y})^2 + \sum_t^k \sum_i^n (y_{ti} - \bar{y}_i - \bar{y}_t + \bar{y})^2 \quad (7B.8)$$

which is the decomposition

$$S = S_A + \quad S_B \quad + \quad S_T \quad + \quad S_R \quad\quad (7B.9)$$

(Notice that the cross-product terms vanish because the corresponding vectors are orthogonal.)

REFERENCES AND FURTHER READING

Anscombe, F.J. and Tukey, T.W. (1963). The examination and analysis of residuals, *Techno-metrics*, **5**, 141.

Box, G. E. P., and D. R. Cox (1964). An analysis of transformations, *J. Roy. Stat. Soc.*, Series B, **26**, 211.

Tukey, J. W. (1949). One degree of freedom for non-additivity, *Biometrics*, **5**, 232.

QUESTIONS FOR CHAPTER 7

1. What is a randomized block design?
2. When is it appropriate to use a randomized block design?
3. Can you imagine situations in which you might want to use a randomized block design but would not be able to do so?
4. What is the usual model for a two-way analysis of variance? What are its possible shortcomings? How can diagnostic checks be made to detect possible inadequacies in the model?
5. With data from a randomized block design, how should reference distributions for treatment means and block means be constructed and interpreted?
6. Treating the footwear data of Section 4.2 as a randomized block design, complete the analysis of variance and show its essential equivalence to the paired t approach. Is any aspect of possible interest obtained only from the analysis of variance approach?
7. What is a two-way factorial design? How does it differ from a randomized block design? Describe an experimental situation in which its use would be appropriate.
8. Given a replicated two-way design, how can we use residuals to check (a) constancy of variance and (b) existence of transformable nonadditivity? How may appropriate transformations be found?
9. In which circumstances is transformation of data potentially important? What considerations should enter into the choice of transformation?

CHAPTER 8

Designs with More
Than One Blocking Variable

Fisher pointed out that by using carefully balanced designs it is often possible to eliminate more than one blocking variable. The earliest arrangement of this kind was the *Latin square*, first used to adjust for fertility differences in two different directions. Later *balanced incomplete blocks* and *Youden squares* were introduced, by F. Yates and W. J. Youden, respectively, for use when the number of treatments to be compared is greater than the block size.

8.1. LATIN SQUARE DESIGNS: AUTOMOBILE EMISSIONS AND SYNTHETIC YARN EXAMPLES

Suppose that we wish to determine the effect of five different fertilizers A, B, C, D, and E on the yield of potatoes and that we have a piece of land which can be subdivided into a 5×5 grid of 25 plots (neither the piece of land nor the individual plots need be geometrically square). The Latin square arrangement shown in Figure 8.1 has the property that each row and each column receives each treatment (fertilizer) exactly once. Now, for example, ground moisture may vary across the field in one direction, and a fertility gradient occur in the other direction. Such contributions to the experimental environment can be largely eliminated by using this design.

Example: Reduction of Automobile Emissions

Latin square designs may also be used when the two kinds of blocking variables are of a different nature. For example, suppose that four cars and four drivers are employed in a study of possible differences between four gasoline additives. Even though the cars are identical models, slight systematic differences are likely to occur in their performance, and even though

columns

	1	2	3	4	5
I	A	B	C	D	E
II	C	D	E	A	B
rows III	E	A	B	C	D
IV	B	C	D	E	A
V	D	E	A	B	C

FIGURE 8.1.   5 × 5 Latin square design for fertilizer study.

each driver may do his best to drive the car in the manner required by the test, slight systematic differences can occur from driver to driver. It would be desirable to eliminate both the car-to-car and driver-to-driver differences.

The Latin square arrangement shown in Figure 8.2 allows this to be done on the assumption that the effects of treatments, cars, and drivers are approximately additive or, equivalently, that no appreciable interaction occurs between them.

Exercise 8.1.   Write a 3 × 3 and a 7 × 7 Latin square design.

Answer: See Appendix 8A.

The appropriate additive model for the Latin square is

$$y_{ijt} = \eta + \beta_i + \gamma_j + \tau_t + \epsilon_{ijt} \tag{8.1}$$

where η is the mean, β_i are the $i = 1, 2, \ldots, k$ row effects, γ_j are the $j = 1, 2, \ldots, k$ column effects, and τ_t are the $t = 1, 2, \ldots, k$ treatment effects.

cars

	1	2	3	4	
I	A	B	D	C	
II	D	C	A	B	
drivers III	B	D	C	A	additives A, B, C, D
IV	C	A	B	D	

FIGURE 8.2.   4 × 4 Latin square for automobile exhaust emission study.

A collection of Latin square designs for $k = 3, 4, \ldots, 9$ is given in Appendix 8A. Just as with randomized blocks, it is important that these designs be appropriately randomized. This is done by first randomly selecting one design from the basic set given in Appendix 8A and then randomly assigning the row, column, and letters to, for example, the drivers, cars and treatments. Significance tests and confidence intervals are derived on the assumption that the errors ϵ_{ijt} are IIDN(0, σ^2). Alternatively, they can be approximately justified on randomization theory.

To understand the analysis of variance associated with a Latin square arrangement we use a set of whole numbers constructed for the purpose of illustration. Suppose that the automobile emission study yielded the results shown in Table 8.1, where the response is a coded measure of the reduction in oxides of nitrogen. Table 8.2 shows the orthogonal decomposition of the numbers. The corresponding analysis of variance table is given as Table 8.3. Notice that the mean square for drivers is 27 times, and the mean square for cars is 3 times the residual mean square. Elimination of such factors greatly contributes to the reduction of the experimental error variance.

One should, of course, consider the adequacy of the implied additive model. Unfortunately, for this example with so few degrees of freedom associated with the residuals, only massive discrepancies from assumption would

TABLE 8.1.   Results for reduction in oxides of nitrogen from 4 × 4 Latin square design, automobile emissions example

		car				average		
		1	2	3	4	cars	drivers	additives
	I	A 21	B 26	D 20	C 25	1: 19	I: 23	A: 18
	II	D 23	C 26	A 20	B 27	2: 20	II: 24	B: 22
driver	III	B 15	D 13	C 16	A 16	3: 19	III: 15	C: 21
	IV	C 17	A 15	B 20	D 20	4: 22	IV: 18	D: 19

grand average = 20

TABLE 8.2. Breakup of data for the Latin square design, automobile emissions example

observations

A	21	B	26	D	20	C	25
D	23	C	26	A	20	B	27
B	15	D	13	C	16	A	16
C	17	A	15	B	20	D	20

=

grand average

20	20	20	20
20	20	20	20
20	20	20	20
20	20	20	20

+

rows (drivers)

3	3	3	3
4	4	4	4
-5	-5	-5	-5
-2	-2	-2	-2

+

columns (cars)

-1	0	-1	2
-1	0	-1	2
-1	0	-1	2
-1	0	-1	2

+

treatments (additives)

-2	2	-1	1
-1	1	-2	2
2	-1	1	-2
1	-2	2	-1

+

residuals

1	1	-1	-1
1	1	-1	-1
-1	-1	1	1
-1	-1	1	1

vector	$\mathbf{Y}$	=	$\mathbf{A}$	+	$\mathbf{B}$	+	$\mathbf{C}$	+	$\mathbf{T}$	+	$\mathbf{R}$
sum of squares	6696	=	6400	+	216	+	24	+	40	+	16
degrees of freedom	16	=	1	+	3	+	3	+	3	+	6

248

TABLE 8.3. Analysis of variance table for Latin square design, automobile emissions example

$$\text{model: } y_{ijt} = \eta + \beta_i + \gamma_j + \tau_t + \epsilon_{ijt}$$
$$\epsilon_{ijt} \sim \text{IIDN}(0, \sigma^2)$$

hypotheses tested: (i) all τ_t are zero, (ii) all γ_j are zero, (iii) all β_i are zero

source of variation	sum of squares	degrees of freedom	mean square	expected value mean square	ratio of mean squares
grand average	6400	1	$s_A^2 = 6400.00$		
rows (drivers)	216	3	$s_B^2 = 72.00$	$\sigma^2 + k \sum \beta_i^2/(k-1)$	$s_B^2/s_R^2 = 27.0$
columns (cars)	24	3	$s_C^2 = 8.00$	$\sigma^2 + k \sum \gamma_j^2/(k-1)$	$s_C^2/s_R^2 = 3.0$
treatments (additives)	40	3	$s_T^2 = 13.33$	$\sigma^2 + k \sum \tau_t^2/(k-1)$	$s_T^2/s_R^2 = 5.0$
residuals	16	6	$s_R^2 = 2.67$	σ^2	
total	6696	16			

be apparent. Assuming model adequacy, one obtains for the ratio of treatment and residual mean squares an $F_{3,6}$ value of 5.0, which is significant at the 0.045 level.

A plot of the individual treatment averages, along with their appropriate $t_6\sqrt{2.67/4}$ reference distribution, is shown in Figure 8.3. This display emphasizes that, although the differences are not readily explained by chance, their configuration remains extremely vague. Reference distributions for drivers and for cars are shown in Figure 8.4.

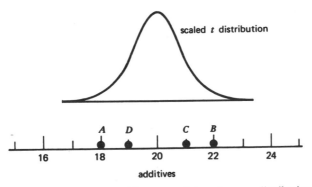

FIGURE 8.3. Treatment averages with appropriate reference distribution, automobile emissions example.

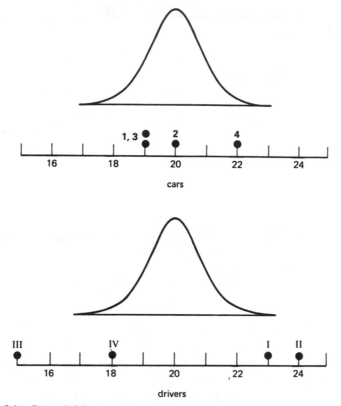

FIGURE 8.4. Car and driver averages with appropriate reference distributions, automobile emissions example.

The 4 × 4 Latin square used in this example yields an error sum of squares based on only six degrees of freedom. The residual number of degrees of freedom is, in general, $(k - 1)(k - 2)$ for a $k \times k$ Latin square, thus:

size of Latin square, k	2	3	4	5	6	7	8
residual degrees of freedom	0	2	6	12	20	30	42

When small Latin squares are used, it is often desirable to replicate them, as in the following example.

Example: Spinning Synthetic Yarn

The data of Table 8.4 were obtained in an experiment to determine to what extent changes in "draw ratio" affected the yarn breaking strength in an

TABLE 8.4. A 3 × 3 Latin square in four replicates: synthetic yarn example

doff		I	II	III	doff total	doff average	doff average minus grand average
			spinneret head				
1	1	A 19.56	B 23.16	C 29.72	72.44	24.14667	2.28195
	2	B 22.94	C 27.51	A 23.71	74.16	24.72000	2.85528
	3	C 25.06	A 17.70	B 22.32	65.08	21.69333	−0.17139
					211.68		
2	4	B 23.24	C 23.54	A 18.75	65.53	21.84333	−0.02139
	5	A 16.28	B 22.29	C 28.09	66.66	22.22000	0.35528
	6	C 18.53	A 19.89	B 20.42	58.84	19.61333	−2.25139
					191.03		
3	7	C 23.98	A 20.46	B 19.28	63.72	21.24000	−0.62472
	8	A 15.33	B 23.02	C 24.97	63.32	21.10667	−0.75805
	9	B 24.41	C 22.44	A 19.23	66.08	22.02667	0.16195
					193.12		
4	10	A 16.65	B 22.69	C 24.94	64.28	21.42667	−0.43805
	11	B 18.96	C 24.19	A 21.95	65.10	21.70000	−0.16472
	12	C 21.49	A 15.78	B 24.65	61.92	20.64000	−1.22472
					191.30		
spinneret total		246.43	262.67	278.03	787.13	$N = 36$	
spinneret average		20.53583	21.88917	23.16917		grand average $\bar{y} = 21.86472$	
spinneret average minus grand average		−1.32889	0.02445	1.30445			

	draw ratio	total	draw ratio average	draw ratio average minus grand average
	A	225.29	18.77417	−3.09055
	B	267.38	22.28167	0.41695
	C	294.46	24.53833	2.67961

251

experimental synthetic fiber. Draw ratio determines the tension applied to the yarn as it is spun. Three spinnerets (I, II, III) were employed in this experimental program. Each spinneret supplied yarn to an individual bobbin, and thus three bobbins were obtained each time the machine was doffed. (To "doff" a spinning machine is to replace completely wound bobbins with empty bobbin spools.) At any given time three different draw ratios were being tested: usual (A), 5% increase (B), and 10% increase (C). At each doff the draw ratios were interchanged according to the plan of Table 8.4, which employs a 3×3 Latin square replicated four times. The table also shows the various totals and averages needed for an analysis based on the additive model of Equation 8.1.

Before proceeding, one should inspect and plot the residuals implied by the model.

With $\bar{y}$, $\bar{y}_i - \bar{y}$, $\bar{y}_j - \bar{y}$, and $\bar{y}_t - \bar{y}$ providing estimates of the overall mean and the row, column, and treatment effects, respectively, the individual estimated values $\hat{y}_{ijt}$ are given by

$$\hat{y}_{ijt} = \bar{y} + (\bar{y}_i - \bar{y}) + (\bar{y}_j - \bar{y}) + (\bar{y}_t - \bar{y}) \tag{8.2}$$

and the residuals are $y_{ijt} - \hat{y}_{ijt}$. Estimated values and residuals for this example are displayed in Table 8.5.

Exercise 8.2. Check the residuals and estimated values in Table 8.5, and apply whatever diagnostics checks of the additive model you consider appropriate.

If the additive model is adequate, we can proceed to interpret the analysis of variance table, Table 8.6. From the F table the ratio $s_T^2/s_R^2 = 19.6$ is significant well beyond the 0.001 level, suggesting that real differences in yarn strength occur when the draw ratio is changed.

A plot of the average breaking strengths for the three different draw ratios, along with their appropriate $t_{20}\sqrt{5.16/12}$ reference distribution, is shown in Figure 8.5. In Figure 8.6 the average breaking strength is plotted against the draw ratio. It appears that over the ranges considered there is a steady increase in breaking strength with increased draw ratio.

The analysis of variance table shows that there was little evidence in this experiment of variation between doffs ($s_B^2/s_R^2 = 1.1$). In the light of hindsight it seems that the experimenter was unnecessarily concerned about this potential source of variability. However, the precautions against the possible effects of doffing incurred only modest additional trouble (the draw ratio-spinneret head combinations had to be changed for each doff). Even though the insurance was unnecessary in this instance, precautions of this kind are generally wise.

TABLE 8.5. Predicted values and residuals, synthetic yarn example

a. Predicted values

19.72723	24.58807	28.12473
23.80806	27.41806	22.93390
23.03805	18.62723	23.41473
20.93139	24.54139	20.05723
17.80056	22.66140	26.19806
20.95805	16.54723	21.33473
22.58472	18.17390	22.96140
16.68723	21.54807	25.08473
21.11473	24.72473	20.24057
17.00723	21.86807	25.40473
20.78806	24.39806	19.91390
21.98472	17.57390	22.36140

b. Residuals

−0.16723	−1.42807	1.59527
−0.86806	0.09194	0.77610
2.02195	−0.92723	−1.09473
2.30861	−1.00139	−1.30723
−1.52056	−0.37140	1.89194
−2.42805	3.34277	−0.91473
1.39528	2.28610	−3.68140
−1.35723	1.47193	−0.11473
3.29527	−2.28473	−1.01057
−0.35723	0.82193	−0.46473
−1.82806	−0.20806	2.03610
−0.49472	−1.79390	2.28860

TABLE 8.6. Analysis of variance table, yarn example

source	sum of squares	degrees of freedom	mean square	ratio of mean squares
grand average	17,210.3788	1		
doffs	63.8844	11	$s_B^2 = 5.8077$	$s_B^2/s_R^2 = 1.1$
spinneret heads	41.6174	2	$s_C^2 = 20.8087$	$s_C^2/s_R^2 = 4.0$
draw ratios	202.4839	2	$s_T^2 = 101.2419$	$s_T^2/s_R^2 = 19.6$
residuals R	103.1972	20	$s_R^2 = 5.1599$	
total	17,621.5617	36		

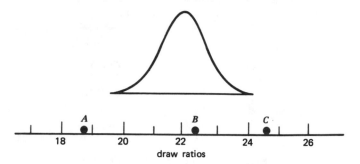

FIGURE 8.5. Reference distribution diagram for draw ratios A, B, C, synthetic yarn example.

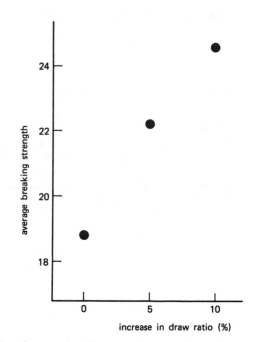

FIGURE 8.6. Plot of average breaking strength versus draw ratio, synthetic yarn example.

The elimination of spinneret heads as a source of variability clearly paid off ($s_C^2/s_R^2 = 4.0$), although the dividend in this example was not particularly large.

Exercise 8.3. Construct reference distribution diagrams for doffs and spinneret heads.

8.2. GRAECO– AND HYPER-GRAECO–LATIN SQUARES: FIRST WEAR TESTING EXAMPLE

To eliminate more than two sources of variability a Graeco–Latin square or hyper-Graeco–Latin square is sometimes useful. A Graeco–Latin square is a $k \times k$ pattern that permits the study of k treatments simultaneously with three different blocking variables. For example, the 4×4 Graeco–Latin square shown in Table 8.7 is an extension of the Latin square design of Table 8.1, in which there is one extra blocking variable, namely. *days* $\alpha, \beta, \gamma, \delta$. It is constructed from the first two 4×4 Latin squares in Appendix 8A.

How to obtain Graeco–Latin and hyper-Graeco–Latin squares more generally is described in Appendix 8A.

Exercise 8.4. Write a 3×3 and a 5×5 Graeco–Latin square.

Answer: See Appendix 8A.

A $k \times k$ hyper-Graeco–Latin square permits the study of k treatments with more than three blocking variables. We illustrate with an experiment on a Martindale wear tester (Table 8.8). This machine is used for testing the wearing quality of types of cloth or other such materials, and it possesses the

TABLE 8.7. A 4 × 4 Graeco–Latin square

		car				
		1	2	3	4	
	I	$A\alpha$	$B\beta$	$C\gamma$	$D\delta$	additives: A, B, C, D
driver	II	$B\delta$	$A\gamma$	$D\beta$	$C\alpha$	days: $\alpha, \beta, \gamma, \delta$
	III	$C\beta$	$D\alpha$	$A\delta$	$B\gamma$	
	IV	$D\gamma$	$C\delta$	$B\alpha$	$A\beta$	

TABLE 8.8. Hyper-Graeco–Latin square replicated twice: first wear testing example

positions

	P_1	P_2	P_3	P_4	*replicate I*
R_1	$\alpha A1$ 320	$\beta B2$ 297	$\gamma C3$ 299	$\delta D4$ 313	
R_2	$\beta C4$ 266	$\alpha D3$ 227	$\delta A2$ 260	$\gamma B1$ 240	treatments: A, B, C, D
R_3	$\gamma D2$ 221	$\delta C1$ 240	$\alpha B4$ 267	$\beta A3$ 252	holders: 1, 2, 3, 4
R_4	$\delta B3$ 301	$\gamma A4$ 238	$\beta D1$ 243	$\alpha C2$ 290	emery paper sheets: $\alpha, \beta, \gamma, \delta$

run

positions

	P_1	P_2	P_3	P_4	*replicate II*
R_5	$\epsilon A1$ 285	$\zeta B2$ 280	$\theta C3$ 331	$\kappa D4$ 311	
R_6	$\zeta C4$ 268	$\epsilon D3$ 233	$\kappa A2$ 291	$\theta B1$ 280	treatments: A, B, C, D
R_7	$\theta D2$ 265	$\kappa C1$ 273	$\epsilon B4$ 234	$\zeta A3$ 243	holders: 1, 2, 3, 4
R_8	$\kappa B3$ 306	$\theta A4$ 271	$\zeta D1$ 270	$\epsilon C2$ 272	emery paper sheets: $\epsilon, \zeta, \theta, \kappa$

run

average

treatments	holders	positions	emery papers	runs	replicates
A 270.000	1 268.875	P_1 279.000	α 276.000	R_1 307.250	R_I 275.8125
B 275.625	2 272.000	P_2 257.375	β 264.500	R_2 248.250	R_{II} 267.1250
C 279.875	3 274.000	P_3 274.375	γ 249.500	R_3 245.000	
D 260.375	4 271.000	P_4 275.125	δ 278.500	R_4 268.000	
			ϵ 256.000	R_5 301.750	
			ζ 265.250	R_6 268.000	
			θ 286.750	R_7 253.750	
grand average = 271.46875			κ 295.250	R_8 279.750	

TABLE 8.9. Analysis of variance table for replicated 4×4 hyper-Graeco–Latin square, first wear testing example

source	degrees of freedom	sum of squares	mean square	ratio of mean squares
average	1	2,358,249.03		
replications	1	603.78	$s_D^2 = 603.78$	$s_D^2/s_R^2 = 5.73$
runs	6	14,770.44	$s_B^2 = 2461.74$	$s_B^2/s_R^2 = 23.35$
positions	3	2,217.34	$s_F^2 = 739.11$	$s_F^2/s_R^2 = 7.01$
emery papers	6	6,108.94	$s_C^2 = 1018.16$	$s_C^2/s_R^2 = 9.66$
holders	3	109.09	$s_E^2 = 36.36$	$s_E^2/s_R^2 = 0.34$
treatments	3	1,705.34	$s_T^2 = 568.45$	$s_T^2/s_R^2 = 5.39$
residuals	9	949.04	$s_R^2 = 105.45$	
total	32 -	2,384,713.00		

feature that four pieces of cloth may be compared simultaneously in one run of the machine.

The response is the weight loss in tenths of a milligram suffered by the test piece when it is rubbed against a standard grade of emery paper for 1000 revolutions of the machine. Specimens of the four different *types of cloth* (treatments) *A*, *B*, *C*, *D* whose wearing qualities are to be compared are mounted in four specimen *holders* 1, 2, 3, 4. Each holder can be in any one of four *positions* P_1, P_2, P_3, P_4 on the machine. Each *emery paper sheet* $\alpha, \beta, \gamma, \delta$ was cut into four quarters, each of which was used to make one observation.

The replicated hyper-Graeco–Latin square design is shown in Table 8.8. In the first square each of the treatments *A*, *B*, *C*, *D* occurs once in every position and every run, together with each of the four sheets of emery paper $\alpha, \beta, \gamma, \delta$ and each of the four holders 1, 2, 3, 4. Since there are four versions of each of the five variables: four treatments, four holders, four positions, four runs, and four sheets of emery paper in a single replication, $5 \times 3 = 15$ degrees of freedom are employed in their comparison and no residual degrees of freedom remain to provide an estimate of experimental error. For this reason the square was repeated* with four additional sheets of emery paper $\epsilon, \xi, \theta, \kappa$ in four further runs. The analysis of variance is given in Table 8.9 (see Appendix 8C for computations).

The design was effective in removing sources of extraneous variation and

---

* A better plan might have been to rearrange randomly the design (while retaining its special properties) in the second square, but this was not done.

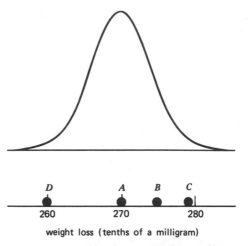

weight loss (tenths of a milligram)

FIGURE 8.7.   Treatment averages with reference distribution, first wear testing example.

in providing some indication of their relative importance. Because of the successful elimination of these disturbances the residual variance was reduced by a factor of about 8, thereby making it possible to detect much smaller differences in the treatments than would otherwise have been possible.

The ratio $s_T^2/s_R^2 = 568.45/105.45 = 5.39$ with three and nine degrees of freedom is significant at about the 0.02 probability level. Figure 8.7 shows the treatment averages for A, B, C, and D with a suitably scaled reference distribution. More replication would evidently be necessary to obtain clear differentiation among these treatments.

8.3.   BALANCED INCOMPLETE BLOCK DESIGNS: SECOND WEAR TESTING EXAMPLE

Suppose that a Martindale wear tester were of a different design which allowed only three, instead of four, samples to be included in one run (block). Suppose, nevertheless, that the experimenter had four treatments A, B, C, and D which he wished to compare. He would then have $t = 4$ treatments but a block size of only $k = 3$, which would be too small to accommodate all the treatments simultaneously. Table 8.10 shows a *balanced incomplete block* design that could then be used. It can alternatively be set out in the manner shown in Table 8.11.

TABLE 8.10. A balanced incomplete block design

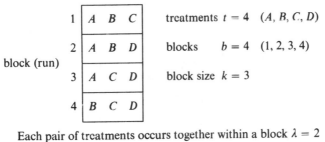

Each pair of treatments occurs together within a block $\lambda = 2$ times.

In general such designs have the property that every pair of treatments occurs together in a block the same number of times λ. Thus in the above design $\lambda = 2$, and we note that A occurs with B twice, with C twice, and with D twice, and likewise for B, C, and D. A listing of a number balanced incomplete block designs appears in Appendic 8C. A numerical example given in Appendix 8D shows how the results from such designs are analyzed. In particular, the analysis shows how the treatment averages need to be adjusted to allow for the fact that not every treatment occurs in every block.

TABLE 8.11. Balanced incomplete block design of Table 8.10 displayed in an alternative way

		treatment			
		A	B	C	D
	1	×	×	×	
block (run)	2	×	×		×
	3	×		×	×
	4		×	×	×

TABLE 8.12. Youden square, second wear testing example

		A	B	C	D	E	F	G
					treatment			
	1		α 627		β 248		γ 563	δ 252
	2	α 344		β 233			δ 442	γ 226
	3			α 251	γ 211	δ 160		β 297
block (run)	4	β 337	δ 537			γ 195		α 300
	5		γ 520	δ 278		β 199	α 595	
	6	γ 369			δ 196	α 185	β 606	
	7	δ 396	β 602	γ 240	α 273			

treatments $t = 7$ blocks $b = 7$ block size $k = 4$
(A, B, C, D, E, F, G) $(1, 2, 3, 4, 5, 6, 7)$
Each pair of treatments occurs together $\lambda = 2$ times.

Youden Squares: Second Wear Testing Example

A less trivial example of a balanced incomplete block is shown in Table 8.12 for comparing seven treatments in seven blocks of size four (ignore for the moment the Greek letters). The data shown within each of the seven blocks (runs) represent the weight loss in tenths of a milligram in 1000 revolutions on a Martindale wear tester, obtained in the comparison of seven different types of cloth (treatments) A, B, C, D, E, F, G. Only four test pieces can be compared simultaneously in the machine, so that we have a fixed block size of $k = 4$.

In this particular balanced incomplete block design the number of blocks happens to equal the number of treatments. When this is so, one often has an opportunity to eliminate a second source of block variation. In the experiment set out in Table 8.12 opportunity was taken to eliminate in the following way possible differences due to machine positions α, β, γ, δ. Each treatment was run in each of the four positions, and each of the four positions was represented in every run. A doubly balanced incomplete block design of this kind is called a *Youden square* after its inventor, W. J. Youden. The method of analysis for this design is set out in Appendix 8D.

APPENDIX 8A. SOME USEFUL LATIN SQUARES AND HOW TO USE THEM TO CONSTRUCT GRAECO–LATIN AND HYPER-GRAECO–LATIN SQUARE DESIGNS

Before running a Latin square or similar design be sure to randomize the design. For example, randomly permute first the rows and then the columns, and finally randomly assign the treatments to the letters.

3×3

$$
\begin{array}{ccc}
A & B & C \\
B & C & A \\
C & A & B
\end{array}
\qquad
\begin{array}{ccc}
A & B & C \\
C & A & B \\
B & C & A
\end{array}
$$

To form the 3×3 Graeco–Latin square superimpose the two designs. Thus, using Greek letter equivalents for the second 3×3 Latin square, we have

$$
\begin{array}{ccc}
A\alpha & B\beta & C\gamma \\
B\gamma & C\alpha & A\beta \\
C\beta & A\gamma & B\alpha
\end{array}
$$

4×4

$$
\begin{array}{cccc}
A & B & C & D \\
B & A & D & C \\
C & D & A & B \\
D & C & B & A
\end{array}
\qquad
\begin{array}{cccc}
A & B & C & D \\
D & C & B & A \\
B & A & D & C \\
C & D & A & B
\end{array}
\qquad
\begin{array}{cccc}
A & B & C & D \\
C & D & A & B \\
D & C & B & A \\
B & A & D & C
\end{array}
$$

These three 4×4 Latin squares may be superimposed to form a hyper-Graeco–Latin square. Superimposing any pair gives a Graeco–Latin square design.

5×5

$$
\begin{array}{ccccc}
A & B & C & D & E \\
B & C & D & E & A \\
C & D & E & A & B \\
D & E & A & B & C \\
E & A & B & C & D
\end{array}
\qquad
\begin{array}{ccccc}
A & B & C & D & E \\
C & D & E & A & B \\
E & A & B & C & D \\
B & C & D & E & A \\
D & E & A & B & C
\end{array}
\qquad
\begin{array}{ccccc}
A & B & C & D & E \\
D & E & A & B & C \\
B & C & D & E & A \\
E & A & B & C & D \\
C & D & E & A & B
\end{array}
\qquad
\begin{array}{ccccc}
A & B & C & D & E \\
E & A & B & C & D \\
D & E & A & B & C \\
C & D & E & A & B \\
B & C & D & E & A
\end{array}
$$

These four 5×5 Latin squares may be superimposed to form a hyper-Graeco–Latin square. Also, superimposing any three gives a hyper-Graeco–Latin square design. Similarly, superimposing any pair gives a Graeco–Latin square design.

6 × 6

```
A B C D E F
B A F E C D
C F B A D E
D C E B F A
E D A F B C
F E D C A B
```

No 6 × 6 Graeco–Latin square exists.

7 × 7

```
A B C D E F G        A B C D E F G
B C D E F G A        C D E F G A B
C D E F G A B        E F G A B C D
D E F G A B C        G A B C D E F
E F G A B C D        B C D E F G A
F G A B C D E        D E F G A B C
G A B C D E F        F G A B C D E
```

These two 7 × 7 Latin squares can be superimposed to give a Graeco–Latin square design.

8 × 8

```
A B C D E F G H        A B C D E F G H
B A D C F E H G        E F G H A B C D
C D A B G H E F        B A D C F E H G
D C B A H G F E        F E H G B A D C
E F G H A B C D        G H E F C D A B
F E H G B A D C        C D A B G H E F
G H E F C D A B        H G F E D C B A
H G F E D C B A        D C B A H G F E
```

These two 8 × 8 Latin squares can be superimposed to give a Graeco–Latin square design.

9 × 9

```
A B C D E F G H I        A B C D E F G H I
B C A E F D H I G        G H I A B C D E F
C A B F D E I G H        D E F G H I A B C
D C F G H I A B C        B C A E F D H I G
E F D H I G B C A        H I G B C A E F D
F D E I G H C A B        E F D H I G B C A
G H I A B C D E F        C A B F D E I G H
H I G B C A E F D        I G H C A B F D E
I G H C A B F D E        F D E I G H C A B
```

These two 9×9 Latin squares can be superimposed to give a Graeco–Latin square design.

For other Latin squares see Fisher and Yates (1963). The designs above are from that source and are reproduced here with permission.

APPENDIX 8B. ANALYSIS OF VARIANCE FOR $k \times k$ LATIN SQUARE DESIGNS WITH r REPLICATES

The usual model used in conjunction with Latin square designs is

$$y_{ijtu} = \eta + \beta_i + \gamma_j + \tau_t + \rho_u + \epsilon_{ijtu} \qquad (8B.1)$$

where η = mean, β_i = row effect, γ_j = column effect, τ_t = treatment effect ($i, j, t = 1, 2, \ldots, k$), ρ_u = replicate effect ($u = 1, 2, \ldots, r$), and the errors are assumed to be IIDN($0, \sigma^2$).

Special Case: $r = 1$ *Replicate*

Consider the data, originally given in Table 8.1, for a Latin square with $k = 4$ and $r = 1$ replicates. The blocking variables are drivers (rows) and cars (columns), and the treatments are gasoline additives A, B, C, D.

		car 1	car 2	car 3	car 4	row total T_i	treatment total T_t
I		A 21	B 26	D 20	C 25	92	A 72
II		D 23	C 26	A 20	B 27	96	B 88
driver III		B 15	D 13	C 16	A 16	60	C 84
IV		C 17	A 15	B 20	D 20	72	D 76
column total T_j		76	80	76	88	320 = G	

$$\sum_i \sum_j y_{ij}^2 = (21)^2 + (26)^2 + \cdots + (20)^2 = 6696$$

$$S_A = \frac{(320)^2}{16} = 6400$$

$$S_B = \frac{(92)^2 + (96)^2 + \cdots + (72)^2}{4} - S_A = 216$$

$$S_C = \frac{(76)^2 + (80)^2 + \cdots + (88)^2}{4} - S_A = 24$$

$$S_T = \frac{(72)^2 + (88)^2 + \cdots + (76)^2}{4} - S_A = 40$$

These values appear in Table 8.3.

Analysis of Variance for Latin Square with $r > 1$ Replicates

When a design is replicated, it is important to note the nature of the replication. For example, suppose that the 4×4 Latin square given above had been repeated r times. The design could have been repeated in the following ways:

(1) using the same drivers and cars in each replicate,
(2) using the same drivers but new cars (or the same cars but new drivers), or
(3) using new cars and new drivers.

The manner in which the degrees of freedom for the blocking variable and for the r replicates are partitioned varies for each case.

Case 1: Latin Square Replicated by Using Same Blocking Variables

The analysis of variance table for the general case in which a Latin square design is replicated r times is given in Table 8B.1. Entries in the sum of squares column for this table can be obtained as follows.

Let G = grand total of all rk^2 observations

T_i = total for ith row, $i = 1, 2, \ldots, k$ (rk observations in each total T_i)
T_j = total for jth column, $j = 1, 2, \ldots, k$ (rk observations in each total T_j)
T_u = total for uth replicate, $u = 1, 2, \ldots, r$ (k^2 observations in each total T_u)
T_t = total for tth treatment, $t = 1, 2, \ldots, k$ (rk observations in each total T_t)

TABLE 8B.1. Analysis of variance table for Latin square design, Case 1

source of variation	sum of squares	degrees of freedom	mean square	ratio of mean squares
average	S_A	1		
rows	S_B	$k-1$	$S_B/(k-1) = s_B^2$	s_B^2/s_R^2
columns	S_C	$k-1$	$S_C/(k-1) = s_C^2$	s_C^2/s_R^2
replicates	S_D	$r-1$	$S_D/(r-1) = s_D^2$	s_D^2/s_R^2
treatments	S_T	$k-1$	$S_T/(k-1) = s_T^2$	s_T^2/s_R^2
residual	S_R	$\nu = (k-1)[r(k+1)-3]$	$S_R/\nu = s_R^2$	
total	$\sum_i \sum_j \sum_u y_{iju}^2$	rk^2		

Then,

$$S_A = \frac{G^2}{rk^2} = \text{the correction factor}$$

$$S_B = \frac{\sum_i T_i^2}{rk} - S_A$$

$$S_C = \frac{\sum_j T_j^2}{rk} - S_A$$

$$S_D = \frac{\sum_u T_u^2}{k^2} - S_A \quad \text{(only for } r > 1\text{)}$$

$$S_T = \frac{\sum_t T_t^2}{rk} - S_A$$

$$S_R = \sum_i \sum_j \sum_u y_{iju}^2 - S_A - S_B - S_C - S_D - S_T$$

Case 2: Latin Square Replicated by Introducing Additional Versions of One Blocking Variable

Consider the data given in Table 8.4 for the wear testing of a fabric made from synthetic yarn. This design consists of $r = 4$ replicates of a 3×3 Latin square. The spinneret heads, one of the blocking variables, remain the same for all replicates. The other blocking variable, doffs, is not constant; each new replicate has three new doffs. Since this blocking variable which introduces new versions for each replicate is identified by rows, the corresponding entries for rows and for residuals are as follows:

source of variation	sum of squares	degrees of freedom	mean square	ratio of mean squares
rows	S_B	$r(k-1)$	$S_B/r(k-1) = s_B^2$	s_B^2/s_R^2
residuals	S_R	$v = (k-1)(rk-2)$	$S_R/v = s_R^2$	

where $S_B = \sum T_i^2/k - S_A - S_D$ ($i = 1, 2, \ldots, rk$), T_i = total for the ith row (with k observations), and $S_R = \sum_i \sum_j \sum_u y_{iju}^2 - S_A - S_B - S_C - S_D - S_T$.

For the data in Table 8.4 the doffs (rows) totals are:

replicate 1	(1) 72.44	(2) 74.16	(3) 65.08
replicate 2	(4) 65.53	(5) 66.66	(6) 58.84
replicate 3	(7) 63.72	(8) 63.32	(9) 66.08
replicate 4	(10) 64.28	(11) 65.10	(12) 61.92

The doff numbers are given in parentheses. We now have

$$\sum_i \sum_j \sum_u y_{iju}^2 = 17,621.5617$$

$$S_A = \frac{G^2}{rk^2} = \frac{(787.13)^2}{4(9)} = 17,210.3788$$

$$S_D = \frac{(211.68)^2 + \cdots + (191.30)^2}{9} - 17,210.3788 = 33.1675$$

$$S_B = \frac{(72.44)^2 + (74.16)^2 + (65.08)^2 + (65.53)^2 + \cdots + (61.92)^2}{3} - S_A - S_D$$

$$= \frac{51822.7897}{3} - 17,210.3788 - 33.1675 = 30.7169$$

$$S_C = \frac{(246.43)^2 + (262.67)^2 + (278.03)^2}{4(3)} - 17,210.3788 = 41.6174$$

$$S_T = \frac{(225.29)^2 + (267.38)^2 + (294.46)^2}{4(3)} - 17,210.3788 = 202.4839$$

If there is no special interest in the differences between replicates, the replicate sum of squares and degrees of freedom can be combined with the row (doff) sum of squares and degrees of freedom as illustrated in the example in Table 8.6. The sum of squares for rows (doffs) given in Table 8B.2 measures the combined variation between individual rows *within* the replicates and is usually identified in the table as "rows within replicates."

Case 3: Latin Square Replicated by Introducing Additional Versions of the Two Blocking Variables

In Case 2 suppose that the second replicate had been composed of three new doffs (rows) *and three new spinnerets* (columns). Here $k = 3$; the corresponding analysis of

TABLE 8B.2. Analysis of variance table for Latin square designs, Case 2

source of variation	sum of squares	degrees of freedom	mean squares	ratio of mean squares
grand average	17,210.3788	1		
replicates	33.1675	3	$s_D^2 = 11.0558$	2.1
doffs within reps	30.7169	8	$s_B^2 = 3.8396$	0.7
spinnerets	41.6174	2	$s_C^2 = 20.8087$	4.0
treatments	202.4839	2	$s_T^2 = 101.2419$	19.6
residuals	103.1972	20	$s_R^2 = 5.1599$	
total	17,621.5617	36		

variance entries for rows, columns, and residuals would then become:

source of variation	sum of squares	degrees of freedom	mean square	ratio of mean squares
factors	S_B	$r(k-1)$	$s_B^2 = S_B/r(k-1)$	s_B^2/s_R^2
columns	S_C	$r(k-1)$	$s_C^2 = S_C/r(k-1)$	s_C^2/s_R^2
residuals	S_R	$v = (k-1)[r(k-1)-1]$	$s_R^2 = S_R/v$	

where $S_B = \sum T_i^2/k - S_A - S_D, i = 1, 2, \ldots, rk$, $S_C = \sum T_j^2/k - S_A - S_D, j = 1, 2, \ldots, rk$, T_i = total for the ith row (with k observations), T_j = total for the jth column (with k observations), and $S_R = \sum \sum \sum y_{iju}^2 - S_A - S_B - S_C - S_D - S_T$.

For example, suppose that the second replicate of the design described in Table 8.4 had consisted of $k = 3$ new doffs (rows) along with $k = 3$ new spinnerets (columns). The spinneret totals within each replicate are:

replicate 1	(1) 67.56	(2) 68.37	(3) 75.75
replicate 2	(4) 58.05	(5) 65.72	(6) 67.26
replicate 3	(7) 63.72	(8) 65.92	(9) 63.48
replicate 4	(10) 57.10	(11) 62.66	(12) 71.54

Spinneret numbers are given in parentheses. The doff (row) sum of squares $S_B = 30.7169$ is computed above. The spinneret (column) sum of squares is

$$S_C = \frac{(67.56)^2 + \cdots + (71.54)^2}{3} - 17,210.3788 - 33.1675 = 66.3717$$

Table 8B.3 is the resulting analysis of variance table for Case 3.

TABLE 8B.3. Analysis of variance table for Latin square design, Case 3

source of variation	sum of squares	degrees of freedom	mean square	ratio of mean squares
grand average	17,210.3788	1		
doffs within replicates	30.7169	8	$s_B^2 = 3.8396$	$s_B^2/s_R^2 = 0.7$
spinnerets within replicates	66.3717	8	$s_C^2 = 8.2925$	$s_C^2/s_R^2 = 1.5$
replicates	33.1675	3	$s_D^2 = 11.0558$	$s_D^2/s_R^2 = 20.0$
treatments	202.4839	3	$s_T^2 = 101.2420$	$s_T^2/s_R^2 = 18.1$
residuals	78.4429	14	$s_R^2 = 5.6031$	
total	17,621.5617	36		

APPENDIX 8C. SOME USEFUL BALANCED INCOMPLETE BLOCK DESIGNS

This standard notation has been adopted for describing balanced incomplete block designs:

t = number of treatments
b = number of blocks
k = block size $(k < t)$
r = number of replicates of each treatment
$N = tr = bk$ = total number of observations

To use the following balanced incomplete block (BIB) designs,

(1) randomly assign numbers 1, 2, 3, ... to the blocks,
(2) randomly assign letters $A, B, C, \ldots$ to the treatments,
(3) randomly assign the treatments within the blocks, and
(4) whenever possible, randomly group blocks within replicates.

In this appendix the term *combinatoric design* frequently appears. A combinatoric design is a balanced incomplete block design in which *all* combinations of t treatments appear in blocks of size k. For example, consider the combinatoric design for $t = 6$, $k = 3$. There are $\binom{6}{3} = 6!/3!3! = 20$ position combinations (blocks). The design is formed by writing down all possible combinations of these letters, drawn from six, as follows:

(A B C)	(A C E)	(B C D)	(B E F)	$t = 6$
(A B D)	(A C F)	(B C E)	(C D E)	$k = 3$
(A B E)	(A D E)	(B C F)	(C D F)	$b = 20$
(A B F)	(A D F)	(B D E)	(C E F)	$r = 10$
(A C D)	(A E F)	(B D F)	(D E F)	$\lambda = 4$

Unfortunately, combinatoric balanced incomplete block designs often require many experimental runs. Smaller balanced incomplete block designs are sometimes possible. For example, as can be seen from the listing below, a design with only $b = 10$ blocks is available for $t = 6, k = 3$.

Some useful balanced incomplete block designs are listed on the following five pages.

The combinatoric designs

Form all possible combinations of the t symbols taken two at a time; each combination composes a block. For example, for $t = 4$:

$$(A, B);\ (A, C);\ (A, D);\ (B, C);\ (B, D);\ (C, D)$$

$k = 2$

$t = 2$	$b = t(t - 1)/2$	
$r = t - 1$	$N = t(t - 1)$	

$t = 6$ $b = 15$
$r = 5$ $N = 30$

Replicate		
I	$(A, B);\ (C, D);\ (E, F)$
II	$(A, C);\ (B, E);\ (D, F)$
III	$(A, D);\ (B, F);\ (C, E)$
IV	$(A, E);\ (B, D);\ (C, F)$
V	$(A, F);\ (B, C);\ (D, E)$

$t = 8$ $b = 28$
$r = 7$ $N = 56$

Replicate	
I	$(1, 2);\ (3, 4);\ (5, 6);\ (7, 8)$
II	$(1, 3);\ (2, 8);\ (4, 5);\ (6, 7)$
III	$(1, 4);\ (2, 7);\ (3, 6);\ (5, 8)$
IV	$(1, 5);\ (2, 3);\ (4, 7);\ (6, 8)$
V	$(1, 6);\ (2, 4);\ (3, 8);\ (5, 7)$
VI	$(1, 7);\ (2, 6);\ (3, 5);\ (4, 8)$
VII	$(1, 8);\ (2, 5);\ (3, 7);\ (4, 6)$

$t = 10$ $b = 45$
$r = 9$ $N = 90$

Replicate	
I	$(1, 2);\ (3, 4);\ (5, 6);\ (7, 8);\ (9, 10)$
II	$(1, 3);\ (2, 7);\ (4, 8);\ (5, 9);\ (6, 10)$
III	$(1, 4);\ (2, 10);\ (3, 7);\ (5, 8);\ (6, 9)$
IV	$(1, 5);\ (2, 8);\ (3, 10);\ (4, 9);\ (6, 7)$
V	$(1, 6);\ (2, 9);\ (3, 8);\ (4, 10);\ (5, 7)$
VI	$(1, 7);\ (2, 6);\ (3, 9);\ (4, 5);\ (8, 10)$
VII	$(1, 8);\ (2, 3);\ (4, 6);\ (5, 10);\ (7, 9)$
VIII	$(1, 9);\ (2, 4);\ (3, 5);\ (6, 8);\ (7, 10)$
IX	$(1, 10);\ (2, 5);\ (3, 6);\ (4, 7);\ (8, 9)$

$k = 3$ | $t > 3$
$b = t(t - 1)(t - 2)/6$
$r = (t - 1)(t - 2)/2$
$N = t(t - 1)(t - 2)/2$

The combinatoric designs
Form all combinations of the t symbols taken three at a time; each combination comprises a block.

| $t = 6$ | $b = 10$ |
| $r = 5$ | $N = 30$ |

(1, 2, 5); (1, 2, 6); (1, 3, 4); (1, 3, 6); (1, 4, 5)
(2, 3, 4); (2, 3, 5); (2, 4, 6); (3, 5, 6); (4, 5, 6)

| $t = 7$ | $b = 7$ |
| $r = 3$ | $N = 21$ |

(1, 2, 5); (3, 4, 5); (1, 3, 6); (2, 4, 6); (1, 4, 7)
(2, 3, 7); (5, 6, 7)

| $t = 8$ | $b = 56$ |
| $r = 21$ | $N = 108$ |

The combinatoric design

| $t = 9$ | $b = 12$ |
| $r = 4$ | $N = 36$ |

Replicate I   (1, 2, 3); (4, 5, 6); (7, 8, 9)
         II   (1, 4, 7); (2, 5, 8); (3, 6, 9)
         III (1, 5, 9); (2, 6, 7); (3, 4, 8)
         IV   (1, 6, 8); (2, 4, 9); (3, 5, 7)

| $t = 10$ | $b = 30$ |
| $r = 9$ | $N = 90$ |

(1, 2, 3); (1, 2, 4); (1, 3, 5); (1, 4, 6); (1, 5, 7); (1, 6, 8)
(1, 7, 9); (1, 8, 9); (1, 9, 10); (2, 3, 6); (2, 4, 10); (2, 5, 8)
(2, 5, 9); (2, 6, 7); (2, 7, 9); (2, 8, 10); (3, 4, 7); (3, 4, 8)
(3, 5, 6); (3, 7, 10); (3, 8, 9); (3, 9, 10); (4, 5, 9); (4, 5, 10)
(4, 6, 9); (4, 7, 8); (5, 6, 10); (5, 7, 8); (6, 7, 10); (6, 8, 9)

(continued on following page)

271

$k = 4$		
	$t = 5$ $b = 5$ $r = 4$ $N = 20$	**The combinatoric design** (1, 2, 3, 4); (1, 2, 3, 5); (1, 2, 4, 5); (1, 3, 4, 5); (2, 3, 4, 5)
	$t = 6$ $b = 15$ $r = 10$ $N = 60$	Replicates I and II (1, 2, 3, 4); (1, 4, 5, 6); (2, 3, 5, 6) III and IV (1, 2, 3, 5); (1, 2, 4, 6); (3, 4, 5, 6) V and VI (1, 2, 3, 6); (1, 3, 4, 5); (2, 4, 5, 6) VII and VIII (1, 2, 4, 5); (1, 3, 5, 6); (2, 3, 4, 6) IX and X (1, 2, 5, 6); (1, 3, 4, 6); (2, 3, 4, 5)
	$t = 7$ $b = 7$ $r = 4$ $N = 28$	Design is a Youden square, as illustrated in Table 8.12. Complementary to $k = 3, t = 7$. (3, 4, 6, 7); (1, 2, 6, 7); (2, 4, 5, 7); (1, 3, 5, 7); (2, 3, 5, 6); (1, 4, 5, 6); (1, 2, 3, 4)
	$t = 8$ $b = 14$ $r = 7$ $N = 56$	Replicate I (1, 2, 3, 4); (5, 6, 7, 8) II (1, 2, 7, 8); (3, 4, 5, 6) III (1, 3, 6, 8); (2, 4, 5, 7) IV (1, 4, 6, 7); (2, 3, 5, 8) V (1, 2, 5, 6); (3, 4, 7, 8) VI (1, 3, 5, 7); (2, 4, 6, 8) VII (1, 4, 5, 8); (2, 3, 6, 7)
	$t = 9$ $b = 18$ $r = 8$ $N = 72$	(1, 2, 3, 4); (1, 2, 5, 6); (1, 2, 7, 8); (1, 3, 5, 7); (1, 4, 6, 8); (1, 3, 6, 9) (1, 4, 8, 9); (1, 5, 7, 9); (2, 3, 8, 9); (2, 4, 5, 9); (2, 6, 7, 9); (2, 3, 4, 7) (2, 5, 6, 8); (3, 5, 6, 8); (4, 5, 7, 9); (3, 4, 5, 6); (3, 6, 7, 8); (4, 5, 7, 8)
	$t = 10$ $b = 15$ $r = 6$ $N = 60$	(1, 2, 3, 4); (1, 2, 5, 6); (1, 3, 7, 8); (1, 4, 9, 10); (1, 5, 7, 9) (1, 6, 8, 10); (2, 3, 6, 9); (2, 4, 7, 10); (2, 5, 8, 10); (2, 7, 8, 9) (3, 5, 9, 10); (3, 6, 7, 10); (3, 4, 5, 8); (4, 5, 6, 7); (4, 6, 8, 9)

$k = 5$	$t = 6$ $b = 6$ $r = 5$ $N = 30$	The combinatoric design $(1, 2, 3, 4, 5)$; $(1, 2, 3, 4, 6)$; $(1, 2, 3, 5, 6)$; $(1, 2, 4, 5, 6)$; $(1, 3, 4, 5, 6)$; $(2, 3, 4, 5, 6)$
	$t = 7$ $b = 21$ $r = 15$ $N = 105$	The combinatoric design
	$t = 8$ $b = 56$ $r = 35$ $N = 280$	The combinatoric design
	$t = 9$ $b = 18$ $r = 10$ $N = 90$	$(1, 2, 3, 5, 8)$; $(1, 2, 3, 6, 9)$; $(1, 2, 4, 5, 9)$; $(1, 2, 4, 6, 7)$; $(1, 2, 7, 8, 9)$ $(1, 3, 4, 5, 8)$; $(1, 3, 4, 7, 9)$; $(1, 3, 6, 7, 8)$; $(1, 4, 5, 6, 7)$; $(1, 5, 6, 8, 9)$ $(3, 4, 5, 6, 8)$; $(2, 3, 5, 6, 7)$; $(2, 3, 5, 7, 9)$; $(2, 4, 6, 7, 8)$; $(2, 4, 6, 8, 9)$ $(3, 4, 5, 6, 9)$; $(3, 4, 7, 8, 9)$; $(5, 6, 7, 8, 9)$
	$t = 10$ $b = 18$ $r = 9$ $N = 90$	$(1, 2, 3, 4, 5)$; $(1, 2, 3, 6, 7)$; $(1, 2, 4, 6, 9)$; $(1, 2, 5, 7, 8)$; $(1, 3, 5, 8, 9)$ $(1, 3, 7, 8, 10)$; $(1, 4, 5, 6, 10)$; $(1, 4, 8, 9, 10)$; $(1, 5, 7, 9, 10)$ $(2, 3, 4, 8, 10)$; $(2, 3, 5, 9, 10)$; $(2, 4, 7, 8, 9)$; $(2, 5, 6, 8, 10)$ $(2, 6, 7, 9, 10)$; $(3, 4, 6, 7, 10)$; $(3, 4, 6, 7, 9)$; $(3, 5, 6, 8, 9)$ $(4, 5, 6, 7, 8)$

(continued on following page)

k = 6	t = 7 r = 6	b = 7 N = 42		The combinatoric design; design may be run as a Youden square (1, 2, 3, 4, 5, 6); (1, 2, 3, 4, 5, 7); (1, 2, 3, 4, 6, 7); (1, 2, 3, 5, 6, 7) (1, 2, 4, 5, 6, 7); (1, 3, 4, 5, 6, 7); (2, 3, 4, 5, 6, 7)
	t = 8 r = 21	b = 28 N = 168		The combinatoric design
	t = 9 r = 8	b = 12 N = 72	Replicates I and II III and IV V and VI VII and VIII	(1, 2, 3, 4, 5, 6); (1, 2, 3, 7, 8, 9); (4, 5, 6, 7, 8, 9) (1, 2, 4, 6, 8, 9); (1, 3, 5, 6, 7, 8); (2, 3, 4, 5, 7, 9) (1, 2, 4, 5, 7, 8); (1, 3, 4, 6, 7, 9); (2, 3, 5, 6, 8, 9) (1, 2, 5, 6, 7, 9); (1, 3, 4, 5, 8, 9); (2, 3, 4, 6, 7, 8)
	t = 10 r = 9	b = 15 N = 90		(1, 2, 3, 4, 7, 10); (1, 2, 3, 8, 9, 10); (1, 2, 4, 5, 8, 9); (1, 2, 4, 6, 7, 8); (1, 2, 5, 7, 9, 10) (1, 3, 4, 5, 6, 10); (1, 3, 4, 6, 7, 9); (1, 3, 5, 6, 8, 9); (1, 4, 5, 7, 8, 10); (2, 3, 4, 5, 7, 9) (2, 3, 4, 6, 8, 10); (2, 3, 5, 6, 7, 8); (2, 4, 5, 6, 9, 10); (3, 4, 7, 8, 9, 10); (5, 6, 7, 8, 9, 10)

Additional balanced incomplete block designs for $k > 6$ and for $t > 10$ are given in Cochran and Cox (1957). The listing above, which is an abridgment of the tables in that reference, is reproduced here by permission.

APPENDIX 8D. ANALYSIS OF VARIANCE AND COMPUTATION OF ADJUSTED TREATMENT AVERAGES FOR BALANCED INCOMPLETE BLOCK DESIGNS

The usual model for data from a balanced incomplete block design is

$$y_{ij} = \eta + \beta_i + \tau_j + \epsilon_{ij} \tag{8D.1}$$

where y_{ij} = observation on treatment j in block i, η = grand mean, β_i = block effect ($i = 1, 2, \ldots, b$), τ_j = treatment effect ($j = 1, 2, \ldots, t$), and the errors ϵ_{ij} are assumed to be IIDN($0, \sigma^2$). Since all treatments do not appear in every block, special care must be taken to eliminate block effects when introducing treatment effects. The method of analysis is as follows.

1. The total is obtained for each treatment T_j, $j = 1, 2, \ldots, t$. Then obtain the total of all blocks containing the jth treatment; call this total B_j. Compute the quantity

$$Q_j = kT_j - B_j \tag{8D.2}$$

2. The estimated effect of the jth treatment is given by

$$\hat{\tau}_j = \frac{Q_j}{t\lambda} \tag{8D.3}$$

Recall that λ, the number of times an individual pair of treatments appears together in the same block, is a constant and a characteristic of a BIB design. Note that $\sum_{j=1}^{t} \hat{\tau}_j = 0$.

3. The adjusted treatment averages are then

$$\bar{y}_{j(\text{adjusted})} = \bar{y} + \hat{\tau}_j \tag{8D.4}$$

where $\bar{y}$ is the grand average $\bar{y} = \sum_i \sum_j y_{ij}/N$.

4. The variance of an adjusted treatment average is

$$V(\bar{y}_{j(\text{adjusted})}) = \frac{k\sigma^2}{t\lambda} \tag{8D.5}$$

5. The variance of the difference between two adjusted treatment averages is $2k\sigma^2/t\lambda$.

6. An estimate of σ^2 can be obtained from the analysis of variance table, Table 8D.1.

TABLE 8D.1. Analysis of variance table for balanced incomplete block design

source of variation	sum of squares	degrees of freedom	mean square	ratio of mean squares
grand average	S_A	1		
blocks	S_B	$b - 1$	$s_B^2 = S_B/(b - 1)$	s_B^2/s_R^2
treatments adjusted for blocks	S_T	$t - 1$	$s_T^2 = S_T/(t - 1)$	s_T^2/s_R^2
residuals	S_R	$v = N - t - b + 1$		
total	$\sum\sum y_{ij}^2$	N		

Here $S_A = G^2/N$, where $G = \sum y_{ij}$ is the grand total

$\quad S_B = \sum_{i=1}^{b} T_i^2/k - S_A$, where T_i is the total for block i

$\quad S_T = \sum_{j=1}^{t} Q_j^2/\lambda kt$, where $Q_j = kT_j - B_j$ as defined in step 1 above

To illustrate these computations we use the data given in Table 8.12. *Ignoring the Greek letters*, we have a balanced incomplete block design with

$t = 7$ treatments $\quad r = 4$ replicates (of treatments)
$k = 4$ block size $\quad \lambda = 2$
$b = 7$ blocks

The design is displayed in random order in Table 8D.2; the capital letter with each observation identifies the treatment. Using these observations, the quantities T_j, B_j, and Q_j, necessary to compute the adjusted treatment averages, are calculated in Table 8D.3.

To compute the analysis of variance displayed in Table 8D.4, we proceed as follows:

total $\quad \sum_j \sum_i y_{ij}^2$ $\qquad = 3{,}974{,}162.00$

average $\quad S_A = \dfrac{G^2}{N} = \dfrac{(9682)^2}{28}$ $\qquad = 3{,}347{,}897.28$

blocks $\quad S_B = \dfrac{(1690)^2 + \cdots + (1511)^2}{4} - S_A = \quad 97{,}394.72$

treatments $\quad S_T = \dfrac{(303)^2 + \cdots + (-923)^2}{20(4)(7)} \quad = \quad 506{,}798.57$

In this example the hypothesis that the treatment effects are zero is contradicted by the data, since $\Pr(F_{6,15} \geq 57.4) < 0.001$. The adjusted treatment averages should be plotted with their appropriate reference distribution, a t_{15} distribution with scale factor $\sqrt{ks^2/t\lambda} = \sqrt{4(1471.42)/(7)(2)} = 20.50$.

TABLE 8D.2. Youden square, second wear testing example in original random order

					block total
1	F 563	D 248	G 252	B 627	1690
2	C 233	A 344	G 226	F 442	1245
3	G 297	D 211	E 160	C 251	919
4	E 195	G 300	B 537	A 337	1369
5	B 520	E 199	C 278	F 595	1592
6	D 196	A 369	E 185	F 606	1356
7	D 273	C 240	B 602	A 396	1511

blocks

$9682 = G; \bar{y} = 345.78$

TABLE 8D.3. Computation of adjusted treatment averages

treatment total T_j	B_j	$Q_j = 4T_j - B_j$	$\hat{t}_j = Q_j/7(2)$	adjusted treatment average
A 1446	5481	303	21.64	367.43
B 2286	6162	2982	213.00	558.79
C 1002	5267	− 1259	− 89.93	255.86
D · 928	5476	− 1764	− 126.00	219.79
E 739	5236	− 2280	− 162.86	182.93
F 2206	5883	2941	210.07	555.86
G 1075	5223	− 923	− 65.93	279.86

TABLE 8D.4.   Analysis of variance table for balanced incomplete block design, second wear testing example

source of variation	sum of squares	degrees of freedom	mean square	ratio of mean squares
grand average	$S_A = 3{,}347{,}897.28$	1		
blocks (runs)	$S_B = \quad 97{,}394.72$	6	$s_B^2 = 16{,}232.33$	$s_B^2/s_R^2 = 11.03$
treatments (adjusted)	$S_T = \quad 506{,}798.57$	6	$s_T^2 = 84{,}466.43$	$s_T^2/s_R^2 = 57.4$
residual	$S_R = \quad 22{,}071.43$	15	$s_R^2 = \quad 1{,}471.42$	
total	3,974,162.00	28		

Analysis of Variance Table for a Youden Square

A Youden square is a balanced incomplete block design with two blocking variables. The model for the Youden square is therefore the model for the balanced incomplete block with one added term:

$$y_{ilj} = \eta + \beta_i + \alpha_l + \tau_j + \epsilon_{ilj}$$

where α_l represents the effects of the second blocking variable, $l = 1, 2, \ldots, k$. The analysis of variance table will be identical to that for the balanced incomplete block design except for the additional sum of squares and degrees of freedom for the second blocking variable, given by

$$S_C = \frac{\sum_{l=1}^{k} T_l^2}{t} - S_A$$

which has $k - 1$ degrees of freedom, where T_l = total for the lth block with t observations. For example, consider the data used above to describe the analysis of a balanced incomplete block. These were originally given in Table 8.12, in which there is a second blocking variable (positions) identified by Greek letters α, β, γ, δ. The position totals are

$$T_\alpha = 2575, \qquad T_\beta = 2522, \qquad T_\gamma = 2324, \qquad T_\delta = 2261$$

The quantities S_A, S_B, S_T in the analysis of variance table are identical to those given above in Table 8D.4 for the balanced incomplete block example. In addition we now have, for positions,

$$S_C = \frac{(2575)^2 + \cdots + (2261)^2}{7} - S_A = 9846.43$$

Table 8D.5 is the analysis of variance table for the Youden square example.

On the assumption that the treatment effects are zero the ratio of the mean squares s_T^2/s_R^2 is distributed as $F_{6,12}$. The probability that $F_{6,12} \geq 82.9$ is less than 0.001,

TABLE 8D.5. Analysis of variance table for Youden square, second wear testing example

source of variation	sum of squares	degrees of freedom	mean square	ratio of mean squares
average	3,347,897.28	1		
blocks (runs)	97,394.72	6	$s_B^2 = 16,232.45$	$s_B^2/s_R^2 = 15.93$
blocks (positions)	9,846.43	3	$s_C^2 = 3,282.14$	$s_C^2/s_R^2 = 3.22$
treatments adjusted	506,798.57	6	$s_T^2 = 84,466.43$	$s_T^2/s_R^2 = 82.9$
residual	12,225.00	12	$s_R^2 = 1,018.75$	
total	3,974,162.00	28		

which casts doubt on the hypothesis of zero treatment effects. As in the above example, the adjusted treatment averages should now be displayed along with their appropriate reference distribution, this time a t_{12} distribution with scale factor $\sqrt{ks^2/t\lambda} = \sqrt{4(1018.75)/7(2)} = 17.06$.

REFERENCES AND FURTHER READING

For further discussion of designs of the type discussed in this chapter see:

Cochran, W. G., and G. M. Cox (1957). *Experimental Designs*, 2nd ed., Wiley.

Tables for more extensive Latin squares than those in Appendix 8A are published in:

Fisher, R. A., and F. Yates (1938, 1963). *Statistical Tables for Biological, Agricultural, and Medical Research*, Oliver and Boyd.

For the important work of Frank Yates, who made fundamental contributions to the design and analysis of experiments, see:

Yates, F. (1970). *Experimental Design: Selected Papers of Frank Yates, C.B.E., F.R.S.*, Hafner (Macmillan).

For a discussion and listing of *partially* balanced incomplete block designs see:

Bose, R. C., W. H. Clatworthy, and S. S. Shrikhande (1954). *Tables of Partially Balanced Designs with Two Associate Classes*, Technical Bulletin 107, North Carolina Agricultural Experiment Station.

QUESTIONS FOR CHAPTER 8

1. What is a Latin square design? Describe a situation in which it could be used, preferably one in your own field. What model is ordinarily used in analyzing data from such a design, and what are its possible shortcomings? How can residuals and reference distributions be plotted, and what should one look for in these plots?

2. Answer the questions above for Graeco–Latin and hyper-Graeco–Latin square designs.

3. What is a balanced incomplete block design? Explain what is meant by the three words *balanced*, *incomplete*, and *block*. Describe a situation, preferably in your own field, in which one might properly use such a design.

4. How can the use of blocking reduce the adverse effects of time trends?

5. What is the usual "analysis of variance model" that is employed in analyzing data from block designs? What are the assumptions associated with this model, and how realistic do you think they are?

6. What is a Youden square design?

7. How can a Latin square design be randomized? A balanced incomplete block design?

8. How would you construct a 7×7 Graeco–Latin square design?

9. Does a balanced incomplete block design exist for $t = 7$ treatments, $b = 7$ blocks, $r = 3$ replicates, and a block size of $k = 3$? Is it always true for a balanced incomplete block design that $tr = bk$?

10. In the analysis of balanced incomplete block designs, why are the treatment averages adjusted?

Problems for Part II

Whether or not specifically asked, the reader should always (1) plot the data in any potentially useful way, (2) state the assumptions made, (3) comment on the appropriateness of these assumptions, and (4) consider alternative analyses.

1. A break-test machine is the subject of a study. In particular we want to find out whether four operators have obtained consistent results, and also to estimate the intrinsic error of the test. From each of five batches of material four ingots taken and randomly allocated to the testers yielded the following strengths:

raw material batch	operator				batch average
	A	B	C	D	
1	89	88	97	94	92
2	84	77	92	79	83
3	81	87	87	85	85
4	87	92	89	84	88
5	79	81	80	88	82
operator average	84	85	89	86	

Make appropriate plots of data and residuals, and carry out an analysis, stating your assumptions. Draw conclusions.

2. Repeat problem 1 after adding 10 to each of the readings in column D.

3. Repeat problem 1 after adding 10 to each of the readings in row 3.

4. Repeat problem 1 after adding 10 to the observation in row 3 and column D.

5. A random sample of 48 devices was selected from those in a warehouse. This sample, in turn, was randomly divided into three samples (A, B, and C), each of size 16. Each of these three samples received a different treatment. To the devices in A, the control, nothing was done. Each of those in B was immersed in water. Each of those in C was dropped from a certain height. All 48 devices were then tested, and the following performance data were obtained. Carefully plot the data and conduct any analysis you consider appropriate. Does there appear to be any measurable effect of immersing the devices in water or dropping them?

A	B	C
0.38	0.53	0.51
0.26	0.35	0.63
0.41	0.38	0.46
0.33	0.45	0.47
0.33	1.09	0.42
0.37	0.46	0.45
0.54	0.57	0.41
0.76	0.47	0.39
0.51	0.39	0.35
0.55	0.74	0.41
0.53	0.32	0.49
0.41	0.74	0.40
0.47	0.48	0.58
0.49	0.37	0.46
0.42	0.52	0.38
0.34	0.44	0.48

(*Source*: This problem is based on an example described in *6th Army Design of Experiments Conference*, p. 263.)

6. The following data were collected from an interlaboratory calibration check. Each of the laboratories (*A*, *B*, and *C*) was sent a sample of some standard wire. The results obtained from their horizontal tension testing machines were as follows:

A	B	C
48	50	51
49	50	52
50	48	50
	52	

These data can be broken up into the following tables:

$$
\begin{array}{ccc}
 & \text{average} & \\
\end{array}
\quad
\begin{array}{c}
\text{laboratory} \\
\text{deviations}
\end{array}
\quad
\text{residuals}
$$

$$
\begin{bmatrix}
50 & 50 & 50 \\
50 & 50 & 50 \\
50 & 50 & 50 \\
   & 50 &
\end{bmatrix}
\begin{bmatrix}
-1 & 0 & +1 \\
-1 & 0 & +1 \\
-1 & 0 & +1 \\
   & 0 &
\end{bmatrix}
\begin{bmatrix}
-1 & 0 & 0 \\
0 & 0 & +1 \\
+1 & -2 & -1 \\
   & +2 &
\end{bmatrix}
$$

(a) Complete the following analysis of variance table:

source	sum of squares	degrees of freedom
avérage		
between		
laboratories		
within		
laboratories		
total	25,018	

(b) Calculate the mean square for between laboratories (s_T^2) and within laboratories (s_R^2).

(c) Describe in detail how the experiment ought to be conducted so that this analysis will help to answer the question of whether there are appreciable differences in the mean levels of tests conducted in the three laboratories.

7. A botanist determined the following entries in an analysis of variance table for some data he collected from a randomized block design. The treatment averages were 55, 68, 56, 66, and 55.

source of variation	sum of squares	degrees of freedom	mean square
average			
blocks	534	2	
treatments	498		
residual	40		
total		15	

(a) Complete the analysis of variance table. (Are any entries impossible to determine with the information given? Explain.)

(b) Construct a reference distribution for the treatment averages.

(c) How should the experiment be conducted to allow valid conclusions to be drawn?

8. The following experiment was designed to find out to what extent a particular type of fabric gave homogeneous results over its surface for a standard wear test. In a single run the test machine could accommodate four samples of fabric, at positions 1, 2, 3, and 4. On a large sheet of the fabric four areas A, B, C, and D were marked out at random at different

places over the surface. From each area 4 samples were taken, and the 16 samples thus obtained were compared in the machine with the following results, given in milligrams of wear:

		run		
	I	II	III	IV
1	A 26.1	D 20.0	C 32.4	B 20.7
2	B 22.3	A 26.9	D 21.5	C 35.0
position 3	C 29.0	B 22.4	A 26.3	D 22.8
4	D 30.6	C 28.8	B 25.2	A 26.4

What type of design is this? Analyze the data, stating carefully your assumptions.

9. The following results were obtained from an experiment to determine whether different operators obtained different mean results in routine soil analyses for nitrogen:

operator

		A	B	C	D	E
	Tuesday	509	512	532	506	509
day	Wednesday	505	507	542	520	519
	Friday	465	472	498	483	475

On each of the 3 days a sample of soil was selected and then divided into five parts. At random these five parts were assigned to the operators to analyze.

(a) Carry out an appropriate analysis of variance. State any conclusions you reach.
(b) State any assumptions you make.
(c) Construct appropriate reference distributions. State any conclusions you reach.

(d) How would your analysis of variance have been different if the experiment had not been blocked?

(e) Do you think that, in this instance, blocking was worthwhile?

10. The following paired data were obtained from a comparison of two treatments, A and B:

treatment

block		A	B	$d = y_A - y_B$	$\bar{d}$	$d - \bar{d}$
	1	16	12	4	4	0
	2	23	16	7	4	+3
	3	30	29	1	4	−3

(a) Plot vectors $\mathbf{d}$, $\bar{\mathbf{d}}$, and $\mathbf{d} - \bar{\mathbf{d}}$ in 3-space.

(b) Construct a 95 % confidence interval for δ, the mean value of d.

(c) Carry out a paired t test.

(d) Construct an analysis of variance table showing sums of squares corresponding to the mean, treatments, blocks, residuals, and total.

11. The following are data on the number of units produced per day by different operators on different machines. Each operator used each machine on two different days.

operator

machine		A	B	C	D
	1	18(9), 17(76)	16(11), 18(77)	17(22), 20(72)	27(3), 27(73)
	2	17(1), 13(71)	18(3), 18(73)	20(57), 16(70)	28(2), 23(78)
	3	16(3), 17(77)	17(7), 19(70)	20(25), 16(73)	31(33), 30(72)
	4	15(2), 17(72)	21(4), 22(74)	16(5), 16(71)	31(6), 24(75)
	5	17(17), 18(84)	16(10), 18(72)	14(39), 13(74)	28(7), 22(82)

Eighty-four working days were needed to collect all these data. The numbers in parentheses refer to the days on which the results were obtained; for example, on the first day operator A produced 17 units using machine 2, and on the eighty-fourth day operator A produced 18 units using machine 5. On some days (e.g., on the third day) more than one item of data was collected, and on some (e.g., day 40) none was collected. Examine the data in any way you consider appropriate. Make the best analysis you can, stating assumptions.

12. Re-analyze the data from problem 11, supposing that only the data up to day 60 are available.

13. Samples of four different metals were subjected to corrosive solutions, and their corrosion rates were measured. The experimenter calculated the following analysis of variance table for the data:

source of variation	sum of squares	degrees of freedom	mean square	ratio
average	213,363.33	1		
between treatments	77.58	3	25.86⎱	11.38
within treatments	59.09	26	2.27⎰	
total	213,500.00	30		

(a) How should this experiment have been conducted to allow valid inferences to be made from this analysis?

(b) The treatment averages are as follows

metal	average
A	$\bar{y}_A = 86.143$ (7 samples)
B	$\bar{y}_B = 85.750$ (8 samples)
C	$\bar{y}_C = 82.625$ (8 samples)
D	$\bar{y}_D = 82.857$ (7 samples)

Make an appropriate reference distribution diagram.

(c) Metals A and B are of one alloy, and C and D of another. This statistic is calculated: $\frac{1}{2}(\bar{y}_A + \bar{y}_B) - \frac{1}{2}(\bar{y}_C + \bar{y}_D)$. What inferences can you draw about the corresponding parameter? What can you conclude about the performance of the two alloys?

14. (a) Develop a design suitable for evaluating three different instruments (I, II, III) for obtaining bacteria counts on three different types of floor tile (A, B, C). A reasonable trial period is 1 day for setting up

three test tiles, coating them with a standard solution, and obtaining the three bacteria counts.

(b) Set out a *detailed schedule* of how the experiment is to be conducted.

(c) Illustrate with constructed data precisely how the analysis is to be conducted.

(d) List in detail all the assumptions you make and the questions you would need to ask.

15. (a) Write a 6 × 6 Latin square design.

(b) What is the usual model employed in the analysis of data from such a design? State all assumptions.

(c) Describe a situation in which this design could be used.

(d) In what way(s) might the model and assumptions listed in (b) be violated in this situation?

16. With certain designs it is customary to use the residual mean square as an estimate of the experimental error variance σ^2. In certain circumstances, however, this estimate may be misleading. List some of these circumstances, in what you judge to be the order of their importance. Give a short explanation of each.

PART III

Measuring the Effects of Variables

Experiments are frequently performed to measure the effects of one or more variables on a response. Factorial designs are extremely useful for this purpose, especially two-level factorial designs. These designs and the fractional designs derived from them are economical and easy to use and can provide a great deal of valuable information.

In this part of the book we discuss empirical modeling (Chapter 9), the principles governing the construction and analysis of factorial designs (Chapter 10), and fractional factorial designs (Chapter 12). Examples are given (Chapters 11 and 13) that further illustrate the use of these designs in practice.

CHAPTER 9

Empirical Modeling

Models studied in preceding chapters contained mostly *qualitative* (or categorical) variables. The models we now consider also contain *quantitative* variables, such as temperature. Such models can take advantage of the inherent continuity of quantitative variables.

9.1. MATHEMATICAL MODELS

Data have no meaning in themselves; they are meaningful only in relation to a conceptual model of the phenomenon studied. Suppose, for instance, that a clock was observed at midnight on Sunday and thereafter every 12 hours. Suppose that the hands were always found to point to 6 o'clock. A set of such observations is shown in Figure 9.1a. The interpretation of these data would be different, depending on what model was thought appropriate.

One idea that would fit the facts is that the clock had stopped at 6 o'clock. The appropriate mathematical model is then

$$\eta = \beta_0 \qquad (9.1)$$

where η is the observed value of the reading of the hour hand, and β_0 is a constant equal to 6, as illustrated in Figure 9.1b. A second idea, illustrated in Figure 9.1c, is that the clock is correctly making one revolution every 12 hours but is 6 hours fast. The appropriate model is

$$\eta = (\beta_0 + x)_{\text{mod } 12} \qquad (9.2)$$

where x is the elapsed time in hours from the first reading, and $(\beta_0 + x)_{\text{mod } 12}$ means the remainder obtained after dividing $\beta_0 + x$ by 12. A third idea is that the clock is making, not one, but p revolutions every 12 hours, where p is any integer, in which case

$$\eta = (\beta_0 + px)_{\text{mod } 12} \qquad (9.3)$$

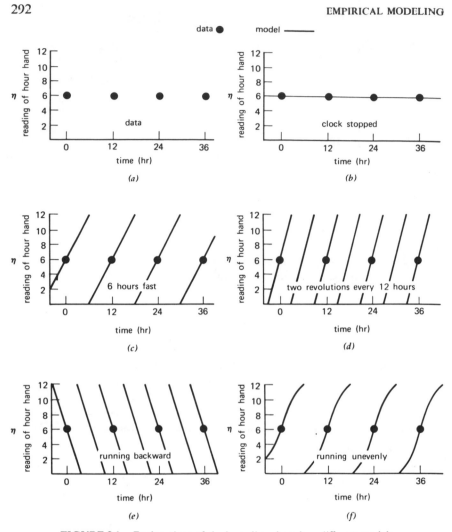

FIGURE 9.1. Explanations of clock readings based on different models.

as illustrated in Figure 9.1d for $p = 2$. In all the above, it is assumed that the hands move forward at a regular rate. Of course, the observations would be equally consistent with a model requiring the hands of the clock to run backward, as in Figure 9.1e, or to speed up at one part of the cycle and to slow down at another part, as in Figure 9.1f.

The possibilities are clearly extensive. In practice, however, some basic knowledge of the phenomenon under study (the clock mechanism) is usually available. This prior knowledge enables an experimenter to classify certain models as plausible and others as implausible. Based on the experimenter's tentative hypothesis regarding which model or models are plausible, an experimental design is chosen. Even when the investigator thinks he knows what the model should be, he must keep in mind reasonable alternatives. Accordingly the experimental design should be constructed so that it can detect inadequacies of the initial model. Model building, an important part of scientific work, is typically accomplished by the iterative process described in Chapter 1; alternative models are exposed to hazard, and survivors and new candidates are scrutinized further.

In general, experimenters are often interested in studying some relationship

$$\eta = f(x_1, x_2, \ldots, x_k) \tag{9.4}$$

between the mean value of a *response*, such as yield, quality, or efficiency, and the levels (or versions) of a number of *variables* $x_1, x_2, \ldots, x_k$, such as time, concentration, pressure, and catalyst type. For conciseness the relationship may be written as

$$\eta = f(\mathbf{x}) \tag{9.5}$$

where $\mathbf{x}$ refers jointly to the k variables $x_1, x_2, \ldots, x_k$.

Theoretical Models

Sometimes the phenomenon under study is well understood, and it is possible, from *theoretical* considerations, to write down a plausible functional form. The required physical laws are often expressed most directly by differential equations. For example, suppose that in a chemical reaction substance A is the reactant, substance B is the product, and first-order kinetics apply. Then the rate of formation of B at any instant of time is proportional to the concentration of unreacted substance A. If the mean value of the concentration of B at time x is indicated by η, the relationship between η and x can be shown* to be

$$\eta = \beta_1(1 - e^{-\beta_2 x}) \tag{9.6}$$

---

* Equation 9.6 results from solving a differential equation, which expresses mathematically the statement "The rate of formation of B is proportional to the concentration of unreacted substance A" as

$$\frac{d\eta}{dx} = \beta_2(\beta_1 - \eta)$$

It is assumed that 1 mole of B is formed from 1 mole of A and that at $x = 0$ the concentration of A is β_1.

where β_1 is the ultimate concentration of B, and β_2 is the rate constant of the reaction. This equation is called a *theoretical* or *mechanistic model* because it is based directly on an appreciation of physical or mechanistic theory governing the system—in this particular case, chemical kinetic theory. Mechanistic modeling is discussed further in Chapter 16.

Empirical Models

Frequently the mechanism underlying a process is not understood sufficiently well, or is too complicated, to allow an exact model to be postulated from theory. In such circumstances an empirical model may be useful, particularly if it is desired to approximate the response only over *limited ranges* of the variables. For example, suppose that a chemical reaction is being studied in two different chemical reactors. In both reactors measurements of yield are made over a range of reaction temperatures from 170 to 190°C. Over this range the experimenter guesses that the relationships between the mean value η of the yield and temperature x can be approximated by straight lines. Thus the empirical model contemplated is:

$$\eta = \alpha_1 + \beta_1 x \qquad \text{for reactor 1}$$
$$\eta = \alpha_2 + \beta_2 x \qquad \text{for reactor 2} \tag{9.7}$$

where the α's and β's are parameters measuring the intercepts and slopes of two straight lines. Equation 9.7 makes no claim to do more than *locally approximate* the true functions over a limited region. These two straight lines may be perfectly adequate for this purpose, even though the experimenter is quite certain that the true relationships cannot possibly be straight over wider ranges (see Figure 9.2).

The degree of complexity that should be incorporated in an empirical model can seldom be guessed with certainty a priori. One approach is to allow for a fairly general model that can be simplified in a variety of ways once the experimental results indicate what simplifications are reasonable. For example, the experimenter might hope that the straight line functions of Equation 9.7 would be adequate but, at the same time, make a provision for fitting quadratic relationships should they prove necessary:

$$\eta = \alpha_1 + \beta_1 x + \gamma_1 x^2 \qquad \text{for reactor 1}$$
$$\eta = \alpha_2 + \beta_2 x + \gamma_2 x^2 \qquad \text{for reactor 2} \tag{9.8}$$

where the γ's are parameters that measure curvature.

In practice, it would not be surprising if, over the temperature range studied, one or more of the following simplifications occurred:

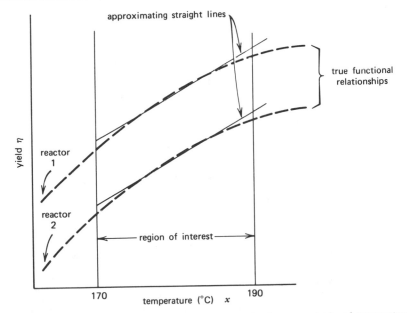

FIGURE 9.2. Straight lines approximating relationships between yield and temperature in two reactors.

1. Straight line relationships of the form of Equation 9.7 provided an adequate approximation (i.e., the coefficients γ_1 and γ_2 were zero).
2. To an adequate approximation the difference between the yields of the reactors was constant. In this case the shape of the yield–temperature curves would be the same ($\beta_1 = \beta_2$ and $\gamma_1 = \gamma_2$), but they would be displaced from one another on the yield axis ($\alpha_1 \neq \alpha_2$). This is the situation illustrated in Figure 9.2.
3. The two reactors gave identical results ($\alpha_1 = \alpha_2$, $\beta_1 = \beta_2$, and $\gamma_1 = \gamma_2$).

A good design for this particular experimental situation will allow contemplation of these various possibilities. The most economical arrangement will be to use three levels of temperature for each of the two reactors in accordance with the scheme shown in Table 9.1.

This arrangement is a 2×3 factorial design. The six sets of conditions it contains are just sufficient to determine the six constants in Equation 9.8. In addition the design allows each of the possible simplifications to be considered, whether by visual inspection of the results or by numerical computation. In practice, of course, each of the six conditions might be replicated several times.

TABLE 9.1.   A 2 × 3 factorial design

	temperature (°C)		
	170	180	190
reactor 1	—	—	—
reactor 2	—	—	—

9.2.   GEOMETRIC REPRESENTATION OF EMPIRICAL RELATIONSHIPS

In trying to understand what is being attempted in planning and analyzing experiments, it is helpful to have in mind a geometric representation. A graph like Figure 9.3 can be used to map out a relationship between a mean response η and a single quantitative variable x. A relationship between η and two quantitative variables x_1 and x_2 may be represented by a curved *surface* like that shown in Figure 9.4a. Alternatively, a contour representation may be used, as shown in Figure 9.4b. The relationship between a mean response η and three quantitative variables $x_1, x_2,$ and x_3 can be represented by contour surfaces plotted in the (x_1, x_2, x_3)-space, as illustrated in Figure 9.5.

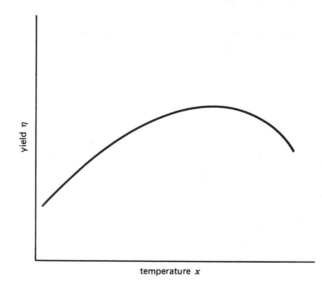

FIGURE 9.3.   Graphical representation of relationship between mean yield and temperature.

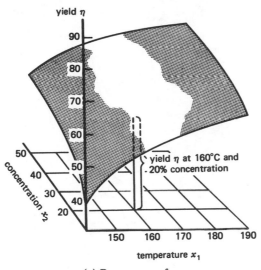

(a) Response surface.

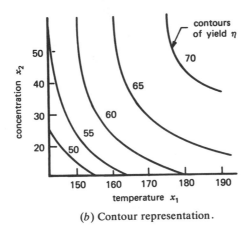

(b) Contour representation.

FIGURE 9.4. Graphical representations of relationship between mean yield, and temperature and concentration.

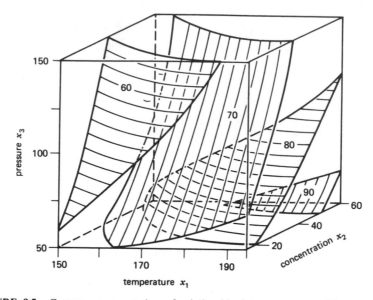

FIGURE 9.5. Contour representation of relationship between mean yield, temperature, concentration, and pressure.

When qualitative variables and quantitative variables occur together, the relationships can be represented by a series of graphs or contour diagrams such as those in Figure 9.6. Here two qualitative variables (reactor and catalyst) and two quantitative variables (temperature and concentration) are represented by four two-dimensional contour diagrams.

9.3. THE PROBLEM OF EXPERIMENTAL DESIGN

The basic problem of experimental design is deciding what pattern of design points will best reveal aspects of the situation of interest. The runs that the experimenter decides to perform can be represented by points on a diagram like Figure 9.6. For example, the point labeled P in that figure represents a potential experimental run: at 170°C and 20% concentration, with catalyst 1 in reactor 1.

The question of where the points should be placed is a circular one in the sense that, if we knew what the response function was like, we could decide where the points should be. But to find out what the response function is like is precisely the object of the investigation. Fortunately this circularity is not crippling, particularly when experiments may be conducted sequentially so

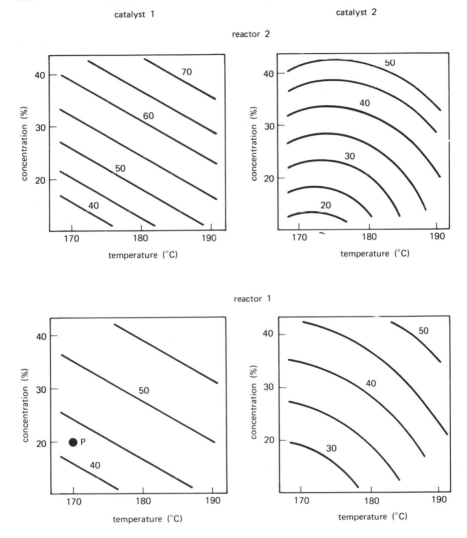

FIGURE 9.6. Example of yield–temperature–concentration relationships for two reactors and two catalysts.

that information gained in one set directly influences the choice of experiments in the next. We discuss this point further in Section 9.4.

The experimenter is like a person attempting to map the depth of the sea by making soundings at a limited number of places. If some theory existed about the nature of the sea bed in a particular area, based perhaps on geological considerations and knowledge of currents and tides, it might allow the

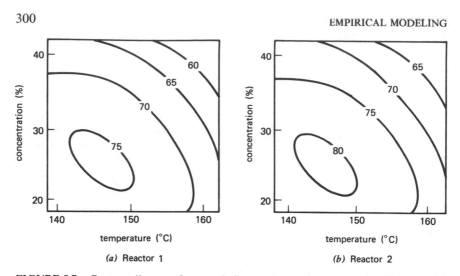

FIGURE 9.7. Contour diagrams for two similar reactors with a systematic difference of 5 units in mean yield.

experimenter to work with a fairly definite theoretical model in which the only uncertainties were the values of certain parameters. The appropriate strategy might be quite different, however, if no theory was available. This book is primarily concerned with the situation in which no very definite theory *is* available and an empirical approach must be followed.

Sometimes it is said that an empirical approach is appropriate when *nothing* is known about the system to be studied. But in practice something is always known. In particular, we can usually be fairly certain of some degree of continuity in the relationship studied.* This is fortunate because, with any reasonable number of experimental runs, mapping a surface resembling a nest of stalagmites or the back of a porcupine would be impossible. Furthermore for such a surface sequential experimentation would be useless, since characteristics of the surface at one point would not be related to characteristics elsewhere.

Smoothness and Similarity

Functions met in practice usually show fairly *smooth* relationships like those represented in Figures 9.3 to 9.6. Ignoring for the time being the question of the experimental error, we can say that the simpler the relationship, the fewer the number of experimental points needed to explore it. For example, if it

---

* Or more precisely when discontinuity is likely to occur (e.g., because a liquid product becomes gaseous at a certain temperature.)

were *certain* that the relation between a response η and a single variable x could be represented by a straight line, in the absence of experimental error only two points would be required to determine it exactly.

A further simplifying circumstance is *similarity*, which is often found in response functions when certain *qualitative* changes are made. For instance, for two reactors of basically the same design, we would not be surprised to find the situation illustrated in Figure 9.7. Over the region of interest, reactor 1 behaves very much the same as reactor 2 but with 5% lower yield.

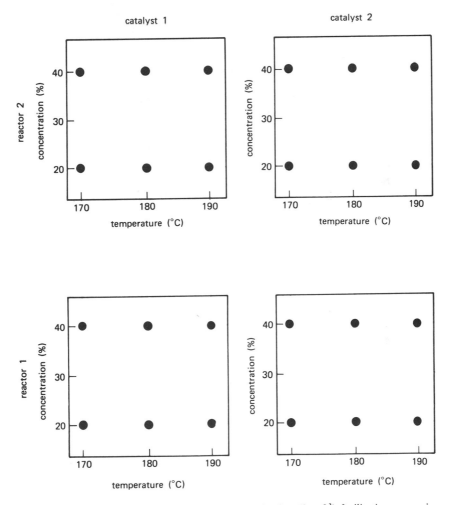

FIGURE 9.8. Factorial arrangement of experimental points (3×2^3), facilitating comparison and model identification.

Factorial designs facilitate the discovery of similarities and simplifications and thus assist the process of model building. These experimental designs can also provide estimates of the "effects" of the changes, which are infected as little as possible by experimental error. For the four variables of Figure 9.6, for example, the investigator might employ three levels of temperature (170, 180, 190°C) and two levels of concentration (20 and 40%) for each of the two catalysts and two reactors. The experimental points would then follow the pattern shown in Figure 9.8. The resulting arrangement is a $3 \times 2 \times 2 \times 2$

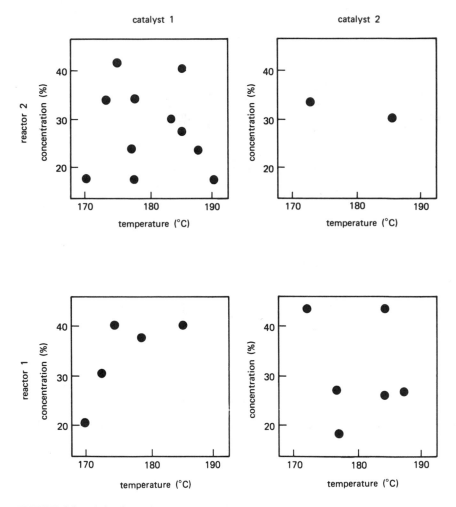

FIGURE 9.9. A haphazard arrangement of experimental points, making comparison and model identification difficult.

factorial design, which requires 24 experimental runs. With this design the behavior of each variable at a variety of levels of the other variables can be readily appreciated, greatly facilitating appropriate choice of a model form.

By contrast a haphazard arrangement of the same number of experimental points like that of Figure 9.9 renders it difficult or impossible to make comparisons and to find the similarities necessary for the development of an appropriate model. One sometimes hears it argued that in this age of computers one can just as easily fit a response function to a haphazard set of points as to the points of a systematic experimental design. It must be remembered, however, that initially the investigator usually does not know *what* function to fit. At the model identification (specification) stage, visualization of the principal characteristics of the system which is aided by the factorial arrangement, is of great help in deciding what *form of function is worth considering*.

9.4. COMPREHENSIVE VERSUS SEQUENTIAL APPROACH TO EXPERIMENTAL INVESTIGATIONS

In exploring a functional relationship it might appear reasonable at first sight to adopt a comprehensive approach in which the entire range of every factor was investigated. The resulting design might contain all combinations of several levels of all factors. However, when runs can be made in successive groups, this is an inefficient way to organize experimental programs. The situation relates to the paradox that the best time to design an experiment is after it is finished, the converse of which is that the worst time is at the beginning, when least is known. If the entire experiment was designed at the outset, the following would have to be assumed as known: (1) which variables were the most important, (2) over what ranges the variables should be studied, (3) in what metrics the variables and the responses should be considered (e.g., linear, logarithmic, or reciprocal scales), and (4) what multivariable transformations should be made (perhaps the effects of variables x_1 and x_2 would be most simply expressed in terms of their ratio x_1/x_2 and their sum $x_1 + x_2$).

The experimenter is least able to answer such questions at the outset of an investigation but gradually becomes more able to do so as a program evolves.

Advantages of Sequential Approach

All the above arguments point to the desirability of a sequence of moderately sized designs and reassessment of the results as each group of experiments becomes available. Thus, as an investigation proceeds, the following may occur:

1. The general location of the experiments in the space of the variables can change to a more promising neighborhood.
2. Some of the variables initially included may be dropped, and others substituted—the subspace of the factor space being explored can change.
3. Evidence collected may indicate that some variables would be best considered in other transformations—the coordinate system can change.
4. The objective of the investigation can change (in particular, if we are specifically looking for silver and we strike gold, the discovery ought not to be ignored).

The 25% Rule

As a rough general rule, not more than one quarter of the experimental effort (budget) should be invested in a first design. Of course many types of investigations are conducted in countless different circumstances, so there will be many exceptions. Nevertheless, in most circumstances it is unwise to plan too comprehensive a design at the outset. When the first part of an investigation has been completed, the experimenter will usually know considerably more than when he started and consequently will be able to plan a better second part, which in turn will lead to improved planning of a third part, and so on.

It is sometimes argued that the sequential approach customarily applied in, for example, engineering, chemical, biological, medical, and psychological research is not applicable to agricultural research, because there, data become available only once a year. On the longer view, however, such research is, of course, also sequential.

In recent years mathematical theories of so-called *optimal* design have been pursued (see also the end of Section 14.2). These tend to greatly oversimplify the experimental situation. In such studies it is usually implicit that the variables of interest, the region of experimentation, and the metrics and transformations in which the variables are to be considered are all known, and the only problem is to find at what points, in a known region of a defined factor space, experiments should be run. This oversimplification is further oversimplified by supposing that, except for experimental error, the mathematical model that describes the physical phenomenon is known exactly. Finally, it is assumed that the experimenter's purposes, which are actually multifaceted, can be summarized in terms of a single quantity called the *design criterion*, which, when "optimized," produces the ideal experimental arrangement. Conclusions reached from mathematical exercises of this kind must be applied with some caution.

Hindsight and Foresight

At the end of an investigation, on looking back at the first experiments, one is frequently impressed with (and even embarrassed about) how odd and pathetic they appear. Initially some of the wrong variables may have been studied, or some of the right variables may have been investigated over inappropriate ranges (too high, low, wide, or narrow) and in what are now known to be inappropriate metrics or transformations. It is rather like looking at an old movie of a swimmer, who can now do back flips from a high diving board, when he was a young child making his first feeble attempts to keep his head above water. It is unrealistic to expect to dive off the high board right away and neurotic to say, "If I can't dive off the high board immediately I'm not going to start learning how to swim." The investigator must learn from the swimmer, who was prepared to begin by putting his foot in the water and was not afraid of getting wet.

QUESTIONS FOR CHAPTER 9

1. What is an empirical model?
2. What is a theoretical or mechanistic model?
3. Can you describe an experimental situation, preferably in your own field, in which an empirical approach would be preferable? In which a theoretical approach would be preferable?
4. Plot contours of surfaces with the following shapes: an inclined plane, an upturned canoe, a pyramid, a doughnut.
5. For most experimental investigations, why is a sequential approach better than a comprehensive one?
6. Think of a specific experimental investigation. Imagine that an experimenter involved in this investigation asks you to help with the statistical design. What questions would you want to ask?

CHAPTER 10

Factorial Designs
at Two Levels

A class of designs is now discussed that are of great practical importance—two-level factorial designs. In this chapter the uses, generation, and analysis of these designs are considered.

10.1. GENERAL FACTORIAL DESIGNS AND DESIGNS AT TWO LEVELS

To perform a general factorial design, an investigator selects a fixed number of "levels" (or "versions") for each of a number of variables (factors) and then runs experiments with all possible combinations. If there are l_1 levels for the first variable, l_2 for the second, ..., and l_k for the kth, the complete arrangement of $l_1 \times l_2 \times \cdots \times l_k$ experimental runs is called an $l_1 \times l_2 \times \cdots \times l_k$ factorial design. For example, a $2 \times 3 \times 5$ factorial design requires $2 \times 3 \times 5 = 30$ runs, and a $2 \times 2 \times 2 = 2^3$ factorial design 8 runs. An application and analysis of a 3×4 factorial design in four replications was given in Section 7.7. In the present chapter we discuss designs in which each variable occurs at only two levels. These designs are of importance for a number of reasons.

1. They require relatively few runs per factor studied; and although they are unable to explore fully a wide region in the factor space, they can indicate major trends and so determine a promising direction for further experimentation.
2. We see in Chapter 15 that, when a more thorough local exploration is needed, they can be suitably augmented to form composite designs.

3. In Chapter 12 we see that they form the basis for two-level *fractional* factorial designs. These fractional designs are often of great value at an early stage of an investigation, when it is frequently good practice to use a preliminary experimental effort to look at a large number of factors superficially rather than a small number (which may or may not include the important ones) thoroughly.
4. These designs and the corresponding fractional designs may be used as building blocks so that the degree of complexity of the finally constructed design can match the sophistication of the problem.
5. The interpretation of the observations produced by the designs can proceed largely by using common sense and elementary arithmetic.

In all these applications the designs fit naturally into the sequential strategy discussed in Chapter 1, which is an essential feature of the scientific process.

10.2. AN EXAMPLE OF A 2^3 FACTORIAL DESIGN: PILOT PLANT INVESTIGATION

Table 10.1a shows a 2^3 factorial experiment in which there are two quantitative variables—temperature and concentration—and a single qualitative variable—catalyst. The response is the chemical yield. The original data were part of a pilot plant investigation of a process and have been simplified somewhat for illustrative purposes. Table 10.1b shows the recorded data with the levels coded so that for the quantitative variables a minus sign represents the low level and a plus sign the high level. For a qualitative variable the two versions or "levels" can also be conveniently coded by minus and plus signs. It does not matter here which is associated with the plus as long as the labeling is consistent. A display of levels to be run in a design such as is given in Table 10.1 is called a *design matrix*.

Exercise 10.1. How many variables and how many runs are there in a $2 \times 4 \times 3 \times 2$ factorial design? *Answer*: 4 variables, 48 runs.

Exercise 10.2. If four variables are to be studied using a three-level factorial design (all variables are studied at three levels), how will we designate the design and how many runs will it require? *Answer*: 3^4 design, 81.

There are other notations in common use for the design matrix of a two-level factorial. One identifies the "upper" level of each factor by the use of the corresponding lower case letter. Another notation uses a 0 and 1 in place of our − and + signs. The three alternative notations for a 2^3 factorial design are as follows:

Run	T	C	K		T	C	K
1	−	−	−	1	0	0	0
2	+	−	−	t	1	0	0
3	−	+	−	c	0	1	0
4	+	+	−	tc	1	1	0
5	−	−	+	k	0	0	1
6	+	−	+	tk	1	0	1
7	−	+	+	ck	0	1	1
8	+	+	+	tck	1	1	1

TABLE 10.1.   Data from a 2^3 factorial design, pilot plant example

test condition number	temperature (°C) T	concentration (%) C	catalyst (A or B) K	yield (grams) $\bar{y}$
		a. Original units of variables		
1	160	20	A	60
2	180	20	A	72
3	160	40	A	54
4	180	40	A	68
5	160	20	B	52
6	180	20	B	83
7	160	40	B	45
8	180	40	B	80
		b. Coded units of variables		
1	−	−	−	60
2	+	−	−	72
3	−	+	−	54
4	+	+	−	68
5	−	−	+	52
6	+	−	+	83
7	−	+	+	45
8	+	+	+	80

temperature (°C)		concentration (%)		catalyst	
−	+	−	+	−	+
160	180	20	40	A	B

We prefer the $\pm$ notation because it relates to a geometric view of the design and, as we will see later, is also readily applicable to regression analysis and to the construction of fractional factorial designs.

10.3. CALCULATION OF MAIN EFFECTS

Averaging Individual Measures of Effects

What can we find from this factorial design? For example, what does it tell us about the *effect* of temperature on yield? (By the "effect" of a factor we mean the change in the response as we move from the $-$ to the $+$ version of that factor, here from the low to the high level of temperature.) Consider the first two tests in Table 10.1. Aside from experimental error the corresponding yields (60 and 72) differ only because of temperature. The concentration (20%) and the catalyst (A) are the same for both of these conditions. Altogether there are four measures of the temperature effect at each of the four combinations of conditions of the other variables as listed below.

	condition at which comparison is made	
individual measure of the effect of changing temperature from 160 to 180°C	concentration C	catalyst K
$y_2 - y_1 = 72 - 60 = 12$	20	A
$y_4 - y_3 = 68 - 54 = 14$	40	A
$y_6 - y_5 = 83 - 52 = 31$	20	B
$y_8 - y_7 = 80 - 45 = \underline{35}$	40	B
main effect of temperature $\quad T = 23$		

The average of these four measures ($+23$ for this example) is called the *main effect* of temperature and is denoted by T. It measures the *average* effect of temperature over all conditions of the other variables.

Because of the general symmetry of the design (see Figure 10.1), there is a similar set of four measures for the effect of concentration, for each of which the levels of the remaining variables are constant.

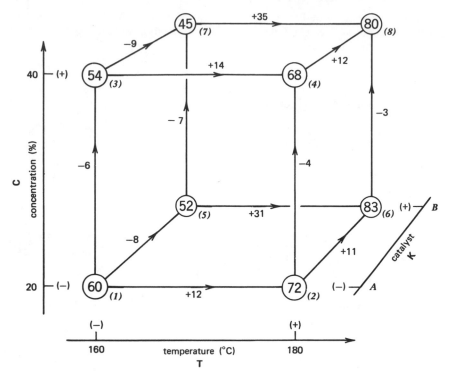

FIGURE 10.1. 2^3 factorial design, pilot plant example.

	condition at which comparison is made	
individual measure of the effect of changing concentration from 20 to 40%	temperature **T**	catalyst **K**
$y_3 - y_1 = 54 - 60 = -6$	160	A
$y_4 - y_2 = 68 - 72 = -4$	180	A
$y_7 - y_5 = 45 - 52 = -7$	160	B
$y_8 - y_6 = 80 - 83 = \underline{-3}$	180	B

main effect
of concentration $C = -5$

Finally (see Figure 10.1), there are four measures of the effect of catalyst.

individual measures of the effect of changing from catalyst A to catalyst B	condition at which comparison is made	
	temperature **T**	concentration **C**
$y_5 - y_1 = 52 - 60 = -8$	160	20
$y_6 - y_2 = 83 - 72 = 11$	180	20
$y_7 - y_3 = 45 - 54 = -9$	160	40
$y_8 - y_4 = 80 - 68 = 12$	180	40
main effect of catalyst $K = 1.5$		

Difference between Two Averages

The main effect for each of the variables is seen to be the difference between two averages:

$$\text{main effect} = \bar{y}_+ - \bar{y}_- \tag{10.1}$$

where $\bar{y}_+$ is the average response for the plus level of the variable and $\bar{y}_-$ is the average response for the minus level. Thus (see Figure 10.2a)

temperature effect

$$T = \frac{72 + 68 + 83 + 80}{4} - \frac{60 + 54 + 52 + 45}{4}$$

$$= 75.75 - 52.75 = 23$$

concentration effect

$$C = \frac{54 + 68 + 45 + 80}{4} - \frac{60 + 72 + 52 + 83}{4}$$

$$= 61.75 - 66.75 = -5$$

catalyst effect

$$K = \frac{52 + 83 + 45 + 80}{4} - \frac{60 + 72 + 54 + 68}{4}$$

$$= 65.0 - 63.5 = 1.5$$

Notice that (1) *all* the observations are being used to supply information on *each* of the main effects, and (2) each effect is determined with the precision of a fourfold replicated difference.

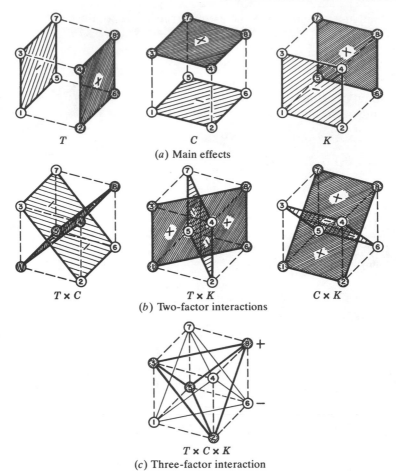

(a) Main effects

(b) Two-factor interactions

(c) Three-factor interaction

FIGURE 10.2. Geometric representation of contrasts corresponding to main effects and interactions.

Advantages over the "One-Factor-at-a-Time" Method

Suppose that in the above investigation, instead of a factorial arrangement, the "one-factor-at-a-time" method had been used. This method, in which experimental factors are varied one at a time, with the remaining factors held constant, was formerly regarded as the only correct way to conduct research (see Section 15.1 for further discussion of this procedure). The method provides an estimate of the effect of a single variable at selected *fixed* conditions of the other variables. However, for such an estimate to have general relevance it is necessary to assume that the effect would be the same at other settings of the other variables—that, over the ranges of current interest, the variables act on the response additively. However, (1) if the variables *do* act additively, the factorial does the job with more precision; and (2) if the variables do *not* act additively, the factorial, unlike the one-factor-at-a-time design, can detect and estimate interactions that *measure* the nonadditivity.

Gain in Precision If Variables Act Additively

To secure the same precision for the estimate of the temperature effect the one-factor-at-a-time experiment would need to employ eight runs, four at each level of temperature, with all the observations made at some arbitrarily *fixed* levels of concentration and catalyst. In a similar manner two further sets of eight runs would be required to study concentration and catalyst. Thus to obtain estimates of the main effects of three variables with the same precision as is provided by the 2^3 factorial design, the one-factor-at-a-time method would require 24 runs—a threefold increase. In general, for k factors a k-fold increase would be required. Some economy can be introduced by using a single experimental condition from which to make all changes; even with this arrangement, however, the one-factor-at-a-time design requires $(k + 1)/2$ times as many runs as the factorial.

10.4. INTERACTION EFFECTS

Two-Factor Interactions

In the example, the average effect of temperature is 23. It is obvious from the data, however, that the temperature effect is much greater with catalyst B than with catalyst A. The variables temperature and catalyst do not behave additively and are therefore said to "interact." A measure of this interaction is supplied by the difference between the average temperature effect with catalyst A and the average temperature effect with catalyst B. By convention, *half* the difference is called the *temperature by catalyst interaction* or, in symbols, the $T \times K$ interaction.

| | average |
catalyst	temperature effect
$(+)B$	33
$(-)A$	13
difference	20

$$T \times K \text{ interaction} = \frac{20}{2} = 10 \qquad (10.2)$$

The $T \times K$ interaction may equally well be thought of as one-half the difference in the average *catalyst* effects at the two levels of *temperature*.

temperature (°C)	average catalyst effect
(+)180	11.5
(−)160	−8.5
difference	20.0

$$T \times K \text{ interaction} = \frac{20}{2} = 10 \qquad (10.3)$$

That the results must be equivalent is seen as soon as we consider the way the interaction employs the observations. Consider the calculations leading to Equation 10.2.

catalyst	average temperature effect
(+)B	$33 = \dfrac{31 + 35}{2} = \frac{1}{2}(y_6 - y_5 + y_8 - y_7)$
(−)A	$13 = \dfrac{12 + 14}{2} = \frac{1}{2}(y_2 - y_1 + y_4 - y_3)$

$$T \times K \text{ interaction} = \tfrac{1}{2}\text{ difference} = \frac{33 - 13}{2} = 10$$

$$= \tfrac{1}{4}(y_1 - y_2 + y_3 - y_4 - y_5 + y_6 - y_7 + y_8)$$

$$= \frac{y_1 + y_3 + y_6 + y_8}{4} - \frac{y_2 + y_4 + y_5 + y_7}{4} \qquad (10.4)$$

The reader should confirm that exactly the same result is obtained by carrying through the same argument for Equation 10.3.

Like the main effects, the interaction effect is seen to be a difference between two averages, half of the eight results being included in one average and half in the other. Just as main effects may be viewed as a *contrast* between observations on parallel faces of the cube, as shown in Figure 10.2a, the $T \times K$ interaction is a contrast between results on two *diagonal planes*, as shown in Figure 10.2b. The $T \times C$ and the $C \times K$ interactions are obtained in a similar way.

concentration	average temperature effect	catalyst	average concentration
$(+)\,40\%$	24.5	$(+)B$	-5.0
$(-)\,20\%$	21.5	$(-)A$	-5.0
difference	3.0	difference	0.0

$$T \times C \text{ interaction} = \frac{3}{2} = 1.5 \qquad C \times K \text{ interaction} = \frac{0}{2} = 0 \quad (10.5)$$

Also, as illustrated in Figure 10.2b,

$$T \times C \text{ interaction} = \frac{y_1 + y_4 + y_5 + y_8}{4} - \frac{y_2 + y_3 + y_6 + y_7}{4}$$

$$C \times K \text{ interaction} = \frac{y_1 + y_2 + y_7 + y_8}{4} - \frac{y_3 + y_4 + y_5 + y_6}{4} \tag{10.6}$$

Three-Factor Interaction

Consider the temperature by concentration $(T \times C)$ interaction. Two measures of the $T \times C$ interaction are available from the experiment, one for each catalyst.

$T \times C$ interaction with catalyst $B\,(+)$:

$$\frac{(y_8 - y_7) - (y_6 - y_5)}{2} = \frac{(80 - 45) - (83 - 52)}{2} = \frac{35 - 31}{2} = 2 \tag{10.7}$$

$T \times C$ interaction with catalyst $A\,(-)$:

$$\frac{(y_4 - y_3) - (y_2 - y_1)}{2} = \frac{(68 - 54) - (72 - 60)}{2} = \frac{14 - 12}{2} = 1 \tag{10.8}$$

The difference measures the consistency of the *temperature by concentration interaction* for the two catalysts. Half this difference is defined as the *three-factor interaction* of temperature, concentration, and catalyst, denoted as the $T \times C \times K$ interaction. Thus

$$T \times C \times K \text{ interaction} = \frac{2 - 1}{2} = 0.5 \tag{10.9}$$

As before, this interaction is symmetric in all the variables. For example, it could equally well have been defined as half the difference between the temperature by catalyst interactions at each of the two concentrations. The estimate of this three-factor interaction can again be reduced to the difference between two averages.* If the experimental points contributing to the two averages are isolated, they define the vertices of the two tetrahedra in Figure 10.2c, which together comprise the cube. It is helpful to remember that for any main effect or interaction the $\bar{y}_+$ average *always* contains the observation from the run in which all variables are at their plus levels, for example, for a 2^3 factorial, the $(+ + +)$ run.

Exercise 10.3.   Calculate the main effects and interactions for the following data:

test condition number	brand of popcorn 1	ratio of popcorn to oil 2	batch size (cup) 3	yield of popcorn y
1	− (ordinary)	− (low)	− ($\frac{1}{3}$)	$6\frac{1}{4}$
2	+ (gourmet)	− (low)	− ($\frac{1}{3}$)	8
3	− (ordinary)	+ (high)	− ($\frac{1}{3}$)	6
4	+ (gourmet)	+ (high)	− ($\frac{1}{3}$)	$9\frac{1}{2}$
5	− (ordinary)	− (low)	+ ($\frac{2}{3}$)	8
6	+ (gourmet)	− (low)	+ ($\frac{2}{3}$)	15
7	− (ordinary)	+ (high)	+ ($\frac{2}{3}$)	9
8	+ (gourmet)	+ (high)	+ ($\frac{2}{3}$)	17

Answer: $1 = 5.1$, $2 = 1.1$, $3 = 4.8$, $1 \times 2 = 0.7$, $1 \times 3 = 2.4$, $2 \times 3 = 0.4$,
$$1 \times 2 \times 3 = -0.2.$$

Notation

In what follows italic type is used to denote the effects of factors (variables). Bold face type is used to denote the factors themselves and the columns of signs that determine their levels. Thus T is the main effect of variable **T**.

* $T \times C \times K$ interaction $= \frac{1}{4}(-y_1 + y_2 + y_3 - y_4 + y_5 - y_6 - y_7 + y_8)$

$$= \frac{y_2 + y_3 + y_5 + y_8}{4} - \frac{y_1 + y_4 + y_6 + y_7}{4}$$

10.5. INTERPRETATION OF RESULTS

The results of the factorial analysis are collected in Table 10.2. A point we have not yet mentioned about the data is that each of the eight yield values in Table 10.1 was actually an average of two replicate runs. We show in Section 10.6 how, by using this fact, we can obtain the standard errors shown in Table 10.2. For the moment we use these standard errors to complete an

TABLE 10.2. Calculated effects and standard errors for the 2^3 factorial design, pilot plant example

effect	estimate $\pm$ standard error
average	64.25 ± 0.7
main effects	
temperature T	23.0 ± 1.4
concentration C	-5.0 ± 1.4
catalyst K	1.5 ± 1.4
two-factor interactions	
$T \times C$	1.5 ± 1.4
$T \times K$	10.0 ± 1.4
$C \times K$	0.0 ± 1.4
three-factor interaction	
$T \times C \times K$	0.5 ± 1.4

analysis of the data. Comparison of the estimates with their standard errors (see also Figure 10.3) suggests that the circled items T, C, and the interaction $T \times K$ require interpretation, while the apparent effects remaining could be generated by the noise.

The main effect of a variable should be individually interpreted only if there is no evidence that the variable interacts with other variables. When

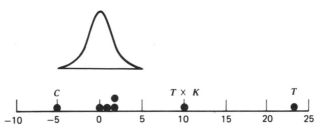

FIGURE 10.3. Main effects and interactions in relation to a reference t distribution with eight degrees of freedom and scale factor 1.4, pilot plant example.

there is evidence of one or more such interaction effects, the interacting variables should be considered jointly.

In Table 10.2 there is a large temperature effect, 23.0 ± 1.4. But since temperature interacts with catalyst type (the $T \times K$ interaction is 10.0 ± 1.4), we make no statement about the effect of temperature alone. The main effect of concentration is -5.0 ± 1.4, and in this case there is no evidence of any interactions involving concentration. Thus we draw the following tentative conclusions:

1. The effect of concentration (**C**) is to reduce the yield by about five units, and this is approximately so irrespective of the tested levels of the other variables.
2. The effects of temperature (**T**) and catalyst (**K**) cannot be interpreted separately because of the large $T \times K$ interaction, and can best be considered using the two-way table shown in Figure 10.4. The interaction evidently arises from a difference in sensitivity to temperature change for the two catalysts. With catalyst A the temperature effect is 13 units, but with catalyst B it is 33 units.

The result of most practical interest was the very different behaviors of the two "catalyst types" in response to temperature. The effect was unexpected, for although obtained from two different suppliers, the catalysts

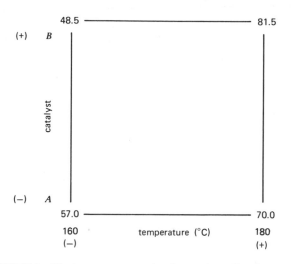

FIGURE 10.4.   The temperature–catalyst interaction, pilot plant example.

were supposedly identical. Also, the yield from catalyst B at 180°C was the highest that, up to that time, had been seen. The finding led to a very careful study of the catalyst in further iterations of the investigation.

10.6. CALCULATION OF STANDARD ERRORS FOR EFFECTS USING REPLICATED RUNS

When genuine run replicates are made under a given set of experimental conditions, the variation between their associated observations may be used to estimate the standard deviation of a single observation and hence the standard deviation of the effects. By *genuine* run replicates we mean that variation between runs made at the same experimental conditions is a reflection of the *total* variability afflicting runs made at different experimental conditions. This point requires careful consideration.

Randomization of run order usually ensures that replicates are genuine. Consider the pilot plant example. A pilot plant run consisted of (1) cleaning the reactor, (2) inserting the appropriate catalyst charge, (3) running the apparatus at a given temperature and a given feed concentration for 3 hours to allow the process to settle down at the chosen experimental conditions, (4) sampling the output every 15 minutes during the final hour of running, and (5) combining chemical analyses made on these samples. A genuine run replicate must involve the taking of all these steps again. In particular, several chemical analyses from a single run would provide only an estimate of *analytical* variance, usually only a small part of the run-to-run variance. Similarly, several samples from the same run could provide only an estimate of sampling plus analytical variance. Generally this problem of wrongly assessing experimental error variance has been particularly troublesome. It has led, for instance, to gross underestimates of the errors associated with such quantities as the astronomical unit (see Youden, 1972, and the references therein).

In general, if g sets of experimental conditions are genuinely replicated and the n_i replicate runs made at the ith set yield an estimate s_i^2 of σ^2 having $v_i = n_i - 1$ degrees of freedom, the pooled estimate of run variance is

$$s^2 = \frac{v_1 s_1^2 + v_2 s_2^2 + \cdots + v_g s_g^2}{v_1 + v_2 + \cdots + v_g} \tag{10.10}$$

with $v = v_1 + v_2 + \cdots + v_g$ degrees of freedom.

With only $n_i = 2$ replicates at each of the g sets of conditions, the formula for the ith variance reduces to $s_i^2 = d_i^2/2$ with $v_i = 1$, where d_i is the difference between the duplicate observations for the ith set of conditions. Thus

TABLE 10.3.   Estimation of the variance, pilot plant example

average response value (previously used in the analysis)	T	C	K	results from individual runs*		difference of duplicate	estimated variance at each set of conditions $s_i^2 = (\text{difference})^2/2$
60	−	−	−	$59^{(6)}$	$61^{(13)}$	−2	2
72	+	−	−	$74^{(2)}$	$70^{(4)}$	4	8
54	−	+	−	$50^{(1)}$	$58^{(16)}$	−8	32
68	+	+	−	$69^{(5)}$	$67^{(10)}$	2	2
52	−	−	+	$50^{(8)}$	$54^{(12)}$	−4	8
83	+	−	+	$81^{(9)}$	$85^{(14)}$	−4	8
45	−	+	+	$46^{(3)}$	$44^{(11)}$	2	2
80	+	+	+	$79^{(7)}$	$81^{(15)}$	−2	2

total   64

s^2 = pooled estimate of σ^2 = average of estimated variances
= $\frac{64}{8}$ = 8 with v = 8 degrees of freedom

* Superscripts give the order in which the runs were made.

Equation 10.10 yields $s^2 = \sum d_i^2/2g$. Using the replicated pilot plant data displayed in Table 10.3, where $g = 8$, we obtain $s^2 = 128/(2 \times 8) = 8$ and $s = 2.8$ with $v = 8$ degrees of freedom.

Since each main effect and interaction is a statistic of the form $\bar{y}_+ - \bar{y}_-$, where each average contains eight observations, the variance of each effect (assuming independent errors) is given by

$$V(\text{effect}) = V(\bar{y}_+ - \bar{y}_-) = (\tfrac{1}{8} + \tfrac{1}{8})\sigma^2 = \tfrac{1}{4}\sigma^2$$

In general, if a total of N runs is made in conducting a two-level factorial or replicated factorial design, then

$$V(\text{effect}) = \frac{4}{N}\sigma^2$$

For the pilot plant data, on substituting for σ^2 the estimate $s^2 = 8$, the estimated variance of an effect is $(\tfrac{1}{8} + \tfrac{1}{8})8 = 2$ (or, equivalently, $\tfrac{4}{16} \times 8 = 2$). Thus the estimated standard error of an effect is $\sqrt{2.0} = 1.4$, the value used in Table 10.2.

From equation 3.16, the variance of a mean is σ^2/N. Thus the estimated standard error of the mean is $s/\sqrt{N} = 2.8/\sqrt{16} = 0.7$, the value used in Table 10.2.

Exercise 10.4. For the following data calculate the main effects and interactions and their standard errors:

test condition number	depth of planting (in.) 1	watering (times daily) 2	type of lima bean 3	yield replication 1	replication 2	replication 3
1	$-$ ($\frac{1}{2}$)	$-$ (once)	$-$ (baby)	6	7	6
2	$+$ ($\frac{3}{2}$)	$-$ (once)	$-$ (baby)	4	5	5
3	$-$ ($\frac{1}{2}$)	$+$ (twice)	$-$ (baby)	10	9	8
4	$+$ ($\frac{3}{2}$)	$+$ (twice)	$-$ (baby)	7	7	6
5	$-$ ($\frac{1}{2}$)	$-$ (once)	$+$ (large)	4	5	4
6	$+$ ($\frac{3}{2}$)	$-$ (once)	$+$ (large)	3	3	1
7	$-$ ($\frac{1}{2}$)	$+$ (twice)	$+$ (large)	8	7	7
8	$+$ ($\frac{3}{2}$)	$+$ (twice)	$+$ (large)	5	5	4

Answer: $1 = -2.2$, $2 = 2.5$, $3 = -2.0$, $1 \times 2 = -0.3$, $1 \times 3 = -0.2$, $2 \times 3 = 0.2$, $1 \times 2 \times 3 = 0.0$, standard error of effect $= 0.3$.

Exercise 10.5. Repeat Exercise 10.4, assuming that data for replication 3 are unavailable.
Answer: $1 = -2.1$, $2 = 2.6$, $3 = -1.9$, $1 \times 2 = -0.4$, $1 \times 3 = 0.1$, $2 \times 3 = -0.1$, $1 \times 2 \times 3 = -0.1$, standard error of effect $= 0.28$.

Exercise 10.6. Suppose that in addition to the eight runs in Exercise 10.3 the following popcorn yields were obtained from genuine replicate runs made at the center conditions $1 =$ a mixture of 50% ordinary and 50% gourmet popcorn, $2 =$ medium, $3 = \frac{1}{2}$ cup: 9, 8, $9\frac{1}{2}$, and 10 cups. Use these extra runs to calculate standard errors for main effects and interactions. Plot the effects in relation to an appropriate reference distribution, and draw conclusions.
Answer: Main effects and interactions remain unchanged. They are given in Exercise 10.3. Estimate of σ^2 from center points $= s^2 = 0.73$ with three degrees of freedom. Standard error of an effect $= 0.60$. Reference t distribution has three degrees of freedom, scale factor $= 0.60$.

Exercise 10.7. Show that the sum of squares in the analysis of variance table associated with any effect from any two-level factorial design containing (including possible replication) a total of N runs $= N \times$ (estimated effect)$^2/4$.

Exercise 10.8. Given the data in Table 10.3, an alternative way to estimate the variance is to set up a one-way classification analysis of variance table. For the 16 observations and 8 treatments of this example, complete the analysis of variance table and show that $s^2 = 8$ with $v = 8$ degrees of freedom.

Economy in Experimentation

In most situations there are more factors to be investigated than can conveniently be accommodated with the time and budget available. Rather than duplicate a 2^3 factorial as was done in the pilot plant study, it is usually better to include a fourth variable and run an unreplicated 2^4 design. Or, as we shall see in Chapter 12, it may be even better to run a half-replicated 2^5 design and use the 16 runs to study five factors. The reader at this point may be worried about obtaining an estimate of error if there is no replication. We show in Section 10.8 how it is usually possible to overcome this difficulty.

10.7. QUICKER METHODS FOR CALCULATING EFFECTS

It would be extremely tedious if effects had to be calculated from first principles whenever a factorial design was analyzed. Fortunately this is unnecessary. We now describe two quicker methods: one employs a table of contrast coefficients; the other, Yates's algorithm.

Table of Contrast Coefficients

The calculations performed to obtain the various effects can be characterized by the table of signs in Table 10.4.

TABLE 10.4.   Signs for calculating effects from the 2^3 factorial design, pilot plant example

mean	T	C	K	TC	TK	CK	TCK	yield averages
+	−	−	−	+	+	+	−	60
+	+	−	−	−	−	+	+	72
+	−	+	−	−	+	−	+	54
+	+	+	−	+	−	−	−	68
+	−	−	+	+	−	−	+	52
+	+	−	+	−	+	−	−	83
+	−	+	+	−	−	+	−	45
+	+	+	+	+	+	+	+	80
divisor 8	4	4	4	4	4	4	4	

Thus the estimate of the mean is calculated from the first column,

$$\frac{+60 + 72 + 54 + 68 + 52 + 83 + 45 + 80}{8} = \frac{514}{8} = 64.25 \quad (10.11)$$

The T main effect is calculated from the second column,

$$\frac{-60 + 72 - 54 + 68 - 52 + 83 - 45 + 80}{4} = \frac{92}{4} = 23.0 \quad (10.12)$$

and the $T \times K$ interaction from the sixth column,

$$\frac{+60 - 72 + 54 - 68 - 52 + 83 - 45 + 80}{4} = \frac{40}{4} = 10.0 \quad (10.13)$$

The remaining effects are obtained similarly to yield the estimates already displayed in Table 10.2.

The signs for the interactions reveal a remarkable fact: they can be obtained by directly multiplying the signs of their respective variables! Thus the array of signs for the $T \times K$ interaction is obtained by multiplying together the signs for T and K. The method is quite general, and effects for any 2^k factorial design may be calculated using a table of signs like Table 10.4. We note further that each effect is a contrast, and that they are all mutually orthogonal.

However, although such a table of signs has many uses, the most rapid way of calculating effects is by means of an algorithm due to Yates.

Yates's Algorithm

Yates's algorithm is applied to the observations after they have been re-arranged* in what is called *standard order*. A 2^k factorial design is in standard order when, as in Table 10.1, the first column of the design matrix consists of successive minus and plus signs, the second column of successive pairs of minus and plus signs, the third column of four minus signs followed by four plus signs, and so forth. In general, the kth column consists of 2^{k-1} minus signs followed by 2^{k-1} plus signs.

The Yates calculations for the pilot plant data are shown in Table 10.5. In this table the design matrix gives the experimental conditions in standard order. Column y contains the corresponding average yields for each run. (Of course, if the runs have not been replicated, each average would be the single observation recorded for that run.) These averages are now considered in successive pairs. The first four entries in column (1) are obtained by adding the pairs together. Thus $60 + 72 = 132$, $54 + 68 = 122$, and so on. The second four entries in column (1) are obtained by subtracting the *top number from the bottom number* of each pair. Thus $72 - 60 = 12$, $68 - 54 = 14$, and so on.

In just the same way that column (1) is obtained from column y, column (2) is obtained from column (1). Finally, column (3) is obtained from column

---

* The order of actual running should, of course, be random.

TABLE 10.5. Yates's algorithm, pilot plant example

test condition number	design matrix variables			run average y	(1)	(2)	(3)	divisor	estimate	identification
	T	C	K		algorithm					
1	−	−	−	60	132	254	514	8	64.25	average
2	+	−	−	72	122	260	92	4	23.0	T
3	−	+	−	54	135	26	− 20	4	− 5.0	C
4	+	+	−	68	125	66	6	4	1.5	TC
5	−	−	+	52	12	− 10	6	4	1.5	K
6	+	−	+	83	14	− 10	40	4	10.0	TK
7	−	+	+	45	31	2	0	4	0.0	CK
8	+	+	+	80	35	4	2	4	0.5	TCK

(2) in the same manner. The entries in column (3) are precisely the values that are obtained by combining the averages with the appropriate column of signs in Table 10.4. To obtain the effects one has only to divide as before by the appropriate divisor, which is 8 for the first entry and 4 for the others. The first estimate is the grand average of all the observations. The remaining effects are identified by locating the plus signs in the design matrix. Thus in the second row a plus sign occurs only in the **T** column, so that the effect in that row is the T effect. In the seventh row plus signs occur in both the **C** and **K** columns, so that the effect in that row is the $C \times K$ interaction.

In general, for a 2^k factorial design, whether working with individual observations or averages of observations, k columns (1), (2), ..., (k) will be generated by adding and subtracting appropriate pairs of numbers. The first divisor will be 2^k, and the remaining divisors will be 2^{k-1}. Yates's algorithm is explained in more detail in Appendix 10A, which includes a description of ways to check the calculations as they are done.

Exercise 10.9.   For the data in Exercise 10.3 and 10.4, compute the main effects and interactions, using the table of contrast coefficients and Yates's algorithm.

Answer: See answers for Exercises 10.3 and 10.4.

10.8.   A 2^4 FACTORIAL DESIGN: PROCESS DEVELOPMENT STUDY

Results from a 2^4 design employed in a process development study are shown in Table 10.6. A visual display of the data is given in Figure 10.5, which repays careful study, particularly after the analysis has been carried out.

TABLE 10.6.   Data obtained in a process development study arranged in standard (Yates) order

observation number	1	2	3	4	conversion (%)	order of runs	variable	−	+
1	−	−	−	−	71	(8)	1 catalyst charge (lb)	10	15
2	+	−	−	−	61	(2)	2 temperature (°C)	220	240
3	−	+	−	−	90	(10)	3 pressure (psi)	50	80
4	+	+	−	−	82	(4)	4 concentration (%)	10	12
5	−	−	+	−	68	(15)			
6	+	−	+	−	61	(9)			
7	−	+	+	−	87	(1)			
8	+	+	+	−	80	(13)			
9	−	−	−	+	61	(16)			
10	+	−	−	+	50	(5)			
11	−	+	−	+	89	(11)			
12	+	+	−	+	83	(14)			
13	−	−	+	+	59	(3)			
14	+	−	+	+	51	(12)			
15	−	+	+	+	85	(6)			
16	+	+	+	+	78	(7)			

Table 10.7 is obtained by multiplying the elements of the columns of the design matrix, first two at a time, next three at a time, then four at a time.

Calculation of Effects

The effects calculated from the table of signs are shown in Table 10.8. Note that, for conciseness, we now denote an interaction as *12* rather than *1 × 2*.

Exercise 10.10.   Recalculate the effects using Yates's algorithm. Check your calculations by the methods described in Appendix 10A.

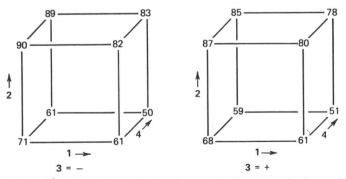

FIGURE 10.5.   A 2^4 design with data displayed geometrically, process development example.

TABLE 10.7. Signs for calculating effects for a 2^4 factorial, process development example

I	1	2	3	4	12	13	14	23	24	34	123	124	134	234	1234	conversion (%)
+	−	−	−	−	+	+	+	+	+	+	−	−	−	−	+	71
+	+	−	−	−	−	−	−	+	+	+	+	+	+	−	−	61
+	−	+	−	−	−	+	+	−	−	+	+	+	−	+	−	90
+	+	+	−	−	+	−	−	−	−	+	−	−	+	+	+	82
+	−	−	+	−	+	−	+	−	+	−	+	−	+	+	−	68
+	+	−	+	−	−	+	−	−	+	−	−	+	−	+	+	61
+	−	+	+	−	−	−	+	+	−	−	−	+	+	−	+	87
+	+	+	+	−	+	+	−	+	−	−	+	−	−	−	−	80
+	−	−	−	+	+	+	−	+	−	−	−	+	+	+	−	61
+	+	−	−	+	−	−	+	+	−	−	+	−	−	+	+	50
+	−	+	−	+	−	+	−	−	+	−	+	−	+	−	+	89
+	+	+	−	+	+	−	+	−	+	−	−	+	−	−	−	83
+	−	−	+	+	+	−	−	−	−	+	+	+	−	−	+	59
+	+	−	+	+	−	+	+	−	−	+	−	−	+	−	−	51
+	−	+	+	+	−	−	−	+	+	+	−	−	−	+	−	85
+	+	+	+	+	+	+	+	+	+	+	+	+	+	+	+	78
divisor 16	8	8	8	8	8	8	8	8	8	8	8	8	8	8	8	

TABLE 10.8. Estimated effects from a 2^4 factorial design, process development example

effects	estimated effects ± standard error
average	72.25 ± 0.27
1	(−8.00)± 0.55
2	(24.00)± 0.55
3	−2.25 ± 0.55
4	(−5.50)± 0.55
12	1.00 ± 0.55
13	0.75 ± 0.55
14	0.00 ± 0.55
23	−1.25 ± 0.55
24	(4.50)± 0.55
34	−0.25 ± 0.55
123	−0.75
124	0.50
134	−0.25
234	−0.75
1234	−0.25

Calculation of Standard Errors for Effects Using Higher-Order Interactions

No direct estimate of σ^2 is available from these 16 runs since there were no replicates. However we can obtain such an estimate if certain assumptions are made. In particular, if all three- and four-factor interactions are supposed negligible (an assumption made plausible by the earlier discussion of smoothness and similarity of response functions) these higher-order interactions would measure differences arising principally from experimental error.

They could thus provide an appropriate reference set for the remaining effects. We find:

	effect	effect^2
123	−0.75	0.5625
124	0.50	0.2500
134	−0.25	0.0625
234	−0.75	0.5625
1234	−0.25	0.0625
	sum	1.5000

Accordingly an estimated value for the variance of an effect, having five degrees of freedom, is $1.50/5 = 0.30$. The estimated standard error of an effect is therefore $\sqrt{0.30} = 0.55$, which is the value used in Table 10.8.

Exercise 10.11. The following are data in standard order from a 2^4 factorial design in which the variables were (1) brand of tape deck ($- = A$, $+ = B$), (2) bass level (low, high), (3) treble level (low, high), and (4) synthesizer (with, without), and the response was judged quality of sound: 58, 44, 55, 45, 55, 42, 56, 46, 51, 45, 58, 44, 60, 46, 54, 45. Calculate the effects by means of Yates's algorithm, and, using the three and four-factor interactions, calculate the standard error of an effect. *Answer*: 1.64.

A diagram showing the effects, together with a reference t distribution centered at zero and having five degrees of freedom and a scale factor of 0.55, is shown in Figure 10.6. It appears that main effects *1*, *2*, and *4* and interaction *24* (circled in Table 10.8) are distinguishable from the noise. Possibly main effect *3* is also in this category.

Interpretation of the Process Development Data

Proceeding as before, we find:

1. An increase in catalyst charge (variable **1**) from 10 to 15 pounds reduces conversion by about 8 %, and the effect is consistent over the levels of the other factors tested.
2. With much less certainty it seems that an increase in pressure (variable **3**) from 50 to 80 psi may possibly reduce conversion by about 2 %.
3. Since there is an appreciable interaction between temperature (variable **2**) and concentration (variable **4**), the effects of these variables must be considered jointly. The nature of the interaction is indicated by the two-way table in Figure 10.7, obtained by averaging over levels of the other variables. Evidently high temperature produces high conversion. The interaction occurs because at the low temperature an increase in concentration reduces conversion, whereas an increase in concentration at a higher temperature produces no appreciable effect.

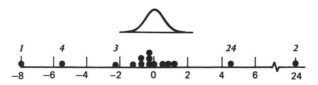

FIGURE 10.6. Main effects and interactions in relation to a reference t distribution with five degrees of freedom and scale factor 0.55, process development example.

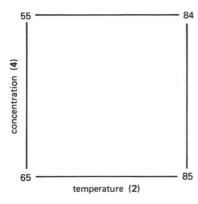

FIGURE 10.7. Two-way table for process development data.

10.9. ANALYSIS OF FACTORIALS USING NORMAL PROBABILITY PAPER

Two problems arise in the assessment of effects from unreplicated factorials: (1) occasionally real and meaningful high-order interactions occur, and (2) it is necessary to allow for selection. The assessment of effects using a reference distribution (Figure 10.6) scaled by an error estimate obtained from higher order interactions does not confront the first of these problems. However, a method (Daniel, 1959) by which *effects* are plotted on normal probability paper often provides an effective way of overcoming both difficulties.

The reader is urged to study Daniel's book (1976) *Applications of Statistics in Industrial Experimentation,* for a penetrating discussion and criticism of factorial designs in general. In particular, this book takes an appropriately skeptical attitude toward mechanical analysis of data and provides many interesting methods for diagnostic checking, including the plotting of *residuals* on normal probability paper.

Normal Probability Paper

A normal distribution is shown in Figure 10.8a. The percentage probability of the occurrence of some value less than X is given by the shaded area P. If we plot P against X, we obtain the sigmoid cumulative normal curve shown in Figure 10.8b. Normal probability paper is obtained by adjusting the vertical scale in the manner shown in Figure 10.8c, so that P versus X plots as a straight line. Suppose that the dots in Figure 10.8a represent a random sample of 10 observations from the normal distribution. Since the sample size is $n = 10$, the observation at the extreme left can be taken to represent the first 10% of the cumulative distribution. We plot that first

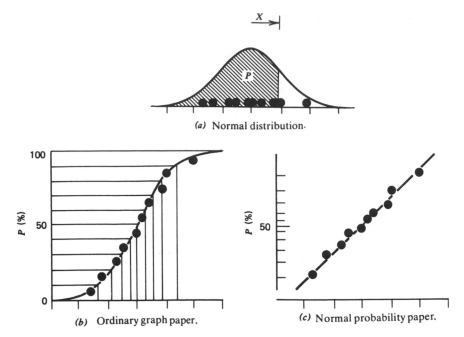

FIGURE 10.8. Normal probability plots.

observation, therefore, in Figure 10.8*b* midway between zero and 10%, that is, at the value 5%. Similarly, the second observation can be taken to represent the second 10% of the cumulative distribution, between 10 and 20%, and is plotted at the intermediate value 15%. The sample values approximately trace out the sigmoid curve, as expected. Furthermore, when these same points are plotted on normal probability paper in Figure 10.8*c*, they plot roughly as a straight line. Scales for making your own normal probability plots are given in Table E at the end of the volume. Intercepts such that $P_i = 100(i - \frac{1}{2})/m$ for $i = 1, 2, \ldots, m$ are given for the frequently needed values $N = 16, 32, 64, 15, 31, 63$.

Process Development Data

For illustration refer to the results from the 2^4 process development experiment considered before. Suppose that these data had occurred simply as the result of random (roughly normal) variation about a fixed mean, and the changes in levels of the variables had had no real effect at all on the percent conversion. Then the $m = 15$ effects (main effects and interactions), rep-

TABLE 10.9. The 15 ordered effects and the probability points P, process development example

order number i	1	2	3	4	5	6	7	8	9	10	11	12	13	14	15
effect	-8.0	-5.5	-2.25	-1.25	-0.75	-0.75	-0.25	-0.25	-0.25	0.00	0.50	0.75	1.00	4.50	24.00
identity of effects	1	4	3	23	123	234	34	134	1234	14	124	13	12	24	2
$P = 100(i - \frac{1}{2})/15$	3.3	10.0	16.7	23.3	30.0	36.7	43.3	50.0	56.7	63.3	70.0	76.7	83.3	90.0	96.7

331

resenting 15 contrasts between pairs of averages containing eight observations each, would have been roughly normal and would have been distributed about zero. They would therefore plot on normal probability paper as a straight line. To see whether they do, we order the 15 effects as in Table 10.9 and plot with the appropriate scale from Table E for $m = 15$. It happens that for these data the estimated main effect of factor 1 represents the first $\frac{1}{15} = 6.7\%$ of the cumulative distribution and should therefore be plotted against the value $\frac{1}{2}/15 = 3.3\%$. The next largest estimate is the main effect of factor 4, which is plotted at $1\frac{1}{2}/15 = 10\%$. Beginning with effects with magnitudes close to zero, 11 of the estimates fit reasonably well on a straight line. Those corresponding to 1, 4, 24, and 2 do not. As before, we conclude that these effects are not easily explained as chance occurrences.

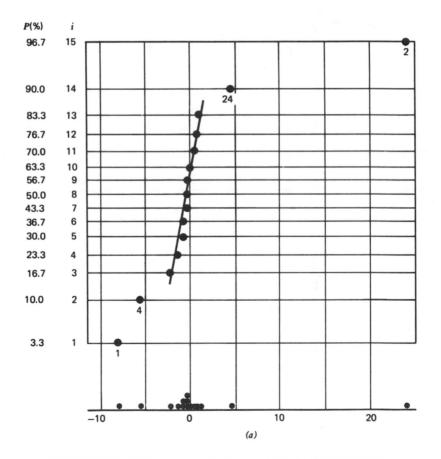

FIGURE 10.9. (a) Normal plot of effects, process development example.

Exercise 10.12. Add nine units to observations 2, 3, 6, 7, 9, 12, 13, 16 in Table 10.6. Recalculate the effects, and construct a normal plot.
Answer: Notice that *124* is the only effect whose numerical value is changed. This three-factor interaction is now picked out as deviating from the line.

Diagnostic Checks

Normal plotting of *residuals* provides a diagnostic check for any tentatively entertained model. For example, the plot in Figure 10.9a suggests that all effects, with the exception of the *average* 72.25, $1 = -8.0$, $2 = 24.0$, $4 = -5.5$, and $24 = 4.5$, can be explained by noise. If this is true, the estimated percent

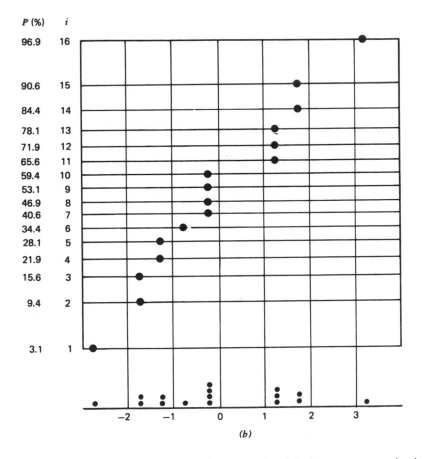

FIGURE 10.9. (*b*) Normal plot of residuals, on normal probability paper, process development example.

conversion for the process development data are given at the vertices of the design by*

$$\hat{y} = 72.25 + \left(\frac{-8.0}{2}\right)x_1 + \left(\frac{24.0}{2}\right)x_2 + \left(\frac{-5.5}{2}\right)x_4 + \left(\frac{4.5}{2}\right)x_2 x_4 \quad (10.14)$$

where x_1, x_2, x_4 take the value -1 or $+1$ according to the columns of signs in Tables 10.6 and 10.7. Notice that the coefficients that appear in the equations are *half* the calculated effects. This is so because a change from $x = -1$ to $x = +1$ is a change of *two* units along the x axis.

The values of y, $\hat{y}$, and $y - \hat{y}$ are then as follows:

y	71	61	90	82	68	61	87	80
$\hat{y}$	69.25	61.25	88.75	80.75	69.25	61.25	88.75	80.75
$y - \hat{y}$	1.75	-0.25	1.25	1.25	-1.25	-0.25	-1.75	-0.75
y	61	50	89	83	59	51	85	78
$\hat{y}$	59.25	51.25	87.75	79.75	59.25	51.25	87.75	79.75
$y - \hat{y}$	1.75	-1.25	1.25	3.25	-0.25	-0.25	-2.75	-1.75

(Alternatively, these residuals may be obtained by using the reverse Yates algorithm described in Appendix 10A.)

The model may now be checked by plotting these residuals on normal probability paper (see Figure 10.9b) for $m = 16$. Unlike the original plot of the effects, all the points from this residual plot now lie close to the line, confirming the conjecture that effects other than *1, 2, 4*, and *24* are readily explained by random noise. This residual check is valuable provided that the number of effects eliminated (four in this case) is fairly small compared to m.

10.10. TRANSFORMATION OF DATA FROM FACTORIAL DESIGNS

When y_{max}/y_{min} is large, the possibility of simplified and more efficient representation as a result of appropriate transformation (see Section 7.8) should always be kept in mind. A striking example of simplification and

---

* The occurrence of a cross-product term in this expression suggests that quadratic terms (involving x_2^2 and x_4^2) might be needed in a model which adequately represented the conversion at *interpolated* levels of the variables. Such terms cannot be estimated using two-level designs. The fitting and use of approximating functions called *response surfaces* are outlined in Chapter 15 where methods are discussed for augmenting two-level designs to allow estimation of quadratic effects.

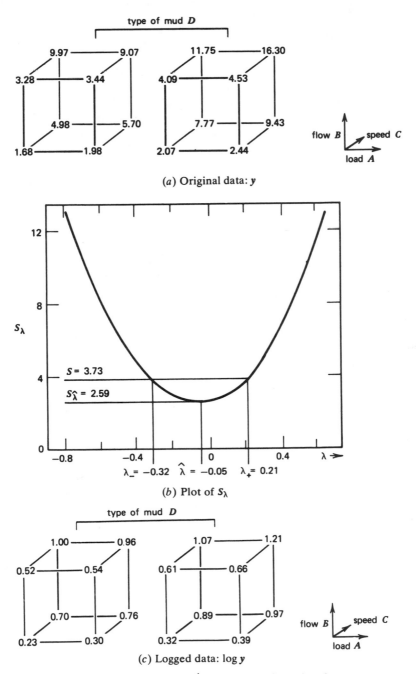

(a) Original data: y

(b) Plot of S_λ

(c) Logged data: log y

FIGURE 10.10. Daniel's 2^4 data with transformation plot.

increased sensitivity achieved by transformation of data from a 3^3 factorial design is given, for example, by Box and Cox (1964).

The data plotted in Figure 10.10a are from a 2^4 factorial given by Daniel (1976, p. 72). The factors were A, the load on a small stone drill; B, the flow rate through it; C, its rotational speed; and D, the "type of mud used in drilling." The response y was the rate of advance of the drill. The plot of the data y in Figure 10.10a shows that the elementary comparisons made along the edges of the cubes are remarkably consistent in sign.

There is, however, a clear tendency for these comparisons to increase in magnitude as the level of the response increases. This will produce interactions (nonadditivity) which might be removed by transformation. In a very interesting and thorough treatment of these data, Daniel shows how such a transformation may be chosen using normal plotting techniques. To illustrate the likelihood approach discussed in Section 7.8 we tentatively assume that, for some suitable power transformation $Y = y^\lambda$ of the original data, only main effects are needed to account for the response. As before, supposing that the tentative assumption is true, we carry through an analysis of $y^{(\lambda)} = (y^\lambda - 1)/\lambda \dot{y}^{\lambda - 1}$ for various values of λ. The sum of squares of residuals S_λ after the four main effects have been eliminated has 11 degrees of freedom. It is plotted against λ in Figure 10.10b. We see that, among the class of models considered, $\hat{\lambda}$ is close to zero, suggesting a transformation to logarithms or to some small negative power. As would be expected, this is in agreement with Daniel's analysis, and the log plot in Figure 10.10c suggests that a remarkably succinct summary of the major influences of the variables is indeed possible in this metric. Confidence limits (95%) obtained as before are $\lambda_- = -0.32$, $\lambda_+ = 0.21$.

In a more extensive likelihood analysis other models and other transformations could be entertained if it was thought worthwhile to do so. As always, residuals from the various transformations should be studied, and the possibility of bad values kept in mind. We will not take this matter further here, but instead refer the reader to the comprehensive analysis in Daniel's book.

10.11. BLOCKING

Suppose that a trial is to be conducted using a 2^3 factorial design, and, to make the eight runs under conditions as homogeneous as possible, it is desirable that batches of raw material sufficient for the complete experiment be blended together. Suppose, however, that the available blender is only large enough to accommodate material for four runs. This means that two different blends will have to be used. Figure 10.11 shows how the 2^3 factorial

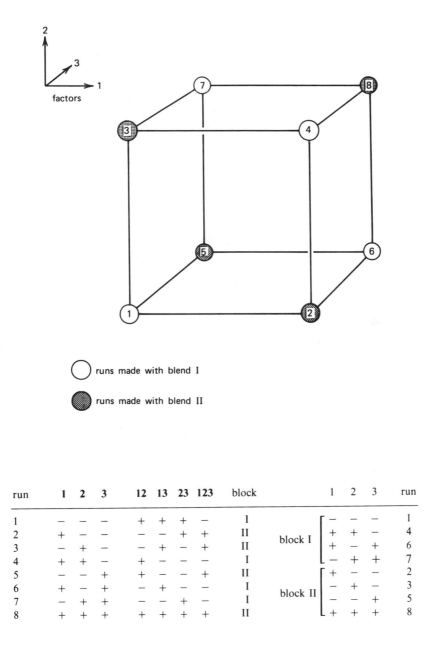

2

3

1

factors

7

⑧

③

4

⑤

6

1

②

◯ runs made with blend I

● runs made with blend II

run	1	2	3	12	13	23	123	block		1	2	3	run
1	−	−	−	+	+	+	−	I		−	−	−	1
2	+	−	−	−	−	+	+	II	block I	+	+	−	4
3	−	+	−	−	+	−	+	II		+	−	+	6
4	+	+	−	+	−	−	−	I		−	+	+	7
5	−	−	+	+	−	−	+	II		+	−	−	2
6	+	−	+	−	+	−	−	I		−	+	−	3
7	−	+	+	−	−	+	−	I	block II	−	−	+	5
8	+	+	+	+	+	+	+	II		+	+	+	8

FIGURE 10.11. Arranging a 2^3 factorial design in two blocks of size four.

337

design can be arranged in two blocks of four runs to neutralize the effect of possible blend differences. Runs indicated by open dots and numbered 1, 4, 6, 7 use blend I, and runs indicated by black dots and numbered 2, 3, 5, 8 use blend II.

The main effects of the factors are contrasts between averages on opposite faces of the cube. But two black dots and two white dots occur on each face, so that any additive effect associated with blends is eliminated from each of the main effects. Now look again at Figure 10.2, and consider the diagonal contrasts, which correspond to the two-factor interactions. Again two black dots and two white dots appear on each side of the contrast. Consequently any additive effect associated with blends is eliminated from each of the two-factor interactions.

The design is blocked in this way by placing all runs in which **123** is minus in one block and all the other runs, in which **123** is plus, in the other block. If all the runs in block II were higher by amount h than they would have been if they had been performed in block I, then, whatever the value of h, it will sum out in the calculation of effects *1*, *2*, *3*, *12*, *13*, and *23*.

Notice that we have had to give up something to get something. We have deliberately *confounded* (i.e., confused) the three-factor interaction and the blend difference. Therefore with this design we cannot estimate the three-factor interaction separately from the blend effect. However, we would usually expect this interaction to be unimportant. In exchange we have been able to run the design in two blocks, which can ensure that main effects and two-factor interactions are much more precisely measured than would otherwise be the case. This important idea of using confounding in the design of experiments is again due to Fisher.

In the 2^3 factorial example, suppose that we give the block variable the identifying number **4**. Then we could think of our experiment as containing four variables, the last of which is assumed to possess the rather special property that it does not interact with the others. If this new variable is introduced by making its levels coincide exactly with the plus and minus signs attributed to the **123** interaction, the blocking may be said to be "generated" by the relationship **4 = 123**. This idea may be used to derive more sophisticated blocking arrangements.

A 2^3 in Four Blocks of Size Two

Suppose that in the 2^3 experiment the blends had been large enough for only *two* runs each. How could the design be arranged in four blocks of two runs so as to do as little damage as possible to the estimates of the important effects?

How Not To Do It. Let us introduce two block factors, which we will call **4** and **5**. At first glance it seems reasonable to associate block factor **4** with the **123** interaction and block factor **5** with some expendable two-factor interaction, say the **23** interaction. The consequences are illustrated in Table 10.10. We assign runs to the different blocks, depending on the signs of the block variables in the columns for **4** and **5**. Thus runs for which these signs are $(- -)$ go in one block, and those that have $(- +)$ signs in another. The $(+ -)$ runs go in a third block, and the $(+ +)$ runs in a fourth.

TABLE 10.10. A 2^3 design in blocks of size two, an undesirable arrangement

	experimental variable			block variable		experiment arranged in four blocks				
run	1	2	3	4 = 123	5 = 23	block	1	2	3	run
1	−	−	−	−	+	I	+	+	−	4
2	+	−	−	+	+		+	−	+	6
3	−	+	−	+	−					
4	+	+	−	−	−	II	−	−	−	1
5	−	−	+	+	−		−	+	+	7
6	+	−	+	−	−					
7	−	+	+	−	+	III	−	+	−	3
8	+	+	+	+	+		−	−	+	5
						IV	+	−	−	2
							+	+	+	8

		−	+
variable **5**	+	1, 7	2, 8
	−	4, 6	3, 5
		−	+
		variable **4**	

There is a serious weakness in the design in Table 10.10. We have confounded block variables **4** and **5** with interactions **123** and **23**. But there are three degrees of freedom between four blocks. With what contrast is the third degree of freedom associated? Inspection of Table 10.10 shows that it is associated with the **45** interaction, but this is the "diagonal" contrast between runs 2, 4, 6, 8 and runs 1, 3, 5, 7 measuring the main effect of variable **1** as well as the contrast of blocks (I and IV) with blocks (II and III). Thus the arrangement we have chosen results in confounding of the main effect for factor **1** with block differences!

Clearly caution is necessary. Fortunately a simple calculus is available to show immediately the consequences of any proposed blocking arrangement. If the elements of any column in Table 10.10 are multiplied by themselves, a column of plus signs is obtained. We denote a column of all plus signs by I. Thus we write

$$I = 11 = 22 = 33 = 44 = 55 \tag{10.15}$$

where, for example, by 33 we mean the product of the elements in column 3 with themselves. The effect of multiplying the elements in any column by the elements in column I is to leave those elements unchanged. Now in the blocking arrangement just considered

$$4 = 123, \qquad 5 = 23 \tag{10.16}$$

The 45 column is thus

$$45 = 123 \cdot 23 = 12233 = 1II = 1 \tag{10.17}$$

which shows at once that column 45 is identical to column 1. That means that interaction 45 and main effect 1 are confounded.

How To Do It. A better arrangement is obtained by confounding the two block variables with any two of the two-factor interactions. The third degree of freedom between blocks is then confounded with the third two-factor interaction. Thus, for

$$4 = 12, \qquad 5 = 13 \tag{10.18}$$

interaction 45 is confounded with interaction 23 since

$$45 = 1123 = 23 \tag{10.19}$$

The organization of the experiment in four blocks using this arrangement is indicated in Table 10.11, the blocks being typified as before by the pairs of observations for which 4 and 5 take the signatures $(- -), (- +), (+ -)$, and $(+ +)$.

The two runs comprising each block in the above example are complementary in the sense that the plus and minus levels of one run are exactly reversed in the second. Each block is said to consist of a *fold-over* pair. For example, in block I, the plus and minus signs for the pair of runs are $(+ - -)$ and $(- + +)$. Any 2^k factorial may be broken into 2^{k-1} blocks of size two by this method. *Such blocking arrangements leave the main effects of the k variables unconfounded with block variables.* All the two-factor interactions, however, are confounded with blocks.

Partial Confounding

When, to achieve sufficient accuracy, replication is necessary, an opportunity is presented to confound different effects in different replicates. Suppose, for example, that four replicates of the 2^3 factorial were to be run in 16 blocks of size two. Then we might run the pattern shown in Table 10.12.

This arrangement would estimate main effects with greatest precision, providing less precision in the estimates of two-factor interactions and still less for three-factor interaction. The reader may find it entertaining to invent other schemes that place different degrees of emphasis on the various effects.

TABLE 10.11. A better arrangement for a 2^3 factorial design in blocks of size two

run	experimental variable			block variable		experiment arranged in four blocks				
	1	**2**	**3**	**4 = 12**	**5 = 13**	**block**	**1**	**2**	**3**	**run**
1	−	−	−	+	+	I	+	−	−	2
2	+	−	−	−	−		−	+	+	7
3	−	+	−	−	+					
4	+	+	−	+	−	II	−	+	−	3
5	−	−	+	+	−		+	−	+	6
6	+	−	+	−	+					
7	−	+	+	−	−	III	+	+	−	4
8	+	+	+	+	+		−	−	+	5
						IV	−	−	−	1
							+	+	+	8

Recovery of Interblock Information

In designs of this kind, estimates confounded with blocks, rather than being regarded as lost, can, following Yates, be thought of merely as having a different (usually larger) variance. Now, from a design such as that in Table 10.12, separate variances for unconfounded and confounded effects can be estimated from the replication of the calculated effects. Suppose that these estimates are s_u^2 and s_c^2 and that for a particular effect we have two estimates: E_u, based on an average of n_u unconfounded estimates, and E_c, based on an average of n_c confounded estimates. Then a combined estimate, using all the information,

TABLE 10.12. Partial confounding: 2^3 design in four replicates

Effects confounded with blocks are indicated by c.
Effects not confounded with blocks are indicated by a u.

replicate	**1**	**2**	**3**	**12**	**13**	**23**	**123**
first	u	u	u	c	c	c	u
second	c	u	u	u	u	c	c
third	u	c	u	u	c	u	c
fourth	u	u	c	c	u	u	c
number of unconfounded replicates	3	3	3	2	2	2	1
number of confounded replicates	1	1	1	2	2	2	3

is provided by the weighted average of the estimates $E = (w_u E_u + w_c E_c)/(w_u + w_c)$, where the weights are given by $w_u = n_u/s_u^2$, $w_c = n_c/s_c^2$. For a more general discussion of the recovery of interblock information see Cochran and Cox (1957).

10.12. SUMMARY

Two-level factorial designs estimate not only main effects, but also interactions, with maximum precision. The significance of effects may be judged from: an estimate of variance obtained by genuine replication when available; from higher order interactions (assumed due to noise); or by plotting effects on normal probability paper. Rapid calculation of effects is possible using Yates's algorithm. The designs may be run in blocks by associating effects of supposedly lesser importance with block differences.

APPENDIX 10A. YATES'S ALGORITHM

This appendix provides more details on Yates's algorithm, introduced in Section 10.7. Table 10A.1 shows diagramatically the addition and subtraction steps required to generate columns (1), (2), and (3). It also shows how the calculations can be checked after each new column of numbers has been computed. This check is provided by the sum of squares of the entries in each column. The total sum of squares of the eight average results is 34,342. The sum of squares of the entries in column (1) is 2(34,342); in column

TABLE 10A.1.   Calculations for carrying out and checking Yates's algorithm: pilot plant example (see Table 10.5)

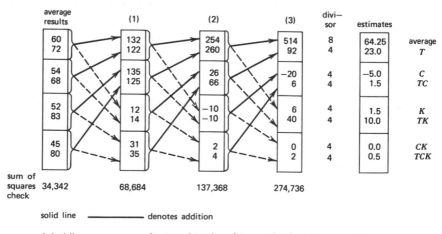

solid line ———————— denotes addition

dashed line — — — — — — — denotes subtraction of <u>top number from bottom number</u>

TABLE 10A.2. Reverse Yates's algorithm for obtaining estimated values $\hat{y}$ and residuals $y - \hat{y}$

effect	modified column (3) from forward Yates	(1)	(2)	(3)	divisor	$\hat{y}$	y	$y - \hat{y}$	identification T	C	K
TCK	0	0	40	626	8	78.25	80	1.75	+	+	+
CK	0	40	586	362	8	45.25	45	−0.25	−	+	+
TK	40	−20	−40	666	8	83.25	83	−0.25	+	−	+
K	0	606	402	402	8	50.25	52	1.75	−	−	+
TC	0	0	40	546	8	68.25	68	−0.25	+	+	−
C	−20	−40	626	442	8	55.25	54	−1.25	−	+	−
T	92	−20	−40	586	8	73.25	72	−1.25	+	−	−
average	514	422	442	482	8	60.25	60	−0.25	−	−	−
sum of squares	274,660	549,320	1,098,640	2,197,280		34,332.5	34,342	9.5			

343

(2), $2^2(34, 342)$; in column (3), $2^3(34,342)$. In general, for a 2^k factorial design, in column (*m*) the sum of squares should be $2^m \sum y^2$.

Note that the sum of the entries in column (2) is 592. The sum of every second entry in column (1) is $122 + 125 + 14 + 35 = 296$, which, if multiplied by 2, gives 592, that is, $2 \times 296 = 592$. If the calculations are correct, it will be true in general that twice the sum of every second entry in column (*i*) is equal to the sum of the entries in column (*i* + 1). This provides an additional way to check calculations as they are done.

Yates's algorithm can be used for a replicated 2^k factorial design by starting with *totals* rather than averages, that is, for this example with $59 + 61 = 120$ for the first test condition $(- - -)$, $74 + 70 = 144$ for the second test condition $(+ - -)$, and so forth. (Recall from Table 10.3 that each test condition was run twice.) When this is done, it will always be true that the main effects and interactions can be obtained by dividing the entries in column (*k*) by $N/2$, except the first entry, which is divided by N, where N is the total number of runs performed (e.g., in this example $N = 16$ and $k = 3$).

Reverse Yates's Algorithm to Give Estimated Values $\hat{y}$ and residuals $y - \hat{y}$

Once again following Cuthbert Daniel, we may obtain estimated values $\hat{y}$ and hence residuals for the 2^k factorials by reversing Yates's algorithm. For example, suppose that in the pilot plant example we wished to obtain estimated values $\hat{y}$ on the supposition that the effects K, TC, CK, and TCK arose only from noise. Then in column (3) of Table 10A.1 these entries would be replaced by zeros. Now, if to this column of entries in *reversed standard order* [see column (1) of Table 10A.2] Yates's algorithm is applied, the $\hat{y}$'s, multiplied by 8, are obtained in reverse standard order at the end of the calculation.

In general, for a 2^k design (*a*) the *k*th column of the forward calculation is written in reverse order, appropriately modified by substitution of zeros or other values of interest; (*b*) the entries are operated on k times with the standard Yates algorithm; and (*c*) the quantities produced are $2^k \times \hat{y}$ arranged in reverse standard order. Notice once again that the sum of squares checks for the columns. Thus $2^k \sum \hat{y}^2 = 8 \sum \hat{y}^2$ $= 8 \times 34,332.5 = 274,660$, which is the sum of squares of the initial entries. Also, since $\sum y^2 = \sum \hat{y}^2 + \sum (y - \hat{y})^2$, $34,342 = 34,332.5 + 9.5$. An example in which the reverse Yates was employed to uncover a maverick observation in a 2^4 factorial is given in Hunter (1966).

Exercise 10A.1.　Recompute the estimated values and residuals for the process development example in Table 10.6, using the reverse Yates algorithm.

APPENDIX 10B.　MORE ON BLOCKING FACTORIAL DESIGNS

We have used the 2^3 factorial design to illustrate the principle of blocking designs by means of confounding. These principles, however, are equally applicable to larger examples. Table 10B.1 provides a list of particularly useful arrangements. To understand how this table is used consider a complete 2^6 factorial design containing 64 runs.

Suppose that the experimenter wishes to arrange this program in eight blocks of eight runs each. The eight blocks, for example, might be associated with eight periods of time or eight blends of raw material. We choose to confound three three-factor or higher order interactions among the experimental variables with the block variables B_1, B_2, and B_3. In addition, we require that the interactions between the block variables be themselves confounded with three-factor or higher order interactions among the experimental variables. The table suggests that we use the arrangement

$$B_1 = 135, \qquad B_2 = 1256, \qquad B_3 = 1234$$

Using the rule $11 = 22 = 33 = 44 = 55 = 66 = I$, we obtain the following relationships after multiplication of the block variables by one another:

$$
\begin{aligned}
B_1 &= 135 \\
B_2 &= 1256 \\
B_3 &= 1234 \\
B_1 B_2 &= 236 \\
B_1 B_3 &= 245 \\
B_2 B_3 &= 3456 \\
B_1 B_2 B_3 &= 146
\end{aligned}
$$

which show that only high-order (three-factor and higher order) interactions are confounded with the seven degrees of freedom associated with the eight blocks.

To allocate the experiments to the eight blocks we follow the same procedure as before: we write down the plus and minus levels corresponding to B_1, B_2, and B_3 and then combine the individual runs into blocks that have the same signs for the block variables. As illustrated in Table 10B.2, the signs of (B_1, B_2, B_3) are associated with the eight blocks according to the following scheme:

block	1	2	3	4	5	6	7	8
(B_1, B_2, B_3)	$(---)$	$(+--)$	$(-+-)$	$(++-)$	$(--+)$	$(+-+)$	$(-++)$	$(+++)$

REFERENCES AND FURTHER READING

Yates, F. (1937). *The Design and Analysis of Factorial Experiments*, Bulletin 35, Imperial Bureau of Soil Science, Harpenden, Herts, England, Hafner (Macmillan).

Yates, F. (1970). *Selected Papers*, Hafner (Macmillan).

Daniel, C. (1976). *Applications of Statistics to Industrial Experimentation*, Wiley. (This excellent text carries considerably further than is done here the analysis of 2^k factorial designs and, in particular, the detection and interpretation of discrepant results.)

Davies, O. L. (Ed.) (1971). *The Design and Analysis of Industrial Experiments*, Hafner (Macmillan).

Cochran, W. G. and G. M. Cox (1957). *Experimental Designs*, 2nd ed., Wiley.

Youden, W. J. (1972). Enduring values, *Technometrics*, **14**, 1.

Box, G. E. P. and D. R. Cox (1964). An analysis of transformations, *J. Roy. Stat. Soc.* Series B, **26**, 211.

Daniel, C. (1959). Use of half-normal plot in interpreting factorial two-level experiments, *Technometrics*, **1**, 149.

Hunter, J. S. (1966). Inverse Yates algorithm, *Technometrics*, **8**, 177.

TABLE 10B.1. Blocking arrangements for 2^k factorial designs

k = number of variables	block size	block generator	interactions confounded with blocks
3	4	$B_1 = 123$	123
	2	$B_1 = 12, B_2 = 13$	12, 13, 23
4	8	$B_1 = 1234$	1234
	4	$B_1 = 124, B_2 = 134$	124, 134, 23
	2	$B_1 = 12, B_2 = 23, B_3 = 34$	12, 23, 34, 13, 1234, 24, 14
5	16	$B_1 = 12345$	12345
	8	$B_1 = 123, B_2 = 345$	123, 345, 1245
	4	$B_1 = 125, B_2 = 235, B_3 = 345$	125, 235, 345, 13, 1234, 24, 145
	2	$B_1 = 12, B_2 = 13, B_3 = 34, B_4 = 45$	12, 13, 34, 45, 23, 1234, 1245, 14, 1345, 35, 24, 2345, 1235, 15, 25 i.e., all 2fi and 4fi*
6	32	$B_1 = 123456$	123456
	16	$B_1 = 1236, B_2 = 3456$	1236, 3456, 1245
	8	$B_1 = 135, B_2 = 1256, B_3 = 1234$	135, 1256, 1234, 236, 245, 3456, 146
	4	$B_1 = 126, B_2 = 136, B_3 = 346, B_4 = 456$	126, 136, 346, 456, 23, 1234, 1245, 14, 1345, 35, 246, 23456, 12356, 156, 25
	2	$B_1 = 12, B_2 = 23, B_3 = 34, B_4 = 45, B_5 = 56$	all 2fi, 4fi, and 6fi

7 factors

Block size	Blocks	
	$B_1 = 1234567$	*1234567*
64	$B_1 = 12367, B_2 = 34567$	*12367, 34567, 1245*
32	$B_1 = 123, B_2 = 456, B_3 = 167$	*123, 456, 167, 123456, 2367, 1457, 23457*
16	$B_1 = 1234, B_2 = 567, B_3 = 345, B_4 = 147$	*1234, 567, 345, 147, 1234567, 125, 237, 3467, 1456, 1357, 1267, 2356, 2457, 136, 246*
8	$B_1 = 127, B_2 = 237, B_3 = 347, B_4 = 457, B_5 = 567$	*127, 237, 347, 457, 567, 13, 1234, 1245, 1256, 24, 2345, 2356, 3456, 35, 1234567, 46, 147, 13457, 13567, 12467, 12357, 257, 24567, 23467, 367, 15, 1456, 1346, 1236, 26, 167*
4		
2	$B_1 = 12, B_2 = 23, B_3 = 34, B_4 = 45, B_5 = 56, B_6 = 67$	*all 2fi, 4fi, and 6fi*

---

* "fi" is an abbreviation for "factor interaction"; thus, for example, 2fi means two-factor interaction.

TABLE 10B.2. A 2^6 factorial design in eight blocks

experiment number	experimental variable						block variable			block number	block number	experimental variable						experiment number
	1	2	3	4	5	6	B1 135	B2 1256	B3 1234			1	2	3	4	5	6	
1	−	−	−	−	−	−	−	+	+	7	1	−	+	−	−	−	−	3
2	+	−	−	−	−	−	+	−	−	2		+	−	+	+	−	−	14
3	−	+	−	−	−	−	−	−	−	1		−	−	+	−	+	−	21
4	+	+	−	−	−	−	+	+	+	8		+	+	−	+	+	−	28
5	−	−	+	−	−	−	+	+	−	4		+	+	+	−	−	+	40
6	+	−	+	−	−	−	−	−	+	5		−	−	−	+	−	+	41
7	−	+	+	−	−	−	+	−	+	6		+	−	−	−	+	+	50
8	+	+	+	−	−	−	−	+	−	3		−	+	+	+	+	+	63
9	−	−	−	+	−	−	−	+	−	3	2	+	−	−	−	−	−	2
10	+	−	−	+	−	−	+	−	+	6		−	+	+	+	−	−	15
11	−	+	−	+	−	−	−	−	+	5		+	+	+	−	+	−	24
12	+	+	−	+	−	−	+	+	−	4		−	−	−	+	+	−	25
13	−	−	+	+	−	−	+	+	+	8		−	−	+	−	−	+	37
14	+	−	+	+	−	−	−	−	−	1		+	+	−	+	−	+	44
15	−	+	+	+	−	−	+	−	−	2		−	+	−	−	+	+	51
16	+	+	+	+	−	−	−	+	+	7		+	−	+	+	+	+	62
17	−	−	−	−	+	−	+	−	+	6	3	+	+	+	−	−	−	8
18	+	−	−	−	+	−	−	+	−	3		−	−	−	+	−	−	9
19	−	+	−	−	+	−	+	+	−	4		+	−	−	−	+	−	18
20	+	+	−	−	+	−	−	−	+	5		−	+	+	+	+	−	31

348

Design table (two-level fractional factorial contrast signs). Entries are `+` and `−`.

Block 1

21	−	+	+	−	−	+	−	1	−	−	+	+	−	35
22	+	+	+	−	+	−	+	8	+	−	+	+	+	46
23	+	+	−	+	−	+	−	7	+	+	−	+	−	53
24	−	−	+	−	+	−	+	2	+	+	+	−	+	60

Block 2 (4)

25	−	−	+	+	+	+	−	2	−	−	−	+	−	5
26	+	+	−	−	+	+	+	7	−	−	+	−	+	12
27	−	+	+	−	+	+	−	8	−	+	+	−	−	19
28	+	+	−	+	+	+	+	1	+	−	−	+	+	30
29	−	−	+	−	+	+	−	5	−	+	−	+	−	34
30	+	−	−	+	+	+	+	4	+	−	+	−	+	47
31	−	−	+	+	+	+	−	3	+	+	−	+	−	56
32	+	+	−	−	+	+	+	6	+	+	+	−	+	57

Block 3 (5)

33	+	+	−	−	−	−	+	5	−	−	+	+	+	6
34	−	+	+	−	−	−	−	4	−	−	+	−	−	11
35	−	+	−	+	−	−	+	3	−	+	+	−	+	20
36	+	+	+	+	−	−	−	6	+	−	−	+	+	29
37	−	−	+	+	−	−	+	2	−	+	−	+	−	33
38	+	−	−	+	−	−	−	7	+	−	+	+	+	48
39	+	−	+	+	−	−	+	8	+	+	−	+	−	55
40	−	+	−	−	−	−	−	1	+	+	+	−	−	58

Block 4 (6)

41	−	+	+	−	−	+	+	1	−	−	+	+	+	7
42	+	+	−	+	−	+	−	8	−	−	+	−	+	10
43	−	+	+	−	−	+	+	7	−	+	+	−	−	17
44	+	+	−	+	−	+	−	2	+	−	−	+	+	32
45	−	−	+	+	−	+	+	6	−	+	−	+	−	36
46	+	+	−	−	−	+	−	3	+	−	+	−	+	45
47	−	+	+	−	−	+	+	4	+	+	−	+	−	54
48	+	+	−	+	−	+	−	5	+	+	+	−	−	59

(continued on following page)

349

TABLE 10B.2. (*continued*)

experiment number	experimental variable 1	2	3	4	5	6	B₁ 135	B₂ 1256	B₃ 1234	block number	block number	experimental variable 1	2	3	4	5	6	experiment number
49	−	−	−	−	+	+	+	+	+	8	7	−	−	−	−	−	−	1
50	+	−	−	−	+	+	−	+	−	1		+	−	−	−	−	−	16
51	−	+	−	−	+	+	+	−	−	2		−	+	+	+	+	−	23
52	+	+	−	−	+	+	−	+	+	7		+	−	+	+	+	−	26
53	−	−	+	−	+	+	−	−	−	3		−	+	+	−	−	+	38
54	+	−	+	−	+	+	+	−	+	6		+	+	−	+	−	+	43
55	−	+	+	−	+	+	−	−	+	5		−	+	−	+	+	+	52
56	+	+	+	−	+	+	+	+	−	4		−	−	+	+	+	+	61
57	−	−	−	+	+	+	+	+	−	4	8	+	+	−	−	−	−	4
58	+	−	−	+	+	+	−	−	+	5		−	−	+	+	−	−	13
59	−	+	−	+	+	+	+	−	+	6		+	−	+	−	+	−	22
60	+	+	−	+	+	+	−	+	−	3		−	+	−	+	+	−	27
61	−	−	+	+	+	+	−	+	+	7		−	+	+	−	−	+	39
62	+	−	+	+	+	+	+	−	−	2		+	−	−	−	−	+	42
63	−	+	+	+	+	+	−	−	−	1		−	−	+	−	+	+	49
64	+	+	+	+	+	+	+	+	+	8		+	+	+	+	+	+	64

block variable

experiment arranged in eight blocks

350

QUESTIONS FOR CHAPTER 10

1. What is a factorial design?
2. Why are factorial designs well suited to empirical studies? Suggest an application in your own field.
3. What is a $4^2 \times 3^3 \times 2$ factorial design? How many runs are there in this design? How many variables does it accommodate?
4. What is a two-level factorial design?
5. How many runs are contained in a two-level factorial design for four variables?
6. How many runs does a 2^6 design have? How many variables? How many levels for each variable?
7. How would you write the design matrix for a 2^8 factorial design? How many rows would it have? How many columns? How would you randomize this design?
8. What are four different but equivalent ways of calculating the main effects and interactions from a 2^5 factorial design? How many main effects will there be? Two-factor interactions? Three-factor interactions? Four-factor interactions? Five-factor interactions?
9. If each run of a 2^4 design has been performed twice, how can the standard error of an effect be calculated? Will the size of the standard error be the same for all the effects?
10. If each run of a 2^7 factorial design has been performed only once, describe and compare two methods that might be used to distinguish real effects from noise.
11. Set out schemes for running a 2^4 factorial design in (a) two blocks of eight runs, (b) four blocks of four runs. Mention any assumptions you make.

CHAPTER 11

More Applications
of Factorial Designs

Once principles are understood, the practitioner must be ready to face details and the conflicting issues that sometimes accompany actual applications. The examples in the present chapter are accounts of investigations as they actually occurred.

The reader may find surprises in these examples. In some instances things turn out more simply than might be expected, and in others not. Sometimes, as in most human enterprises, things go wrong but useful results emerge anyway. Occasionally experimenters take shortcuts and rely on hunches to a greater extent than most statisticians would prefer.

11.1. EXAMPLE 1: THE EFFECTS OF THREE VARIABLES ON CLARITY OF FILM

Figure 11.1 shows the clarity or cloudiness of a floor wax when certain changes were introduced into the formula for its preparation. The three variables studied were the amount of emulsifier A, the amount of emulsifier B, and the catalyst concentration. Eight formulations were prepared in a 2^3 factorial design. A portion of each formulation was spread thinly on a glass microscope slide, and the clarity of the film observed. The display of the data in Figure 11.1 leaves little doubt that over the ranges of the variables studied it was the presence of emulsifier B that produced cloudiness. These conclusions, though simply obtained, represented important findings and illustrate the value of geometric display of experimental results.

In this example only one response, albeit an important one, was considered. In the next example there are three important responses, and again valuable conclusions can be reached by a simple inspection of the data.

EXAMPLE 2   353

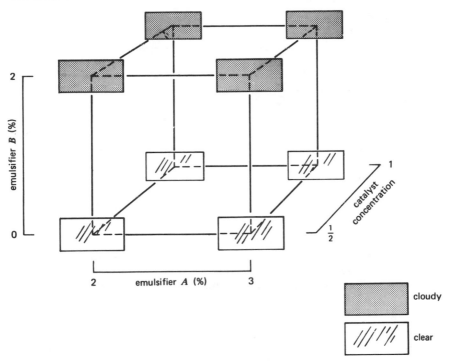

FIGURE 11.1.   Appearance of cast films on glass slides, Example 1.

11.2.   EXAMPLE 2: THE EFFECTS OF THREE VARIABLES ON PHYSICAL PROPERTIES OF A POLYMER SOLUTION

Eight liquid polymer formulations were prepared according to a 2^3 factorial design. The three variables were the amount of a reactive monomer, the type of chain length regulator, and the amount of chain length regulator. The design and the results are shown in Table 11.1. Responses observed were (1) whether the resulting solution was milky, (2) whether it was viscous, and (3) whether it had a yellow color.

The most obvious result when bottles containing the solution were lined up next to one another in standard order was the regular alternation between milky and clear solutions. This phenomenon is directly correlated with variable **1**, the amount of reactive monomer present. Similarly, variable **3**, the amount of chain length regulator, seems to control viscosity. Finally, yellowness apparently results from an interaction between variables **1** and **2**; it is necessary to have *both* a high level (30 %) of reactive monomer and the

TABLE 11.1. Experimental design and results, Example 2

formulation	variable			response		
	1	**2**	**3**	milky?	viscous?	yellow?
1	−	−	−	yes	yes	no
2	+	−	−	no	yes	no
3	−	+	−	yes	yes	no
4	+	+	−	no	yes	slightly
5	−	−	+	yes	no	no
6	+	−	+	no	no	no
7	−	+	+	yes	no	no
8	+	+	+	no	no	slightly

variable	−	+
1 amount of reactive monomer (%)	10	30
2 type of chain length regulator	A	B
3 amount of chain length regulator (%)	1	3

type-B chain length regulator for this effect to be produced. If the variables had been studied in customary one-variable-at-a-time fashion, this inter-action effect could have escaped detection.

11.3. EXAMPLE 3: DEVELOPMENT OF SCREENING FACILITY FOR STORM WATER OVERFLOWS

Illustrating the use of a sequence of simple factorial designs, this example shows how iterations occur in tactical objectives as well as in the nature and levels of the variables. It also indicates how unexpected features of the problem can become dominant and how experimental difficulties can occur so that certain planned experiments cannot be run at all. Most of all, this example shows the importance of common sense in the conduct of any experimental investigation. The reader may rightly conclude from this example that the course of a real investigation, like that of true love, seldom runs smooth although the eventual outcome may be quite satisfactory.

Motivation for the Study

During heavy rainstorms the total flow coming to a sewage treatment plant may exceed its capacity, making it necessary to bypass the excess flow around

EXAMPLE 3   355

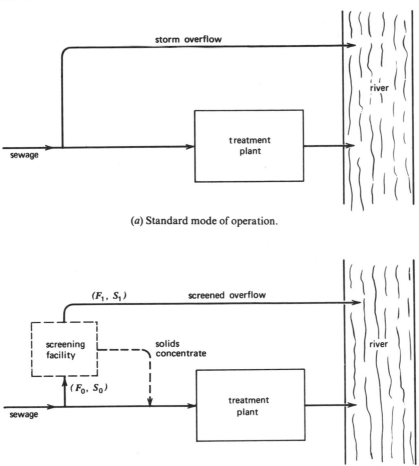

(*a*) Standard mode of operation.

(*b*) Modified mode of operation, with screening facility. F = flow, S = settleable solids.

FIGURE 11.2.   Operation of sewage treatment plant, Example 3.

the treatment plant as shown in Figure 11.2*a*. Unfortunately the storm over-flow of untreated sewage causes pollution of the receiving body of water. A possible alternative, sketched in Figure 11.2*b*, is to screen most of the solids out of the overflow in some way and return them to the plant for treatment. Only the less objectionable screened overflow is discharged directly to the river.

To determine whether it was possible economically to construct and operate such a screening facility, the Federal Water Pollution Control Administration of the Department of the Interior sponsored a research project at the Sullivan Gulch pump station in Portland, Oregon. Usually the flow to the pump station was about 20 million gallons per day (mgd), but during storms the flow could exceed 50 mgd.

Figure 11.3 shows the original version of the experimental screening unit, which could handle approximately 1000 gallons per minute (gpm). Figure 11.3a is a perspective view, and Figure 11.3b is a simplified schematic diagram. A single unit was about 7 feet high and 7 feet in diameter. The flow of raw sewage struck a rotating collar screen at a velocity of 5 to 15 feet per second. This speed was a function of the flow rate into the unit and hence the diameter of the influent pipe. Depending on the speed of rotation of this screen and its fineness, up to 90% of the feed penetrated the collar screen. The rest of the feed dropped to the horizontal screen, which vibrated to remove excess water. The solids concentrate, which passed through neither screen, was sent to the sewage treatment plant. Unfortunately during operation the screens became clogged with solid matter, not only sewage but also oil, paint, and fish packing wastes. Backwash sprays were therefore installed for both screens to permit cleaning during operation.

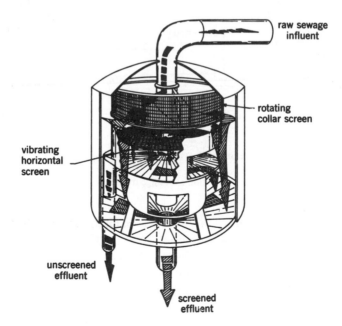

FIGURE 11.3. (a) Original version of screening unit, Example 3. Detailed diagram.

EXAMPLE 3   357

The Investigation

The objective of the investigation was to determine good operating conditions.

Changing Criteria

What are *good* operating conditions? Initially it was believed they were those resulting in the highest possible removal of solids. In Figure 11.2*b* settleable solids in the influent are denoted by S_0 and settleable solids in the effluent by S_1. The *percent solids removed* by the screen is therefore $y = 100(S_0 - S_1)/S_0$. Thus initially it was believed that good operation meant achieving a high value for y. However, it became evident after the first set of runs had been

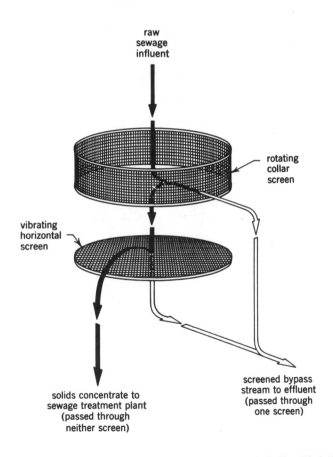

FIGURE 11.3.   (*b*) Original version of screening unit, Example 3. Simplified diagram.

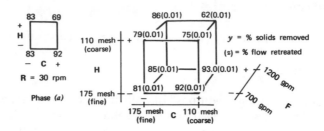

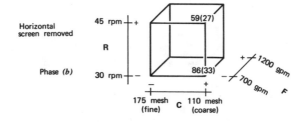

83   69
+
H
−
83   92
− C +

R = 30 rpm

Phase (a)

86(0.01)   62(0.01)
79(0.01)   75(0.01)   y = % solids removed
110 mesh + +   (z) = % flow retreated
(coarse)
                    85(0.01)——93.0(0.01) + 1200 gpm
H
                    81(0.01)   92(0.01) − 700 gpm   F
175 mesh + −
(fine)   −   +
        175 mesh C 110 mesh
        (fine)   (coarse)

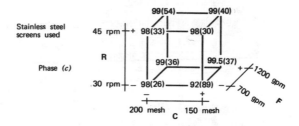

Horizontal
screen removed

                45 rpm + +   59(27)

                R

Phase (b)   30 rpm + −   86(33)
                                        + 1200 gpm
                                        − 700 gpm   F
                −   +
                175 mesh C 110 mesh
                (fine)   (coarse)

Stainless steel
screens used

                            99(54)——99(40)
                45 rpm + + 98(33)——98(30)

                R   99(36)   99.5(37)

Phase (c)   30 rpm + − 98(26)——92(89) − 1200 gpm
                                            − 700 gpm   F
                        −   +
                        200 mesh C 150 mesh

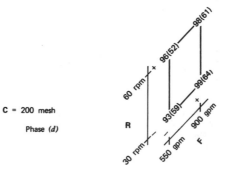

C = 200 mesh

Phase (d)

                98(61)
        96(52)
60 rpm + +   99(84)
                93(59) +
R   900 gpm
30 rpm + −   F
        550 gpm

FIGURE 11.4. Summary of experimental program, Example 3.

EXAMPLE 3   359

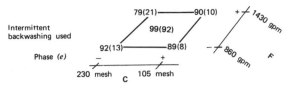

analysis, phase (*a*) for solids removed		analysis, phase (*c*) for		
			solids removed	flow retreated
	y		**y**	**z**
C	−2 ± 6	*C*	−1	−1
H	−12 ± 6	*R*	1	5
F	0 ± 6	*F*	3	10
CH	−12 ± 6	*CR*	1	2
CF	−6 ± 6	*CF*	2	6
HF	−3 ± 6	*RF*	−2	−6
CHF	−4 ± 6	*CRF*	−2	0

FIGURE 11.4. (*continued*)

made that the *percentage of flow retreated* (flow returned to treatment plant), which we denote by z, also had to be taken into account. In Figure 11.2*b* influent flow to the screens is denoted by F_0, and effluent flow from the screens to the river by F_1. Thus $z = 100(F_0 - F_1)/F_0$. The various phases of the investigation are summarized in Figure 11.4.

Phase (a)

In this initial phase a 2^3 design was run in which the variables studied and their levels were as follows:

collar screen mesh size		horizontal screen mesh size		flow rate (gpm)	
C		**H**		**F**	
−	+	−	+	−	+
175	110	175	110	700	1200
(fine)	(coarse)	(fine)	(coarse)		

In Figure 11.4a percentage of solids removed (y) is shown at the vertices of the cube, with percentage of flow retreated (z) in parentheses. The effects for y with standard errors are also shown. At this stage:

1. The experimenters were encouraged by the generally high values achieved for y.
2. Highest values for y were apparently achieved by using a horizontal screen with coarse mesh and a collar screen with fine mesh.
3. Contrary to expectation, flow rate did not show up as an important variable affecting y.
4. Most importantly, the experiment was unexpectedly dominated by the z values, which measure the flow retreated. These were uniformly very low, with about 0.01 % of the flow being returned to the treatment plant and 99.99 % leaving the screen for discharge into the river. Although it was desirable that the retreated flow be small, these values were embarrassingly low. As the experimenters remarked, "[T]he horizontal screen produced a solids concentrate... dry enough to shovel.... This represented a waste of concentrating effort because the concentrated solids were intended to *flow* from the units."

Phase (b)

It was now clear (1) that z as well as y was important and (2) that z was too low. It was conjectured that matters might be improved by removing the horizontal screen altogether. A further 2^3 design was therefore performed with no horizontal screen. The speed of rotation of the collar screen was introduced as a new variable. The levels of the variables were as follows:

collar screen mesh size C		collar screen speed (rpm) R		flow rate (gpm) F	
−	+	−	+	−	+
175	110	30	45	700	1200

Unfortunately after only two runs this particular phase had to be terminated because of excessive tearing of the cloth screens. The results are shown in Figure 11.4b. From the scanty results obtained it appeared, however, that with no horizontal screen high solids removal rates *could* be achieved with

EXAMPLE 3   361

higher portions of the flow retreated. It was therefore decided to repeat these runs with screens made of stainless steel instead of cloth.

Phase (c)

With stainless steel collar screens of two mesh sizes, a third 2^3 design similar to that attempted in phase (b) was performed with these levels:

collar screen mesh size C		collar screen speed (rpm) R		flow rate (gpm) F	
−	+	−	+	−	+
200	150	30	45	700	1200

The results given in Figure 11.4c show that, with a stainless steel collar screen, high removal rates y were possible at all eight sets of conditions. However, these high y values were obtained with retreated flow z at undesirably *high* values (before, they had been too low). The object was to get reasonably small values for z, but not so small as to make shoveling necessary; values between 5 and 20% were desirable. It was believed that, by varying flow rate and speed of rotation of the collar screen, this objective could be achieved without sacrificing solids removal.

Phase (d)

Using a stainless steel collar screen, therefore, the investigators set up a 2^2 factorial design with these levels:

collar screen speed (rpm) R		flow rate (gpm) F	
−	+	−	+
30	60	550	900

The results are shown in Figure 11.4d. High values for solids removal were maintained, but unfortunately flow retreated values were even higher than before.

Phase (e)

It was now conjectured that intermittent back washing could overcome the difficulties. This procedure was now introduced with influent flow rate and collar screen mesh size varied as follows:

collar screen mesh size C		flow rate (gpm) F	
−	+	−	+
230	105	860	1430

The results are shown in Figure 11.4*e*. It will be seen, in particular, that a removal efficiency of 89 % was achieved with a retreated flow of only 8 %. This was regarded as providing a satisfactory and practical solution, and the investigation was terminated at that point.

Commentary

Notice how wise use of technical knowledge and common sense, employed in conjunction with simple experimental design, moved this investigation iteratively toward a satisfactory solution. Notice also how the objective changed as the investigation proceeded.

11.4. EXAMPLE 4: SIMPLE FACTORIALS USED SEQUENTIALLY IN EVOLUTIONARY OPERATION— PETROCHEMICAL PLANT

It is seldom true that industrial processes are run at the best possible operating conditions. Often small-scale work, on which many process designs are based, provides no more than a rather crude approximation of the best conditions to use for full-scale production. Evolutionary operation (EVOP) is an ongoing mode of using an operating full-scale process so that information on how to improve the process is generated from a simple experimental design while production is under way. To avoid appreciable changes in the characteristics of the product only small changes are made in the levels of the process variables. To determine the effects of these changes it is therefore necessary to repeat the runs a number of times and average the observations.

Evolutionary operation is designed to be run, by *process operators*, on a *full-scale* manufacturing process while it continues to produce *satisfactory*

EXAMPLE 4   363

product. Hence the circumstances are very different from those in the laboratory or pilot plant, where (a) additional money and time must be expended to make the runs, but (b) many factors can be simultaneously changed in carefully conducted runs by skilled technicians, and (c) the manufacture of unsalable material is not a problem. In the latter situation it is usually good strategy to pack as many factors as possible into each design, thus keeping the (factors studied)/(runs made) ratio as high as possible. If 16 runs are to be made at a given stage of an investigation, it is frequently better to use them to study four factors in a 2^4 factorial, or, even better, to study five factors in a half-fraction of a 2^5 factorial (see Chapter 12), than to study three factors in a duplicated 2^3 design.

By contrast consider the situation in which evolutionary operation is used. Here (a) because the signal/noise ratio must be kept low, a large number of runs are usually necessary to reveal the effects of changes, but (b), these are manufacturing runs that must be made anyway and result in very little additional cost; (c) in the manufacturing environment things must be kept very simple, and usually it is practical to vary only two or three factors in any given phase of the investigation. In these circumstances it makes sense to use replicated 2^2 or 2^3 factorial designs, often with an added center point, in the manner we describe.

As results become available, averages and estimates of effects are continually updated and displayed on an information board, which can be an ordinary chalk board. It must be located where it is clearly visible to those responsible for running the process and, in particular, to the process superintendent. In consultation with an EVOP committee the process superintendent uses the information board to guide him to better process conditions.

In a study described by Jenkins (1969) the object was to decrease the cost per ton of a certain product of a petrochemical plant. At one stage of the investigation two variables believed to be important were:

1. The reflux ratio of a distillation column.
2. The ratio of recycle flow to purge flow.

It was expected that changes of the magnitude planned would introduce transient effects that would die out in about 6 hours. A further 18 hours of steady operation was then allowed to make the necessary measurements for each run.

The appearance of the information board at the end of each of three phases is shown in Table 11.2. The response was the average cost per ton, recorded only to the nearest unit. The design employed was a 2^2 factorial with a center point. The results at the end of phase I, shown on the left of Table 11.2, are

TABLE 11.2. Appearance of the information board at the end of phases I, II, and III, petrochemical plant data. Response is average cost per ton.

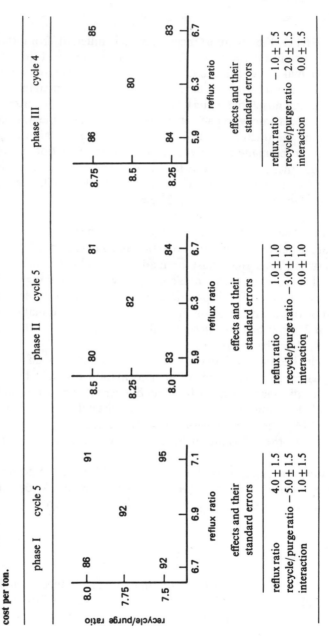

EXAMPLE 5 365

averages obtained after five repetitions (cycles). At this point it was decided that there was sufficient evidence to justify a move to the lower reflux ratio and higher recycle/purge ratio used in phase II. The results from phase II, which was terminated after five further cycles, confirmed that lower costs indeed resulted from this move and suggested that still higher values of the recycle/purge ratio should be tried, leading to phase III. This phase, brought to an end after four cycles, led to the conclusion that the lowest cost was obtained with the reflux ratio close to 6.3 and the recycle/purge ratio about 8.5, where a cost per ton of about £80 was achieved.

Exercise 11.1. Plot the results from all three phases of the petrochemical plant study on a single grid, with reflux ratio and recycle/purge ratio as coordinates. Do you think that shifts in the general level of the response may have occurred during the course of this study?

The program described took $4\frac{1}{2}$ months to complete at a cost of about £6,000. The savings resulting from reducing the cost per ton from £92 to £80 was about £100,000 per year. Patience has its rewards.

11.5. EXAMPLE 5: SIMPLE FACTORIALS USED SEQUENTIALLY IN EVOLUTIONARY OPERATION— POLYMER UNIT

The chemical product in this case was a polymer latex. The *initial* objective was to determine the conditions that, subject to a number of other constraints, gave the *lowest optical densities* for a given batch of raw material. Later, when it was realized that shorter addition times were possible, resulting in great savings because of increased throughput, the objective shifted to *increasing plant throughput* subject to obtaining satisfactory optical densities (less than 40).

Three process variables were studied in this EVOP program: addition time (t) of one of the reactants, temperature (T), and stirring rate (S). Although it would have been possible to use all three variables simultaneously, it was decided for the sake of simplicity of operation to study these factors in different combinations two at a time. The data below have been coded for proprietary reasons.

Initially the standard operating conditions were t = 3 hours, T = 120°F, S = 88 rpm. In phase I temperature was held at its standard condition, and time and stirring rate were varied as in Table 11.3. The ranges adopted for these two factors were, however, extremely narrow because production

TABLE 11.3. Summary of results, Example 5

optical density reading

phase I (stirring rate vs. addition time)

```
stirring rate
   35.2    36.0
        ┌──────┐
        │ 36.6 │
        └──────┘
   36.2    35.2
        → addition time
```

phase II (stirring rate vs. addition time)

```
stirring rate
   40.8    41.0
        ┌──────┐
        │ 38.8 │
        └──────┘
   38.8    39.8
        → addition time
```

phase III (temperature vs. addition time)

```
temperature
   44.0    44.8
        ┌──────┐
        │ 41.8 │
        └──────┘
   39.2    40.0
        → addition time
```

phase IV (temperature vs. stirring rate)

```
temperature
   45.2    43.6
        ┌──────┐
        │ 42.0 │
        └──────┘
   39.6    40.8
        → stirring rate
```

phase V (temperature vs. addition time)

```
temperature
   39.8    40.2
        ┌──────┐
        │ 37.6 │
        └──────┘
   36.4    37.6
        → addition time
```

	phase I	phase II	phase III	phase IV	phase V
settings					
temperature	120	120	118 120 122	116 118 120	114 116 118
stirring rate	83 88 93	78 88 98	88	68 78 88	78
addition time	2:50 3:00 3:10	2:35 3:00 3:25	2:30 2:45 3:00	2:45	2:15 2:30 2:45
effects with standard errors					
temperature			4.8 ± 1.2	4.2 ± 0.7	3.0 ± 0.4
stirring rate	-0.4 ± 0.5	1.6 ± 0.7		-0.2 ± 0.7	
addition time	0.2 ± 0.5	0.6 ± 0.7	0.8 ± 1.2		0.8 ± 0.4
interaction	0.6 ± 0.5	-0.4 ± 0.7	0.0 ± 1.2	-1.4 ± 0.7	-0.4 ± 0.4

EXAMPLE 5   367

personnel feared that substandard product might be produced.* Twelve cycles were performed in this first phase, during which a total of $12 \times 5 = 60$ batches were manufactured. It will be seen that even after this long period of production no effects of any kind were detected. This result, at first disappointing, was actually quite helpful. It convinced those in charge of production that changing the two selected variables over wider ranges would be unlikely to produce substandard material. Wider ranges were therefore allowed in phase II, and after 18 cycles a stirring effect was apparent (1.6 $\pm$ 0.7), indicating that lower optical densities might be obtained at lower stirring rates. The engineers in charge, however, did not choose to lower the stirring rate at this stage. For them the most important finding was the unexpected one that no deleterious effect resulted from reducing the addition time. The unit had been believed to be operating at full capacity. A saving of 15 minutes in addition time would result in an 8% increase in throughput and a considerable rise in profitability. To further investigate the possibilities for increased throughput, addition time and temperature were varied simultaneously in phase III. Continuing to contradict established belief, the results confirmed that increased throughput did not result in an inferior product and suggested that optical density might be reduced by a reduction in temperature. Phase IV used slightly lower temperatures, while stirring rate was varied about a lower value and addition time was held cautiously at the reduced value of 2 hours 45 minutes (2:45).

In phase V both temperature and addition time were varied about lower values (116°F, 2:30), without deleterious effect, and the stirring rate was held at 78. The results indicated that temperature could be reduced still further.

Commentary

The net result of the EVOP program up to this point was that throughput was increased by a factor of 25%, stirring rate was reduced from 88 to 78, and temperature was lowered from 120 to 114°F. All of these changes, but especially the first, resulted in economies.

One iteration that occurred in this example, as it did in Example 3, concerned a change in the objective itself. This was first stated as improvement

---

* Although it is essential for management to exercise control over operating conditions, attitudes should not be excessively rigid. In particular, we believe that licensers of processes are not operating in their own best interest, or in that of the licensee, by prohibiting *any* departures from standard operating conditions established at the time the plant was designed. In particular, if a plant is constructed in a new environment, somewhat modified operating conditions will usually be required to achieve best performance. In these circumstances EVOP is a safe, systematic, proved method for "tuning up" a plant to the benefit of all parties.

of the optical density. What was actually achieved was the profitable but unexpected discovery that the unit, while yielding a product of about the same optical density, could give dramatically increased production.

Exercise 11.2. Sketch a three-dimensional grid with coordinates covering appropriate ranges of temperature, stirring rate, and addition time. Plot all the results from Table 11.3 on the grid. Report any apparent inconsistencies, and discuss how they might have occurred.

11.6. SUMMARY

The examples show the value of even very simple factorial experiments. Conclusions from such designs are often best comprehended when the results are displayed geometrically. These designs used in combination with geometric visualization are valuable building blocks in the natural sequential process of learning by experiment. They allow the structure of our thinking to be appropriately modified as investigation proceeds. Evolutionary operation applies these ideas to manufacture and has produced substantial improvement in operating processes.

APPENDIX 11A. A SUGGESTED EXERCISE

One must learn by doing the thing; for though you think you know it, you have no certainty until you try.

SOPHOCLES

You may not be currently involved in a formal project to which you can apply experimental design. In that case we believe you will find it rewarding at this point to plan and perform a home-made factorial experiment and collect and analyze the data. We have regularly assigned such a project to our classes when teaching this material. The students have enjoyed the experience and have learned a great deal from it. We have always left it to the individual to decide what he or she wants to study.

What follows is one of the reports we received. It is by Norman Miller of the University of Wisconsin.

Description of Design

I ride a bicycle a great deal and thought that it might be interesting to see how some of the various parts of the bike affect its performance. In particular, I was interested in the effects of varying the seat height, varying the tire pressure, and using or not using the generator.

In deciding on a design, a 2^3 factorial design immediately came to mind. I expected a large amount of variability, so I decided to replicate the eight data points in order to be able to estimate the variance. Since I wished to do the experiment in a reasonable length of time, I decided to make four runs a day. Although a long run might be desirable in that various types of terrain could be covered, I felt that four long runs a day would be out of the question. I did not have time to make them, I feared that I would tire as the day progressed, making the runs late in the day incompatible with the earlier runs, and I also feared that with the increased exercise my own strength and endurance would increase as the experiment progressed. As a result of the above considerations I decided to do a rather short one and a half block uphill run. My test hill was on Summit Avenue in Madison, starting at Breese Terrace. This hill is of such an inclination that it all must be taken in first gear, thus eliminating gear changes as a possible source of experimental error. I tried to follow a schedule of 9:00 A.M., 11:30 A.M., 2:00 P.M., and 4:30 P.M. As can be seen from the data sheet [Table 11A.1], I deviated from the schedule quite a bit; however, I did avoid making runs right after meals. Note that on April 6 I was able to make only three runs because of bad weather. I made up the lost run on April 8. I assigned the variable levels as shown on the data sheet. The seat height of 30 inches is the maximum elevation of the seat on my bike. I could ride with it higher, so my bike is in effect too small for me.

I measured two responses. First, I measured the speed in seconds to climb the hill. I naturally was trying to ride as fast as possible. Second, I hoped to get some idea of the amount of energy I was expending during the run by measuring my pulse right before and right after the run. I always rode around two blocks at a slow pace before each run to get warmed up. I then stopped at the bottom of the hill and let my pulse settle to the range of 75 to 80 per minute. I would then make my run, timing it on a watch with a large sweep second hand. At the top of the hill, I took my pulse again. I randomized the experiment by writing the numbers 1 through 16 on pieces of cardboard, putting them in a small box, shaking up the box, and drawing out one number before each experiment to determine the setup of the variables.

Two possible criticisms of the experiment came to mind. First, since I knew the setup of the variables, I might bias the experiment to "come out right." I could not think of a way around this. It would have been quite an imposition to ask a friend to set up the bicycle 16 times! In addition, even if a friend had changed the variables each time, I could have heard the generator operating (or seen that the light was burning) and could have seen or felt the height of the seat. Another rider would probably be just as biased as I would be. In addition, who would ride up a hill 16 times for me? So I just decided to try to be as unbiased as possible. The second criticism involves the rather large temperature range encountered during the experiment. Naturally oil tends to be more viscous at low temperatures. I considered that this problem was probably less acute than it might have been, however, since I used silicone oil to lubricate the bike.

The original data sheet appears [in Table 11A.1]. Notice that there is a problem with the data on pulse. Setup 9, the first run, shows my pulse at the top of the hill to be 105 per minute. Setup 15, the second run, has it at 122 per minute. Setup 14, the third run, has it at 125 per minute, and setup 6, the fourth run, at 133 per minute. What was happening was the following. When I arrived at the top of the hill, my heart was beating quite fast, but it started slowing down quickly as soon as I stopped. With practice, I was

TABLE 11A.1. Data from bicycle experiment

setup	seat height	generator	tire pressure	pulse before	pulse after	time (secs)	date and hour	notes
1	−	−	−	76	135	51	4/8/72; 11:40 A.M.	36°F
2	−	−	−	76	133	54	4/9/72; 9:40 A.M.	37°F
3	+	−	−	76	132	41	4/7/72; 12:35 A.M.	23°F, snow on road
4	+	−	−	79	137	43	4/7/72; 4:05 P.M.	25°F, snow melting
5	−	+	−	77	134	54	4/8/72; 2:40 P.M.	36°F
6	−	+	−	77	(133)	(60)	4/7/72; 9:50 A.M.	fourth run, 23°F, snow
7	+	+	−	76	130	44	4/8/72; 9:10 A.M.	26°F, road clear
8	+	+	+	79	139	43	4/9/72; 3:35 P.M.	40°F
9	−	−	+	76	(105)	50	4/6/72; 9:15 A.M.	first run
10	−	−	+	80	144	48	4/9/72; 6:05 P.M.	40°F
11	+	−	+	77	139	39	4/8/72; 6:35 P.M.	32°F
12	+	−	+	78	139	39	4/9/72; 11:40 A.M.	41°F
13	−	+	+	78	137	53	4/8/72; 4:50 P.M.	35°F, rain
14	−	+	+	79	(125)	51	4/6/72; 4:25 P.M.	third run
15	+	+	+	80	(122)	41	4/6/72; 11:40 A.M.	second run
16	+	+	+	77	133	44	4/7/72; 6:35 P.M.	21°F

	seat height (inches from center of crank)	generator	tire pressure (psi)
−	26	off	40
+	30 (maximum height)	on	55

370

simply starting to measure my pulse sooner. Since the pulse data are damaged by this obvious error, I have chosen to ignore them in the following analysis. The readings taken at the bottom of the hill did prove valuable since they served as a gauge of when my body was in a "normal" state.

Analysis of Data

First, I plotted the average times on a cube [Figure 11A.1] and looked at them. Then I calculated:

1. Estimated variance of each of 16 observations = 4.19 square seconds with eight degrees of freedom.
2. Main effects and interactions.
3. Estimated variance of effect = 4.19/4 = 1.05.
4. Individual 95% confidence intervals for true values of effects.

1 (seat)	=	-10.9 ± 2.4
2 (generator)	=	3.1 ± 2.4
3 (pressure)	=	-3.1 ± 2.4
12 (seat × generator)	=	-0.6 ± 2.4
13 (seat × pressure)	=	1.1 ± 2.4
23 (generator × pressure)	=	0.1 ± 2.4
123 (seat × generator × pressure)	=	0.9 ± 2.4

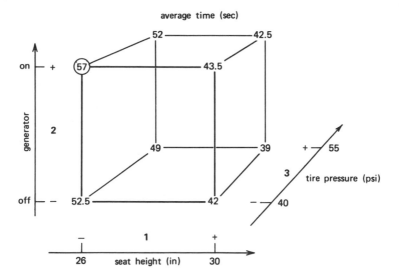

FIGURE 11A.1. Results from bicycle experiment. Response is average time to ride bicycle over set course.

I believe that the value of 57 circled on the cube is probably too high. Notice that it is the result of setup 6, which had a time of 60 seconds. Its comparison run (setup 5) was only 54 seconds. This is the largest amount of variation in the whole table, by far. I suspect that the correct reading for setup 6 was 55 seconds, that is, I glanced at my watch and thought that it said 60 instead of 55 seconds. Since I am not sure, however, I have not changed it for the analysis. The conclusions would be the same in either case.

Obviously a good stopwatch would have helped the experiment, but I believe that the results are rather good considering the adverse weather and poor timer.

A slightly better design might include a downhill and a level stretch to provide more representative terrain. Probably only two runs could be safely made each day in such a case.

Conclusions

Raising the seat to the high position cuts the time to climb the hill by about 10 seconds. Turning on the generator increases the time by about one third of that amount. Inflating the tires to 55 psi reduces the time by about the same amount that the generator increases it. The variables appear to be substantially free of interactions.

Since I could ride with the seat at an even higher position, I am planning to modify the bike to achieve this. I also plan to inflate my tires to 60 psi, which is the maximum for which they are designed.

REFERENCES AND FURTHER READING

Examples 1 and 2 in this chapter are from this article:

Hunter, W. G., and M. E. Hoff (1967). Planning experiments to increase research efficiency, *Ind. Eng. Chem.*, **59**, 43. (The final example in this paper illustrates the sequential use of simple factorial designs in a successful project in which the experimenter's intuition and subject-matter knowledge played a dominant role in the decisions at each stage about the location of the next statistically designed experiment, the factors to be studied, and their ranges.)

Example 3 is based on this report:

Rotary Vibratory Fine Screening of Combined Sewer Overflows: Primary Treatment of Storm Water Overflows: Primary Treatment of Storm Water Overflow from Combined Sewers by High-Rate, Fine-Mesh Screens, by CH2M Hill, Inc., Engineers, Planners, Economists, Scientists, Corvallis, Oregon 97330, Federal Water Pollution Control Administration, Department of the Interior, Contract 14-12-128, March 1970. (The quotations in this chapter are from this report. We are indebted to Mr. Donald M. Marske, who directed this research, for bringing it to our attention.)

A paper based on the above report is:

Marske, D. M. (1970). High-rate, fine-mesh screening of combined wastewater flows, *J. Water Pollut. Control Fed.*, **42**, 1476.

Example 4 is from the following:

Jenkins, G. M. (1969). A systems study of a petrochemical plant, *J. Syst. Eng.*, **1**, 90.

Example 5 is from this article:

Hunter, W. G., and E. Chacko (1971). Increasing industrial productivity in developing countries, *International Development Review*, **13**, 3, 11.

For further reading on evolutionary operation see:

Box, G. E. P., and N. R. Draper (1969). *Evolutionary Operation: A Statistical Method for Process Improvement*, Wiley.

For further examples of applications of factorial designs, see:

Daniel, C. (1976). *Applications of Statistics to Industrial Experimentation*, Wiley.
Davies, O. L. (Ed.) (1954). *Design and Analysis of Industrial Experiments*, Hafner (Macmillan).
Hunter, W. G. (1976). Some ideas about teaching design of experiments, with 2^5 examples of experiments conducted by students, *Am. Stat.*, **31**, 12. (Appendix 11A describes one of the 32 experiments listed in this paper.)

QUESTIONS FOR CHAPTER 11

1. In the cast film and polymer solution examples (Examples 1 and 2), why was not any mathematical analysis of the results necessary?
2. In the example about storm water overflows (Example 3), what was the object of the investigation? Was it accomplished?
3. In terms of the description of the iterative nature of experimentation given in Chapter 1, can you identify the hypothesis, design, experiment, and analysis steps at each of the several stages involved in Example 3? Example 4? Example 5?
4. What is evolutionary operation? Make a list of characteristics that differ in laboratory or pilot plant experimentation and in evolutionary operation. How do these decide strategy in the two cases?
5. How could you make a geometric display of the results from a 2^5 factorial?

CHAPTER 12

Fractional Factorial Designs at Two Levels

The number of runs required by a full 2^k factorial design increases geometrically as k is increased. It turns out, however, that when k is not small the desired information can often be obtained by performing only a fraction of the full factorial design. This chapter describes how suitable fractions can be generated and discusses their advantages and limitations.

12.1. REDUNDANCY

Consider a two-level design in seven variables. A complete factorial arrangement requires $2^7 = 128$ runs. From these runs 128 statistics can be calculated, which estimate the following effects:

			interactions				
average	main effects	2-factor	3-factor	4-factor	5-factor	6-factor	7-factor
1	7	21	35	35	21	7	1

Now the fact that all these effects can be estimated does not imply that they all are of appreciable size. There tends to be a certain hierarchy. In terms of absolute magnitude, main effects tend to be larger than two-factor interactions, which in turn tend to be larger than three-factor interactions, and so on. This fact relates directly to the properties of smoothness and similarity discussed earlier. (In particular, for quantitative variables the main effects and interactions can be associated with the terms of a Taylor series expansion of a response function. Ignoring, say, three-factor interactions corresponds to ignoring terms of third order in the Taylor expansion.)

It is often true, then, that at some point higher order interactions tend to become negligible and can properly be disregarded. Also, when a moderately large number of variables is introduced into a design, it often happens that some have no distinguishable effects at all. We can encompass both these ideas by saying that there tends to be redundancy in a 2^k design if k is not small—redundancy in terms of an excess number of interactions that can be estimated and sometimes in an excess number of variables that are studied. Fractional factorial designs exploit this redundancy. We begin by considering what effects can be estimated using only a half-fraction of a 2^5 factorial design.

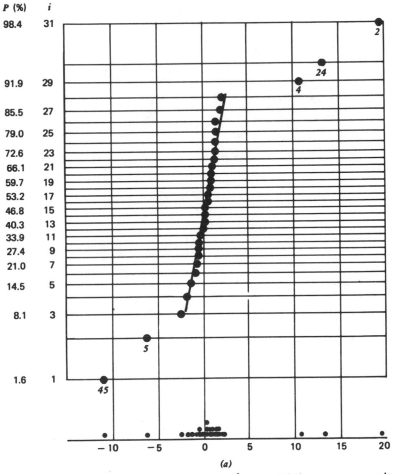

FIGURE 12.1. (a) Normal plot of effects from 2^5 factorial design, reactor example.

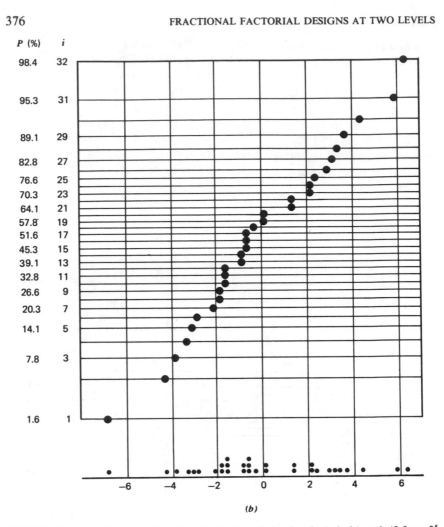

FIGURE 12.1. (*b*) Normal plot of residuals after eliminating *2, 4, 5, 24,* and *45* from 2^5 factorial design, reactor example.

12.2.   A HALF-FRACTION OF A 2^5 DESIGN: REACTOR
EXAMPLE

Table 12.1*a* shows data from a complete 2^5 factorial design analyzed in Table 12.1*b*. Normal plots (Figure 12.1) indicate that over the ranges of the variables studied the main effects *2, 4,* and *5* and interactions *24* and *45* are the only effects distinguishable from noise.

TABLE 12.1a. Results from 2^5 factorial design, reactor example

variable	−	+
1 feed rate (liters/min)	10	15
2 catalyst (%)	1	2
3 agitation rate (rpm)	100	120
4 temperature (°C)	140	180
5 concentration (%)	3	6

	variable					response (% reacted)
run	1	2	3	4	5	y
1	−	−	−	−	−	61
*2	+	−	−	−	−	53
*3	−	+	−	−	−	63
4	+	+	−	−	−	61
*5	−	−	+	−	−	53
6	- +	−	+	−	−	56
7	−	+	+	−	−	54
*8	+	+	+	−	−	61
*9	−	−	−	+	−	69
10	+	−	−	+	−	61
11	−	+	−	+	−	94
*12	+	+	−	+	−	93
13	−	−	+	+	−	66
*14	+	−	+	+	−	60
*15	−	+	+	+	−	95
16	+	+	+	+	−	98
*17	−	−	−	−	+	56
18	+	−	−	−	+	63
19	−	+	−	−	+	70
*20	+	+	−	−	+	65
21	−	−	+	−	+	59
*22	+	−	+	−	+	55
*23	−	+	+	−	+	67
24	+	+	+	−	+	65
25	−	−	−	+	+	44
*26	+	−	−	+	+	45
*27	−	+	−	+	+	78
28	+	+	−	+	+	77
*29	−	−	+	+	+	49
30	+	−	+	+	+	42
31	−	+	+	+	+	81
*32	+	+	+	+	+	82

TABLE 12.1b. Analysis of 2^5 factorial design, reactor example

---

estimates of effects

---

average = 65.5

$1 = -1.375$		$123 = 1.50$
$2 = 19.5$		$124 = 1.375$
$3 = -0.625$		$125 = -1.875$
$4 = 10.75$		$134 = -0.75$
$5 = -6.25$		$135 = -2.50$
		$145 = 0.625$
$12 = 1.375$		$235 = 0.125$
$13 = 0.75$		$234 = 1.125$
$14 = 0.875$		$245 = -0.250$
$15 = 0.125$		$345 = 0.125$
$23 = 0.875$		
$24 = 13.25$		$1234 = 0.0$
$25 = 2.0$		$1245 = 0.625$
$34 = 2.125$		$2345 = -0.625$
$35 = 0.875$		$1235 = 1.5$
$45 = -11.0$		$1345 = 1.0$

$12345 = -0.25$

The full 2^5 factorial requires 32 runs. Suppose that the experimenter had chosen to make only the 16 runs marked with asterisks in Table 12.1, so that only the data of Table 12.2 were available. When the 15 main effects and two-factor interactions are calculated from the reduced set of data in Table 12.2, they produce the estimates listed there, which are not very different from those obtained from the complete factorial design. Furthermore the normal plots of Figure 12.2 call attention to precisely the same effects: 2, 4, 24, 45 and 5. Thus the essential information could have been obtained with only half the effort.

The 16-run design in Table 12.2 is called a *half-fraction*. It is often designated as a 2^{5-1} fractional factorial design since

$$\tfrac{1}{2}2^5 = 2^{-1}2^5 = 2^5 2^{-1} = 2^{5-1}$$

The notation tells us that the design accommodates five variables, each at two levels, but that only $2^{5-1} = 2^4 = 16$ runs are employed.

TABLE 12.2 Analysis of a half-fraction of the full 2^5 design: a 2^{5-1} fractional factorial design, reactor example

	variable	−	+
1	feed rate (liters/min)	10	15
2	catalyst (%)	1	2
3	agitation rate (rpm)	'100	120
4	temperature (°C)	140	180
5	concentration (%)	3	6

run	1	2	3	4	5	12	13	14	15	23	24	25	34	35	45	response (% reacted) y
17	−	−	−	−	+	+	+	+	−	+	+	−	+	−	−	56
2	+	−	−	−	−	−	−	−	−	+	+	+	+	+	+	53
3	−	+	−	−	−	−	+	+	+	−	−	−	+	+	+	63
20	+	+	−	−	+	+	−	−	+	−	−	+	+	−	−	65
5	−	−	+	−	−	+	−	+	+	−	+	+	−	−	+	53
22	+	−	+	−	+	−	+	−	+	−	+	−	−	+	−	55
23	−	+	+	−	+	−	−	+	−	+	−	+	−	+	−	67
8	+	+	+	−	−	+	+	−	−	+	−	−	−	−	+	61
9	−	−	−	+	−	+	+	−	+	+	−	+	−	+	−	69
26	+	−	−	+	+	−	−	+	+	+	−	−	−	−	+	45
27	−	+	−	+	+	−	+	−	−	−	+	+	−	−	+	78
12	+	+	−	+	−	+	−	+	−	−	+	−	−	+	−	93
29	−	−	+	+	+	+	−	−	−	−	−	−	+	+	+	49
14	+	−	+	+	−	−	+	+	−	−	−	+	+	−	−	60
15	−	+	+	+	−	−	−	−	+	+	+	−	+	−	−	95
32	+	+	+	+	+	+	+	+	+	+	+	+	+	+	+	82

estimates of effects
(assuming that three-factor and higher order interactions are negligible)

average =	65.25		12 =	1.5
1 =	− 2.0		13 =	0.5
2 =	20.5		14 =	−0.75
3 =	0.0		15 =	1.25
4 =	12.25		23 =	1.50
5 =	− 6.25		24 =	10.75
			25 =	1.25
			34 =	0.25
			35 =	2.25
			45 =	−9.50

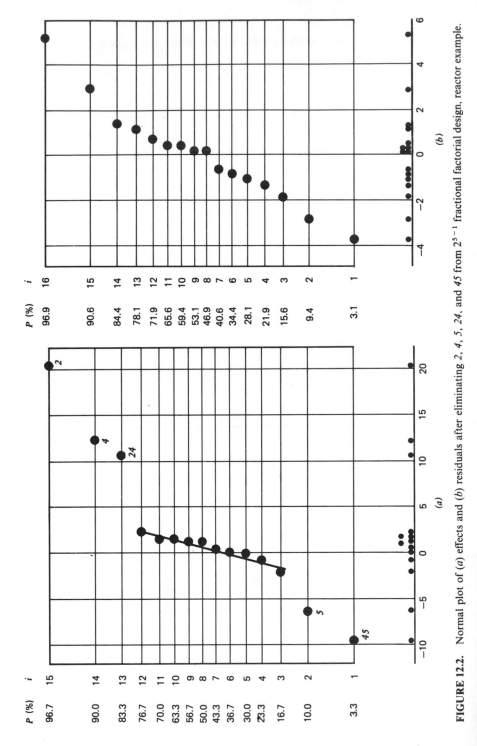

FIGURE 12.2. Normal plot of (*a*) effects and (*b*) residuals after eliminating *2, 4, 5, 24,* and *45* from 2^{5-1} fractional factorial design, reactor example.

12.3. CONSTRUCTION AND ANALYSIS OF HALF-FRACTIONS: REACTOR EXAMPLE

How Were the 16 Runs Chosen?

The 2^{5-1} design in Table 12.2 was constructed as follows:

1. A full 2^4 design was written for the four variables **1, 2, 3,** and **4.**
2. The column of signs for the **1234** interaction was written, and these were used to define the levels of variable **5.** Thus we made **5 = 1234.**

Exercise 12.1.   By using this procedure, verify that the design obtained is the one given in Table 12.2.

The Anatomy of the Half-Fraction

At this point we seem to have gained something for nothing. It is natural to ask, Have we lost anything? Look again at the fractional factorial design of Table 12.2. We have made 16 runs and estimated 16 quantities: the mean, the 5 main effects, and the 10 two-factor interactions. But what happened to the remaining 16 effects we were able to estimate with the full factorial design — the 10 three-factor interactions, the 5 four-factor interactions, and the 1 five-factor interaction?

Let us try to estimate the value of the three-factor interaction *123.* Multiplying the signs in columns **1, 2,** and **3,** we obtain the sequence (which, to save space, we write as a row rather than a column)

$$\mathbf{123} = -++-+--+-++-+--+$$

We notice that this is identical to

$$\mathbf{45} = -++-+--+-++-+--+$$

Thus **123 = 45,** and as a consequence the *123* and *45* interactions are confounded. Equivalently, in the fractional design the individual interactions *123* and *45* are said to be *aliases* of each other. Now suppose that we use the symbol l_{45} to denote the linear function of the observations which we used to estimate the *45* interaction:

$$l_{45} = \tfrac{1}{8}(-56 + 53 + 63 - 65 + 53 - 55 - 67 + 61 - 69 + 45 + 78$$
$$-93 + 49 - 60 - 95 + 82) = -9.50 \tag{12.1}$$

We can call this the l_{45} *contrast* since it is the *difference* between two averages of eight results. Properly speaking, contrast l_{45} estimates the *sum* of the mean values of effects *45* and *123*. We indicate this by the notation $l_{45} \rightarrow 45 + 123$. If the columns of signs corresponding to all the other three-factor, four-factor, and five-factor interactions are obtained by multiplying signs, we get the results shown in Table 12.3.

TABLE 12.3. Confounding pattern and estimates from 2^{5-1} design
of Table 12.2

relationship between column pairs	confounding pattern	estimate
1 = 2345	$l_1 \rightarrow 1 + 2345$	$l_1 = -2.0$
2 = 1345	$l_2 \rightarrow 2 + 1345$	$l_2 = 20.5$
3 = 1245	$l_3 \rightarrow 3 + 1245$	$l_3 = 0.0$
4 = 1235	$l_4 \rightarrow 4 + 1235$	$l_4 = 12.25$
5 = 1234	$l_5 \rightarrow 5 + 1234$	$l_5 = -6.25$
12 = 345	$l_{12} \rightarrow 12 + 345$	$l_{12} = 1.5$
13 = 245	$l_{13} \rightarrow 13 + 245$	$l_{13} = 0.5$
14 = 235	$l_{14} \rightarrow 14 + 235$	$l_{14} = -0.75$
15 = 234	$l_{15} \rightarrow 15 + 234$	$l_{15} = 1.25$
23 = 145	$l_{23} \rightarrow 23 + 145$	$l_{23} = 1.5$
24 = 135	$l_{24} \rightarrow 24 + 135$	$l_{24} = 10.75$
25 = 134	$l_{25} \rightarrow 25 + 134$	$l_{25} = 1.25$
34 = 125	$l_{34} \rightarrow 34 + 125$	$l_{34} = 0.25$
35 = 124	$l_{35} \rightarrow 35 + 124$	$l_{35} = 2.25$
45 = 123	$l_{45} \rightarrow 45 + 123$	$l_{45} = -9.50$
(I = 12345)	[$l_1 \rightarrow$ average $+ \frac{1}{2}(12345)$]	($l_1 = 65.25$)

Exercise 12.2. As was done for columns **45** and **123**, verify that columns **24** and **135** are identical. Verify the identity of the other column pairs in Table 12.3.

A Justification for the Analysis

Evidently our earlier analysis would be justified if it could be assumed that effects of third and fourth order (represented by three-factor and four-factor interactions) could be ignored. In the reactor example the assumption was apparently justified. We shall see later that the analysis could also be justified on different and somewhat more subtle grounds (see the subsection entitled "An Alternative Rationale for the Half-Fraction Design in the Reactor Experiment").

How to Find the Confounding Patterns

In manipulating fractional factorials it is important to be able to obtain the confounding pattern for any given design. The method of associating like sign sequences is extremely tedious. Fortunately a much more expeditious route is available. To understand it remember the following four points:

1. Boldface numerals (e.g., **3** and **12**) refer to *columns* of plus and minus signs.
2. A product column is obtained by multiplication of the individual elements in the columns that make up that product. (The product column **124**, for instance, is obtained by multiplication of the individual elements in the corresponding columns, **1**, **2**, and **4**.)
3. Multiplying the elements in any column by a column of identical elements gives a column of plus signs, which is designated by the letter **I**, that is, $1 \times 1 = 1^2 = I$, $2^2 = I$, $3^2 = I$, $4^2 = I$, and so forth.
4. A contrast like l_{45} in Equation 12.1 is obtained by multiplying the observations by the appropriate plus and minus signs in column **45** and dividing by $N/2 = 8$ where N is the number of observations (16 in this case). Each quantity l is thus a contrast between two averages, each of $N/2$ observations. The single exception is $l_I = \bar{y}$, which is obtained by multiplying the observations by the column **I** of plus signs (i.e., summing the observations) and dividing the result by N (in this example $N = 16$).

Generator and Defining Relation

The 2^{5-1} design in Table 12.2 was constructed by setting

$$5 = 1234 \qquad\qquad (12.2)$$

This relation is called the *generator* of the design. Multiplying both sides by **5**, we obtain

$$5 \times 5 = 1234 \times 5 \qquad\qquad (12.3)$$

or

$$5^2 = 12345 \qquad\qquad (12.4)$$

Thus the generator for the design can equivalently (and more conveniently) be written as

$$I = 12345 \qquad\qquad (12.5)$$

This version of the identity is readily confirmed by multiplying together the elements in columns **1**, **2**, **3**, **4**, and **5**, and noting that a column of plus signs,

I, is actually obtained. The half-fraction is defined* by a single generator, so that the relation **I** = **12345** also provides the *defining relation* of the design.

This defining relation is the key to the confounding pattern. For example, multiplying the defining relation on both sides by **1** yields

$$1 = 2345$$

In a similar way multiplying by **2** gives **2** = **1345** and so on to produce all the identities in the first column of Table 12.3.

The Complementary Half-Fraction

In the above example the generator **5** = **1234**, or, equivalently, **I** = **12345**, produced the defining relation for the design. In other words, by generating a new column **5** = **1234** we obtained the half-fraction corresponding to the runs marked with asterisks in Table 12.1. The defining relation **I** = **12345** provided by this generator immediately yields the confounding pattern of Table 12.3. The complementary half-fraction is generated by putting **5** = −**1234**. We then obtain the half-fraction corresponding to the runs of the original 2^5 that are *not* marked with asterisks in Table 12.1. The defining relation for this design may be written as

$$I = -12345$$

In practice either half-fraction can equally well be used. For the data of Table 12.1 the complementary half-fraction would have given, for example,

$$l'_1 = -0.75 \rightarrow 1 - 2345$$
$$l'_2 = 18.50 \rightarrow 2 - 1345$$

Exercise 12.3. For the 16 runs in Table 12.1 that do *not* have asterisks, calculate the average and the 15 contrasts $l'_1, l'_2, \ldots, l'_{45}$. Show by making a normal plot that the conclusions that would result from this fraction would be similar to those obtained from the other one.

Answer: (average, *1, 2. 3, 4, 5, 12, 13, 14, 15, 23, 24, 25, 34, 35, 45*) = (65.75, −0.75, 18.5, −1.25, 9.25, −6.25, 1.25, 1.0, −1.0, −1.0, 0.25, 15.75, 2.75, 4.0, −0.5, −12.5).

Combining the Two Half-Fractions

Suppose that after completing one of the half-fractions the other was subsequently added, so that the whole factorial was available. Unconfounded estimates of all effects

---

* When higher fractions are employed, there is more than one generator. For example, a quarter-fraction is defined by two generators. For more complicated fractions see Appendix 12A.

could then be obtained by analyzing the 32 runs as a full 2^5 factorial design run in two blocks of 16. The same result would be obtained by suitably adding and subtracting estimates from the two individual fractions. For example, we have

first fraction	second fraction
$l_2 = 20.5 \rightarrow 2 + 1345$	$l'_2 = 18.5 \rightarrow 2 - 1345$

whence

$$\tfrac{1}{2}(l_2 + l'_2) = \tfrac{1}{2}(20.5 + 18.5) = 19.5 \rightarrow 2$$
$$\tfrac{1}{2}(l_2 - l'_2) = \tfrac{1}{2}(20.5 - 18.5) = 1.0 \rightarrow 1345$$

(12.6)

These values for *2* and *1345* agree with those given in Table 12.1 for the complete 2^5 design.

12.4. THE CONCEPT OF DESIGN RESOLUTION: REACTOR EXAMPLE

The 2^{5-1} fraction is called a *resolution V* design. Looking at the confounding pattern in Table 12.3, we see, for example, that $l_1 \rightarrow 1 + 2345$ and $l_{12} \rightarrow 12 + 345$. Thus main effects are confounded with four-factor interactions, and two-factor interactions with three-factor interactions.

In general, a design of resolution R is one in which no p-factor effect is confounded with any other effect containing less than $R - p$ factors. The resolution of a design is denoted by the appropriate Roman letter appended as a subscript. Thus we could refer to the design of Table 12.2 as a 2_V^{5-1} design. To illustrate:

1. A design of resolution $R =$ III does not confound main effects with one another but does confound main effects with two-factor interactions.
2. A design of resolution $R =$ IV does not confound main effects and two-factor interactions but does confound two-factor interactions with other two-factor interactions.
3. A design of resolution $R =$ V does not confound main effects and two-factor interactions with each other, but does confound two-factor interactions with three-factor interactions, and so on.

In general, the *resolution* of a two-level fractional design is *the length of the shortest word in the defining relation.*

Resolutions of Some Half-Fractions

For any half-fraction the number of symbols on the right of the defining relation denotes the resolution of the design. Thus a 2^{5-1} half-fraction with defining relation $\mathbf{I} = \pm\mathbf{12345}$ has resolution V. In Table 12.4 the 2^{3-1} half-fractions with defining relations $\mathbf{I} = \pm\mathbf{123}$ have resolution III, and the 2^{4-1} fractions with defining relations $\mathbf{I} = \pm\mathbf{1234}$ have resolution IV.

Half-Fractions of Highest Resolution

At the beginning of Section 12.3 we gave a procedure for constructing a 2^{5-1} design. In fact, it would have been possible to use *any* interaction or main effect column to accommodate the fifth variable. The choice we made yields a half-fraction with *highest possible resolution*. In general, to construct a 2^{k-1} fractional factorial design of highest possible resolution:

1. Write a full factorial design for the first $k - 1$ variables.
2. Associate the kth variable with plus or minus the interaction column **123...(k − 1)**.

Table 12.4 gives examples of 2_{III}^{3-1}, 2_{IV}^{4-1}, and 2_{V}^{5-1} half-fractions of this kind. The two 2^{3-1} half-fractions obtained by the above rule are shown geometrically in Table 12.4.

Exercise 12.4. Obtain the confounding pattern for a 2^{5-1} design generated by setting $\mathbf{5} = \mathbf{123}$. Discuss its properties. What is its resolution? Can you imagine circumstances in which it might be preferred to the resolution V design?

Partial answer: $l_1 \rightarrow 1 + 235, l_2 \rightarrow 2 + 135, R = \mathrm{IV}$.

An Alternative Rationale for the Half-Fraction Design in the Reactor Experiment

Consider the 2_{V}^{5-1} half-fraction with $\mathbf{I} = \mathbf{12345}$ given in Tables 12.2 and 12.4. Obviously (from its mode of construction), if we omit the fifth column of plus and minus signs from this design, we have a complete factorial in variables **1, 2, 3,** and **4**. But try omitting column **1** instead. There is now a complete factorial in variables **2, 3, 4,** and **5**! Indeed, a *complete* factorial in the remaining variables is obtained *whichever column is omitted*. We have already seen that the experimenter could justify the 2^{5-1} half-fraction on the assumption that three-factor, four-factor, and five-factor interactions could be ignored. An alternative justifying assumption is that at most only four of the five variables will produce detectable effects and the other will be essentially

TABLE 12.4. Best half-fractions for $k = 3$, $k = 4$, and $k = 5$

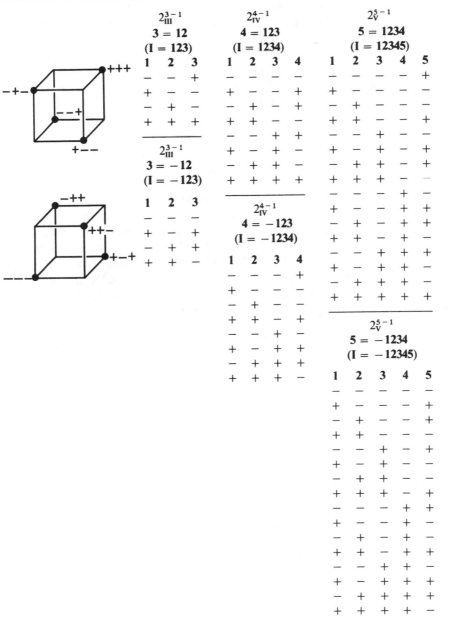

2_{III}^{3-1}
3 = 12
(I = 123)

1	2	3
−	−	+
+	−	−
−	+	−
+	+	+

2_{III}^{3-1}
3 = −12
(I = −123)

1	2	3
−	−	−
+	−	+
−	+	+
+	+	−

2_{IV}^{4-1}
4 = 123
(I = 1234)

1	2	3	4
−	−	−	−
+	−	−	+
−	+	−	+
+	+	−	−
−	−	+	+
+	−	+	−
−	+	+	−
+	+	+	+

2_{IV}^{4-1}
4 = −123
(I = −1234)

1	2	3	4
−	−	−	+
+	−	−	−
−	+	−	−
+	+	−	+
−	−	+	−
+	−	+	+
−	+	+	+
+	+	+	−

2_{V}^{5-1}
5 = 1234
(I = 12345)

1	2	3	4	5
−	−	−	−	+
+	−	−	−	−
−	+	−	−	−
+	+	−	−	+
−	−	+	−	−
+	−	+	−	+
−	+	+	−	+
+	+	+	−	−
−	−	−	+	−
+	−	−	+	+
−	+	−	+	+
+	+	−	+	−
−	−	+	+	+
+	−	+	+	−
−	+	+	+	−
+	+	+	+	+

2_{V}^{5-1}
5 = −1234
(I = −12345)

1	2	3	4	5
−	−	−	−	−
+	−	−	−	+
−	+	−	−	+
+	+	−	−	−
−	−	+	−	+
+	−	+	−	−
−	+	+	−	−
+	+	+	−	+
−	−	−	+	+
+	−	−	+	−
−	+	−	+	−
+	+	−	+	+
−	−	+	+	−
+	−	+	+	+
−	+	+	+	+
+	+	+	+	−

inert—it will have no detectable main effect or interaction with any other variable. On the assumption of one or more inert variables, the 2_V^{5-1} design will generate complete factorials in the remaining variables, *no matter which variables these are.*

In fact, our analysis for the reactor example suggests that only three of the variables had detectable effects: **2, 4**, and **5** (catalyst, temperature, and concentration). Since variables **1** and **3** were effectively inert, we had a *replicated* 2^3 factorial in variables **2, 4**, and **5**, and the results can be assembled as in Figure 12.3.

Factorials Embedded in Fractions:
The General Importance of the Concept of Resolution

In general, it can be shown that a fractional factorial design of resolution R contains complete factorials (possibly replicated) in every set of $R - 1$ variables. Suppose, then, that the experimenter has a number of candidate variables but believes that all but $R - 1$ of them (specific identity unknown) may have no detectable effects. Then, if he employs a design of resolution R and his conjecture is justified, he will have a complete factorial design in the effective variables. This idea is illustrated with the 2_{III}^{3-1} design in Figure 12.4, which projects a 2^2 pattern in every subspace of two dimensions.

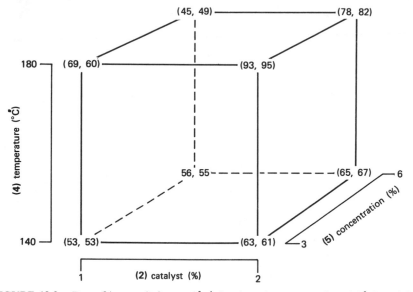

FIGURE 12.3. Data (% reacted) from a 2^{5-1} fraction, shown as replicated 2^3 factorial in variables *2, 4*, and *5*, reactor example.

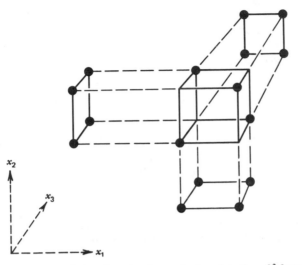

FIGURE 12.4. A 2_{III}^{3-1} design, showing projections into three 2^2 factorials.

Exercise 12.5. If a resolution R design gives a full factorial in every set of $R - 1$ variables, is it necessarily true that a full factorial is obtained in every subset containing fewer than $R - 1$ variables?   *Answer*: Yes.

Exercise 12.6. A 2_V^{5-1} design gives full factorials in every subset of q variables. What is the value of q?   *Answer*: 4, 3, 2, or 1 (for an example of $q = 3$ see Figure 12.3).

Economy in Experimentation Arising from the
Sequential Use of Fractional Designs

Suppose that an experimenter who can make his runs sequentially wishes to investigate five factors, each at two levels, and is contemplating a 2^5 design involving 32 runs. It is almost always better for him to run a half-fraction containing 16 runs first, analyze the results, and think about them. If necessary, he can always run the second fraction later to complete the full design. Frequently, however, the first half-fraction itself will allow him to proceed to the next stage of experimental iteration, which may involve, for example, the introduction of new variables or different levels of the old ones. Use of this sequential approach can thus greatly accelerate progress. It is worth noting that:

1. The experimenter should randomize within each fraction.
2. If eventually it is decided to run both fractions, these fractions will be randomized orthogonal blocks of the complete design.

3. No information will be "lost" except that concerning the interaction which is actually confounded with the block contrast.
4. The design run as two randomized fractions can give greater precision than the whole design run in random order because the block difference is eliminated.

Recapitulation

We began the chapter by discussing redundancy. It was pointed out that, for moderate k, a full factorial design frequently makes possible the estimation of many more effects than are detectably different from the noise. Sometimes these nondetectable effects are high-order interactions and sometimes they are all the effects associated with some inert variable or variables.

The fractional factorials discussed in this chapter are ideally suited to exploiting the probable existence of redundancy of one or both of these kinds for the following reason:

1. It can be arranged so that the confounding that occurs is between effects of high and low order,
2. A complete factorial design is available for whichever subset of $R - 1$ variables turns out to have appreciable effects.

In sequential experimentation, unless the total number of runs for a full or replicated factorial is needed to achieve sufficient precision, it is usually better to run fractional factorial designs. The fractions, used as building blocks, can build up to the full factorial design if this is necessary.

We now illustrate these ideas for designs of resolution III.

12.5. RESOLUTION III DESIGNS: BICYCLE EXAMPLE*

Suppose that the hypothetical data of Table 12.5 are times in seconds for a particular person to complete eight trial bicycle runs up a hill between fixed marks. These runs were performed in random order on eight successive days. The design is of resolution III and is a $\frac{8}{128} = \frac{1}{16}$ fraction of the full 2^7 factorial. Thus it is a 2_{III}^{7-4} design. (Note that $2^{7-4} = 2^7 2^{-4} = 2^{-4} 2^7 = \frac{1}{16} 2^7$.)

Table 12.6 gives the calculated contrasts. For example,

$$l_1 = \tfrac{1}{4}(-69 + 52 - 60 + 83 - 71 + 50 - 59 + 88) \qquad (12.7)$$

---

* This hypothetical example is an extension of the real one in Appendix 11A, but it is assumed now that both the rider and the bicycle are different.

TABLE 12.5. An eight-run experimental design for studying how time to cycle up a hill is affected by seven variables (I = 124, I = 135, I = 236, I = 1237).

run	seat up/down 1	dynamo off/on 2	handlebars up/down 3	gear low/medium 4 12	raincoat on/off 5 13	breakfast yes/no 6 23	tires hard/soft 7 123	time to climb hill (sec) y
1	−	−	−	+	+	+	−	69
2	+	−	−	−	−	+	+	52
3	−	+	−	−	+	−	+	60
4	+	+	−	+	−	−	−	83
5	−	−	+	+	−	−	+	71
6	+	−	+	−	+	−	−	50
7	−	+	+	−	−	+	−	59
8	+	+	+	+	+	+	+	88

TABLE 12.6. Calculated contrasts and abbreviated confounding pattern for data and design in Table 12.5

seat	$l_1 =$	3.5	$\rightarrow 1 + 24 + 35 + 67$
dynamo	$l_2 =$	⎛12.0⎞	$\rightarrow 2 + 14 + 36 + 57$
handlebars	$l_3 =$	1.0	$\rightarrow 3 + 15 + 26 + 47$
gear	$l_4 =$	⎛22.5⎞	$\rightarrow 4 + 12 + 56 + 37$
raincoat	$l_5 =$	0.5	$\rightarrow 5 + 13 + 46 + 27$
breakfast	$l_6 =$	1.0	$\rightarrow 6 + 23 + 45 + 17$
tires	$l_7 =$	2.5	$\rightarrow 7 + 34 + 25 + 16$
	$(l_1 =$	66.5	$\rightarrow$ average)

```
                        69              83
         medium        71_____88
                        |               |
         gear 4         |               |
                        |               |
           low         52_____60
                       50              59

                       off             on
                          dynamo 2
```

The table also gives an abbreviated* confounding pattern in which interactions between three or more factors have been ignored. Suppose that previous experience suggested that the standard deviation for repeated runs up the hill under the same conditions is about 3 seconds. Thus the calculated effects $l_1, l_2, \ldots, l_7$ have a standard error of about

$$\sqrt{\frac{3^2}{4} + \frac{3^2}{4}} = 2.1$$

Evidently only two contrasts, l_2 and l_4, are distinguishable from noise. Their values are circled in Table 12.6. The simplest interpretation of the results is that only two of the seven factors, the dynamo (2) and gear (4), exert a detectable influence, and they do so by way of their main effects. Having the dynamo on adds about 12 seconds to the time, and using medium gear instead of low gear adds about 22 seconds. On this interpretation we have in effect

* The method by which the confounding pattern has been obtained is given in Appendix 12A.

a replicated 2^2 design in the variables **2** and **4**, as indicated at the bottom of Table 12.6. There is, of course, some ambiguity in these conclusions. It is possible, for example, that l_4 is large, not because of a large main effect *4*, but because one or more of the interactions *12, 56, 37* are large. We see in Appendix 12B how sequential addition of further runs can resolve such ambiguities. However, for this example we suppose that the experimenter's knowledge of the nature of his bicycle suggests that the simpler explanation is likely to be right. The experimenter might well decide to proceed to the next stage of the investigation at this point.

Because one use of resolution III designs is to determine the main effects of each of the factors, assuming that they do not interact, these arrangements have sometimes been called "main effect plans."

Embedded 2^2 Factorials in Resolution III designs

A resolution R design has a complete factorial (possibly replicated) in every subset of $R - 1$ variables. For the resolution III design of Table 12.5, for example, whichever two columns of the design are chosen, they form a complete 2^2 factorial replicated twice. Also notice what happens to the confounding pattern in Table 12.6 supposing that two variables, say **2** and **4**, are effective, and the rest, that is, **1, 3, 5, 6**, and **7**, are essentially inert. Then all interactions and main effects containing these numbers vanish, $l_2 \to 2$, $l_4 \to 4$, and $l_1 \to 24$, and the remaining *l*'s measure experimental error only.

Exercise 12.7. For the examples in Table 12.4, verify that any subset of $R - 1$ variables from a design of resolution R produces a full factorial design.

Construction of 2_{III}^{7-4} Design

The 2^{7-4} design in Table 12.5 can be constructed as follows:

1. Write a full factorial design for the three variables, **1, 2,** and **3**.
2. Associate additional variables **4, 5, 6,** and **7** with all the interaction columns **12, 13, 23,** and **123**, respectively.

The design is obtained by associating every available contrast with a variable and is therefore sometimes called a *saturated design.**

---

* It is actually possible to construct supersaturated designs, but we do not recommend them in ordinary circumstances.

Other 2^{7-4} Fractions

In Table 12.5 a one-sixteenth fraction of a full 2^7 factorial design is shown. How can the other one-sixteenth fractions that make up the full factorial design be generated? The first design was generated by setting

$$4 = +12 \quad 5 = +13 \quad 6 = +23 \quad 7 = +123 \tag{12.8}$$

but, for example, we could equally well have used

$$4 = -12 \quad 5 = +13 \quad 6 = +23 \quad 7 = +123 \tag{12.9}$$

This gives a different one-sixteenth fraction, which is shown in Table 12.7 with further hypothetical data on times to cycle up the hill. Note that none of the runs in this new design is the same as any of those in the preceding design. Calculated contrasts for this design are shown in Table 12.8.

TABLE 12.7. A second 2_{III}^{7-4} fractional factorial design with times to cycle up a hill ($I = -124, I = 135, I = 236, I = 1,237$).

run	seat 1	dynamo 2	handlebars 3	gear 4 −12	raincoat 5 13	breakfast 6 23	tires 7 123	time to climb hill (sec)
9	−	−	−	−	+	+	−	47
10	+	−	−	+	−	+	+	74
11	−	+	−	+	+	−	+	84
12	+	+	−	−	−	−	−	62
13	−	−	+	−	−	−	+	53
14	+	−	+	+	+	−	−	78
15	−	+	+	+	−	+	−	87
16	+	+	+	−	+	+	+	60

What is the confounding pattern for the new fraction? Notice that the new fraction was obtained by switching signs for variable **4** in the first design (variable **4** was associated with −**12** instead of +**12**). The abbreviated confounding pattern for this new fraction may be obtained, therefore, by switching signs in the confounding pattern of Table 12.6. This gives the confounding pattern in Table 12.8.

For this set of data the contrasts calculated from the second fraction confirm the conclusions from the first fraction.

TABLE 12.8.   Calculated contrasts and abbreviated confounding pattern for second design in bicycle experiment

$l'_1 = \quad 0.8 \rightarrow 1 - 24 + 35 + 67$

$l'_2 = \quad 10.2 \rightarrow 2 - 14 + 36 + 57$

$l'_3 = \quad 2.7 \rightarrow 3 + 15 + 26 - 47$

$l'_4 = \quad 25.2 \rightarrow 4 - 12 - 56 - 37$ (i.e., $l'_{-4} = -25.2 \rightarrow$

$$-4 + 12 + 56 + 37)$$

$l'_5 = -1.7 \rightarrow 5 + 13 - 46 + 27$

$l'_6 = \quad 2.2 \rightarrow 6 + 23 - 45 + 17$

$l'_7 = -0.7 \rightarrow 7 - 34 + 25 + 16$

The Sixteen Different Fractions

In all there are 16 different ways of allocating signs to the four generators:

$$4 = \pm 12, \qquad 5 = \pm 13, \qquad 6 = \pm 23, \qquad 7 = \pm 123 \qquad (12.10)$$

Thus appropriate sign switching in columns* **4, 5, 6,** and **7** of Table 12.5 produces 16 fractional factorial designs which together make up the complete 2^7 factorial design. Corresponding sign switching in Table 12.6 produces the 16 different confounding patterns.

Designing Two Fractions

Consider again the bicycle example. Suppose that the 16 results from the two 2_{III}^{7-4} fractionals were considered together. What conclusions could be drawn? Combining the results from Tables 12.6 and 12.8, we obtain Table 12.9.

Conclusions would now be somewhat more certain. In particular, the large main effect of factor **4** (gear) is now estimated free of bias from two-factor interactions, and has a value close to that conjectured earlier. The joint effect of the string of interactions *12 + 56 + 37* can now be estimated separately from the main effect *4*, and it is shown to be small. Most interestingly, all the two-factor interactions involving the important variable 4 are now *free of aliases.* (Of course we continue to assume all three-factor and higher order interactions to be zero.) For this particular set of data, however, none of these two-factor interactions is distinguishable from noise. Factor **2** (dynamo), somewhat less aliased than before, is showing an effect similar to that previously conjectured.

---

* The reader can confirm by experimentation that switching signs in other columns of the design only produces one or another of these basic 16 fractions. However, the *order* in which the runs appear can be different.

TABLE 12.9. Analysis of complete set of 16 runs, combining the results of the two fractions, bicycle example

seat	$\frac{1}{2}(l_1 + l_1') = \frac{1}{2}(3.5 + 0.8)$	$=$	$2.2 \to 1 + 35 + 67$
dynamo	$\frac{1}{2}(l_2 + l_2') = \frac{1}{2}(12.0 + 10.2)$	$=$	$11.1 \to 2 + 36 + 57$
handlebars	$\frac{1}{2}(l_3 + l_3') = \frac{1}{2}(1.0 + 2.7)$	$=$	$1.9 \to 3 + 15 + 26$
gear	$\frac{1}{2}(l_4 + l_4') = \frac{1}{2}(22.5 + 25.2)$	$=$	$23.9 \to 4$
raincoat	$\frac{1}{2}(l_5 + l_5') = \frac{1}{2}(0.5 - 1.7)$	$=$	$-0.6 \to 5 + 13 + 27$
breakfast	$\frac{1}{2}(l_6 + l_6') = \frac{1}{2}(1.0 + 2.2)$	$=$	$1.8 \to 6 + 23 + 17$
tires	$\frac{1}{2}(l_7 + l_7') = \frac{1}{2}(2.5 - 0.7)$	$=$	$0.9 \to 7 + 25 + 16$
	$\frac{1}{2}(l_1 - l_1') = \frac{1}{2}(3.5 - 0.8)$	$=$	$1.3 \to 24$
	$\frac{1}{2}(l_2 - l_2') = \frac{1}{2}(12.0 - 10.2)$	$=$	$0.9 \to 14$
	$\frac{1}{2}(l_3 - l_3') = \frac{1}{2}(1.0 - 2.7)$	$=$	$-0.9 \to 47$
	$\frac{1}{2}(l_4 - l_4') = \frac{1}{2}(22.5 - 25.2)$	$=$	$-1.4 \to 12 + 56 + 37$
	$\frac{1}{2}(l_5 - l_5') = \frac{1}{2}(0.5 + 1.7)$	$=$	$1.1 \to 46$
	$\frac{1}{2}(l_6 - l_6') = \frac{1}{2}(1.0 - 2.2)$	$=$	$-0.6 \to 45$
	$\frac{1}{2}(l_7 - l_7') = \frac{1}{2}(2.5 + 0.7)$	$=$	$1.6 \to 34$

Sequential Use of Highly Fractionated Designs

The preceding example illustrates a useful application of highly fractionated designs as sequential building blocks. Additional fractions may be selected to resolve ambiguities, which knowledge of the variables and data available so far suggest may be of importance. We explore two important applications of this idea. The reader can devise others to suit particular circumstances.

Addition of a Second Fraction to De-alias Any One Main Effect and All Its Associated Two-Factor Interactions

Consider the two fractions used in the bicycle experiment. The largest effect obtained from the first set of eight runs was associated with the choice of gear (variable 4). It might have been argued, therefore, that if further runs were to be made, they could best be employed to de-alias 4 and all the interactions of other variables with 4.

Table 12.9 shows that by adding a second fraction in which the sign of variable 4 has been switched, a design of 16 runs possessing the desired property is obtained. This ability to de-alias one effect and all its two-factor interactions by adding a second fraction with the appropriate column of signs switched is a handy device for the sequential use of these designs.

Adding a Second Fraction to De-alias All Main Effects

Consider Table 12.5 again, and suppose that a different second fraction is added in which signs are switched in *all* the columns. Then for the new fraction the first two rows in the confounding pattern (obtained by switching signs in Table 12.6) are

$$l'_1 \rightarrow 1 - 24 - 35 - 67 \qquad (l'_{-1} \rightarrow -1 + 24 + 35 + 67)$$
$$l'_2 \rightarrow 2 - 14 - 36 - 57 \qquad (l'_{-2} \rightarrow -2 + 14 + 36 + 57) \tag{12.11}$$

By combining this second fraction with the original fraction, we obtain

$$\tfrac{1}{2}(l_1 + l'_1) \rightarrow 1, \qquad \tfrac{1}{2}(l_1 - l'_1) \rightarrow 24 + 35 + 67$$
$$\tfrac{1}{2}(l_2 + l'_2) \rightarrow 2, \qquad \tfrac{1}{2}(l_2 - l'_2) \rightarrow 14 + 36 + 57 \tag{12.12}$$

and so on.

This way of augmenting the design yields *all* main effects clear of all two-factor interactions, but the two-factor interactions themselves are still confounded in groups of three. An example of the use of this sequence is given in Section 13.3.

Exercise 12.8. Show that the second fraction obtained above by switching all signs may also be obtained (with runs in a different order) by switching signs in columns **4**, **5**, **6**, and **7** only. Can you find other ways to reproduce the second fraction? Explain the equivalences you find.

General Construction of Resolution III Designs

Resolution III designs for $2^k - 1$ variables may be obtained by saturating a 2^k factorial with additional variables. For example, to construct a saturated 16-run design in 15 variables first write a full factorial design for four variables and then associate the extra variables **5**, **6**, ..., **15** with the 11 interaction columns **12**, **13**, **14**, **23**, **24**, **34**, **123**, **124**, **134**, **234**, and **1234**, respectively. The resulting design is a 2_{III}^{15-11} fractional factorial design for 15 variables in 16 runs.

Exercise 12.9. Construct a two-level fractional factorial design for 31 variables in 32 runs. This is a 2^{k-p} design; what values do k and p have? *Answer*: $k = 31, p = 26$.

Exercise 12.10. Indicate how you could construct a 2^{63-57} fractional factorial design. Is this a saturated design?   *Answer*: Yes.

Useful designs may be obtained by appropriately deleting columns from the saturated designs. For example, dropping columns **4** and **7** from the design matrix for a 2^{7-4} design yields a 2^{5-2} design, the defining relation for which can be obtained from that for the 2^{7-4} design by deleting all words containing **4** and **7**. The variables to be dropped are selected so as to obtain the most satisfactory alias arrangement.

Plackett and Burman Saturated Designs

The saturated fractional factorial designs have the following orthogonal*
property: if we take any two columns, then, corresponding to the $N/2$ plus
signs in the first column, there will be $N/4$ plus and $N/4$ minus signs in the
second column, and similarly for the minus signs in the first column. Provided
that all interactions are negligible, designs with this property allow unbiased
estimation of all main effects of $N - 1$ variables with smallest possible
variance. The fractional factorials so far discussed are available only if N is
a power of 2. Plackett and Burman (1946) have obtained arrangements with
this same orthogonal property when N is a multiple of 4. For example, their
design for $k = 11$ factors in $N = 12$ runs is shown in Table 12.10. The fashion
in which two-factor interactions confound main effects for most Plackett and
Burman designs is complicated. However, fold-over pairs of any such
orthogonal design are of resolution IV (see Box and Wilson, 1951).

**TABLE 12.10. Plackett and Burman design for study of
11 factors in 12 runs**

run	\multicolumn{11}{c}{variable}
	1	2	3	4	5	6	7	8	9	$\overline{10}$	$\overline{11}$
1	+	−	+	−	−	−	+	+	+	−	+
2	+	+	−	+	−	−	−	+	+	+	−
3	−	+	+	−	+	−	−	−	+	+	+
4	+	−	+	+	−	+	−	−	−	+	+
5	+	+	−	+	+	−	+	−	−	−	+
6	+	+	+	−	+	+	−	+	−	−	−
7	−	+	+	+	−	+	+	−	+	−	−
8	−	−	+	+	+	−	+	+	−	+	−
9	−	−	−	+	+	+	−	+	+	−	+
10	+	−	−	−	+	+	+	−	+	+	−
11	−	+	−	−	−	+	+	+	−	+	+
12	−	−	−	−	−	−	−	−	−	−	−

12.6. RESOLUTION IV DESIGNS: INJECTION MOLDING EXAMPLE

We have seen that for designs of resolution V main effects are confounded
only with four-factor interactions, and two-factor interactions only with
three-factor interactions. Full factorial designs are generated by every subset

---

* If the level of the ith variable is represented by $x_i = \pm 1$ and that of the jth variable by $x_j = \pm 1$,
then $\Sigma x_i = 0$, $\Sigma x_j = 0$, and $\Sigma x_i x_j = 0$ for every i and j.

of four variables. Designs of resolution III introduce much more serious confounding, with main effects having two-factor interactions as aliases. For these designs full factorial designs exist for every subset of two variables. Designs of resolution IV occupy an intermediate position. No main effect is confounded with any two-factor interaction, but two-factor interactions *are* confounded with each other. For these designs full factorial designs exist for every subset of three variables.

An Experiment on Injection Molding

In an injection molding experiment (Table 12.11) eight variables were studied in a 2_{IV}^{8-4} (a $\frac{1}{16}$ replicate of a 2^8 factorial of resolution IV). The normal plots, shown in Figure 12.5, suggest that the linear contrasts l_3, l_{15}, and l_5 are distinguishable from the noise. The largest remaining effect is l_8. The confounding pattern, assuming negligible interaction between three or more factors, is shown in Table 12.12. It seems likely that main effects associated with holding pressure (3) and booster pressure (5) exist. Also, the interactions most likely to explain the large size of l_{15} are perhaps *15* and *38*, since these involve factors 3 and 5, which have large main effects. It is, however, possible that interactions exist between factors that have no main effects. Without further information the situation is uncertain. One way to proceed is to choose a further fraction of eight or 16 runs designed to resolve the ambiguity. However, in this particular example the large size of l_{15} suggested that the problem might be resolved with even fewer than eight runs. We show in Appendix 12B how four additional runs were chosen and used to discover and estimate the responsible interaction.

Construction of the Resolution IV Design by "Folding Over" a Resolution III Design

The sixteen-run 2_{IV}^{8-4} design in Table 12.11 was constructed as follows. The eight-run 2_{III}^{7-4} design was first written as in Table 12.5 for the seven variables 1, 2, 3, ..., 7. A further column labeled 8 and consisting entirely of plus signs was then added. The remaining eight runs were obtained by switching all signs in the first set of eight runs. Thus run 9 was obtained by switching all signs in run 1 and so on.

The Alias Pattern

The alias pattern for the folded-over design given in Table 12.11 can be obtained from that of the resolution III design (Table 12.6) by the following argument. Suppose that we compute for the first set of eight runs

$$l_1 = \tfrac{1}{4}(-y_1 + y_2 \ldots + y_8)$$

and for the second set of eight runs

$$-l'_1 = \tfrac{1}{4}(-y_9 + y_{10} \cdots + y_{16})$$

Then using Table 12.6

$$l_1 \rightarrow 1 + 18 + 24 + 35 + 67 \quad \text{and} \quad -l'_1 = -1 + 18 + 24 + 35 + 67$$

Now the contrast l_1 for the complete set of 16 runs is

$$l_1 = \tfrac{1}{8}(-y_1 + y_2 + \cdots + y_8 + y_9 - y_{10} \cdots - y_{16}) = \tfrac{1}{2}(l_1 + l'_1)$$

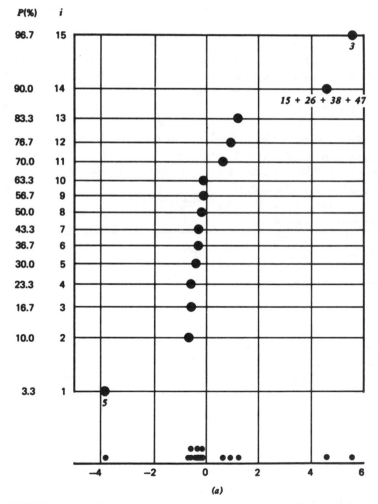

FIGURE 12.5. (a) Normal plot of contrasts, injection molding example.

Similarly for the contrast associated with the interaction 18 it is

$$l_{18} = \tfrac{1}{8}(-y_1 + y_2 + \cdots - y_8 - y_9 + y_{10} \cdots + y_{16}) = \tfrac{1}{2}(l_1 - l_1').$$

Thus $l_1 \rightarrow 1$ and $l_{18} \rightarrow 18 + 24 + 35 + 67$. The same argument applied to the remaining contrasts yields the confounding pattern of Table 12.12. A more complete discussion is given in Appendix 12A.

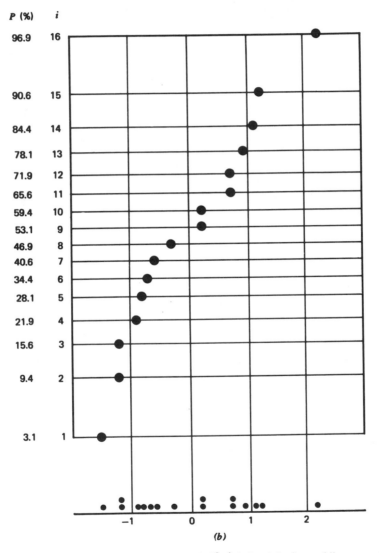

FIGURE 12.5. (*b*) Normal plot of residuals 2_{IV}^{8-4} design, injection molding example.

TABLE 12.11. A 2^{8-4} resolution IV design, molding example ($I = 1248, I = 1358, I = 2368, I = 1237$).

run	mold temperature 1	moisture content 2	holding pressure 3	cavity thickness 4	booster pressure 5	cycle time 6	gate size 7	screw speed 8	shrinkage y
1	−	−	−	+	+	+	−	+	14.0
2	+	−	−	−	−	+	+	+	16.8
3	−	+	−	−	+	−	+	+	15.0
4	+	+	−	+	−	−	−	+	15.4
5	−	−	+	+	−	−	+	+	27.6
6	+	−	+	−	+	−	−	+	24.0
7	−	+	+	−	−	+	−	+	27.4
8	+	+	+	+	+	+	+	+	22.6
9	+	+	+	−	−	−	+	−	22.3
10	−	+	+	+	+	−	−	−	17.1
11	+	−	+	−	−	+	−	−	21.5
12	−	−	+	−	+	+	+	−	17.5
13	+	+	−	−	+	+	−	−	15.9
14	−	+	−	+	−	+	+	−	21.9
15	+	−	−	+	+	−	+	−	16.7
16	−	−	−	−	−	−	−	−	20.3

TABLE 12.12. Calculated contrasts with their expected values: interactions between three or more factors ignored, molding example

$$l_1 = -0.7 \rightarrow 1$$
$$l_2 = -0.1 \rightarrow 2$$
$$l_3 = 5.5 \rightarrow 3$$
$$l_4 = -0.3 \rightarrow 4$$
$$l_5 = -3.8 \rightarrow 5$$
$$l_6 = -0.1 \rightarrow 6$$
$$l_7 = 0.6 \rightarrow 7$$
$$l_8 = 1.2 \rightarrow 8$$
$$l_{12} = -0.6 \rightarrow 12 + 37 + 48 + 56$$
$$l_{13} = 0.9 \rightarrow 13 + 27 + 46 + 58$$
$$l_{14} = -0.4 \rightarrow 14 + 28 + 36 + 57$$
$$l_{15} = 4.6 \rightarrow 15 + 26 + 38 + 47$$
$$l_{16} = -0.3 \rightarrow 16 + 25 + 34 + 78$$
$$l_{17} = -0.2 \rightarrow 17 + 23 + 68 + 45$$
$$l_{18} = -0.6 \rightarrow 18 + 24 + 35 + 67$$

average = 19.75

Alternative 2^{8-4} Fractions

Sixteen different 2_{IV}^{8-4} fractions are members of the family making up the complete 2^8 design. Individual members of the family may be generated by sign switching. Exactly as with the resolution III designs, the switching of signs in one or more columns will always yield a member of the family, and the associated confounding pattern is obtained by making the corresponding sign changes in the alias patterns of Table 12.12.

Building Blocks

Resolution IV designs may be used sequentially as were the resolution III designs. As before, sign switching may be used to eliminate particular confounding links.

General Construction of Resolution IV Designs

The construction of a resolution IV design containing 2^k variables follows exactly the pattern given for the 2_{IV}^{8-4} design:

1. Write a complete 2^k factorial with added columns for all interaction terms.
2. Generate a resolution III design for $2^k - 1$ variables by saturating this design with variables.
3. Add a further variable as a column of plus signs.
4. Repeat the design with all signs reversed to give a resolution IV design for 2^k variables in 2^{k+1} runs.

An alternative general method is given in Appendix 12A.

12.7. ELIMINATION OF BLOCK EFFECTS IN FRACTIONAL DESIGNS

Fractional designs may be run in blocks, with suitable contrasts used as "block variables." A design in 2^q blocks is defined by q independent contrasts. All effects (including aliases) associated with these basic contrasts *and all their interactions* are confounded with blocks.

Example 2_V^{5-1} Design in Two Blocks of Eight

Consider again the 2^{5-1} design of Table 12.2. Suppose the investigator decided that interaction between feed rate and catalyst concentration was likely to be negligible. This interaction *13* could then be used for blocking. The eight runs 2, 20, 5, ..., 15, having a minus sign in the **13** column, would be run in one block, and the eight runs 17, 3, 22, ..., 32 in the other. Notice that in this design the alias *245* (here assumed negligible) of *13* is also confounded with blocks.

Example: 2_V^{5-1} Design in Four Blocks of Four Runs

Suppose that, in the 2^{5-1} design of Table 12.2, columns **13** and **23** are confounded with blocks. Then the interaction between these columns $\mathbf{13} \times \mathbf{23} = \mathbf{123}^2 = \mathbf{12}$ is also confounded. The design would thus be appropriate if we were prepared to confound with blocks all two-factor interactions between variables **1**, **2**, and **3** and their aliases. To achieve this arrangement, runs 20, 5, 12, and 29, for which the **13** and **23** columns have signs $(--)$, could be put in the first block, runs 2, 23, 26, and 15, for which columns **13** and **23** have signs $(-+)$ in the second block, and so on. Thus in terms of a two-way table the arrangement would be as follows:

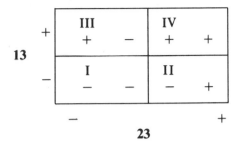

The Resolution IV Designs as Main Effect Plans in Blocks of Two

It occasionally happens that we must work with very small block sizes. A remarkable class of such designs based on the resolution IV arrangement provides economical main effect plans with a block size of only two. In one investigation the subject of study was an effluent impurity that tended to vary slowly with time. Runs made consecutively were thus much more comparable than those made further apart. It was possible to run the design in blocks of 2-hour periods, one experimental condition being run in the first hour and one in the second. At one stage of the investigation a 16-run main effect plan was used to study the main effects of eight variables based on a blocked 2_{IV}^{8-4} design. The plan is shown in Table 12.13. To see how this is derived, consider the original design given in Table 12.11 and the aliasing strings in Table 12.12. For the blocking scheme suppose that we use any two-factor interaction contrast, say l_{12}, to accommodate $\mathbf{B_1}$, and a second, say l_{13}, to accommodate $\mathbf{B_2}$; then l_{17} cannot be used for $\mathbf{B_3}$ since it can be obtained by multiplying the signs of **12** and **13**. Suppose, therefore, we use l_{14} for $\mathbf{B_3}$. (The reader may confirm that any other remaining two-factor interaction contrast can equally well be employed.) Then the seven columns of signs obtained for $\mathbf{B_1}, \mathbf{B_2}, \mathbf{B_3}, \mathbf{B_1B_2}, \mathbf{B_1B_3}, \mathbf{B_2B_3}, \mathbf{B_1B_2B_3}$ exactly correspond to the contrasts $l_{12}, l_{13}, l_{14}, l_{15}, l_{16}, l_{17}, l_{18}$, in some order. They thus involve only the strings of interactions and not the main effects. When the design is rearranged in the eight blocks as on the right of Table 12.13, it is seen that the second run in each block is the mirror image or "fold-over" of the first run, that is, the signs in one run are exactly reversed in the other.

In designs of this kind, both the ordering within pairs and the sequence in which the pairs (blocks) are run should be random.

Rather than regard all between-block information as lost, the design can be analyzed on the basis that there are two different error variances. The within-block variance is appropriate for inferences about main effects, and the between-block variance for inferences about the strings of two-factor

TABLE 12.13. 2_{IV}^{8-4} design in eight blocks of size two

	2_{IV}^{8-4} design								block variable		
run	1	2	3	4	5	6	7	8	B_1 12	B_2 13	B_3 14
1	−	−	−	+	+	+	−	+	+	+	−
2	+	−	−	−	−	+	+	+	+	−	−
3	−	+	−	−	+	−	+	+	−	+	+
4	+	+	−	+	−	−	−	+	+	−	+
5	−	−	+	+	−	−	+	+	+	−	−
6	+	−	+	−	+	−	−	+	−	+	−
7	−	+	+	−	−	+	−	+	−	+	+
8	+	+	+	+	+	+	+	+	+	+	+
9	+	+	+	−	−	−	+	−	+	+	−
10	−	+	+	+	+	+	−	−	−	−	−
11	+	−	+	−	−	+	−	−	−	+	+
12	−	−	+	+	+	+	+	−	+	−	+
13	+	+	−	−	+	+	+	−	+	−	−
14	−	+	−	+	−	+	−	−	+	+	−
15	+	−	−	+	+	−	+	−	−	+	+
16	−	−	−	−	−	−	−	−	+	+	+

design rearranged in eight blocks

block	1	2	3	4	5	6	7	8	B_1	B_2	B_3	run
1	+	−	−	−	−	+	+	+	−	−	−	2
	−	+	+	+	+	−	−	−	−	−	−	10
2	−	−	+	+	−	−	+	+	+	−	−	5
	+	+	−	−	+	+	−	−	+	−	−	13
3	+	−	+	−	+	−	−	+	−	+	−	6
	−	+	−	+	−	+	+	−	−	+	−	14
4	−	−	−	−	+	+	−	+	+	+	−	1
	+	+	+	+	−	−	+	−	+	+	−	9
5	−	+	+	−	−	+	−	+	−	−	+	7
	+	−	−	+	+	−	+	−	−	−	+	15
6	+	−	+	−	−	+	−	+	+	−	+	4
	−	+	−	+	+	−	+	−	+	−	+	12
7	−	+	−	−	+	+	+	+	−	+	+	3
	+	−	+	+	−	−	−	−	−	+	+	11
8	+	+	+	+	+	+	+	+	+	+	+	8
	−	−	−	−	−	−	−	−	+	+	+	16

406

interactions. For large designs two *separate* plots can be made on normal probability paper.

Exercise 12.11. Above we used l_{12}, l_{13}, and l_{14} as three independent interaction contrasts. Confirm that the same final blocking plan is obtained whichever three independent interactions are used.

Exercise 12.12. Suppose that $k = 2^q$ ($q = 1, 2, 3, \ldots$). Show (*a*) that a 2_V^{k-p} design may be obtained with $p = k - q - 1$, and (*b*) that the design can always be run in blocks of two as main effect plans in k variables, and that the "mirrored pair" property always holds.

12.8. DESIGNS OF RESOLUTION V AND HIGHER

At the beginning of this chapter we introduced the 2_V^{5-1} design. Like other resolution V designs, this arrangement has the property that no main effect or two-factor interaction is confounded with any other main effect or two-factor interaction. Table 12.14 lists some other designs of resolution V and higher and shows how they may be blocked so that *no main effect or two-factor interaction is confounded with any other main effect or two-factor interaction*.

For illustration consider the 2_V^{11-4} designs for studying 11 variables in 128 runs. To obtain the design, write a complete factorial in the $11 - 4 = 7$ variables **1, 2, 3, ..., 7**; then associate new variables **8, 9, $\overline{10}$, $\overline{11}$** with interactions as shown in column (5) of the table. To arrange in eight blocks of 16 runs write columns $\mathbf{B_1} = \mathbf{149}$, $\mathbf{B_2} = \mathbf{12\overline{10}}$, $\mathbf{B_3} = \mathbf{89\overline{10}}$. The 16 runs for which $(\mathbf{B_1 B_2 B_3})$ are $(- - -)$ are put in one block, the 16 runs for which $(\mathbf{B_1 B_2 B_3})$ are $(+ - -)$ in another, and so on.

Application of Yates's Algorithm to Fractional Designs

Yates's algorithm can be used in analyzing data from any 2^{k-p} fractional factorial design. The algorithm is applied in the usual way to any embedded complete factorial in k-p factors. For example one way to compute and identify the effects for the 2_{IV}^{8-4} design of Table 12.11 is as follows:

(i) rearrange the 16 runs in Yates order as a complete factorial in variables **1, 2, 3** and **8**

(ii) calculate the effects using Yates algorithm

(iii) associate the calculated effects with their appropriate aliases.

Exercise 12.13. Make an analysis of the data in Table 12.11, using Yates's algorithm.

TABLE 12.14. Construction and blocking of some designs of resolution V and higher so that no main effect or interaction is confounded with any other main effect or interaction

(1) number of variables	(2) number of runs	(3) degree of fractionation	(4) type of design	(5) method of introducing "new" factors	(6) blocking (with no main effect or interaction confounded)	(7) method of introducing blocks
5	16	$\frac{1}{2}$	2_V^{5-1}	$\pm 5 = 1234$	not available	
6	32	$\frac{1}{2}$	2_{VI}^{6-1}	$\pm 6 = 12345$	two blocks of 16 runs	$B_1 = 123$
7	64	$\frac{1}{2}$	2_{VII}^{7-1}	$\pm 7 = 123456$	eight blocks of 8 runs	$B_1 = 1357$ $B_2 = 1256$ $B_3 = 1234$
8	64	$\frac{1}{4}$	2_V^{8-2}	$\pm 7 = 1234$ $\pm 8 = 1256$	four blocks of 16 runs	$B_1 = 135$ $B_2 = 348$
9	128	$\frac{1}{4}$	2_{VI}^{9-2}	$\pm 8 = 13467$ $\pm 9 = 23567$	eight blocks of 16 runs	$B_1 = 138$ $B_2 = 129$ $B_3 = 789$
10	128	$\frac{1}{8}$	2_V^{10-3}	$\pm 8 = 1237$ $\pm 9 = 2345$ $\pm \overline{10} = 1346$	eight blocks of 16 runs	$B_1 = 149$ $B_2 = 121\overline{0}$ $B_3 = 891\overline{0}$
11	128	$\frac{1}{16}$	2_V^{11-4}	$\pm 8 = 1237$ $\pm 9 = 2345$ $\pm \overline{10} = 1346$ $\pm \overline{11} = 1234567$	eight blocks of 16 runs	$B_1 = 149$ $B_2 = 121\overline{0}$ $B_3 = 891\overline{0}$

408

12.9. SUMMARY

Estimation redundancy often occurs in data from 2^k factorials. Many higher order interactions may be negligible, and some of the factors may be without detectable effects of *any* kind. Utilization of fractional factorials can then reduce experimental effort. In general, increase in the degree of fractionation lowers the resolution of the best fraction and increases confounding between effects of various orders. Fractional designs may be employed as building blocks in the iterative acquisition of knowledge. In this evolution, designs can be augmented so that ambiguities revealed at one stage of experimentation can be resolved in the next. A summary of some useful fractional designs is given in Table 12.15.

APPENDIX 12A. STRUCTURE OF THE FRACTIONAL DESIGNS*

Confounding Patterns for Resolution III Designs

The 2^{7-4} design of Table 12.5 was obtained by setting

$$4 = 12, \quad 5 = 13, \quad 6 = 23, \quad 7 = 123 \qquad (12A.1)$$

Multiplying both sides of each of these identities by **4, 5, 6**, and **7**, respectively, provides the four *generating relations* in the form

$$I = 124, \quad I = 135, \quad I = 236, \quad I = 1237 \qquad (12A.2)$$

Combinations such as **124** and **135** may be referred to as "words." The *defining relation* includes *all* words that are equal to the identity **I**. These are the *generators* **124, 135, 236, 1237** themselves and all other words that can be obtained by multiplying these generators together. Multiplying two at a time gives

$$I = 2345 = 1346 = 347 = 1256 = 257 = 167 \qquad (12A.3)$$

three at a time† gives

$$I = 456 = 1457 = 2467 = 3567 \qquad (12A.4)$$

and four at a time gives

$$I = 1234567 \qquad (12A.5)$$

The complete defining relation is therefore

$$\begin{aligned} I &= 124 = 135 = 236 = 1237 = 2345 = 1346 = 347 \\ &= 1256 = 257 = 167 = 456 = 1457 = 2467 = 3567 \\ &= 1234567 \end{aligned} \qquad (12A.6)$$

---

* Further discussion will be found in Box and Hunter (1961).
† For example, $124 \times 135 \times 236 = 1^2 2^2 3^2 456 = 456$.

TABLE 12.15. Two-level fractional factorial designs for k variables and N runs (numbers in parentheses represent replication)

number of variables k

N	3	4	5	6	7	8	9	10	11
4	2_{III}^{3-1} $\pm 3 = 12$								
8	2^3	2_{IV}^{4-1} $\pm 4 = 123$	2_{III}^{5-2} $\pm 4 = 12$ $\pm 5 = 13$	2_{III}^{6-3} $\pm 4 = 12$ $\pm 5 = 13$ $\pm 6 = 23$	2_{III}^{7-4} $\pm 4 = 12$ $\pm 5 = 13$ $\pm 6 = 23$ $\pm 7 = 123$				
16	2^3 2 times	2^4 2 times	2_V^{5-1} $\pm 5 = 1234$	2_{IV}^{6-2} $\pm 5 = 123$ $\pm 6 = 234$	2_{IV}^{7-3} $\pm 5 = 123$ $\pm 6 = 234$ $\pm 7 = 134$	2_{IV}^{8-4} $\pm 5 = 234$ $\pm 6 = 134$ $\pm 7 = 123$ $\pm 8 = 124$	2_{III}^{9-5} $\pm 5 = 123$ $\pm 6 = 234$ $\pm 7 = 134$ $\pm 8 = 124$ $\pm 9 = 1234$	2_{III}^{10-6} $\pm 5 = 123$ $\pm 6 = 234$ $\pm 7 = 134$ $\pm 8 = 124$ $\pm 9 = 1234$ $\pm \overline{10} = 12$	2_{III}^{11-7} $\pm 5 = 123$ $\pm 6 = 234$ $\pm 7 = 134$ $\pm 8 = 124$ $\pm 9 = 1234$ $\pm \overline{10} = 12$ $\pm \overline{11} = 13$
32	2^3 4 times	2^4 4 times	2^5 2 times	2_{VI}^{6-1} $\pm 6 = 12345$	2_{IV}^{7-2} $\pm 6 = 1234$ $\pm 7 = 1245$	2_{IV}^{8-3} $\pm 6 = 123$ $\pm 7 = 124$ $\pm 8 = 2345$	2_{IV}^{9-4} $\pm 6 = 2345$ $\pm 7 = 1345$ $\pm 8 = 1245$ $\pm 9 = 1235$	2_{IV}^{10-5} $\pm 6 = 1234$ $\pm 7 = 1235$ $\pm 8 = 1245$ $\pm 9 = 1345$ $\pm \overline{10} = 2345$	2_{IV}^{11-6} $\pm 6 = 123$ $\pm 7 = 234$ $\pm 8 = 345$ $\pm 9 = 134$ $\pm \overline{10} = 145$ $\pm \overline{11} = 245$
64	2^3 8 times	2^4 8 times	2^5 4 times	2^6 2 times	2_{VII}^{7-1} $\pm 7 = 123456$	2_V^{8-2} $\pm 7 = 1234$ $\pm 8 = 1256$	2_{IV}^{9-3} $\pm 7 = 1234$ $\pm 8 = 1356$ $\pm 9 = 3456$	2_{IV}^{10-4} $\pm 7 = 2346$ $\pm 8 = 1346$ $\pm 9 = 1245$ $\pm \overline{10} = 1235$	2_{IV}^{11-5} $\pm 7 = 345$ $\pm 8 = 1234$ $\pm 9 = 126$ $\pm \overline{10} = 2456$ $\pm \overline{11} = 1456$
128	2^3 16 times	2^4 16 times	2^5 8 times	2^6 4 times	2^7 2 times	2_{VIII}^{8-1} $\pm 8 = 1234567$	2_{VI}^{9-2} $\pm 8 = 13467$ $\pm 9 = 23567$	2_V^{10-3} $\pm 8 = 1237$ $\pm 9 = 2345$ $\pm \overline{10} = 1346$	2_{IV}^{11-4} $\pm 8 = 1237$ $\pm 9 = 2345$ $\pm \overline{10} = 1346$ $\pm \overline{11} = 1234567$

number of runs N

$(\frac{1}{32})$ $(\frac{1}{64})$ $(\frac{1}{128})$ $(\frac{1}{16})$ $(\frac{1}{8})$ $(\frac{1}{4})$ $(\frac{1}{2})$

This defining relation provides the confounding pattern for the whole design. For example, multiplying through by 1 gives

$$1 = 24 = 35 = 1236 = 237 = 12345 = 346 = 1347 = 256$$
$$= 1257 = 67 = 1456 = 457 = 12467 = 13567 = 234567 \qquad (12A.7)$$

Thus interactions 24, 35, 1236, etc., are confounded with (are aliases of) main effect 1. By repeatedly using the defining relation, and omitting words with three or more letters, we obtain the abbreviated version of the confounding pattern of Table 12.6, which is appropriate on the assumption that all interactions among three or more variables are negligible.

Note that the defining relation for the 2^{7-4} design contains 16 words and each main effect and interaction has 15 aliases. In general, a 2^{k-p} design is produced by p generators and has a defining relation containing 2^p words.

The 16 possible combinations of $\pm$ signs for the four generators

$$I = \pm 124, \qquad I = \pm 135, \qquad I = \pm 236, \qquad I = \pm 1237 \qquad (12A.8)$$

determine the 16 separate fractions. In composing the defining relations for each of these different fractions, we employ the usual rules of algebraic multiplication to determine the signs in the defining relation and hence in the confounding pattern. For example, if the individual generators had been -124, $+135$, $+236$ and -1237 the complete defining relation would be $I = -124 = 135 = 236 = -1237 = -2345 = -1346 = 347 = 1256 = -257 = -167 = -456 = 1457 = 2467 = -3567 = 1234567$.

Exercise 12A.1. Obtain the generators and defining relation for the 2_{IV}^{7-4} fraction obtained by setting $4 = -12$, $5 = 13$, $6 = 23$, $7 = 123$. Use the defining relation to obtain the aliases of all main effects. By omitting interactions between more than three factors, confirm the entries in Table 12.8.

Resolution

The resolution R of a fractional design is the *length of the shortest word in the defining relation*. It should be clear from this definition that the saturation method for generating fractionals must always give designs of resolution III.

Confounding Patterns for Resolution IV Designs

In Section 12.6 we described the generation of a 2_{IV}^{8-4} design by "fold-over." We now consider more carefully the confounding pattern for this design. The two component groups of eight runs can be regarded as separate 2^{8-5} designs with generating relations

$$I_8 = 8 = 124 = 135 = 236 = 1237 \qquad (12A.9)$$

and

$$I_8 = -8 = -124 = -135 = -236 = 1237 \qquad (12A.10)$$

respectively, where the notation I_8 refers to a column of eight plus signs. A set of four generators, and hence the defining relation, for the combined 2^{8-4} design can now be obtained from these expressions as follows. The relation $I_8 = 1237$ holds for both sets of eight runs. Consequently, if I_{16} represents the column of 16 plus signs associated with the combined design,

$$I_{16} = 1237$$

and 1237 is a generator for the combined design. Also, for the first part of the design $I_8 = (8)(124) = 1248$, and for the second part $I_8 = (-8)(-124) = 1248$. Thus for the complete design $I_{16} = 1248$.

By a similar argument $I_{16} = 1358 = 2368$. Thus the complete set of four generators is

$$I_{16} = 1237, \qquad I_{16} = 1248, \qquad I_{16} = 1358, \qquad I_{16} = 2368 \qquad (12A.11)$$

By multiplication as before we obtain the defining relation:

$$
\begin{aligned}
I_{16} &= 1237 = 1248 = 1358 = 2368 \\
&= 3478 = 2578 = 1678 = 2345 = 1346 = 1256 \\
&= 1457 = 2467 = 3567 = 4568 \\
&= 12345678
\end{aligned}
\qquad (12A.12)
$$

Since the shortest word is of length four, the combined design is of resolution IV as required.

Exercise 12A.2.   Verify the confounding pattern shown in Table 12.13, and extend it to include aliases of all orders.

Exercise 12A.3.   Although the defining relation for any fractional factorial is unique, the generators for a 2^{k-p} factorial design with $p > 1$ are not unique. Any set of $k - p$ words that generate the defining relation is a set of generators. Show that an alternative set of generators for the 2_{IV}^{8-4} design of Table 12.11 is 3478, 2578, 1457, and 2467.

An Alternative Method for Generating Resolution IV Designs

As an alternative to the fold-over method, any 2_{IV}^{k-p} design may be constructed as follows:

1. Write the design matrix for a full factorial design for $k - p$ variables.
2. Associate extra variables with all interaction columns containing an *odd* number of numerals.

We demonstrate by obtaining anew the 2_{IV}^{8-4} design of Table 12.11, whose confounding was discussed above. To do this, write down a 2^4 factorial for the variables 1, 2, 3, and 8. The four three-factor interaction columns are then 128, 138, 238, and 123. Now associate these with the four "new" variables 4, 5, 6, and 7 to obtain a set of four generators:

$$
\begin{array}{cccc}
1 & 2 & & 8 & 4 \\
1 & & 3 & 8 & 5 \\
& 2 & 3 & 8 & 6 \\
1 & 2 & 3 & & 7
\end{array}
$$

The design thus constructed is identical to that given in Table 12.11. The only reason for starting with variables termed **1, 2, 3**, and **8** instead of **1, 2, 3**, and **4** is to make it easy to see the identity between this method and the preceding one.

The defining relation for this design is $I = 1248 = 1358 = 2368 = 1237 = 2345 = 1346 = 3478 = 1256 = 2578 = 1678 = 4568 = 1457 = 2467 = 3567 = 1234567$. The identical design could be obtained, for example, using the generators **1237, 1248, 1346, 2345**, that is, by first writing down the 16 runs of a 2^4 factorial in variables **1, 2, 3** and **4**, and then associating the extra variables with the three factor interactions. This method for constructing the 2_{IV}^{8-4} design is that presented in Table 12.15.

APPENDIX 12B. CHOOSING ADDITIONAL RUNS TO RESOLVE AMBIGUITIES FROM FRACTIONAL FACTORIALS

In the injection molding example of Section 12.6 the linear contrast $l_{15} = 4.6$ is not easily explained by system noise. But there is ambiguity in its interpretation, since it estimates the sum of the effects $15 + 26 + 38 + 47$. Three is the smallest number of additional runs that could allow separate estimation of the true values of these two-factor interactions. However, since the additional runs have to be made at a different time from the first 16 runs, we should also allow for a general change in level (a block effect). Thus the minimum number of additional runs we need is 4. We now consider how 4 such runs might be chosen.

For the runs made so far, the columns of signs corresponding to the interactions 15, $26, 38$, and 47 are identical. We need 4 additional runs that will permit separate estimation of the mean values of these interactions and will also allow for a possible block effect (a change in level between the first 16 runs and the additional 4 runs). One sensible possibility is to employ additional runs that yield signs for the interaction columns as follows:

15	26	47	38
+	−	−	+
+	+	−	−
+	−	+	−
+	+	+	+

Four additional runs that will do this are:

1	2	3	4	5	6	7	8
+	+	+	+	+	−	−	+
+	+	+	+	+	+	−	−
+	+	+	+	+	−	+	−
+	+	+	+	+	+	+	+

This four-run design is obtained by writing a plus sign for each element in the **1, 2, 3, 4** columns and then choosing the remaining signs to satisfy the requirements of the previous table. The choice is not unique. In particular, the signs in one or more rows and/or in one or more columns may be switched, and the resulting design will still possess the desired characteristics. The experimenter has many choices. Suppose that he decides to perform a design in which variables **3** and **5**, which apparently have the largest main effects, are not maintained at the same levels but are varied in a 2^2 factorial design. Such a sequence is easily obtained as before by assigning the necessary signs to columns **3** and **5** and then arranging the other columns to satisfy the required condition. One such arrangement for the injection molding example is given in Table 12B.1, which also contains new data.

TABLE 12B.1. Four additional runs with data, injection molding example

run	1	2	3	4	5	6	7	8	shrinkage
17	−	+	+	+	−	−	−	+	29.4
18	−	+	−	−	−	+	+	+	19.7
19	+	+	−	−	+	−	−	+	13.6
20	+	+	+	+	+	+	+	+	24.7

Incorporating the New Data

In general, the incorporation of design fragments can always be achieved by use of the method of least squares (see Chapter 14 and references given there). For designs of the kind here considered, where the number of extra constants is exactly equal to the number of additional runs, the least squares analysis simplifies. It then corresponds exactly to a commonsense analysis that is illustrated below for the data of Tables 12.11 and 12.B.1.

From the analysis of the first 16 runs it appears that, apart from the effect of noise, the data may be explained by a mean level plus the main effects of variables **3** and **5** and the effect of one more of interactions *15, 26, 47,* and *38*. The main effect of variable **8** is the next largest effect and has also been treated as real in this analysis. We denote the mean level in the second block by M. Now the response expected when an effect (main effect or interaction) is at the plus level is the mean plus one half of that effect. Similarly the response expected at the minus level is the mean minus one half of that effect. Thus we can make a table of the following kind, which summarizes all of the information available about the four interactions both from the new runs and from the previous 16 runs.

run	M	$\frac{1}{2}l_3 = 2.75$	$\frac{1}{2}l_5 = -1.9$	$\frac{1}{2}l_8 = 0.6$	$\frac{1}{2}(15)$	$\frac{1}{2}(26)$	$\frac{1}{2}(47)$	$\frac{1}{2}(38)$	
17	+	+	−	+	+	−	−	+	$29.4 = y_{17}$
18	+	−	−	+	+	+	−	−	$19.7 = y_{18}$
19	+	−	+	+	+	−	+	−	$13.6 = y_{19}$
20	+	+	+	+	+	+	+	+	$24.7 = y_{20}$
					+	+	+	+	$2.3 = \frac{1}{2}l_{15}$

In the first row of this table, for example, $\frac{1}{2}l_3 = 2.75$ is the best available estimate (taken from Table 12.12) of one half of the main effect of factor **3**. The second row of the table tells us that the result $y_{17} = 29.4$ should be explained by the equation

$$M + 2.75 - (-1.9) + 0.6 + \tfrac{1}{2}(15 - 26 - 47 + 38) = 29.4 \qquad (12\text{B}.1)$$

This yields

$$M + \tfrac{1}{2}(15 - 26 - 47 + 38) = 24.15 \qquad (12\text{B}.2)$$

The last row of the table presents all the information provided by the first 16 runs about the interaction effects, that is, that half their sum is equal to $\frac{1}{2}l_{15} = 2.3$. Putting the reduced equations together, we have

M	$\frac{1}{2}(15)$	$\frac{1}{2}(26)$	$\frac{1}{2}(47)$	$\frac{1}{2}(38)$	
+	+	−	−	+	24.15
+	+	+	−	−	19.95
+	+	−	+	−	17.65
+	+	+	+	+	23.25
	+	+	+	+	2.3

From the last two equations M is estimated as 20.95. Substituting this value in each of the preceding equations, we obtain

$\frac{1}{2}(15)$	$\frac{1}{2}(26)$	$\frac{1}{2}(47)$	$\frac{1}{2}(38)$	
+	−	−	+	3.2
+	+	−	−	−1.0
+	−	+	−	−3.3
+	+	+	+	2.3

From this table the effect of interest seems mainly associated with interaction *38*. More precisely, by solving these last four equations we obtain

$$0.6 \to 15, \qquad 0.7 \to 26, \qquad -1.6 \to 47, \qquad 4.9 \to 38$$

It seems very likely, therefore, that a considerable interaction between holding pressure and screw speed accounts for the majority, if not all, of the interaction effects found.

The nature of this interaction may be comprehended from the two-way table below, which shows the average shrinkage at all combinations of low and high values of holding pressure and screw speed using data from the original 16 runs.

		−	+
screw speed **(8)**	+	15.3	25.4
	−	18.7	19.6

holding pressure **(3)**

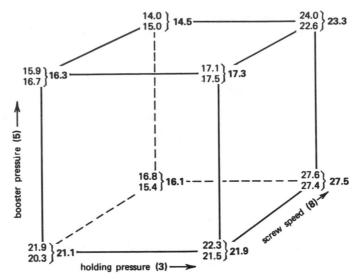

FIGURE 12B.1.   The 2_{IV}^{8-4} design as a replicated 2^3 factorial in variables **3**, **5**, and **8**.

We see that, whereas at low holding pressure increasing screw speed reduces shrinkage from 19 to 15 units, at high holding pressure increasing screw speed increases shrinkage from 20 to 25 units.

A Replicated 2^3 Design Encapsulated in the 2_{IV}^{8-4} Fractional Factorial

The results appear to be explained by main effects and interactions of factors **3**, **5**, and **8**. Supposing the remaining factors to be essentially inert, we rearrange the data as a

TABLE 12B.2.   The 2_{IV}^{8-4} design as a replicated 2^3 factorial in variables 3, 5, and 8

runs	3	5	8	shrinkage		average from first 16 runs
14,16	−	−	−	21.9	20.3	21.1
9, 11	+	−	−	22.3	21.5	21.9
13, 15	−	+	−	15.9	16.7	16.3
10, 12	+	+	−	17.1	17.5	17.3
2, 4 (18*)	−	−	+	16.8	15.4 (18.5*)	16.1
5, 7 (17*)	+	−	+	27.6	27.4 (28.2*)	27.5
1, 3 (19*)	−	+	+	14.0	15.0 (12.4*)	14.5
6, 8 (20*)	+	+	+	24.0	22.6 (23.5*)	23.3

replicated 2^3 factorial in factors **3**, **5**, and **8** in Table 12B.2 and Figure 12B.1, taking advantage of the property that the design provides duplicated 2^3 factorials in all possible subsets of three variables.

In Table 12B.2 the results from the additional runs (17, 18, 19, 20), indicated by asterisks, have been adjusted for differences associated with blocks, that is, $20.95 - 19.75 = 1.2$ has been subtracted from each value. These adjusted values are seen to agree quite well with the original shrinkage values *on the assumption that* **3**, **5**, *and* **8** *are the important factors.*

REFERENCES AND FURTHER READING

For further discussion of fractional factorial designs see the following and the references listed therein:

Daniel, C. (1976). *Applications of Statistics to Industrial Experimentation*, Wiley.
Plackett, R. L., and J. P. Burman (1946). The design of optimum multifactorial experiments, *Biometrika*, **33**, 305.
Box, G. E. P., and J. S. Hunter (1961). The 2^{k-p} fractional factorial designs, *Technometrics*, **3**, 311, 449.

For augmentation of designs see Daniel's book and the following articles:

Davies, O. L., and W. A. Hay (1950). Construction and uses of fractional factorial designs in industrial research, *Biometrics*, **6**, 233.
Box, G. E. P., and K. B. Wilson (1951). On the experimental attainment of optimum conditions, *Roy. Stat. Soc., Ser. B*, **13**, 1.
Daniel, C. (1962). Sequences of fractional replicates in the 2^{p-q} series, *J. Am. Stat. Soc.*, **58**, 403.
Box, G. E. P., (1966). A note on augmented designs, *Technometrics*, **8**, 184.

QUESTIONS FOR CHAPTER 12

1. What is a fractional factorial design?
2. What is a half-fraction, and how can you construct such a design?
3. What is a saturated design, and how can you construct such a design?
4. Discuss the sequential use of fractional designs.
5. A 2^{8-3} design has how many runs? How many variables? How many levels for each variable? Answer the same questions for a 2^{k-p} design.
6. All other things being equal, why would a resolution IV design be preferred to a resolution III design?
7. Is it possible to construct a 2^{8-3} design of resolution III? Resolution IV? Resolution V?
8. In what situations is it useful to employ fractional factorial designs?

9. How can fractional factorial designs be blocked? How should they be randomized?

10. How might you analyze data from a 2^{7-1} design? A 2^{7-4} design?

11. What is a defining relation? A generator? A confounding pattern? Why is it necessary to know the confounding pattern for a fractional factorial design?

12. Construct, starting with the 12-run Plackett and Burman design, a 12-variable resolution IV design.

13. Design a 2^{5-1} design in eight blocks of size two so that main effects are clear of block effects.

CHAPTER 13

More Applications of
Fractional Factorial Designs

The purpose of this chapter is to give further examples of the use of fractional factorial designs. It is a companion to Chapter 11, which served the same purpose for factorial designs.

13.1. EXAMPLE 1: EFFECTS OF FIVE VARIABLES ON SOME PROPERTIES OF CAST FILMS

In this example five variables were studied: the catalyst concentration, the amount of a certain additive, and the amounts of three emulsifiers A, B, and C. In an initial 2^{5-2} fraction eight polymer solutions were prepared, each was spread as a film on a microscope slide, and the properties of the films were recorded after they dried. The results for six different responses are shown in Table 13.1.

Surprisingly many conclusions can be drawn from these results by mere visual inspection. On the assumption (reasonable for this particular application) of dominant main effects, the important variable affecting haziness is emulsifier A (variable **3**). The important variable affecting adhesion is catalyst concentration (variable **1**). The important variables affecting the remaining responses of grease on top of the film, grease under the film, dullness of the film when the pH was adjusted, and dullness of the film when the original pH was used are **4, 5, 4**, and **4**, respectively.

Commentary

If experiments are carefully planned, a great deal of information can sometimes be obtained without much mathematical analysis (no computation was needed in this case). The converse of this statement, however, is not

419

TABLE 13.1. 2^{5-2} fraction (4 = 23, 5 = 123) with results for six responses of interest, Example 1

| | variable | | | | | | | | response | | |
run	1	2	3	4	5	hazy?	adheres?	grease on top of film?	grease under film?	dull (adjusted pH)?	dull (original pH)?
1	−	−	−	+	−	no	no	yes	no	slightly	yes
2	+	−	−	+	+	no	yes	yes	yes	slightly	yes
3	−	+	−	+	+	no	no	no	yes	no	no
4	+	+	−	−	−	no	yes	no	no	no	no
5	−	−	+	−	+	yes	no	no	yes	no	slightly
6	+	−	+	−	−	yes	yes	no	no	no	no
7	−	+	+	+	−	yes	no	yes	no	slightly	yes
8	+	+	+	+	+	yes	yes	yes	yes	slightly	yes

variable	−	+
1 catalyst (%)	1	$1\frac{1}{2}$
2 additive (%)	$\frac{1}{4}$	$\frac{1}{2}$
3 emulsifier A (%)	2	3
4 emulsifier B (%)	1	2
5 emulsifier C (%)	1	2

420

EXAMPLE 1   421

true. If experiments are not carefully planned, it may be impossible to obtain much useful information even with extensive and careful analysis. This is the reason why design is more important than analysis. In this case, after the initial set of eight runs further fractions could have been run to confirm the tentative findings, but the experimenter judged this unnecessary and moved successfully to the next part of his investigation.

Exercise 13.1. Show that the generating relations for the design of Table 13.1 are $I = 234$ and $I = 1235$. Use these relations to obtain the defining relation and the alias structure.   *Partial answer*: $I = 234 = 1235 = 145$, $l_1 \rightarrow 1234 + 235 + 45$.

Exercise 13.2. Show that the alias structure may be obtained alternatively by thinking of the design as a saturated 2_{III}^{7-4} design in which two variables have been omitted.

The reader may ask how the experimenter could be certain that the effects were due to main effects and not to interactions in the alias strings. The answer is that he was not certain. From his knowledge of these variables, however, he thought that to press on was the best bet. In doing so he took a calculated risk of having to turn back at a later stage, if his bet did not come off. Decisions of this kind by experimenters are not peculiar to the use of fractional factorial designs. The truth is that the experimenter, in deciding what to do next, is never *certain* of what is best. He always guesses or, to use a more palatable term, uses judgment, and he always runs the risk of being wrong. This apparently dangerous mode of life may come as a surprise, but it is easy to see that efficiency requires it.

Imagine three experimenters, one who was very cautious, one who took carefully calculated risks, and one who jumped to conclusions on very thin evidence. Bear in mind that time and resources are really *always* limited. Consequently to repeat a set of experiments unnecessarily is to deny that effort to the investigation of other variables; on the other hand, frequent doubling back, because false trails are followed, wastes effort. It is clear, then, that the experimenter who takes some suitable intermediate position between the ultraconservative and the foolhardy is likely to be most successful.

"But," some may say, "I thought statistics made everything objective." If by that they mean that statistics ought to lead along a unique, painless route to the truth, they are mistaken. What statistics does, for a given amount of effort, is to lessen the risks of being wrong, either through missing important facts or through giving credence to phenomena that have no reality. Statistics allows the investigator to play the calculated-risk game most efficiently.

13.2. EXAMPLE 2: STABILITY OF NEW PRODUCT

A chemist in an industrial development laboratory was trying to formulate a household liquid product using a completely new process. He was able to demonstrate fairly quickly that the product could be manufactured and that it possessed a number of attractive properties. Unfortunately it could not be marketed because it was unstable.

When the statistician first met him, the chemist had for months been trying many different ways of synthesizing the product in the hopes of hitting on conditions that would give stability, but without success. He had succeeded, however, in identifying four variables that had important influences on stability: (1) acid concentration, (2) catalyst concentration, (3) temperature, and (4) monomer concentration.

With his budget almost expended, he agreed somewhat reluctantly to perform his first statistically planned experiment, the 2^{4-1} fractional factorial design shown in Table 13.2. In these tests, which were performed in random order, the chemist was trying to achieve a stability value of at least 25. His initial reaction to the data was disgust, since none of the individual observations reached the desired stability level.

Using the analysis shown in Table 13.3, the statistician suggested the simplest explanation of the data was that only two variables, **1** and **2**,

TABLE 13.2. **Results for Example 2**

test	variable 1	2	3	4	stability (R)
1	−	−	−	−	20
2	+	−	−	+	14
3	−	+	−	+	17
4	+	+	−	−	10
5	−	−	+	+	19
6	+	−	+	−	13
7	−	+	+	−	14
8	+	+	+	+	10

variable	−	+
1 acid concentration (%)	20	30
2 catalyst concentration (%)	1	2
3 temperature (°C)	100	150
4 monomer concentration (%)	25	50

EXAMPLE 2   423

TABLE 13.3. Analysis of data for Example 2

---

main effects and three-factor interactions

$$l_1 = -5.8 \rightarrow 1 + 234$$
$$l_2 = -3.8 \rightarrow 2 + 134$$
$$l_3 = -1.2 \rightarrow 3 + 124$$
$$l_4 = 0.8 \rightarrow 4 + 123$$

two-factor interactions

$$l_{12} = 0.2 \rightarrow 12 + 34$$
$$l_{13} = 0.8 \rightarrow 13 + 24$$
$$l_{14} = -0.2 \rightarrow 14 + 23$$

average and four-factor interaction

$$l_{1234} = 14.6 \rightarrow \text{average} + \tfrac{1}{2}(1234)$$

---

influenced stability, the other two being inert. If this hypothesis were true, the design could be viewed as a duplicated 2^2 factorial design in acid concentration (**1**) and catalyst concentration (**2**) as shown in Figure 13.1. The chemist was asked, therefore, whether he would have been surprised if he had obtained discrepancies in duplicate runs similar to those in Figure 13.1. He said that he would not.

The possible implications of the results if the statistician's hypothesis was true were demonstrated by roughly sketching in by eye contour lines of a "stability plane" as shown in Figure 13.1. This picture suggested that experiments should be performed in the direction of the arrow. The contour lines shown are actually those obtained by the method of least squares (see Section 14.1). The direction at right angles to these contours is called the *direction of steepest ascent* (see Section 15.2). Refinement is, however, unnecessary in this example; "eyeball analysis" is all that is needed. A few exploratory runs performed in the general direction indicated produced, for the first time since the beginning of the investigation, a product with stability greater than the goal of 25.

Commentary

This example illustrates the following:

1. How a fractional factorial design was used for screening purposes to isolate two important variables from the original set of four proposed by the experimenter.
2. How a desirable *direction* in which to carry out further experiments was discovered.

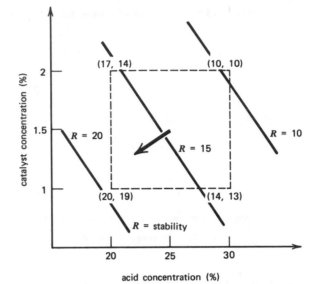

FIGURE 13.1. Results of 2^{4-1} fractional factorial design viewed as a duplicated 2^2 factorial design with approximate stability contours, Example 2.

An additional result of this investigation was that the experimenter became an evangelist for experimental design.

13.3. EXAMPLE 3: BOTTLENECK AT THE FILTRATION STAGE OF AN INDUSTRIAL PLANT

A number of similar chemical plants had been successfully operating for several years in different locations. In the older plants the time to complete a particular filtration cycle was about 40 minutes, but in a newly constructed plant filtration took almost twice as long, resulting in serious delays. What was the cause of the difficulty?

Seven Variables

To begin to solve this problem a meeting was called, at which a number of possibilities were considered.

EXAMPLE 3 425

1. Source of water supply. The plant engineer explained that the water for the new plant, which came from the city reservoir some 30 miles away, differed somewhat in mineral content from that available at the other locations. Some well water was available at the new site that more closely resembled the water supply at the older plants. The engineer suggested, therefore, that some tests be run with well water.
2. Origin of raw material. The process superintendant pointed out that the raw material, which was manufactured on site, was not identical in all respects to that used in the older plants. Consequently he proposed that some of the raw material from one of the older plants be shipped to the new plant and used in some tests.
3. Level of temperature. The temperature of filtration in the new plant was slightly lower than in the older plants. The plant chemist thought that this might be the cause of the problem.
4. Presence of recycle. A major difference between the new plant and the older ones was a recycle device absent in the latter. It was suggested that the inclusion of this device could increase filtration time.
5. Rate of addition of caustic soda. The rate of caustic soda addition in the new plant was higher than in the older plants. The process foreman suggested that the rate be decreased in the new plant.
6. Type of filter cloth. A new type of filter cloth was being used in the new plant. The process superintendent pointed out that it would be a relatively simple matter to get some filter cloths from the older plants and make some test runs with them.
7. Length of holdup time. In the new plant the holdup time was lower than in the older plants, and the quality control engineer gave reasons for believing that this might be the cause of the problem.

Much disagreement was expressed at the meeting about these factors. Some participants even argued that changes proposed by others were ridiculous.

The Design and the Results

To sort out these ideas the 2^{7-4} screening design shown in Table 13.4 was performed on the plant. At the outset the attitude of the person responsible for the investigation was that out of these seven factors perhaps one or two would be found to be important. The chance was small, he thought, that there would be as many as three or more important variables; in fact, it was judged quite possible that none of those selected for investigation would have any effect at all. The order of the eight tests was randomized, and the data shown in Table 13.4 were obtained.

TABLE 13.4. Results of Example 3

variable			−	+
1 water supply			town reservoir	well
2 raw material			on site	other
3 temperature			low	high
4 recycle			yes	no
5 caustic soda			fast	slow
6 filter cloth			new	old
7 holdup time			low	high

				12	13	23	123	filtration time (min)
test	1	2	3	4	5	6	7	y
1	−	−	−	+	+	+	−	68.4
2	+	−	−	−	−	+	+	77.7
3	−	+	−	−	+	−	+	66.4
4	+	+	−	+	−	−	−	81.0
5	−	−	+	+	−	−	+	78.6
6	+	−	+	−	+	−	−	41.2
7	−	+	+	−	−	+	−	68.7
8	+	+	+	+	+	+	+	38.7

Four Tentative Interpretations of Results

In Table 13.5 three of the calculated effects (l_1, l_3, and l_5) are large in absolute value and have been circled. There are several possible interpretations. Four of the most likely are:

1. Main effects *1*, *3*, and *5* are producing the effects.
2. Main effects *1* and *3* and interaction *13* are producing the effects.
3. Main effects *1* and *5* and interaction *15* are producing the effects.
4. Main effects *3* and *5* and interaction *35* are producing the effects.

The Second Design

To reduce these ambiguities a selected set of eight additional tests (see Table 13.6) was run, converting the original resolution III design to one of resolution IV. This was done by "fold-over" (Chapter 12), that is, by arranging that the added fraction had signs opposite to those in the original design.

EXAMPLE 3   427

TABLE 13.5.   Calculated values and abbreviated confounding pattern for eight-run filtration
experiment, Example 3

$$l_1 = \boxed{-10.9} \rightarrow 1 + 24 + 35 + 67$$

$$l_2 = -2.8 \rightarrow 2 + 14 + 36 + 57$$

$$l_3 = \boxed{-16.6} \rightarrow 3 + 15 + 26 + 47$$

$$l_4 = 3.2 \rightarrow 4 + 12 + 37 + 56$$

$$l_5 = \boxed{-22.8} \rightarrow 5 + 13 + 27 + 46$$

$$l_6 = -3.4 \rightarrow 6 + 17 + 23 + 45$$

$$l_7 = 0.5 \rightarrow 7 + 16 + 25 + 34$$

TABLE 13.6.   Results of second filtration experiment, Example 3

test	1	2	3	−12 4	−13 5	−23 6	123 7	filtration time (min) y
9	+	+	+	−	−	−	+	66.7
10	−	+	+	+	+	−	−	65.0
11	+	−	+	+	−	+	−	86.4
12	−	−	+	−	+	+	+	61.9
13	+	+	−	−	+	+	−	47.8
14	−	+	−	+	−	+	+	59.0
15	+	−	−	+	+	−	+	42.6
16	−	−	−	−	−	−	⌐	67.6

Analysis of Sixteen Results

Combining the data from both eight-run designs yields the estimates given
in Table 13.7. The three largest effects in absolute value are l_1, l_5, and l_{15}.
It now seems likely that 1 and 5 are the two most important variables,
with not only large main effects but also a large interaction. On this interpretation, 2, 3, 4, 6, and 7 are inert variables and the 16 tests are essentially
a 2^2 factorial design in variables 1 and 5 replicated four times (see Figure
13.2).

TABLE 13.7.   **Calculated values and abbreviated confounding pattern for 16-run filtration experiment, Example 3**

$$l_1 = \boxed{-6.7} \to 1$$
$$l_2 = -3.9 \to 2$$
$$l_3 = -0.4 \to 3$$
$$l_4 = 2.8 \to 4$$
$$l_5 = \boxed{-19.2} \to 5$$
$$l_6 = 0.1 \to 6$$
$$l_7 = -4.4 \to 7$$
$$l_{12} = 0.5 \to 12 + 37 + 56$$
$$l_{13} = -3.6 \to 13 + 27 + 46$$
$$l_{14} = 1.1 \to 14 + 36 + 57$$
$$l_{15} = \boxed{-16.2} \to 15 + 26 + 47$$
$$l_{16} = 4.9 \to 16 + 25 + 34$$
$$l_{17} = -3.4 \to 17 + 23 + 45$$
$$l_{24} = -4.2 \to 24 + 35 + 67$$

Although this was the simplest and hence the most appealing hypothesis, the investigative team did not feel completely confident that it was correct. Subsequent testing was therefore undertaken in which all variables except **1** and **5** were set at their normal levels. These tests verified that the problem of excessive filtration time could be eliminated by appropriately adjusting the source of the water supply (**1**) and the rate of addition of caustic soda (**5**).

Commentary

This example again illustrates how fractional factorial designs may be used sequentially as building blocks to produce a design of suitable resolution to answer questions at issue.

Substantial noise was present. The key to solving this problem lay in screening a large number of variables to discover effects involving only a few of them. Note from Figure 13.2 that the discovery of the large *interaction* between variables **1** and **5** was essential to the solution of this problem. A one-factor-at-a-time approach would have failed.

EXAMPLE 4   429

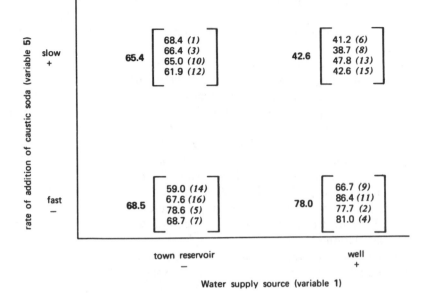

FIGURE 13.2. Results of the 16 trials in relation to two variables, water supply source (**1**) and rate of addition of caustic soda (**5**), Example 3. The average result under each one of the four conditions is in bold type. The test condition numbers are given in parentheses.

13.4. EXAMPLE 4: SENSITIVITY ANALYSIS OF A SIMULATION MODEL—CONTROLLER–AIRCRAFT SYSTEM

The present system for air traffic control has evolved over many years, and any recommendations for changing it are carefully reviewed before they are adopted. Ideally one would like to experiment with an operating system to determine whether alternative modes of operation might be preferable. In the case of air traffic control, however, experimentation is almost impossible since it involves questions of safety and of protocols and modes of operation rooted in practice and often prescribed by law. In such situations a simulation model may be constructed to serve as a stand-in for the actual system. Experiments performed on the simulation model can often provide valuable insights with respect to the actual operating system.

A simulation model for a controller–aircraft system was gradually evolved over many years of experience. It was verified using actual data. The model was quite complicated, and its output was a function of many factors, for example, the number of lengths of tracks within a sector, the number of adjacent high-altitude sectors, the number and location of navigation beacons, and the mix of jumbo versus standard jets. Other factors were parameters that characterized the probability distributions of various events, such as aircraft arrivals to the sector, the number of communication transactions between aircraft and controller, the lengths of such transactions, and the intertransaction gap time experienced by controllers.

Important responses, in addition to measures of safety and cost, were the number of aircraft present per second, the proportion of communication channel time absorbed by certain message types, and the proportion of time the communication channel was utilized. One important response was the amount of time a pilot consumed while waiting to communicate with a sector controller. If another aircraft was using the communications channel, the pilot had to wait until the channel was free before he could speak to the controller. Obviously a wait of even a few seconds could be of great importance. Both the frequency and the duration of these delays were studied with the simulation model.

Once the simulation model was available, it was natural to perform a sensitivity analysis, that is, to ask what effect small changes in the factors

TABLE 13.8.   Results from sensitivity analysis of simulation model, Example 4

data									result of analysis		
factor in simulation								response	calculated	identity of
1	2	3	4	5	6	7	8	y	value	effects
−	−	−	−	−	−	−	−	65.81	60.62	→ average
+	−	−	−	+	+	+	−	58.49	−3.78	→ 1
−	+	−	−	+	+	−	+	62.51	2.65	→ 2
+	+	−	−	−	−	+	+	60.19	0.37	→ 12 + 35 + 46 + 78
−	−	+	−	+	−	+	+	60.22	0.51	→ 3
+	−	+	−	−	+	−	+	59.20	0.37	→ 13 + 25 + 47 + 68
−	+	+	−	−	+	+	−	66.58	0.43	→ 23 + 15 + 48 + 67
+	+	+	−	+	−	−	−	61.68	−1.79	→ 5
−	−	−	+	−	+	+	+	59.01	−2.43	→ 4
+	−	−	+	+	−	−	+	53.71	0.11	→ 14 + 26 + 37 + 58
−	+	−	+	+	−	+	−	62.43	0.84	→ 24 + 16 + 38 + 57
+	+	−	+	−	+	−	−	60.77	0.09	→ 6
−	−	+	+	+	+	−	−	60.44	0.34	→ 34 + 17 + 28 + 56
+	−	+	+	−	−	+	−	57.48	−0.56	→ 7
−	+	+	+	−	−	−	+	63.08	−2.18	→ 8
+	+	+	+	+	+	+	+	58.32	0.43	→ 45 + 36 + 27 + 18

EXAMPLE 4 431

would have on a response. Given that no analytic solution is possible, one may approach the problem experimentally using the computer. But even here, because the simulation involves statistical sampling from waiting time distributions, repetition of a simulation under a fixed set of factor conditions produces variable results. Thus we have a situation in which (1) we have a large number of variables to consider and (2) the variation cannot be ignored. This is exactly the situation for which experimental designs were invented.

It was decided to change eight of the factors in the system, each at two levels. Even in this age of fast computers, for this example individual simulation trials were expensive, and a full 2^8 design would have been too prodigal. Instead a 2^{8-4} fraction with generators **1235, 1246, 1347**, and **2348** was run. The data and estimated effects are displayed in Table 13.8. The 15 estimated effects are identified, assuming that third-order and higher order effects are negligible. The effects, listed in order of their absolute size, are as follows:

effect	estimate	effect	estimate
1	−3.78	45 + 36 + 27 + 18	0.43
2	2.65	23 + 15 + 48 + 67	0.43
4	−2.43	13 + 25 + 37 + 68	0.37
8	−2.18	12 + 35 + 46 + 78	0.37
5	−1.79	34 + 17 + 28 + 56	0.34
24 + 16 + 38 + 57	0.84	14 + 26 + 37 + 58	0.11
7	−0.56	6	0.09
3	0.51		

Using evidence from previous simulation studies it could be assumed that the standard deviation of an estimated effect was about 0.35. On this basis it was confirmed that the changes introduced into factors **1, 2, 4, 5** and **8** had effects of the kind expected. The apparent small effect of factor **6**, on the other hand, was contrary to the intuition of the system experts and led directly to questioning of the structure of the simulation model. Further simulation runs were performed to check this.

Exercise 13.3. Make a plot on normal probability paper of the estimated effects listed in Table 13.8. Make any other plots or analysis you think appropriate. Comment on your results and particularly on the assumption that the standard deviation of an effect is 0.35.

Commentary

This example shows how a fractional factorial design was used with a simulation model. The iterative nature of scientific investigation is illustrated once again, as the results suggested that modifications be made in the model itself. The work described above prompted further "experimentation" on the computer.

13.5. SUMMARY

Fractional factorial designs are one means available to experimenters to increase research efficiency. This is illustrated in this chapter with examples that show the basic simplicity of the designs and how they fit naturally into iterative scientific method.

Areas of Application

There are three main areas of application for fractional factorial designs:

1. In *screening situations*, in which only a subset of the variables is expected to be important, but *which* subset is unknown.
2. In situations where *certain interactions can be tentatively assumed to be negligible*.
3. In situations where *groups of experiments are to be performed sequentially*, ambiguities being resolved as an investigation evolves.

REFERENCES AND FURTHER READING

Examples 1 and 2 are from

Hunter, W. G., and M. E. Hoff (1967). Planning experiments to increase research efficiency, *Ind. Eng. Chem.*, **59**, 43.

Example 3 is from

Box, G. E. P., and J. S. Hunter (1961). The 2^{k-p} fractional factorial designs, *Technometrics*, **3**, 311.

Example 4 is based on work described in

Hunter, J. S., and D. A. Hsu (1975). *Simulation Model for New York Air Traffic Control Communications*, Department of Transportation, FAA Report RD-74-203, February.

For another computer simulation example, similar to Example 4, see:

Close, D. J. (1967). A design approach for solar processes, *Solar Energy*, **11**, 112.

For case studies involving the use of factorial and fractional factorial designs, see:

Hill, W. J., and R. A. Wiles (1975). Plant Experimentation, *J. Qual. Technol.*, 7, 115.
Hunter, J. S., and T. H. Naylor (1971). Experimental Designs for Computer Simulation Experiments, *Management Science*, 16, 422.
Shepler, P. R. (1975). *Fractional Factorial Plans for Diesel Engine Oil Control and for Seals*, National Conference on Fluid Power, Chicago (see also references listed therein).

QUESTIONS FOR CHAPTER 13

1. Why were no computations done in the example on cast films (Example 1)?
2. What was the experimental problem facing the chemist in the example concerning the development of a new product (Example 2)? What were the key factors in finding a solution to this problem?
3. In the filtration problem (Example 3) why would a one-variable-at-a-time approach have failed?
4. In Example 3 an alternative eight-run design that could have been used at the second stage would have been a 2^3 factorial design in variables **1, 3,** and **5**, all other variables being held at their normal levels. Compare this design to the one used; list advantages and disadvantages for each.
5. In Examples 2 and 3 can you identify the iterative cycle described in Chapter 1?
6. What are three main areas of application of fractional factorial designs? Do these areas overlap? Which of these areas relate to each of the four examples described in this chapter?

Problems for Part III

Whether or not specifically asked, the reader should always (1) plot the data in any potentially useful way, (2) state the assumptions made, (3) comment on the appropriateness of these assumptions, and (4) consider alternative analyses.

1. Find the main effects and interaction from the following data:

nickel (%)	manganese (%)	breaking strength (ft-lb)
0	1	35
3	1	46
0	2	42
3	2	40

The purpose of these four trials was to discover the effects of these two alloying elements (nickel and manganese) on the ductility of a certain product. Comment on the possible implications of these results.

2. Stating any assumptions you make, plot and analyze the following data:

temperature	concentration	catalyst	yield
−	−	−	60
+	−	−	77
−	+	−	59
+	+	−	68
−	−	+	57
+	−	+	83
−	+	+	45
+	+	+	85

The ranges for these variables are as follows:

	temperature (°F)	concentration (%)	catalyst
−	160	20	1
+	180	40	2

434

3. A chemist used a 2^4 factorial design to study the effects of temperature, pH, concentration, and agitation rate on yield (measured in grams). If the standard deviation of an individual observation is 6 grams, what is the variance of the temperature effect?

4. A metallurgical engineer is about to begin a comprehensive study to determine the effects of six variables on the strength of a certain type of alloy.

   (a) If a 2^6 factorial design were used, how many runs would be made?

   (b) If σ^2 is the experimental error variance of an individual observation, what is the variance of a main effect?

   (c) What is the usual formula for a 99% confidence interval for the main effect of variable 1?

   (d) On the basis of some previous work it is believed that $\sigma = 8000$ pounds. If the experimenter wants 99% confidence intervals for the main effects and interactions whose lengths are equal to 4000 pounds (i.e., the upper limit minus the lower limit is equal to 4000 pounds), how many replications of the 2^6 factorial design will be required?

5. Write a design matrix for a 3^2 factorial design and one for a 2^3 factorial design.

6. A study was conducted to determine the effects of individual bathers on the fecal and total coliform bacterial populations in water. The variables of interest were the time since the subject's last bath, the vigor of the subject's activity in the water, and the subject's sex. The experiments were performed in a 100-gallon polyethylene tub, using dechlorinated tap water at 38°C. The bacterial contribution of each bather was determined by subtracting the bacterial concentration measured at 15 and 30 minutes from that measured initially.

   A replicated 2^3 factorial design was used for this experiment. The data obtained are presented below. (*Note*: Because the measurement of bacterial populations in water involves a dilution technique, the experimental errors do not have constant variance. Rather, the variation increases with the value of the mean.) Perform analysis, using a logarithmic transformation of the data.

   (a) Calculate main and interaction effects on fecal and total coliform populations after 15 and 30 minutes.

   (b) Construct analysis of variance tables.

   (c) Do you think the log transformation is appropriate? Can you discover a better transformation?

Experimental variables

code	name	low level	high level
x_1	time since last bath	1 hour	24 hours
x_2	vigor of bathing activity	lethargic	vigorous
x_3	sex of bather	female	male

code	name
y_1	fecal coliform contribution after 15 minutes (organisms per 100 milliliters)
y_2	fecal coliform contribution after 30 minutes (organisms per 100 milliliters)
y_3	total coliform contribution after 15 minutes (organisms per 100 milliliters)
y_4	total coliform contribution after 30 minutes (organisms per 100 milliliters)

| | | | | responses (organisms/100 ml) | | | |
run	x_1	x_2	x_3	y_1	y_2	y_3	y_4
1	−	−	−	1	1	3	7
2	+	−	−	12	15	57	80
3	−	+	−	16	10	323	360
4	+	+	−	4	6	183	193
5	−	−	+	153	170	426	590
6	+	−	+	129	148	250	243
7	−	+	+	143	170	580	450
8	+	+	+	113	217	650	735
9	−	−	−	2	4	10	27
10	+	−	−	37	39	280	250
11	−	+	−	21	21	33	53
12	+	+	−	2	5	10	87
13	−	−	+	96	67	147	193
14	+	−	+	390	360	1470	1560
15	−	+	+	300	377	665	810
16	+	+	+	280	250	675	795

(*Source*: Glen D. Drew, "The Effects of Bathers on the Fecal Coliform, Total Coliform and Total Bacteria Density of Water," Master's Thesis, Department of Civil Engineering, Tufts University, Medford, Mass., October 1971.)

7. Using a 2^3 factorial design, a chemical engineer studied three variables (temperature, pH, and agitation rate) on the yield of a chemical reaction and obtained the following data. Using Yates's algorithm, estimate the main effects and interactions. Plot the data. Say what the results might mean, stating the assumptions you make

x_1	x_2	x_3	y
-1	-1	-1	60
$+1$	-1	-1	61
-1	$+1$	-1	54
$+1$	$+1$	-1	75
-1	-1	$+1$	58
$+1$	-1	$+1$	61
-1	$+1$	$+1$	55
$+1$	$+1$	$+1$	75

where

$$x_1 = \frac{\text{temperature} - 150°C}{10°C}$$

$$x_2 = \frac{\text{pH} - 8.0}{0.5}$$

$$x_3 = \frac{\text{agitation rate} - 30 \text{ rpm}}{5 \text{ rpm}}$$

$$y = \text{yield} \ (\% \text{ theoretical})$$

8. A chemist performed the following experiments, randomizing the order of the runs within each week. Analyze the results.

	variable			yield	
run	temperature	catalyst	pH	week 1	week 2
1	$-$	$-$	$-$	60.4	62.1
2	$+$	$-$	$-$	75.4	73.1
3	$-$	$+$	$-$	61.2	59.6
4	$+$	$+$	$-$	67.3	66.7
5	$-$	$-$	$+$	66.0	63.3
6	$+$	$-$	$+$	82.9	82.4
7	$-$	$+$	$+$	68.1	71.3
8	$+$	$+$	$+$	75.3	77.1

The minus and plus values of the three variables are as follows:

	temperature (°C)	catalyst (%)	pH
−	130	1	6.8
+	150	2	6.9

The object was to obtain increased yield. Write a report for the chemist, *asking appropriate questions* and giving advice on how the investigation might proceed.

9. Some tests were carried out on a newly designed carburetor. Four variables were studied as follows:

variable		−	+
A	tension on spring	low	high
B	air gap	narrow	open
C	size of aperture	small	large
D	rate of flow of gas	slow	rapid

The immediate object was to find the effects of these changes on the amount of unburned hydrocarbons in the engine exhaust gas. The following results were obtained:

A	B	C	D	unburned hydrocarbons
−	+	+	+	8.2
−	−	+	−	1.7
−	−	−	+	6.2
+	−	−	−	3.0
+	−	+	+	6.8
+	+	+	−	5.0
−	+	−	−	3.8
+	+	−	+	9.3

Stating any assumptions, analyze these data. It was hoped to develop a design giving lower levels of unburned hydrocarbons. Write a report, asking appropriate questions and making tentative suggestions for further work.

10. In studying a chemical reaction, a chemist performed a 2^3 factorial design and obtained the following results:

run	temperature (°C)	concentration (%)	stirring rate (rpm)	yield (%)
1	50	6	60	54
2	60	6	60	57
3	50	10	60	69
4	60	10	60	70
5	50	6	100	55
6	60	6	100	54
7	50	10	100	80
8	60	10	100	81

The chemist believes that the standard deviation of each observation is approximately unity. Explaining your assumptions, analyze these data.
(a) What are the main conclusions you draw from the data?
(b) If the chemist wants to perform two further runs and his object is to obtain high yield values, what settings of temperature, concentration, and stirring rate would you recommend?

11. Write a design matrix for a 4^2 factorial design, and draw a geometrical representation of it in the space of the variables. (*Note*: This is *not* a 2^4.)

12. (a) Why do we "block" experimental designs?
    (b) Write a 2^3 factorial design.
    (c) Write a 2^3 factorial design in four blocks of two runs each, such that main effects are not confounded with blocks.

13. A group of experimenters is investigating a certain synthetic leather fabric. Set up for them a factorial design suitable for studying four different levels of filler (5, 10, 15, 20%), two curing agents, and two methods of spreading these agents.

14. Consider the following data:

run	order in which runs were performed	variable: relative humidity (%)	variable: ambient temperature (°F)	variable: pump	response: finish imperfections
1	4	40	70	on	77
2	7	50	70	on	67
3	1	40	95	on	28
4	2	50	95	on	32
5	3	40	70	off	75
6	6	50	70	off	73
7	5	40	95	off	29
8	8	50	95	off	28

What inferences can be made about the main effects and interactions? What do you conclude about the production of defects? State assumptions.

15. A 2^3 factorial design has been run in duplicate for the purpose of studying in a research laboratory the effects of three variables (temperature, concentration, and time) on the amount of a certain metal recovered from an ore sample of a given weight. Analyze these data, and state the practical conclusions you reach.

variable: temperature	variable: concentration	variable: time	response: weight of metal recovered (grams)	
−	−	−	80	62
+	−	−	65	63
−	+	−	69	73
+	+	−	74	80
−	−	+	81	79
+	−	+	84	86
−	+	+	91	93
+	+	+	93	93

code:

	temperature	concentration	time
−	1600°C	30%	1 hour
+	1900°C	35%	3 hours

(a) Do you think that these data contain a discrepant value?
(b) Perform an analysis, using all the data, and plot residuals.
(c) Perform an approximate analysis, omitting the discrepant value.
(d) Report your findings.

16. The following data were obtained from a study dealing with the behavior of a solar energy water-heater system. A commercial, selective surface, single-cover collector with a well-insulated storage tank was theoretically modeled. Seventeen computer runs were made, using this theoretical model. The runs were made in accordance with a 2^4 factorial design with an added center point where the four "variables" were four dimensionless groups corresponding to (1) the total daily insolation, (2) the tank storage capacity, (3) the water mass flow rate through the absorber, and (4) the "intermittency" of the input solar radiation. For each of these 17 runs two response values were calculated, the collection efficiency (y_1) and the energy delivery efficiency (y_2). Analyze the following results, where L, M, and H designate low, medium, and high values, respectively, for the groups.

Case	1	2	3	4	y_1	y_2
1	H	H	H	H	41.6	100.0
2	H	H	H	L	39.9	68.6
3	H	H	L	H	51.9	89.8
4	H	H	L	L	43.0	82.2
5	H	L	H	H	39.2	100.0
6	H	L	H	L	37.5	66.0
7	H	L	L	H	50.2	86.3
8	H	L	L	L	41.3	82.0
9	L	H	H	H	41.3	100.0
10	L	H	H	L	39.7	67.7
11	L	H	L	H	52.4	84.1
12	L	H	L	L	44.9	82.1
13	L	L	H	H	38.4	100.0
14	L	L	H	L	35.0	61.7
15	L	L	L	H	51.3	83.7
16	L	L	L	L	43.5	82.0
17	M	M	M	M	45.3	88.9

The motivation for this work was the fact that the theoretical model was too complicated to allow for ready appreciation of the effects of various variables on the responses of interest. This factorial design approach provided the kind of information that was desired.

[Source: D. J. Close, A design approach for solar processes, *Solar Energy*, **11**, 112 (1967).]

17. In the situation described in problem 16 there was actually a total of 14 dimensionless groups in the original formulation of the model, but it was simplified to one that contained only four groups. If it had been decided to perform some computer "experiments" with the original model with 14 dimensionless groups, list some possible designs that might have been useful.

18. A 2^3 factorial design on radar tracking was replicated 20 times with these results:

1	2	3	$\bar{y}$
−	−	−	7.76 ± 0.53
+	−	−	10.13 ± 0.74
−	+	−	5.86 ± 0.47
+	+	−	8.76 ± 1.24
−	−	+	9.03 ± 1.12
+	−	+	14.59 ± 3.22
−	+	+	9.18 ± 1.80
+	+	+	13.04 ± 2.58

where $\bar{y}$ is the average acquisition time (in seconds), and the number following the plus or minus sign is the calculated standard deviation of $\bar{y}$. The variables were:

1. Modified (−) or standard (+) mode of operation.
2. Short (−) or long (+) range.
3. Slow (−) or fast (+) plane.

Analyze the results. Do you think a transformation of the data should be made?

19. This investigation was concerned with the study of the growth of algal blooms and the effects of phosphorus, nitrogen, and carbon on algae growth. A 2^3 factorial experiment was performed, using laboratory batch cultures of blue–green *Microcystis aeruginosa*. Carbon (as bicarbonate), phosphorus (as phosphate), and nitrogen (as nitrate) were the variable factors. The response was algal population, measured spectrophotometrically.

The levels of the experimental variables were:

variable		$-$	$+$
1	phosphorus (mg/liter)	0.06	0.30
2	nitrogen (mg/liter)	2.00	10.00
3	carbon (mg/liter)	20.0	100.0

The responses (in absorbance units) were:

y_1	algal population after 4 days' incubation
y_2	algal population after 6 days' incubation
y_3	algal population after 8 days' incubation
y_4	algal population after 10 days' incubation

The following data were obtained:

experimental design			response			
1	2	3	y_1	y_2	y_3	y_4
$-$	$-$	$-$	0.312	0.448	0.576	0.326
$+$	$-$	$-$	0.391	0.242	0.309	0.323
$-$	$+$	$-$	0.412	0.434	0.280	0.481
$+$	$+$	$-$	0.376	0.251	0.201	0.312
$-$	$-$	$+$	0.479	0.639	0.656	0.679
$+$	$-$	$+$	0.481	0.583	0.631	0.648
$-$	$+$	$+$	0.465	0.657	0.736	0.680
$+$	$+$	$+$	0.451	0.768	0.814	0.799

Given that $s^2 = 0.003821$ was an independent estimate of variance based on 18 degrees of freedom and appropriate for all four responses, analyze the data from this factorial design.

(Source: R. W. Mann, "Significance of Phosphorus, Nitrogen, and Carbon as Regulators of the Growth of the Alga Microcystis aeruginosa," Master's Thesis, Department of Civil Engineering, Tufts University, Medford, Mass., June 1971.)

20. To reduce the amount of a certain pollutant, a waste stream of a small plastic molding factory has to be treated before it is discharged. State laws stipulate that the daily average of this pollutant cannot exceed 10 pounds. The following 11 experiments were performed to determine the best way to treat this waste stream. Analyze these data, and list the conclusions you reach.

order in which experiments were performed	chemical brand	temperature (°F)	stirring	pollutant (lb/day)
5	A	72	none	5
6	B	72	none	30
1	A	100	none	6
9	B	100	none	33
11	A	72	fast	4
4	B	72	fast	3
2	A	100	fast	5
7	B	100	fast	4
3	AB*	86	intermediate	7
8	AB	86	intermediate	4
10	AB	86	intermediate	3

* AB denotes a 50–50 mixture of both brands.

21. An experiment is run on an operating chemical process in which four variables are changed in accordance with a randomized factorial plan:

	variable	−	+
1	concentration of catalyst (%)	5	7
2	concentration of NaOH (%)	40	45
3	agitation speed (rpm)	10	20
4	temperature (°F)	150	180

order of running	1	2	3	4	impurity
2	−	−	−	−	0.38
6	+	−	−	−	0.40
12	−	+	−	−	0.27
4	+	+	−	−	0.30
1	−	−	+	−	0.58
7	+	−	+	−	0.56
14	−	+	+	−	0.30
3	+	+	+	−	0.32
8	−	−	−	+	0.59
10	+	−	−	+	0.62
15	−	+	−	+	0.53
11	+	+	−	+	0.50
16	−	−	+	+	0.79
9	+	−	+	+	0.75
5	−	+	+	+	0.53
13	+	+	+	+	0.54

(a) Make a table of plus and minus signs from which the various effects and interactions can be calculated.

(b) Calculate the effects, using Yates's algorithm.

(c) Assuming three-factor and higher order interactions to be zero, compute an estimate of the error variance.

(d) Make normal plots.

(e) Write a report, setting out the conclusions to be drawn from the experiment, and make recommendations. (Assume that the present conditions of operation are as follows: $1 = 5\%$, $2 = 40\%$, $3 = 10$ rpm, $4 = 180°F$.)

(f) Show pictorially what your conclusions mean.

22. Estimate effects for the following 2^{5-1} fractional factorial design, stating your assumptions.

A	B	C	D	E	observation y
−	−	−	−	+	14.8
+	−	−	−	−	14.5
−	+	−	−	−	18.1
+	+	−	−	+	19.4
−	−	+	−	−	18.4
+	−	+	−	+	15.7
−	+	+	−	+	27.3
+	+	+	−	−	28.2
−	−	−	+	−	16.0
+	−	−	+	+	15.1
−	+	−	+	+	18.9
+	+	−	+	−	22.0
−	−	+	+	+	19.8
+	−	+	+	−	18.9
−	+	+	+	−	29.9
+	+	+	+	+	27.4

Plot the effects on probability paper, draw tentative conclusions, and verify by plotting residuals.

23. Consider a 2^{8-4} fractional factorial design.

(a) How many variables does this design have?

(b) How many runs are involved in this design?

(c) How many levels are used for each of the variables?

(d) How many (independent) generators are there for this design?

(e) How many words are there in the defining relation (counting I)?

24. Repeat problem 23 for a 2^{9-4} fractional factorial design.

25. Repeat problem 23 for a 2^{6-1} fractional factorial design.

26. Write generators for a 2_{IV}^{8-4} design and for a 2_{IV}^{6-2} design. Does a 2_V^{6-2} design exist?

27. (a) Construct a 2^{6-2} fractional factorial design with as high a resolution as possible.
    (b) What are the generators of your design?
    (c) What is the defining relation of your design?
    (d) What is confounded with the main effect of variable 3 in your design?
    (e) What is confounded with the two-factor interaction 12 in your design?

28. Derive generators for a 2_V^{11-4} design, and show how, knowing these generators, you could set out the design.

29. Analyze the following experimental results from a welding study. Do you notice any relationship among y_1, y_2, and y_3?

order of runs	x_1	x_2	x_3	x_4	x_5	y_1 (V)	y_2 (A)	y_3 heat input (W)
12	−1	−1	−1	−1	1	23.43	141.61	3318
1	1	−1	−1	−1	−1	25.70	161.13	4141
2	−1	1	−1	−1	−1	27.75	136.54	3790
6	1	1	−1	−1	1	31.60	128.52	4061
15	−1	−1	1	−1	−1	23.57	145.55	3431
8	1	−1	1	−1	1	27.68	123.75	3425
7	−1	1	1	−1	1	28.76	121.93	3507
4	1	1	1	−1	−1	31.82	120.37	3765
11	−1	−1	−1	1	−1	27.09	95.24	2580
14	1	−1	−1	1	1	31.28	78.31	2450
3	−1	1	−1	1	1	31.20	74.34	2319
16	1	1	−1	1	−1	33.42	91.76	3067
13	−1	−1	1	1	1	29.51	65.23	1925
10	1	−1	1	1	−1	31.35	78.65	2466
5	−1	1	1	1	−1	31.16	79.76	2485
9	1	1	1	1	1	33.65	72.80	2450

The variables and their settings were as follows:

variable		−1	1
x_1	open-circuit voltage (V)	31	34
x_2	slope	11	6
x_3	electrode melt-off rate (ipm)	162	137
x_4	electrode diameter (in.)	0.045	0.035
x_5	electrode extension (in.)	$\frac{3}{8}$	$\frac{5}{8}$

[*Source*: D. A. J. Stegner, S. M. Wu, and N. R. Braton, Prediction of heat input for welding, *Welding J. Res. Supple.*, 1 (March 1967).]

30. (*a*) Write a design of resolution III in seven variables and eight runs.
    (*b*) From (*a*), obtain a design for eight variables in 16 runs of resolution IV, arranged in eight blocks of two.
    (*c*) How could this design be used to eliminate the effects of time trends?
    (*d*) What are the generators for your design?
    (*e*) If no blocking is to be used, how could this design be employed to study the main effects and interactions of three principal variables and the main effects of eight minor variables not expected to interact with the principal variables and with each other?

31. (*a*) Write an eight-run, two-level design of resolution III in seven variables, say what kind of fraction it is, and give its generators and those of other members of this family.
    (*b*) How may members of this family of designs be combined (i) to isolate the main effect and interactions of a particular factor, and (ii) to separate all main effects from two-factor interactions?
    (*c*) Use such an arrangement to generate a 2_{IV}^{8-4} design, and block it in eight blocks of two so that main effects are unconfounded with blocks.

32. Construct a 2^{7-1} fractional factorial design. Show how the design may be divided into eight blocks of eight runs each so that no main effect or two-factor interaction is confounded with any block effect.

33. Suppose that a chemical engineer wants to study the effects of seven variables on the yield of a chemical reaction and is willing to make 32 runs. He elects to perform a 2^{7-2} fractional factorial design with the generators $I = 1237$ and $I = 126$. The main effect of variable 1 in this design is confounded with what other:
    (*a*) Main effect(s)?
    (*b*) Two-factor interaction(s)?
    (*c*) three-factor interaction(s)?
    (*d*) Four-factor interaction(s)?
    (*e*) Five-factor interaction(s)?
    (If the main effect of variable 1 is unconfounded with any particular item, write "none.")
    (*f*) Suggest an alternative plan that produces less confounding of low-order effects.

34. A consulting firm engaged in road-building work is asked by one of its clients to carry out an experimental study to determine the effects of six variables on the physical properties of a certain kind of asphalt. Call these variables A, B, C, D, E, and F.
    (*a*) If a full two-level factorial design was used, how many runs would be made?

(b) Write a two-level resolution IV fractional factorial design requiring only 16 runs.

(c) Write a set of generators for this design.

(d) What is the defining relation for this design?

(e) In your design, what effects are confounded with the main effect of variable **A**? With the two-factor interaction **BD**?

35. Suppose that, after you have set up a 2_{III}^{14-10} main effect plan for an electrical engineer, he says it is essential to split the design into two blocks of equal size because the tests must be done on 2 days and he fears a day-to-day difference in the results. Make recommendations.

36. An experimenter performs a 2^{5-2} fractional factorial design. The generators for this first design are **I = 1234** and **I = 135**. After analyzing the data from the first design, he decides to perform a second 2^{5-2} fractional factorial exactly the same as the first except that the signs are changed in column **3** of the design matrix.

(a) How many runs does the first design contain?

(b) Give a set of generators for the second design.

(c) What is the resolution of the second design?

(d) What is the defining relation of the combined design? (The combined design is the first and second designs considered together as a single design.)

(e) What is the resolution of the combined design?

(f) The combined design is a 2^{k-p} fractional factorial design where k and p have what values, that is, $k = ?$ and $p = ?$

(g) Give one possible reason why the experimenter might have chosen the second design in the way he did.

37. A mechanical engineer used the following design:

1	2	3	4	5
−	−	−	+	−
+	−	−	−	−
−	+	−	−	+
+	+	−	+	+
−	−	+	−	+
+	−	+	+	+
−	+	+	+	−
+	+	+	−	−

(a) Write a set of generators for this design.

(b) Is this set unique? If so, explain why. If not, write a different set of generators for this design.

38. (a) Write a 2^{6-1} fractional factorial design of the highest possible resolution for variables **1, 2, 3, 4, 5**, and **6**. Start with a full factorial in the first five variables.

   (b) Write the generator, the defining relation, and the resolution for this design.

   (c) Suppose that you analyzed the data from this design using Yates's algorithm, making the calculation as if the design were a 2^5 factorial with variables **1, 2, 3, 4**, and **5**. What effects do the following contrasts estimate: l_2, l_{23}, l_{234}, l_{2345}?

39. Design an eight-run fractional factorial design for an experimenter with the following five variables: temperature, concentration, pH, agitation rate, and catalyst type (A or B). He tells you he is particularly concerned about the two-factor interactions between temperature and concentration, and between catalyst type and temperature. He would like a design, if it is possible to construct one, such that these two two-factor interactions are unconfounded with main effects and unconfounded with one another.

PART IV

Building Models
and Using Them

In the first three parts of this book we studied some simple statistical methods that have very wide applicability. Scientists and engineers in all fields are involved in investigations to compare two treatments (Part I), to compare more than two treatments (Part II), or to measure the effects of variables (Part III).

Building on an understanding of these basic principles, we now provide an introduction to other statistical methods appropriate in situations where experimenters' purposes are somewhat more sophisticated. In this fourth part of the book we give elementary accounts of the method of least squares, or regression analysis (Chapter 14), response surface methods (Chapter 15), mechanistic modeling (Chapter 16), quality control, components of variance, propagation of errors (Chapter 17), and time series analysis (Chapter 18). All these topics are related to the central problem of how data can be used to build models and how these models can then be utilized. We discuss each of them only briefly; however, references are provided for those needing more information, and we hope that this part of the book will serve as a springboard for further study.

CHAPTER 14

Simple Modeling
with Least Squares
(Regression Analysis)

In this chapter some simple modeling problems provide an introduction to the use of the method of *least squares* (also called *regression analysis*).

14.1. ONE-PARAMETER MODEL (STRAIGHT LINE THROUGH THE ORIGIN): AEROSOL EXAMPLE

This first example involves fitting to data a straight line passing through the origin. Although elementary, the example illustrates all the basic principles of least squares.

In studying the decay of an aerosol spray, experimenters obtained the results shown in Table 14.1, where x is the age in minutes of the aerosol and y is its observed dispersion at that time. (Dispersion was measured as the reciprocal of the number of particles in a unit volume.) The $n = 9$ experiments were run in random order.

Physical considerations suggested a proportional relationship $\eta = \beta x$ between η, the mean value of the dispersion, and x, the elapsed time since the spray was initiated. Thus the relationship between η and x should be described by a straight line through the origin.

Experience suggests that x is essentially without error, while the observed response has an error $y - \eta = \epsilon$ with a variance that is roughly constant over the range studied.

Thus for the uth experimental run it is reasonable to entertain the model

$$y_u = \beta x_u + \epsilon_u, \qquad u = 1, 2, \ldots, n \qquad (14.1)$$

TABLE 14.1. Aerosol data

observation number u	order in which experiments were performed	age x_u	dispersion y_u
1	6	8	6.16
2	9	22	9.88
3	2	35	14.35
4	8	40	24.06
5	4	57	30.34
6	5	73	32.17
7	7	78	42.18
8	1	87	43.23
9	3	98	48.76

where β is the constant of proportionality and the ϵ_u are random, independent experimental errors with zero mean and constant variance. Inspection of the plot of the data shown in Figure 14.1 shows no immediate reason to question this model. The unknown constant of proportionality β is variously termed a parameter, a constant, and a coefficient. It measures the slope of the line.

Equation 14.1 is an elementary example of what is called a *regression equation*. In this equation the response or output variable y is sometimes called the *dependent variable*. The input variable x is sometimes called an *independent variable* or a *regressor*.

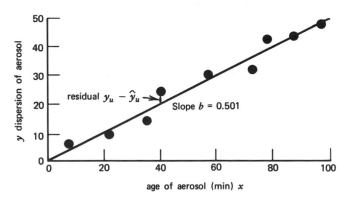

FIGURE 14.1. Plot of data and fitted least squares line, aerosol example.

Sum of Squares S

The method of least squares takes the "best fitting" model to be the one that comes closest to the data in the sense of minimizing the quantity

$$S = \sum_{u=1}^{n} (y_u - \eta_u)^2 \qquad (14.2)$$

which is the sum of squared discrepancies between the observed values y_u and the values given by the model $\eta_u = \beta x_u$. By substituting $\eta_u = \beta x_u$ in Equation 14.2, we see that S is a function of the parameter β. To obtain the best fitting straight line we must find the value of β that minimizes

$$S = S(\beta) = \sum_{u=1}^{n} (y_u - \beta x_u)^2 \qquad (14.3)$$

When there is no danger of ambiguity we shall sometimes omit subscripts to simplify notation. For example, we may write Equation 14.3 as

$$S(\beta) = \sum (y - \beta x)^2$$

Plotting S as a Function of β

One way to find the value of the slope parameter β that will make S as small as possible is to plot S as a function of β. In Figure 14.2, $S(\beta)$ has been calculated for $\beta = 0.26, 0.30, 0.34, \ldots, 0.74$, and a smooth curve drawn through the points. [For illustration the calculation of $S(0.30)$ is given in Table 14.2.]

TABLE 14.2. Calculation of sum of squares S with $\beta = 0.30$

y_u	$\beta x_u = 0.3x_u$	$y_u - \beta x_u$	$(y_u - \beta x_u)^2$
6.16	2.4	3.76	14.1376
9.88	6.6	3.28	10.7584
14.35	10.5	3.85	14.8225
24.06	12.0	12.06	145.4436
30.34	17.1	13.24	175.2976
32.17	21.9	10.27	105.4729
42.18	23.4	18.78	352.6884
43.23	26.1	17.13	293.4369
48.76	29.4	19.36	374.8096
		sum =	1486.8675 = $S(0.30)$

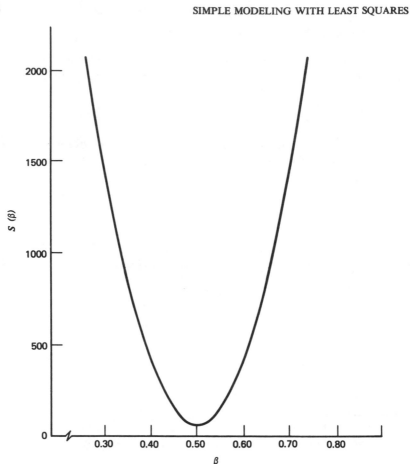

FIGURE 14.2. Sum of squares $S(\beta)$ as a function of β, aerosol example.

The curve is a parabola with the smallest value of S occurring when the parameter is equal to 0.501. This value is denoted by b and is called the *least squares estimate* of β. Thus for this example the least squares estimate is $b = 0.501$. The minimum sum of squares associated with this value is

$$S_{\min} = S(b) = \sum_{u=1}^{n} (y_u - bx_u)^2 = \sum_{u=1}^{n} (y_u - 0.501x_u)^2 = 64.669 \qquad (14.4)$$

Residuals

Having obtained the least squares estimate b of the unknown parameter β, we can calculate an estimated response $\hat{y}_u = bx_u$ for each x_u. (This is the value obtained from the fitted line in Figure 14.1). We may now compare $\hat{y}_u$ with the corresponding observed value y_u. The quantities $y_u - \hat{y}_u$ are the *residuals*.

TABLE 14.3. Calculation of the estimated values $\hat{y}_u$ and residuals $y_u - \hat{y}_u$ for the best fitting line

observation number u	age x_u	dispersion y_u	estimated dispersion $\hat{y}_u = 0.50098x_u$	residual $y_u - \hat{y}_u$
1	8	6.16	4.0079	2.1521
2	22	9.88	11.0216	−1.1416
3	35	14.35	17.5344	−3.1844
4	40	24.06	20.0393	4.0207
5	57	30.34	28.5560	1.7840
6	73	32.17	36.5718	−4.4018
7	78	42.18	39.0767	3.1033
8	87	43.23	43.5855	−0.3555
9	98	48.76	49.0963	−0.3363
		$\sum y_u^2 = 8901.31$	$\sum \hat{y}_u^2 = 8836.64$	$\sum (y_u - \hat{y})^2 = 64.669$

The calculations are illustrated in Table 14.3. Note that the sum of the squares of the residuals S_R is also the minimum value of the sum of squares,

$$S_R = S_{\min} = S(b) = \sum_{u=1}^{n} (y_u - \hat{y}_u)^2 = 64.669 \qquad (14.5)$$

As always, it is important to examine the residuals both individually and jointly for possible inadequacies in the model.

Plot of the Least Squares Line

The residuals from the least squares line in Figure 14.1 appear to exhibit no suspicious peculiarities;* we therefore tentatively adopt the proportional model, estimated by

$$\hat{y} = bx = 0.501x \qquad (14.6)$$

and represented by the straight line through the origin drawn in Figure 14.1, as supplying a useful summary of the data *over the range of ages x_u considered.*

Least Squares Estimation Using the Normal Equation

To obtain the estimate b it is not necessary to plot $S(\beta)$; instead solution of what is called the *normal equation* provides a readily generalized method for direct and exact *calculation*. This uses the fact that the vector of residuals, obtained when the least squares estimate is used, has the unique property of being normal (at right angles) to the vector of x values.

---

* The reader may wish to confirm, in particular, that they show no evidence of dependence on the order in which the runs were made.

The normal equation expressing this fact (see Appendix 6B) is

$$\sum (y - \hat{y})x = 0 \tag{14.7}$$

or, equivalently, since $\hat{y} = bx$,

$$\sum (y - bx)x = 0 \tag{14.8}$$

that is,

$$\sum yx - b \sum x^2 = 0 \tag{14.9}$$

The normal equation may thus be solved to give an explicit expression for the least squares estimate:

$$b = \frac{\sum xy}{\sum x^2} \tag{14.10}$$

For this example, using the data in Table 14.1, we have

$$\sum_{u=1}^{n} x_u y_u = (8)(6.16) + (22)(9.88) + (35)(14.35) + (40)(24.06) + (57)(30.34)$$

$$+ (73)(32.17) + (78)(42.18) + (87)(43.23) + (98)(48.76)$$

$$= 17,638.61$$

and

$$\sum_{u=1}^{n} x_u^2 = 8^2 + (22)^2 + (35)^2 + (40)^2 + (57)^2 + (73)^2 + (78)^2 + (87)^2$$

$$+ (98)^2 = 35,208$$

so that

$$b = \frac{17,638.61}{35,208} = 0.500983 \tag{14.11}$$

As before, to three-decimal accuracy, $b = 0.501$. The fitted model (or regression equation, as it is sometimes called) is thus

$$\hat{y}_u = 0.501 x_u \tag{14.12}$$

which is the equation of the fitted straight line of Figure 14.1.

Exercise 14.1. Weighed amounts of zinc are dissolved in 100-milliliter portions of diluted acid. The solutions are presented to a chemist for analysis, yielding the following data:

| milligrams of zinc dissolved (x) | 0.102 | 0.213 | 0.306 | 0.407 | 0.511 | 0.602 |
| milligrams of zinc found (y) | 0.097 | 0.207 | 0.300 | 0.393 | 0.502 | 0.613 |

Fit the model $y_u = \beta x_u + \epsilon_u$ to the data. Speculate on the meaning of your result.

An Estimate of Experimental Error Variance σ^2

On the supposition that the proportional model (Equation 14.1) is adequate, an estimate s^2 of the experimental error variance σ^2 may be obtained by dividing the residual sum of squares S_R by its number of degrees of freedom v. In general, v will equal the number of observations less the number of parameters estimated. Since in this example only a single parameter β is estimated, the divisor of S_R is $n - 1 = 9 - 1 = 8$. Thus an estimate of σ^2 is

$$s^2 = \frac{S_R}{n - 1} = \frac{64.6692}{8} = 8.0837 \tag{14.13}$$

and consequently the estimated standard deviation is

$$s = 2.8432 \tag{14.14}$$

Precision of the Estimate

Again, on the assumption that the proportional model (Equation 14.1) is adequate, the estimated variance of b is

$$\hat{V}(b) = \frac{s^2}{\sum_{u=1}^{n} x_u^2} = \frac{8.0837}{35208} = 0.0002296 \tag{14.15}$$

(A derivation of this formula is given at the end of this section.) Thus the standard error of b is $\sqrt{0.0002296} = 0.01515$, and

$$b \pm \text{S.E.}(b) = 0.501 \pm 0.015 \tag{14.16}$$

A Confidence Interval for β

Any hypothesis that the true value of the parameter was equal to some specific value β_* could be tested by referring the quantity

$$t = \frac{b - \beta_*}{\text{S.E.}(b)} \tag{14.17}$$

to the table of Student's t with $n - 1 = 8$ degrees of freedom.

The confidence interval for β is bounded, as always, by the values that just produce a significant result at the desired probability level. Thus the $1 - \alpha$ confidence limits for β are given by

$$b \pm [t_{\alpha/2} \times \text{S.E.}(b)] \tag{14.18}$$

For example, the 95% limits are

$$0.501 \pm (2.306 \times 0.015)$$

or

$$0.501 \pm 0.035$$

The 95% confidence interval thus extends from 0.466 to 0.536.

Analysis of Variance

In Table 14.3 each of the columns of elements y_u, $\hat{y}_u$, and $y_u - \hat{y}_u$ is shown together with its corresponding sum of squares. For the linear least squares problems considered in this chapter identities exist among the sums of squares and their degrees of freedom as follows:

$$\sum y_u^2 = \sum \hat{y}_u^2 + \sum (y_u - \hat{y}_u)^2 \qquad (14.19)$$

$$n \;\; = \;\; p \;\; + \;\; n - p \qquad (14.20)$$

where p is the number of parameters estimated by least squares. For the aerosol problem $p = 1$, and the identities yield the analysis of variance shown in Table 14.4, appropriate for testing the null hypothesis that $\beta_* = 0$. If this question were of interest, we could refer the ratio $8836.64/8.08 = 1094$ of mean squares to the tables of the F distribution with one and eight degrees of freedom. The ratio for this example is, of course, overwhelmingly significant.

This test is exactly equivalent to the test using Student's t which was mentioned earlier. If $\beta_* = 0$ in Equation 14.17,

$$t = \frac{b}{\text{S.E.}(b)} = \frac{0.50098}{0.01515} = 33.07 = \sqrt{1094} \qquad (14.21)$$

When there is only a single degree of freedom, as has previously been noted, $t^2 = F$, and the tests yield identical probabilities.

TABLE 14.4. **An analysis of variance, aerosol example**

source	sum of squares	degrees of freedom	mean square
model	$S_M = 8836.64$	1	$\left.\begin{array}{l}8836.64\\ 8.08\end{array}\right\}$ ratio $= 1094$
residual	$S_R = 64.67$	8	
total	$S_T = 8901.31$	9	

An analysis of variance appropriate to any hypothesis $\beta = \beta_*$, with β_* not necessarily zero, is based on the identity

$$\sum (y_u - \beta_* x_u)^2 = \sum (\hat{y}_u - \beta_* x_u)^2 + \sum (y_u - \hat{y}_u)^2$$

with the same allocation of degrees of freedom as before.

Derivation of Equation 14.15

To derive Equation 14.15, rewrite Equation 14.10 as

$$b = \left(\frac{x_1}{\sum x^2}\right) y_1 + \left(\frac{x_2}{\sum x^2}\right) y_2 + \cdots + \left(\frac{x_n}{\sum x^2}\right) y_n \tag{14.22}$$

where each summation $\sum$ goes from $u = 1$ to n. It is thus seen that b is a linear combination of the observations, that is to say, it is of the form

$$b = a_1 y_1 + a_2 y_2 + \cdots + a_n y_n \tag{14.23}$$

where $a_u = x_u / \sum x^2$ is a constant for each u.

Assuming that the variance of each observation is equal to σ^2 and that the errors are uncorrelated, we can then use Equation 3A.9 for the variance of a linear combination of observations to compute the variance of the estimate b:

$$V(b) = (a_1^2 + a_2^2 + \cdots + a_n^2)\sigma^2 \tag{14.24}$$

$$= \left[\left(\frac{x_1}{\sum x^2}\right)^2 + \left(\frac{x_2}{\sum x^2}\right)^2 + \cdots + \left(\frac{x_n}{\sum x_n^2}\right)^2\right]\sigma^2 \tag{14.25}$$

$$= \frac{\sum x^2}{(\sum x^2)^2}\sigma^2 = \frac{\sigma^2}{\sum x^2} \tag{14.26}$$

This formula contains the (unknown) value of the experimental error variance σ^2. However, if the model is adequate, we can take the residual mean square s^2 as an estimate of σ^2. [If the model is inadequate, the residual mean square will tend to be inflated. When genuine replicate runs are available to estimate σ^2, they may be used to check the adequacy of the model in a manner to be discussed in the next section (see Table 14.7). In the aerosol example, although there were no replicate runs, there was no apparent reason to question the adequacy of the model, so we proceed assuming $s^2 = 8.08$ to be a valid estimate of σ^2.] On substituting s^2 for σ^2 in Equation 14.26, we obtain

$$\hat{V}(b) = \frac{s^2}{\sum x^2} \tag{14.27}$$

which is Equation 14.15.

Exercise 14.2.   Add five units to each observation y_u in Table 14.1. Supposing these values to be the actual data on the dispersion of the aerosol, carry out a complete analysis for the model $\eta = \beta x$. Can you detect inadequacy of this model from inspection of residuals?

14.2.   TWO-PARAMETER MODEL: IMPURITY EXAMPLE

The procedures used for the one-parameter model $\eta = \beta x$ can be readily extended. Consider the two-parameter model

$$\eta = \beta_1 x_1 + \beta_2 x_2 \tag{14.28}$$

For clarity we illustrate with the small data set of Table 14.5. It was hypothesized that the mean value η of the initial rate of formation of a chemical impurity causing discoloration was linearly dependent on (1) the concentration x_1 of monomer and (2) the concentration x_2 of dimer, the rate being zero when both components were absent.

Sum of Squares Surface

The sum of squared discrepancies between observations and fitted equation is

$$S = S(\boldsymbol{\beta}) = \sum (y - \beta_1 x_1 - \beta_2 x_2)^2 \tag{14.29}$$

where $\boldsymbol{\beta}$ represents β_1 and β_2.

For example, for the trial values $\beta_1 = 1$ and $\beta_2 = 7$,

$$S(1, 7) = \sum (y - 1x_1 - 7x_2)^2 = 1.9022$$

TABLE 14.5.   Impurity data

observation number	order in which experiments were performed	concentration of monomer x_1	concentration of dimer x_2	initial rate of formation of impurity y
1	3	0.34	0.73	5.75
2	6	0.34	0.73	4.79
3	1	0.58	0.69	5.44
4	4	1.26	0.97	9.09
5	2	1.26	0.97	8.59
6	5	1.82	0.46	5.09

By repeating the calculation for other trial values of the parameters β_1 and β_2, one can construct Figure 14.3, which is a plot of S versus β_1 and β_2. Now that there are two parameters, the sum of squares S is represented, not by a curved line, but by a curved surface, plotted over the space of the parameters. The minimum value on this surface ($S_{min} = 1.3308$) occurs at the point defining the least squares values $b_1 = 1.207$ and $b_2 = 7.123$. Figure 14.3a shows the sum of squares as a bowl-shaped surface, and Figure 14.3b shows a contour representation of the same surface. Although in principle one could obtain the least squares estimates from a grid of trial values, it is easier, as before, to solve the normal equations.

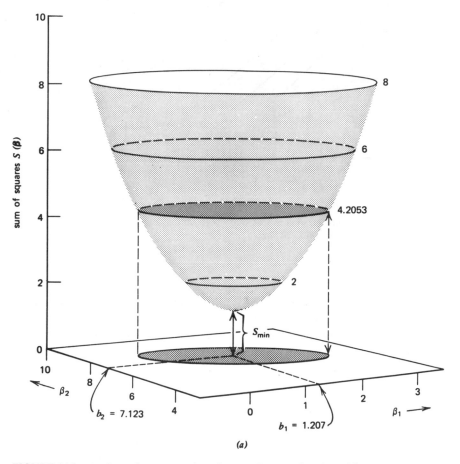

(a)

FIGURE 14.3. (a) Sum of squares surface showing $S(\boldsymbol{\beta})$ as a function of β_1 and β_2, impurity example.

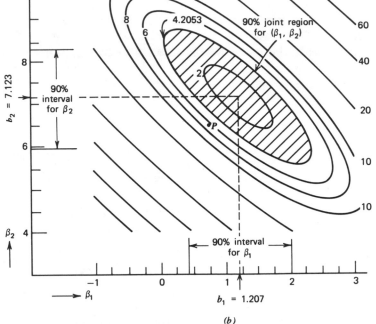

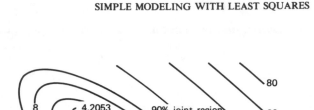

FIGURE 14.3. (*b*) Sum of squares contours, with 90% confidence region and individual (marginal) 90% confidence intervals shown.

Normal Equations

For this model there are two regressors x_1 and x_2. In general, least square estimates are always such that the vector of residuals from the fitted least squares equation is normal to each of the regressor vectors. For this example, therefore, the normal equations are

$$\sum (y - \hat{y})x_1 = 0, \qquad \sum (y - \hat{y})x_2 = 0 \qquad (14.30)$$

that is,

$$\sum (y - b_1 x_1 - b_2 x_2)x_1 = 0, \qquad \sum (y - b_1 x_1 - b_2 x_2)x_2 = 0$$

On multiplying out these two expressions, we obtain the normal equations in the form

$$\sum yx_1 - b_1 \sum x_1^2 - b_2 \sum x_1 x_2 = 0, \qquad \sum yx_2 - b_1 \sum x_1 x_2 - b_2 \sum x_2^2 = 0$$

The pattern becomes clear if we write

$$[ij] \quad \text{for} \quad \sum x_i x_j \quad \text{and} \quad [yi] \quad \text{for} \quad \sum y x_i$$

whereupon the equations may be written as

$$[11]b_1 + [12]b_2 = [y1]$$
$$[12]b_1 + [22]b_2 = [y2] \tag{14.31}$$

Arguing similarly, for three regressors x_1, x_2, x_3 we would have

$$[11]b_1 + [12]b_2 + [13]b_3 = [y1]$$
$$[12]b_1 + [22]b_2 + [23]b_3 = [y2] \tag{14.32}$$
$$[13]b_1 + [23]b_2 + [33]b_3 = [y3]$$

The pattern is readily extended for any number of regressors.

Substituting data values in the summations in Equation 14.31, we obtain

$$7.055b_1 + 4.178b_2 = 38.279$$
$$4.178b_1 + 3.635b_2 = 30.939 \tag{14.33}$$

Solving these linear simultaneous equations for b_1 and b_2 gives the least squares estimates $b_1 = 1.207$ and $b_2 = 7.123$. Thus the fitted equation is

$$\hat{y} = 1.207x_1 + 7.123x_2 \tag{14.34}$$

Residuals

Table 14.6 shows the estimated values $\hat{y}$ and the residuals $y - \hat{y}$. Before proceeding, it is necessary to inspect the residuals for possible indications of model inadequacy. With so small a data set only gross discrepancies would be detectable and none such appear. Nevertheless, we take this opportunity of illustrating a further technique for detecting possible lack of fit of the model.

TABLE 14.6. Calculation of the estimated values $\hat{y}$ and the residuals $y - \hat{y}$ for the impurity data

u	x_{1u}	x_{2u}	$1.207x_{1u}$	$7.123x_{2u}$	$\hat{y}_u$	y_u	$y_u - \hat{y}_u$
1	0.34	0.73	0.410	5.200	5.61	5.75	0.14
2	0.34	0.73	0.410	5.200	5.61	4.79	−0.82
3	0.58	0.69	0.700	4.915	5.62	5.44	−0.18
4	1.26	0.97	1.521	6.909	8.43	9.09	0.66
5	1.26	0.97	1.521	6.909	8.43	8.59	0.16
6	1.82	0.46	2.197	3.277	5.47	5.09	−0.38
				sum of squares =	266.59	267.92	1.33

*Analysis of Variance to Check Adequacy of Fit When Some
Runs Are Replicated*

The following analysis is appropriate when some of the experimental runs
have been genuinely replicated. Recall that we mean by "genuine replicates"
repetitions which are subject to all the sources of error that affect runs made
at different experimental conditions. In Table 14.6 runs 1 and 2 are replicates,
and so are 4 and 5.

At the bottom of Table 14.6 are shown the sums of squares for the observed
values y_u, the estimated values $\hat{y}_u$, and the residuals $y_u - \hat{y}_u$. These provide
the basis for an analysis of variance according to the following relationships:

$$\text{sum of squares} \begin{cases} \sum y^2 = \sum \hat{y}^2 + \sum (y - \hat{y})^2 \\ 267.92 = 266.59 + \qquad 1.33 \end{cases} \tag{14.35}$$

$$\text{degrees of freedom} \begin{cases} n = p + n - p \\ 6 = 2 + \qquad 4 \end{cases} \tag{14.36}$$

Now the sum of squares associated with the replicate runs is

$$\frac{(5.75 - 4.79)^2}{2} + \frac{(9.09 - 8.59)^2}{2} = 0.59 = S_E \tag{14.37}$$

This sum of squares, which has $1 + 1 = 2$ degrees of freedom, is part of the
residual sum of squares and is a measure of pure experimental error. The
remaining part of the residual sum of squares is

$$S_L = S_R - S_E = 1.33 - 0.59 = 0.74 \tag{14.38}$$

The quantity S_L measures experimental error *plus* any contribution from
possible lack of fit of the model. Thus a comparison of mean squares derived
from S_E and S_L can serve to check lack of fit. An appropriate analysis of
variance for this purpose is presented in Table 14.7.

TABLE 14.7.   Analysis of variance for impurity data

source		sum of squares	degrees of freedom	mean square
model		$S_M = 266.59$	2	
residual	$\begin{cases} \text{lack of fit} \\ \text{pure error} \end{cases}$	$S_R = 1.33 \begin{cases} S_L = 0.74 \\ S_E = 0.59 \end{cases}$	$4 \begin{cases} 2 \\ 2 \end{cases}$	$0.33 \begin{cases} 0.37 \\ 0.30 \end{cases}$ ratio = 1.2
total		$S_T = 267.92$	6	

In this example the close agreement between the two mean squares gives no reason to suspect lack of fit, but with so few degrees of freedom this assessment is extremely insensitive. A formal determination of the significance level may be obtained by referring the mean square ratio to the appropriate F distribution. In this instance mean square ratios as great or greater than 1.2 can be expected about 45% of the time, so the result does not approach significance. Whether the data set is large or small, a formal analysis of this kind does not relieve the investigator from the duty of appropriately plotting and carefully considering individual residuals. For the present example we shall proceed, assuming that we have an adequate model.

Plot of the Fitted Surface

The fitted equation $\hat{y} = 1.207x_1 + 7.123x_2$ is represented by a plane passing through the origin as shown in Figure 14.4a. The slope of this plane in the x_1 direction is $b_1 = 1.207$, and in the x_2 direction is $b_2 = 7.123$. The contours of the fitted plane are shown in Figure 14.4b, together with the actual data values. We see that the six points fall in only three principal locations in the (x_1, x_2)-space, so it is now even more apparent that the data are quite unsuitable to either confirm or refute the adequacy of this particular model. However, for purposes of illustration we suppose the model to be appropriate.

Estimate of σ^2

On the assumption that the model is adequate, an estimate of the error variance σ^2 is

$$s^2 = \frac{S_R}{n - p} = \frac{1.3308}{4} = 0.3309 \tag{14.39}$$

Thus the estimated standard deviation for this experiment is $s = \sqrt{0.3309} = 0.5752$.

Precision of Estimates

The formulas for the estimated variances of b_1 and b_2, for this particular model representing a plane passing through the origin, may be written

$$\hat{V}(b_1) = \frac{1}{(1 - \rho^2)\sum x_1^2}s^2 = \frac{1}{0.3193}\frac{0.3309}{7.055} = 0.1469$$

$$\hat{V}(b_2) = \frac{1}{(1 - \rho^2)\sum x_2^2}s^2 = \frac{1}{0.3193}\frac{0.3309}{3.635} = 0.2851$$

$$\tag{14.40}$$

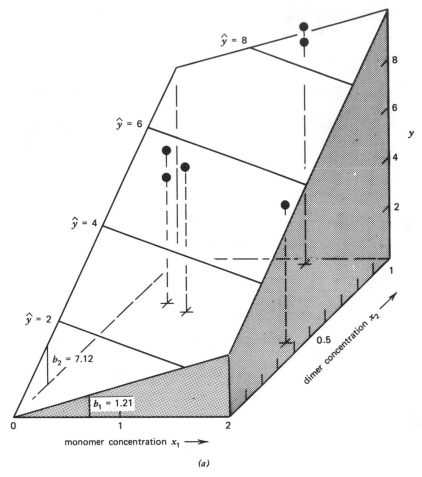

FIGURE 14.4. (*a*) Fitted plane $\hat{y} = 1.21x_1 + 7.12x_2$, impurity example. Three-dimensional diagram.

where ρ, which measures the correlation between the estimates b_1 and b_2, is

$$\rho = \frac{-\sum x_1 x_2}{\sqrt{\sum x_1^2 \sum x_2^2}} = -0.8250 \tag{14.41}$$

Thus the standard errors of b_1 and b_2 are

$$\text{S.E.}(b_1) = 0.383, \qquad \text{S.E.}(b_2) = 0.534$$

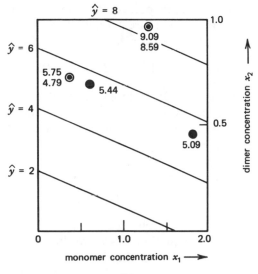

FIGURE 14.4. (*b*) Fitted plane $\hat{y} = 1.21x_1 + 7.12x_2$, impurity example. Contour representation.

The best fitting equation is then conveniently written with standard errors attached as follows:

$$\hat{y} = 1.21x_1 + 7.12x_2$$
$$\pm 0.38 \qquad \pm 0.53 \tag{14.42}$$

Individual 90% confidence limits for the regression coefficients β_1 and β_2 are therefore

$$\begin{array}{llr} \text{For } \beta_1: & 1.21 \pm (2.132 \times 0.38) = 1.21 \pm 0.82 \\ \text{For } \beta_2: & 7.12 \pm (2.132 \times 0.53) = 7.12 \pm 1.13 \end{array} \tag{14.43}$$

where 2.132 is the point of the t distribution, with four degrees of freedom, that leaves 5% in each tail of the distribution. The individual confidence limits are associated with the horizontal and vertical *margins* of Figure 14.3*b* and may therefore be called *marginal* intervals.

A Joint Confidence Region for β_1 and β_2

It is possible to show that confidence regions for different probabilities are given by areas in Figure 14.3*b* contained within specific sum of squares contours. Specifically, on the assumption that the model form is adequate, a $1 - \alpha$ joint confidence region for β_1 and β_2 is bounded by a sum of squares contour S such that

$$S = S_R\left[1 + \frac{p}{n - p} F_\alpha(p, n - p)\right] \tag{14.44}$$

and $F_\alpha(p, n - p)$ is the significance point of the F distribution with p and $n - p$ degrees of freedom. In the present case, for a 90% confidence region we have $S_R = 1.3308$, $p = 2$, $n - p = 6 - 2 = 4$, $F_{0.10}(2, 4) = 4.32$. Thus

$$S = 1.3308[1 + \tfrac{2}{4}(4.32)] = 4.2053 \qquad (14.45)$$

The shaded area in Figure 14.3b enclosed within this contour supplies the required 90% confidence region. Its oblique orientation implies that the estimates are (negatively) correlated. Now it can be shown that if S is the sum of squares at some point β_1, β_2

$$(\beta_1 - b_1)^2 \sum x_1^2 + 2(\beta_1 - b_1)(\beta_2 - b_2) \sum x_1 x_2$$
$$+ (\beta_2 - b_2)^2 \sum x_2^2 = S - S_R \qquad (14.46)$$

Thus, by using Equation 14.44 (solving it for $S - S_R$ and then substituting on the right-hand side of Equation 14.46), we find that the $100(1 - \alpha)\%$ confidence region is enclosed by the ellipse

$$(\beta_1 - b_1)^2 \sum x_1^2 + 2(\beta_1 - b_1)(\beta_2 - b_2) \sum x_1 x_2$$
$$+ (\beta_2 - b_2)^2 \sum x_2^2 = \frac{p S_R}{n - p} F_\alpha(p, n - p) \qquad (14.47)$$

Substituting values for $b_1, b_2, p, S_R, n - p$, and $F_\alpha(p, n - p)$ for the present example, we have for the boundary of the 90% confidence region

$$(\beta_1 - 1.207)^2\, 7.055 + 2(\beta_1 - 1.207)(\beta_2 - 7.123)\, 4.178$$
$$+ (\beta_2 - 7.123)^2\, 3.635 = 2.874$$

which is the equation of the ellipse that defines the shaded region in Figure 14.3b. The nature of a confidence region parallels that of a confidence interval. If we imagine the present experiment repeated many times and a region calculated in the above manner each time, 90% of the calculated regions will include the true parameter point (β_1, β_2) and 10% will exclude it.

Individual (Marginal) Limits and Joint Confidence Regions

One may ask what the relation is between the shaded joint confidence *region* in Figure 14.3b and the individual marginal *limits* also shown in the figure and given by Equation 14.43. Consider the point P with coordinates $\beta_1 = 0.75$, $\beta_2 = 6.50$. The value $\beta_1 = 0.75$ lies well within the marginal 90% confidence limits for β_1, and the value $\beta_2 = 6.50$ lies within the marginal confidence limits for β_2, but P does not lie within the joint confidence region. This means, for example, that, although the value $\beta_1 = 0.75$ is not discredited by the data when *any* value of the other parameter β_2 is allowed, the hypothesized *pair* of values $(\beta_1, \beta_2) = (0.75, 6.50)$ *is* discredited by the data.

A Glance at "Optimal" Design Theory

Consider again Equations 14.40, which give the variances for the estimates b_1 and b_2 in the model $y = \beta_1 x_1 + \beta_2 x_2 + \epsilon$ as follows:

$$\hat{V}(b_1) = \frac{1}{1 - \rho^2} \frac{s^2}{\sum x_{1u}^2}, \qquad \hat{V}(b_2) = \frac{1}{1 - \rho^2} \frac{s^2}{\sum x_{2u}^2} \tag{14.48}$$

with $\rho = -\sum x_{1u} x_{2u} / \sqrt{\sum x_{1u}^2 \sum x_{2u}^2}$ the correlation between b_1 and b_2. Suppose that we were *planning* this experiment and could choose in advance a list of levels to be run for the monomer concentrations x_1 and dimer concentrations x_2. How could we choose the best list, that is, the best *design*?

First we need to consider some general rules implied by the experimental situation. Obviously, for this example the levels x_1 and x_2 must all be positive, and there will be two values, say x_{10}, x_{20}, which are practical upper limits for the two variables, so that

$$0 < x_1 < x_{10}, \qquad 0 < x_2 < x_{20}$$

Now, to estimate the two parameters β_1 and β_2 in the model $\eta = \beta_1 x_1 + \beta_2 x_2$, we need *at least* $n = 2$ runs. For simplicity suppose at first that only two runs are to be made. Then the list of projected runs (called the *design matrix*) looks like this:

$$\begin{array}{c} \\ \text{run 1} \\ \text{run 2} \end{array} \begin{array}{cc} x_1 & x_2 \\ \begin{bmatrix} x_{11} & x_{21} \\ x_{12} & x_{22} \end{bmatrix} \end{array}$$

The problem is how best to choose the four numbers $x_{11}, x_{12}, x_{21}, x_{22}$.

Suppose it is declared that an "optimal" design is one which gives smallest variances $\hat{V}(b_1)$ and $\hat{V}(b_2)$. Then it is possible to show that the design which uniquely satisfies this requirement is

$$\begin{bmatrix} x_{10} & 0 \\ 0 & x_{20} \end{bmatrix}$$

With this arrangement, in the first run the level of monomer x_1 is held at its maximum value with no dimer present at all; in the second run the level of dimer x_2 is held at its maximum value with no monomer present at all. Now suppose that n is even and is larger than 2; then it is easy to show that the "optimal" design consists of $n/2$ replications of this identical arrangement. Now consider equation 14.48. For this arrangement $\sum x_{1u} x_{2u} = 0$ so that $\rho = 0$. Also $\sum x_{1u}^2 = n x_{10}^2$ and $\sum x_{2u}^2 = n x_{20}^2$, which are their maximum values. Thus

$$\hat{V}(b_1) = \frac{s^2}{n x_{10}^2}, \qquad \hat{V}(b_2) = \frac{s^2}{n x_{20}^2}$$

and these minimum variance estimates are uncorrelated.

This kind of approach to the theory of experimental design can provide helpful background to be taken into account with other considerations. Two particularly important points of more general application are illustrated by this example:

1. Highly correlated estimates are usually associated with high variances (and in extreme cases uncertainty as to which of two variables is responsible for a particular effect). Such correlations can often be avoided, as in this example, by arranging that cross-product sums $\sum x_i x_j$ are zero. Notice that for the coded variables of two-level factorial designs all such cross-product sums associated both with main effects and with interactions vanish.

2. Whereas *cross-product* sums should be zero, the *sum of squares* of the x's should be large. We frequently find, therefore, as in this case, that the optimal design has experiments at the boundary conditions. When translated to factorial designs, this implies that the design should cover the widest possible range. Notice, for example, that a doubling of x_{10} would result in a doubling of the range $0 < x_1 < x_{10}$ and a quadrupling of $\sum x_{1u}^2 = n x_{10}^2$. We see that such an enlargement would have the same effect on variances of the b's as a *fourfold reduction* in the experimental error variance or a *fourfold increase* in the number of runs. A similar result applies, for example, to two-level factorial designs.

However, one can sympathize with the experimenter who says, "Well, this design with half runs at $(x_{10}, 0)$ and the other half at $(0, x_{20})$ may be optimal, but I would never run an experiment that way." The reason is, of course, that the experimenter's implied criterion of optimality is much more complex than the one we have used. He would certainly like $\hat{V}(b_1)$ and $\hat{V}(b_2)$ to be small, if the model happened to be appropriate, but certain things about this design would bother him.

Remember that to apply the above theory we would need to have persuaded the experimenter to guess *in advance* not only an appropriate model $y = \beta_1 x_1 + \beta_2 x_2 + \epsilon$ but also an associated square region defined by $0 < x_1 < x_1(\text{max})$, $0 < x_2 < x_2(\text{max})$ over which the guessed model is supposed to apply with sufficient accuracy. We have arrived at our design by treating his guesses about both the model and the size and shape of the appropriate region as exact statements. In particular, the experimenter will almost certainly feel uncomfortable that the points at which the optimal design required all experiments to be run are the very ones that strain his guesses the most. Knowing his conjectures to be fallible, he will almost certainly want to place points in other locations in case he was wrong about the model and the region over which it applies.

The "optimal" design approach is useful provided that the limitations implied by the assumptions and the particular choice of optimality criteria are kept firmly in mind. In particular, rather than say vaguely that a given design is optimal, we should say that it minimizes variances (or whatever it does). In recent years the study of "optimal" design has become separated from real experimentation with the predictable consequence that its limitations have not been stressed or, often, even realized. Some of the other considerations that must be borne in mind in designing real experiments are listed in Table 14.16. A valuable summary of recent work on "optimal" designs has been given by St. John and Draper (1975).

14.3. STRAIGHT LINE MODEL: WELDING EXAMPLE

The least squares results presented earlier may be used to solve a wider class of problems than is at first apparent. As an illustration we consider the problem of fitting to data a straight line not necessarily passing through the origin. The equation of such a line

$$\eta = \beta_0 + \beta x \qquad (14.49)$$

contains two parameters: β, the slope of the line, and β_0, the intercept on the y axis. Here it is *not* assumed (as it was in the proportional model of Section 14.1) that this intercept $\beta_0 = 0$.

We illustrate with the data of Table 14.8, taken from an inertia welding experiment. The data are plotted in Figure 14.5a. In this experiment one part of a workpiece, which was rotating at a certain velocity, was brought to a standstill by forcing it into contact with the other part, which was stationary. The heat generated by friction at the interface produced a hot-pressure weld. Over the region studied, the measured breaking strength y of the weld had a nearly linear relationship with the velocity x (in feet per minute) of the rotating workpiece.

TABLE 14.8. Welding data

velocity (10^2 f/min) x	breaking strength (ksi) y
2.00	89
2.50	97
2.50	91
2.75	98
3.00	100
3.00	104
3.00	97
sum = $\overline{18.75}$	sum = $\overline{676}$
$\bar{x} = 2.679$	$\bar{y} = 96.571$

Use of an Indicator Variable

The straight line model $\eta = \beta_0 + \beta x$ is a special case of the two-parameter model studied in the last section because if we substitute

$$\beta_1 = \beta_0, \qquad x_1 = 1, \qquad \beta_2 = \beta, \qquad x_2 = x$$

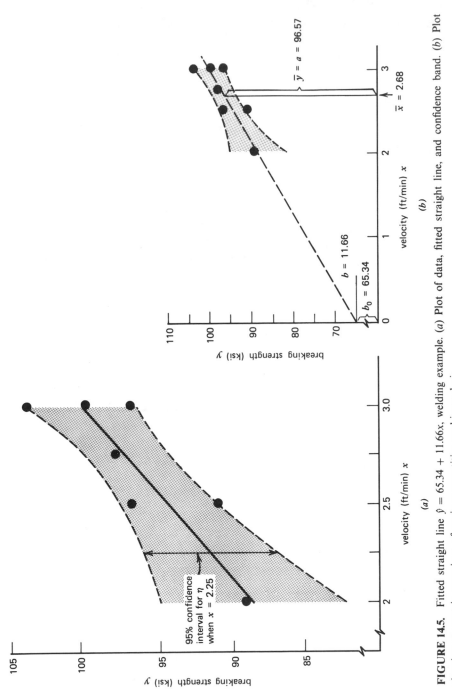

FIGURE 14.5. Fitted straight line $\hat{y} = 65.34 + 11.66x$, welding example. (*a*) Plot of data, fitted straight line, and confidence band. (*b*) Plot showing geometric meanings of various quantities used in analysis.

in the two-parameter model

$$\eta = \beta_1 x_1 + \beta_2 x_2 \tag{14.50}$$

we obtain

$$\eta = \beta_0 \cdot 1 + \beta x \tag{14.51}$$

which is identical to Equation 14.49. The "variable" $x_1 = 1$ is introduced to indicate the presence of the constant β_0 and is therefore called an *indicator variable*. The data then appear as shown in Table 14.9, and the analysis proceeds precisely as before.

TABLE 14.9.   Data with indicator variable

indicator variable $x_1(=1)$	velocity (f/min) $x_2(=x)$	breaking strength (ksi) y
1	2.00	89
1	2.50	97
1	2.50	91
1	2.75	98
1	3.00	100
1	3.00	104
1	3.00	97

The normal Equations 14.31 simplify to

$$b_0 n + b \sum x = \sum y$$
$$b_0 \sum x + b \sum x^2 = \sum xy \tag{14.52}$$

the solutions to which are

$$b = \frac{n \sum xy - \sum x \sum y}{n \sum x^2 - (\sum x)^2} = \frac{\sum (x - \bar{x})(y - \bar{y})}{\sum (x - \bar{x})^2} \tag{14.53}$$

$$b_0 = \frac{\sum x^2 \sum y - \sum x \sum xy}{n \sum x^2 - (\sum x)^2} = \bar{y} - b\bar{x}$$

Exercise 14.3.   Derive Equations 14.53 from Equations 14.52.

For this example

$$n = 7, \qquad \sum x = 18.75, \qquad \sum y = 676,$$

$$\bar{x} = 2.679, \qquad \bar{y} = 96.571 \tag{14.54}$$

$$\sum x^2 = 51.0625, \qquad \sum xy = 1820.5, \qquad \sum y^2 = 65{,}440 \tag{14.55}$$

$$\sum (x - \bar{x})^2 = 0.8393, \qquad \sum (x - \bar{x})(y - \bar{y}) = 9.7857,$$

$$\sum (y - \bar{y})^2 = 157.7143 \tag{14.56}$$

Thus

$$b = \frac{9.7857}{0.8393} = 11.66 \tag{14.57}$$

$$b_0 = 96.571 - (11.66 \times 2.679) = 65.34 \tag{14.58}$$

The fitted straight line is therefore

$$\hat{y} = b_0 + bx \tag{14.59}$$

$$\hat{y} = 65.34 + 11.66x \tag{14.60}$$

This line is plotted in Figure 14.5.

Shifting the Origin from Zero to $\bar{x}$

Since $b_0 = \bar{y} - b\bar{x}$ (see Figure 14.5b), Equation 14.59 can be written as

$$\hat{y} = \bar{y} - b\bar{x} + bx$$

or

$$\hat{y} = \bar{y} + b(x - \bar{x}) \tag{14.61}$$

For the present example

$$\hat{y} = 96.57 + 11.66(x - 2.68) \tag{14.62}$$

which defines the same fitted straight line as does Equation 14.60.

In practice, instead of working with the straight line model in the form $\eta = \beta_0 + \beta x$, it is sometimes a little easier to employ the equivalent form,

$$\eta = \alpha + \beta(x - \bar{x}) \tag{14.63}$$

in which $\alpha = \beta_0 + \beta\bar{x}$. One can estimate this form of the equation directly by setting in Equation 14.50

$$\beta_1 = \alpha, \qquad x_1 = 1, \qquad \beta_2 = \beta, \qquad x_2 = x - \bar{x}$$

If a is the least squares estimate of α, then [since $\sum x_1 x_2 = \sum 1(x - \bar{x}) = 0$], the normal Equations 14.52 simplify to

$$an = \sum y$$

$$b \sum (x - \bar{x})^2 = \sum (x - \bar{x})(y - \bar{y}) \tag{14.64}$$

Thus the estimates (which are uncorrelated) are given by

$$a = \bar{y}$$

$$b = \frac{\sum (x - \bar{x})(y - \bar{y})}{\sum (x - \bar{x})^2} \tag{14.65}$$

Note that Equations 14.64 are easier to solve than Equations 14.52 because the former have no cross-product terms. Inspection of the diagram in Figure 14.5b makes clear that this simplification is associated with shifting the origin from zero to $\bar{x}$. Thus, while b_0 is the intercept at $x = 0$, $a = \bar{y}$ is the intercept at $x = \bar{x}$.

Referring the fitted line to the origin $x = \bar{x}$ has another desirable result: it serves to remind the user that the fitted line is only known to apply approximately *over the range of x actually used*, which is centered at $\bar{x}$. In particular, if the x's used are remote from $x = 0$, as in Figure 14.5, extrapolation to this origin (i.e., assumption of the relevance of the statement $\hat{y} = b_0$ at $x = 0$) could be totally inappropriate. [Indeed, in the present inertia welding example, from a knowledge of the physical setup it is clear that at $x = 0$ (velocity equals zero) the breaking strength must be zero, not 65.34 ksi. Extrapolation to $x = 0$ in this case gives a grossly incorrect result.] If the form $\hat{y} = b_0 + b_1 x$ is used, then, unless the range of the x's includes the origin $x = 0$, b_0 should be regarded merely as a convenient construction point for drawing the line over the actual range of the data. Notice, too, that every fitted straight line passes through the point $\bar{x}, \bar{y}$.

Exercise 14.4. Construct an analysis of variance table, and test the goodness of fit of the straight line, using the data given in Table 14.8. (*Hint*: Refer to Table 14.7.) *Answer*: The ratio of the lack of fit and pure error mean squares is $0.47/14.22 = 0.03$. Lack of fit would be indicated by a significantly *large* value of this ratio. The observed value is actually unusually *small*. (Can you think of reasons that might explain this fact?)

If it could be assumed that the model were adequate, a valid estimate of σ^2 would be provided by

$$s^2 = \frac{S_R}{n - 2} = \frac{43.6130}{5} = 8.7226 \tag{14.66}$$

where

$$S_R = \sum (y - \bar{y})^2 - b^2 \sum (x - \bar{x})^2 = \sum (y - \bar{y})^2 - b \sum (x - \bar{x})(y - \bar{y})$$

$$= 157.7143 \quad - 114.1013 \quad = 43.6130 \tag{14.67}$$

On the same assumption, the standard errors for b_0, b and a would be

$$\text{S.E.}(b_0) = s\left[\frac{1}{n} + \frac{\bar{x}^2}{\sum(x - \bar{x})^2}\right]^{1/2} = 8.64$$

$$\text{S.E.}(b) = \frac{s}{\sqrt{\sum(x - \bar{x})^2}} = 3.22 \qquad (14.68)$$

$$\text{S.E.}(a) = \frac{s}{\sqrt{n}} = 1.12$$

Variance of an Estimated Value and Confidence Limits

If the assumed form of model is adequate, we can use the graph of Figure 14.5a to read off an estimate $\hat{y}_0$ of the breaking strength of a weld for any given velocity x_0. Equivalently, we can calculate $\hat{y}_0$ using the formula

$$\hat{y}_0 = \bar{y} + b(x_0 - \bar{x})$$

It is natural to consider how accurate such an estimate will be. Using the equation for the variance of a linear form, we have, since $\bar{y}$ and b are uncorrelated,

$$V(\hat{y}_0) = V(\bar{y}) + (x_0 - \bar{x})^2 V(b)$$

Substituting estimates for $V(\bar{y})$ and $V(b)$ gives

$$\hat{V}(\hat{y}_0) = \left[\frac{1}{n} + \frac{(x_0 - \bar{x})^2}{\sum(x - \bar{x})^2}\right]s^2$$

For the welding example

$$\hat{V}(\hat{y}_0) = \left[\frac{1}{7} + \frac{(x_0 - 2.679)^2}{0.8393}\right]8.7226$$

A $1 - \alpha$ confidence interval for η_0 at some particular x_0 is now given by

$$\hat{y}_0 \pm t_{\alpha/2}\sqrt{\hat{V}(\hat{y}_0)}$$

where t has 5 degrees of freedom.

Confidence limits calculated by this formula are shown by dotted lines in Figure 14.5a. Such limits are, of course, meaningful only over the range in which a straight line representation is believed to be adequate—most often, this is the range of the data itself.

14.4. GENERAL CASE FOR MODELS LINEAR IN THE PARAMETERS

A general procedure for solving any linear least squares problem is described in Appendix 14A. The linearity referred to here is linearity in the *parameters*. A model is linear in the parameters $\beta_0, \beta_1, \ldots, \beta_p$ if $\eta = E(y)$ can be written in the form

$$\eta = \beta_0 + \beta_1 x_1 + \cdots + \beta_p x_p \qquad (14.69)$$

In this expression the x's are quantities known for each experimental run and are not functions of the β's. Thus the models

$$\eta = \beta x, \qquad \eta = \beta_1 x_1 + \beta_2 x_2, \qquad \eta = \beta_0 + \beta x$$

already discussed in this chapter are all of this linear type. Other examples of model functions that are linear in the parameters are the following:

1. Polynomial models of the form

$$\eta = \beta_0 + \beta_1 x + \beta_2 x^2 + \cdots + \beta_p x^p \qquad (14.70)$$

(Make the substitution $x_j = x^j$ in Equation 14.69.)
2. Sinusoidal models such as

$$\eta = \beta_0 + \beta_1 \sin \theta + \beta_2 \cos \theta \qquad (14.71)$$

where θ is varied in different experimental runs. (Set $x_1 = \sin \theta$ and $x_2 = \cos \theta$ in Equation 14.69.)
3. Models such as

$$\eta = \beta_0 + \beta_1 \log \xi_1 + \beta_2 \frac{e^{\xi_2}}{\xi_3} \qquad (14.72)$$

where ξ_1, ξ_2, ξ_3 are known for each experimental run. (Set $x_1 = \log \xi_1$ and $x_2 = e^{\xi_2}/\xi_3$ in Equation 14.69.)

By contrast, the exponential model

$$\eta = \beta_0(1 - e^{-\beta_1 \xi}) \qquad (14.73)$$

where ξ is known for each experimental run, cannot be written in the form of Equation 14.69. It is therefore an example of a model that is *nonlinear* in the parameters.

In handling the general linear model, it is useful to remember that a model of this type containing a constant term β_0 can always be rewritten as

$$\eta = \alpha + \beta_1(x_1 - \bar{x}_1) + \beta_2(x_2 - \bar{x}_2) + \cdots + \beta_p(x_p - \bar{x}_p) \quad (14.74)$$

as has already been illustrated in Section 14.3 for the case $p = 1$.

It is the form in Equation 14.74 that is most often referred to as a *regression equation*. Computer programs for fitting this regression model are widely available. The ease with which such calculations may thus be carried out has sometimes resulted in their reckless use, particularly in the analysis of "happenstance" data (see Section 14.7). However, properly used, the computer is a well-nigh indispensable tool in model building. This is true not only at the fitting stage; it is valuable also at the preliminary stage of model identification (specification) where graphing and general visual display of data in many different aspects are of special importance. The computer is also of tremendous assistance in the criticism and diagnostic checking of fitted models (see, e.g., Daniel and Wood, 1971).

14.5. POLYNOMIAL MODEL: GROWTH RATE EXAMPLE

The use of the general regression model is now illustrated further with growth rate data (Table 14.10) for experimental rats fed various doses of a dietary supplement. From similar investigations it was believed that over the range

TABLE 14.10.   Growth rate data

observation number	amount of supplement (grams) x	growth rate (coded units) y
1	10 ⎱	73 ⎱
2	10 ⎰	78 ⎰
3	15	85
4	20 ⎱	90 ⎱
5	20 ⎰	91 ⎰
6	25 ⎱	87 ⎱
7	25 ⎰	86 ⎰
8	25 ⎰	91 ⎰
9	30	75
10	35	65

of dosage tested the response would be approximately linear, and a straight line was routinely fitted to the data, yielding

$$\hat{y} = 86.44 - 0.20x \qquad (14.75)$$

Inadequacy of Straight Line Model

An analysis of variance appropriate for checking model adequacy with replicated data was given in Table 14.7. If we had applied that method for the straight line fit to the growth rate data, we would have obtained the analysis of variance shown in Table 14.11. [Note that the model sum of squares S_M has been split into two additive parts: the sum of squares for the mean $N\bar{y}^2 = 67,404.1$, and the extra sum of squares due to the linear term in the model, $67,428.6 - 67,404.1 = 24.5$. The first part is the sum of squares associated with the simple model $\eta = \beta_0$, which contains only the mean β_0 and is represented by a horizontal straight line. The second part is the extra sum of squares that is obtained by fitting the sloping line model $\eta = \beta_0 + \beta x$ containing, in addition, the first-order (or linear) term βx.]

TABLE 14.11.   Analysis of variance for growth rate data: straight line model

source	sum of squares	degrees of freedom	mean square
model	$S_M = 67,428.6 \begin{cases} \text{mean } 67,404.1 \\ \text{extra for linear } 24.5 \end{cases}$	$2 \begin{cases} 1 \\ 1 \end{cases}$	$\begin{array}{c} 67,404.1 \\ 24.5 \end{array}$
residual $\begin{cases} \text{lack of fit} \\ \text{pure error} \end{cases}$	$S_R = 686.4 \begin{cases} S_L = 659.40 \\ S_E = 27.0 \end{cases}$	$8 \begin{cases} 4 \\ 4 \end{cases}$	$85.8 \begin{cases} 164.85 \\ 6.75 \end{cases}$ ratio = 24.42
total	$S_T = 68,115.0$	10	

The value 24.42 for the lack of fit/pure error mean square ratio is large and is significant at the 0.005 level. This analysis, however, does not show the nature of this inadequacy. Such formal lack of fit tests should not make us forget the virtues of simple plotting of data and residuals. From such a plot in Figure 14.6 it is immediately obvious that inadequacy arises from the curvilinear nature of the data. Increasing amounts of the additive up to about 20 grams produce increasing growth rate, but thereafter more additive reduces growth rate.

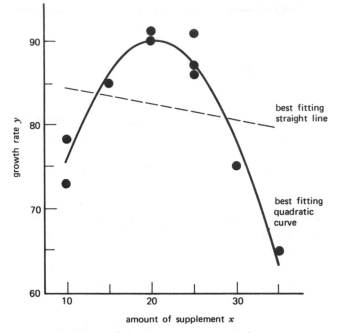

FIGURE 14.6.   Plot of data, best fitting straight line, and best fitting quadratic curve, growth rate example.

A Further Iteration with a Quadratic Model

A better approximation might be expected from the quadratic model

$$\eta = \beta_0 + \beta_1 x + \beta_2 x^2 \tag{14.76}$$

where β_2 is a measure of curvature. As an example of the general regression model, this model is fitted to the data in Appendix 14B, yielding the equation

$$\hat{y} = 35.66 + 5.26x - 0.128x^2 \tag{14.77}$$

which is also plotted in Figure 14.6. The analysis of variance (Table 14.12) for this quadratic model shows no evidence of lack of fit. The model sum of squares is now split into three parts, the third being the extra sum of squares obtained by including the quadratic term $\beta_2 x^2$. Its large size (643.2) in relation to the error mean square (6.75) reflects the important contribution that this term makes in providing an adequate explanation of the data.

   The residuals plotted in Figure 14.7 now appear to display no systematic tendencies in relation to x or $\hat{y}$. Taken together with the fitted curve in

TABLE 14.12. Analysis of variance for growth rate data: quadratic model

source	sum of squares	degrees of freedom	mean square
model	$S_M = 68,071.8$ {mean 67,404.1 / extra for linear 24.5 / extra for quadratic 643.2	3 {1 / 1 / 1	67,404.1 / 24.5 / 643.2
residual	$S_R = 43.2$ {$S_L = 16.2$ / $S_E = 27.0$	7 {3 / 4	{5.40 / 6.75} ratio = 0.80
total	$S_T = 68,115.0$	10	

Figure 14.6 and the analysis of variance table, these results suggest that the quadratic equation supplies an adequate representation over the region studied.

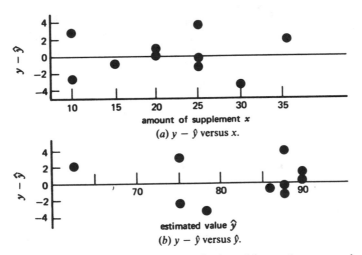

(a) $y - \hat{y}$ versus x.

(b) $y - \hat{y}$ versus $\hat{y}$.

FIGURE 14.7. Plots of residuals from quadratic model, growth rate example.

14.6. NONLINEAR MODEL: BIOCHEMICAL OXYGEN DEMAND EXAMPLE

So far, all the models discussed have been *linear in the parameters*. The following example shows how the method of least squares can also be used to fit models that are nonlinear in the parameters.

TABLE 14.13. Biochemical oxygen demand
(BOD) data

observation number u	BOD (mg/liter) y	incubation (days) x
1	109	1
2	149	2
3	149	3
4	191	5
5	213	7
6	224	10

Biochemical oxygen demand (BOD) is used as a measure of the pollution produced by domestic and industrial wastes. In this test a small amount of the waste is mixed with pure water in a sealed bottle and incubated for a number of days at a fixed temperature. The loss of dissolved oxygen from the water allows calculation of the BOD occurring in a given incubation period. The data given in Table 14.13 and plotted in Figure 14.8 were obtained from six separate sealed bottles tested over a range of incubation periods.

Physical considerations suggest, as an initial speculation, that an exponential model of the form

$$\eta = \beta_1(1 - e^{-\beta_2 x}) \tag{14.78}$$

should describe the phenomenon. The parameter β_2 would then be an overall rate constant, and β_1 would be the ultimate BOD.

Sum of Squares Surface

As before, it is possible to calculate the sum of squares for a grid of values of β_1 and β_2 and to plot contours of the sum of squares

$$S = \sum_{u=1}^{n} [y_u - \beta_1(1 - e^{-\beta_2 x_u})]^2 \tag{14.79}$$

in the (β_1, β_2)-plane as in Figure 14.9. The minimum is at the least squares values $b_1 = 213.8$ and $b_2 = 0.5473$. Substituting these values into Equation 14.78 gives the fitted least squares line shown in Figure 14.8. The residuals

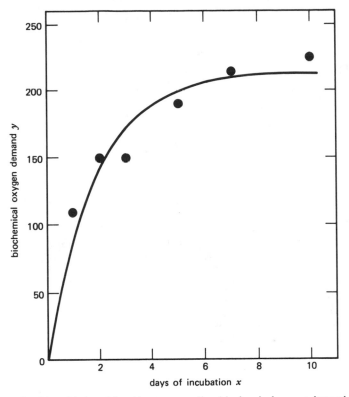

FIGURE 14.8. Plot of data and fitted least squares line, biochemical oxygen demand example.

listed in Table 14.14 provide no reason to suppose that the model is suspect. The sum of squares of these residuals is $S_R = 1168.01$.

In general, a (*very approximate*) $1 - \alpha$ joint confidence region for β_1 and β_2 is given by Equation 14.44 (which is exact for models linear in the parameters). The equation is

$$S = S_R \left[1 + \frac{p}{n - p} F_\alpha(p, n - p) \right] \tag{14.80}$$

where the contour S encloses the region. In this formula, p is the number of parameters and $F_\alpha(p, n - p)$ is the upper $100\alpha\%$ point of the F distribution with p and $n - p$ degrees of freedom. For the BOD data the shaded area in Figure 14.9 is an approximate 90% region for β_1 and β_2. In this case $S_R = 1168.01, p = 2, n = 6$, and $F_\alpha(p, n - p) = F_{0.10}(2, 4) = 4.32$, so the boundary of the region lies on the contour for which $S = 3690.91$.

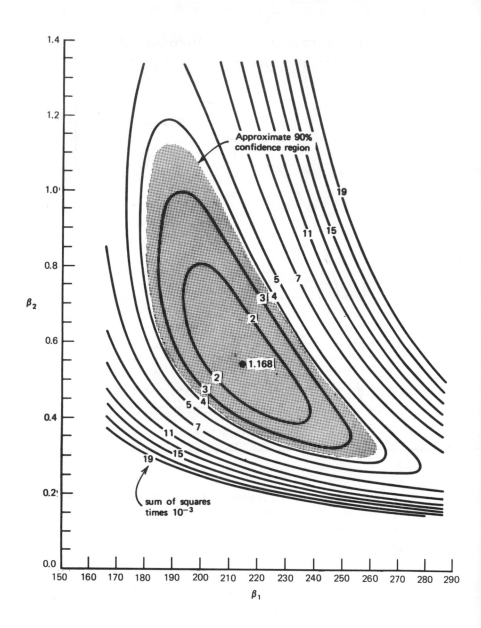

FIGURE 14.9. Contours of sum of squares surface with approximate 90% confidence region, biochemical oxygen demand example.

TABLE 14.14. Calculation of the predicted values y_u and the residuals $y_u - \hat{y}_u$ for BOD data

u	x_u	y_u	$\hat{y}_u = 213.8(1 - e^{-0.5473x_u})$	$y_u - \hat{y}_u$
1	1	109	90.38	18.62
2	2	149	142.53	6.47
3	3	149	172.62	−23.62
4	5	191	200.01	−9.01
5	7	213	209.11	3.89
6	10	224	212.81	11.19

Computer programs exist that carry out iterative calculations to locate the minimum value of the sum of squares for nonlinear models. Using these programs, one supplies the data, the model, and initial guesses for the parameter values. The computer sets out to search for the minimum, which it will usually find. Approximate standard errors and confidence regions for the parameters can also be determined.

14.7. HAZARDS OF FITTING REGRESSION EQUATIONS TO HAPPENSTANCE DATA

The reader must be warned that, should he admit to some knowledge of least squares and multiple regression, he may be expected to extract information from historical records or "happenstance" data. This risk is especially great should he have the temerity to suggest that a designed experiment is needed to solve a particular problem concerning the operation of, say, a chemical plant. Unless he has a strong character, he is likely to find himself outside the door with advice on where he can obtain records of the operation of the process for the last 10 years. "Certainly," the plant superintendent will say while waving him away, "the answer to the problem must be somewhere in these records. This ought to be an excellent opportunity for the use of statistics." Well, it might or it might not.

Happenstance data often take the form of a number of daily or hourly readings on variables such as are indicated in Table 14.15. In that table there are just four independent variables, but in practice there could be a much larger number.

Frequently, without much thought such data are thrown into the regression model

$$y = \alpha + \beta_1 x_1 + \beta_2 x_2 + \cdots + \beta_k x_k + \epsilon \qquad (14.81)$$

TABLE 14.15. Typical appearance of happenstance data

independent (input) variables				dependent (output) variable:
temperature (°C) ξ_1	pressure (psi) ξ_2	concentration (%) ξ_3	air flow (10^5 m³/min) ξ_4	plant yield y
159	173	82.1	10.3	86.2
161	180	89.5	10.4	84.7
160	178	87.9	10.3	85.9
162	181	83.4	10.3	85.3
⋮	⋮	⋮	⋮	⋮

[For convenience the data may be suitably coded, for example, $x_i = (\xi_i - \bar{\xi}_i)/s_i$, with ξ_i the ith independent variable and s_i some suitable scale factor, such as $s_i^2 = \sum (\xi_i - \bar{\xi}_i)^2/(n - 1)$.]

In favor of such a model it is sometimes argued that, for the small changes from the mean that typically occur for both input and output variables, the model function should be approximately linear in the x's. Although it is true that such an analysis is occasionally useful, it is often downright misleading. We now consider some of the hazards involved in fitting equations to happenstance data in this way. An understanding of these possible dangers is perhaps the most telling argument for the use of designed experiments.

1. Inconsistent Data

It is quite rare that a long record of data from any process is consistent and comparable. Typically, for example, standards are modified, instruments change, calibrations drift, operators come and go, changes occur in raw materials, and processes age and in some cases are affected by weather. It is sometimes possible to take at least some of these effects into account, for example, when the times of change in the supplier of raw material are noted. Usually, however, much that is relevant is not recorded.

2. Range of Variables Limited by Control

Figure 14.10 shows a plot of yield against temperature as it might appear if temperature were freely varied. Suppose, however, that during normal operation temperature was carefully controlled to say, 160 ± 2°C. Over the more limited range (open dots) it could well be that no relationship was

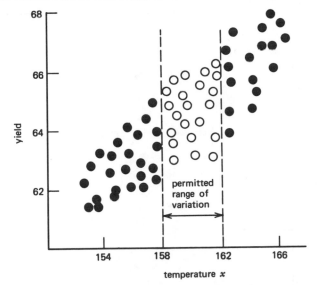

FIGURE 14.10. Plot of yield versus temperature where permissible range of variation in temperature is restricted.

apparent, either visually or statistically. [As usual, common sense and good statistics agree. The variance of the slope b of the fitted line is given by $\sigma^2/\sum(x - \bar{x})^2$, where the denominator is a measure of the squared spread of the x's. Obviously, therefore, the variance of the slope of a line fitted to the restricted data may be many times as large as the variance of the slope of a line fitted to the unrestricted data.] If an investigator in such a case fits a regression equation to the restricted data and informs the process engineer that the temperature has no (statistically) "significant" effect on yield, he may well be laughed to scorn* and told, "But of course it has an effect; *that's the reason we control temperature so carefully!*"

3. Semiconfounding of Effects

Processes often operate so that a particular change in one input variable is usually (sometimes invariably) accompanied by a corresponding change in another. Thus in a closed reactor system an increase in temperature may automatically produce an increase in pressure. Alternatively, such dependence may be *produced* by an automatic controller or by operating policy.

* Here, of course, the process engineer misunderstands the meaning of "statistically significant."

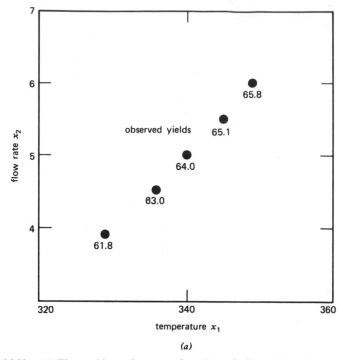

FIGURE 14.11. (a) The problem of near-confounding of effects. Experimental "design" points lying nearly on a straight line in the space of temperature flow rate, with observed yields.

For example, suppose a process operator is told that, whenever the temperature increases by 1° he *must* increase the flow rate by a specified amount. The operator's attempt to do this would produce a highly unsatisfactory "design" in temperature and flow rate like that in Figure 14.11a. At the five different conditions shown, there is a marked increase in yield as temperature and flow rate are increased. Assuming some kind of causal relationship, is the increase produced by higher temperature, by faster flow rate, or by both? Common sense rightly says that the question may be impossible to answer. So does statistics.

The best fitting plane for these data

$$\hat{y} = 63.94 + 0.10(x_1 - \bar{x}_1) + 1.01(x_2 - \bar{x}_2) \qquad (14.82)$$
$$\quad\quad\quad (\pm 0.47) \qquad\qquad (\pm 4.48)$$

(where the quantities in parentheses below the coefficients are standard errors) is shown in Figure 14.11b. Now look at the 95% confidence limits and regions in Figure 14.11c. Notice first of all that the point 0 ($\beta_1 = 0$,

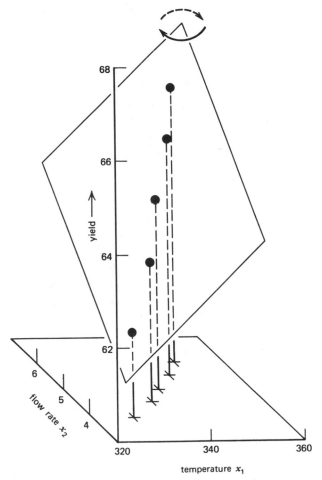

FIGURE 14.11.   (b) The problem of near-confounding of effects. Fitted surface.

$\beta_2 = 0$) falls outside the joint region. The hypothesis that *neither* variable is related to yield is obviously not acceptable. However, both *marginal* confidence intervals cover zero, that is to say, the data can be readily explained by a plane with zero slope in the x_1 direction or by a different plane with zero slope in the x_2 direction.

This state of affairs arises because of the unsatisfactory design, in which x_1 and x_2 are almost linearly related. The fitted plane, Figure 14.11b, sits unstably on the obliquely oriented points and has a great tendency to wobble in the direction shown by the curved arrows. A whole range of planes, in which an increase in the slope β_1 is compensated for by a corresponding

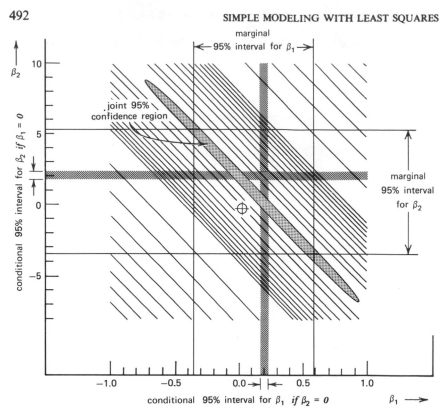

FIGURE 14.11. (c) The problem of near-confounding of effects. Contours of sum of squares surface with joint confidence region and marginal and conditional confidence intervals.

decrease in slope β_2, would fit the data almost equally well. This results in highly correlated estimates b_1 and b_2 with large standard errors. These standard errors would be infinite if x_1 and x_2 were exactly linearly dependent,* and in that case β_1 and β_2 would be perfectly confounded. (We could estimate only some linear combinations $k_1\beta_1 + k_2\beta_2$ of the two coefficients.)

Think how much better off we would be if we had the "four-legged table" on which to balance our plane as provided by the 2^2 factorial design. For such an arrangement small errors in the length of the "legs" have minimum effect on the stability of the table top. The unstable tendency of the present design is reflected in the oblique contours of the sum of squares shown in Figure 14.11c.

Notice that, although neither b_1 nor b_2 is significantly different from zero when *both* temperature and flow rate are included in the regression equation, when only one is included it appears to have a highly significant effect.

---

* That is if the design points lay exactly on a line in the (x_1, x_2) plane of Figure 14.11(a).

Fitting $\eta = \beta_0 + \beta_1 x_1$ to the data gives 0.21 ± 0.02 for the 95% interval for β_1, and fitting $\eta = \beta_0 + \beta_2 x_2$ gives 1.94 ± 0.22 for the β_2 interval. These are called *conditional* intervals in Figure 14.11c because *omission* of one of the variables is equivalent to the assumption that its effect is *exactly zero. On this assumption* the effect of the other variable can be precisely estimated. This is seen if we imagine a straight line being fitted to the plot of y versus x_1 or to the plot of y versus x_2. Also, if, in Figure 14.11c, β_1 is varied, with β_2 set equal to zero, the sum of squares function rapidly changes as β_1 changes and we get a precise estimate of β_1.

In summary, it will be impossible to identify the separate contributions of x_1 and x_2 if the data points in the (x_1, x_2) plane form a highly correlated pattern, yielding a poor "experimental design."

4. Nonsense Correlation—Beware the Lurking Variable!

Inevitably we do not observe all the variables that affect a process. We tacitly admit this fact as soon as we write a model containing an error term ϵ. The error term is a catchall for all the other "lurking" variables that affect the process but are not observed or even known to exist. A serious difficulty with happenstance data is that, when we establish a significant *correlation* between y and x_1, this does not provide evidence that these variables are necessarily *causally* related. (The storks and population example, Figure 1.4, illustrates the problem). In particular, such a correlation can occur if a change in a lurking variable x_2 produces a change in *both* y and x_1.

Impurity, Frothing, and Yield

For illustration consider the hypothetical reactor data relating yield y and pressure x_1 (Figure 14.12a). The 95% confidence interval for the slope of the regression line is -1.05 ± 0.17, so there is no question as to the significance of the correlation. However, the cause of this correlation (unknown to the investigator) might be as follows:

1. There exists an unmeasured and unknown impurity x_2 (a "lurking" variable), which varies from one batch of raw material to another.
2. High levels of the impurity cause low yield.
3. High levels of the impurity also cause frothing.
4. The standard operating procedure is to reduce frothing by increasing reactor pressure. The situation is shown diagrammatically in Figures 14.12b and c.

In Figure 14.12c the causal relationship between the level x_2 of the unknown impurity in the raw material and the yield is shown by the straight

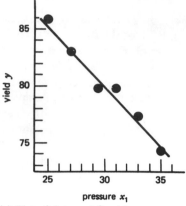

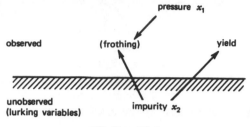

(a) Plot of data and fitted least squares line.

(b) Causal links.

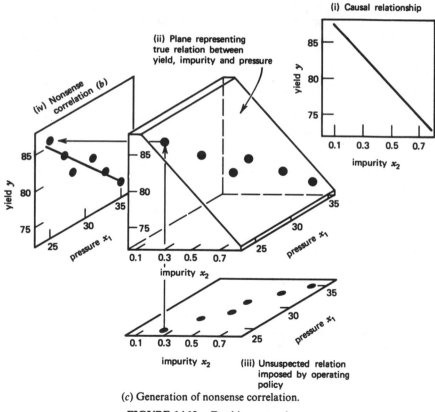

(c) Generation of nonsense correlation.

FIGURE 14.12. Frothing example.

line in (i). Since, over the relevant range, pressure does not affect yield, the plane in (ii) represents the true but unknown functional dependence of yield on impurity and pressure. (Notice that the plane has no slope in the pressure direction, since changing pressure with fixed level of impurity does not change yield.) The unsuspected relation between x_1 and x_2 imposed by operating policy is shown in (iii), and the (nonsense) correlation thus induced between pressure and yield in (iv).

Correlations of this kind are sometimes called *nonsense correlations* because they do not imply direct causation. Just as shooting storks will not reduce the birth rate, decreasing pressure in the present example will not increase yield.

It is important to understand what can and what cannot be done with such relationships. Even though no direct causal relationship between pressure and yield could be inferred, the value of the pressure could nevertheless be used to *predict* the level of *y*, using the fitted line, *provided that the system continued to be operated in the same fashion as when the data were recorded*. It is perfectly true, for example, that, on days when pressure is high, yield tends to be low.

Two Ways to Use Regression Equations

Generalizing from this example, there are two ways in which regression analysis may be used appropriately. The first relates to *correlation*, and the second to *causation*.

1. On the one hand, one may wish *to predict* the response η from *passive* observation of $x_1, x_2, \ldots, x_k$. This can be done if it is assumed that the same correlative relationships that existed during the period when the data were collected also operate during the period when predictions are being made.

2. On the other hand, to *intervene* and change a system one needs causal relationship. With happenstance data correlation *may* indicate causation, but this will not necessarily be the case. The monthly sales of *x* fuel oil in Madison, Wisconsin, may provide a good prediction of the average monthly temperature *y* during the winter months, but experience has shown that it is not possible to induce warm weather by canceling the order for oil. Broadly speaking, *to find out what happens when you change something it is necessary to change it*. To safely infer causality the experimenter cannot rely on natural happenings to choose the design for him; he must choose the design for himself and, in particular, must introduce randomization to break the links with possible lurking variables.

5. Serially Correlated Errors

The method of least squares used in regression analysis is appropriate if the assumed form of model is adequate and, in particular, if the errors ϵ are *independently* distributed with the same variance. But for sets of data that occur as a sequence in time (such as the process data in Table 14.15) the errors are often not independent; in particular, when the error ϵ_t is high, the error ϵ_{t+1} may also tend to be high (in this case ϵ_t and ϵ_{t+1} are said to be positively serially correlated).* When such dependence occurs, the ordinary method of least squares is inappropriate. In particular, the estimates of the standard errors of the regression coefficients can be dramatically wrong. For example, in a study (Coen, Gomme, and Kendall, 1969) on possible relationships between current stock prices and car sales seven quarters earlier, ordinary least squares produced a regression coefficient that was 14 times its standard error and hence tremendously "significant." On the basis of this correlation the authors of this study believed that car sales could help to foretell stock prices. Use of a model that took account of dependence between the errors (in a manner outlined in Chapter 18) showed, however, that in fact no significant effect existed (see Box and Newbold, 1971).

6. Dynamic Relationships

Another problem that sometimes arises in the analysis of serially collected data concerns the proper treatment of dynamic relationships. Consider a simple example. The viscosity of a product is measured every 15 minutes at the outlet of a continuous reactor. There is just one input variable, temperature, which has the value of x_t at time t, the value x_{t+1} at time $t + 1$ one interval later, and so on. If we denote the corresponding values of viscosity by $y_t, y_{t+1}, \ldots$, the usual regression model

$$y_t = \alpha + \beta x_t + \epsilon_t \tag{14.83}$$

says that the viscosity y_t of the product at time t is related only to the temperature at that same time t. Now in this situation the viscosity at time t is, in fact, likely to be dependent, not just on the current temperature, but also on the recent past history of temperature.

Indeed, for the output from a stirred tank reactor, it would be much easier to justify a model of the form

$$y_t = \alpha + \beta(x_{t-1} + \delta x_{t-2} + \delta^2 x_{t-3} + \delta^3 x_{t-4} + \cdots) + \epsilon_t \tag{14.84}$$

where δ is between zero and unity. This is a model in which the output is related via an "exponential memory" to recent past inputs. For a situation

---

* If, alternatively, when ϵ_t is high, ϵ_{t+1} tends to be low, they are negatively serially correlated.

in which Equation 14.84 was applicable, use in its place of the naive regression Equation 14.83 could produce totally misleading results. (See Box and Jenkins, 1970, where more sophisticated "memory functions" and their applications are discussed.)

7. Feedback

Quite frequently data of the kind listed in Table 14.15 arise from processes where some form of feedback control is in operation. For example, suppose that, to maintain the level of viscosity y_t at some desired level L, the temperature x_t is adjusted every 15 minutes in accordance with the following "control equation":

$$x_t = c + k(y_t - L) \tag{14.85}$$

TABLE 14.16. **Experimental design procedures for avoiding the problems that occur in the analysis of happenstance data**

Problems in the analysis of happenstance data	Experimental design procedures for avoiding such problems
1. Inconsistent data	Blocking and randomization
2. Range limited by control	Experimenter makes own choice of ranges for the variables
3. Semiconfounding of effects	Use of designs such as factorial that provide uncorrelated estimates of the separate effects
4. Nonsense correlations due to lurking variables	Randomization
5. Serially correlated errors	Randomization*
6. Dynamic relationships	Where only steady-state characteristics are of interest,† sufficient time is allowed between successive runs for process to settle down
7. Feedback	Temporary disconnection of feedback system‡

* Many time series records are historical data that are necessarily happenstance in nature, for example, various economic series such as unemployment records. Here the use of experimental design (in particular, randomization) is impossible, but the investigator can allow for serial correlation in the statistical model (see Chapter 18 and, e.g., Box and Jenkins, 1970).

† When dynamic characteristics must be estimated, special experimental designs in the time variable may be employed.

‡ When this is impossible, one may proceed by modeling the feedback system (Box and Jenkins, 1970; Box and MacGregor, 1974, 1976).

where c and k are constants. Turning this equation around, we see that control according to Equation 14.85 *guarantees* a relation between y_t and x_t of the form

$$y_t = \alpha + \beta(x_t - c) \tag{14.86}$$

where $\alpha = L$ and $\beta = 1/k$. Notice that this (exact!) relationship would occur in the data whether or not a change in temperature *really* affected viscosity. In this case the regression equation contains only information about the control equation being used (which is known already) and nothing about the causal relationships between the dependent and independent variables.

Designed Experiments Versus Happenstance Data

A good way to appreciate the value of experimental design and the reason for its various principles is to consider the perils of regression analysis of happenstance data. Table 14.16, on the previous page, lists some problems and some means of avoiding them.

APPENDIX 14A.   WHY DO THE NORMAL EQUATIONS YIELD LEAST SQUARES ESTIMATES?

We have made use of the fact that for any model that is linear in the parameters and so can be written in the form

$$\eta = \beta_1 x_1 + \beta_2 x_2 + \cdots + \beta_p x_p \tag{14A.1}$$

the least squares estimates of the parameters $\boldsymbol{\beta}$ can be obtained by substituting $\hat{y} = b_1 x_1 + b_2 x_2 + \cdots + b_p x_p$ in the p normal equations

$$\sum (y - \hat{y})x_1 = 0, \quad \sum (y - \hat{y})x_2 = 0, \quad \ldots, \quad \sum (y - \hat{y})x_p = 0$$

and solving for $b_1, b_2, \ldots, b_p$.

To demonstrate that this is true it is sufficient to consider the $p = 2$ parameter model:

$$\eta = \beta_1 x_1 + \beta_2 x_2$$

The generalization to more parameters involves merely the writing out of longer but precisely similar expressions.

We shall assume that the regressor vectors are not *linearly dependent*. For $k = 2$ this says there are no constants k_1 and k_2 such that

$$k_1 x_{1u} + k_2 x_{2u} = 0 \quad \text{for } u = 1, 2, \ldots, n \tag{14A.2}$$

If this were not true, x_1 and x_2 would be perfectly related and the parameters (β_1, β_2) would be confounded. It would then not be possible to estimate the parameters separately but only some linear combination of them.

Look at the sum of squares surface in Figure 14.3a. This is a plot of a sum of squares function

$$S(\boldsymbol{\beta}) = \sum (y - \eta)^2 = \sum (y - \beta_1 x_1 - \beta_2 x_2)^2 \qquad (14A.3)$$

Let b_1 and b_2 be the values of the parameters obtained by solving the normal equations

$$\sum (y - \hat{y})x_1 = 0, \qquad \sum (y - \hat{y})x_2 = 0 \qquad (14A.4)$$

where

$$\hat{y} = b_1 x_1 + b_2 x_2 \qquad (14A.5)$$

We want to show that substitution of the particular values $\beta_1 = b_1, \beta_2 = b_2$ in the sum of squares function produces its minimum value and hence that b_1 and b_2 are indeed *least squares* estimates.

Introducing the n values $\hat{y}_1, \hat{y}_2, \ldots, \hat{y}_n$, obtained from Equation 14A.5, we can always write

$$y_u - \eta_u = (y_u - \hat{y}_u) + (\hat{y}_u - \eta_u) \qquad \text{for } u = 1, 2, \ldots, n$$

The sum of squares function is therefore

$$\begin{aligned} S(\boldsymbol{\beta}) = \sum (y - \eta)^2 &= \sum [(y - \hat{y}) + (\hat{y} - \eta)]^2 \\ &= \sum (y - \hat{y})^2 + \sum (\hat{y} - \eta)^2 + 2 \sum (y - \hat{y})(\hat{y} - \eta) \end{aligned} \qquad (14A.6)$$

But since

$$\hat{y} - \eta = (b_1 - \beta_1)x_1 + (b_2 - \beta_2)x_2 \qquad (14A.7)$$

the cross-product term in Equation 14A.6 is

$$\sum (y - \hat{y})(\hat{y} - \eta) = (b_1 - \beta_1) \sum (y - \hat{y})x_1 + (b_2 - \beta_2) \sum (y - \hat{y})x_2$$

which from Equation 14A.4 is zero for every value of β. Thus the sum of squares function can *always* be written in the form

$$S(\boldsymbol{\beta}) = \sum (y - \hat{y})^2 + \sum (\hat{y} - \eta)^2$$

or, using Equation 14A.7,

$$S(\boldsymbol{\beta}) = S(\mathbf{b}) + \sum [(b_1 - \beta_1)x_1 + (b_2 - \beta_2)x_2]^2 \qquad (14A.8)$$

where $S(\mathbf{b}) = \sum (y - \hat{y})^2$ is the *fixed* value obtained for the sum of squares when b_1 and b_2 are substituted for β_1 and β_2.

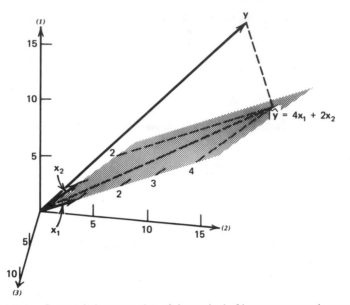

FIGURE 14A.1. Geometric interpretation of the method of least squares and normal equations.

Now the second term on the right of Equation 14A.8 is a sum of squares; and since we have assumed that there exists no set of relations of the form of Equation 14A.2, it is always positive *unless* $\beta_1 = b_1$ and $\beta_2 = b_2$. Thus the values (b_1, b_2) obtained from the normal equations do in fact uniquely minimize the sum of squares $S(\boldsymbol{\beta})$.

Geometry of Least Squares

It is instructive also to consider the problem geometrically, supposing $k = 2$ and $n = 3$. In Figure 14.A1 the two regressor vectors $\mathbf{x}_1$, $\mathbf{x}_2$ and the observation vector $\mathbf{y}$ are shown plotted in a three-dimensional space in the manner discussed in Appendix 6B. The regressor vectors $\mathbf{x}_1$ and $\mathbf{x}_2$ define a plane* passing through them. Furthermore we can draw an (oblique) coordinate system on the plane in which the basic unit is not a square, as with ordinary graph paper, but a parallelogram having $\mathbf{x}_1$ and $\mathbf{x}_2$ for its sides. Then $\hat{\mathbf{y}} = b_1\mathbf{x}_1 + b_2\mathbf{x}_2$ corresponds to some point on this plane with oblique coordinates (b_1, b_2). In the situation illustrated, $b_1 = 4$ and $b_2 = 2$. Now we want to choose b_1 and b_2 and hence $\hat{\mathbf{y}}$ to minimize the sum of squares $\sum (y - \hat{y})^2$, that is, to minimize the length of the line $\mathbf{y} - \hat{\mathbf{y}}$. This requires that $\mathbf{y} - \hat{\mathbf{y}}$ be normal to the plane generated by the vectors $\mathbf{x}_1$ and $\mathbf{x}_2$ and hence that it be normal to the vectors $\mathbf{x}_1$ and $\mathbf{x}_2$ individually.

* The condition that $\mathbf{x}_1$ and $\mathbf{x}_2$ be linearly independent ensures that one vector does not lie on top of the other and consequently that we have a plane and not a line.

Thus

$$\sum (y - \hat{y})x_1 = 0, \qquad \sum (y - \hat{y})x_2 = 0$$

which are the normal equations.

Exercise 14.5. In geometrical terms, using Figure 14A.1, explain for the regression problem (a) the analysis of variance table, (b) the associated F test.

APPENDIX 14B.   MATRIX VERSION OF THE NORMAL EQUATIONS

Matrices (see, e.g., Aitken, 1951) provide a convenient shorthand method for writing the important equations used in least squares calculations. In matrix notation Equations 14A.1 at the n distinct sets of conditions can be written as

$$\boldsymbol{\eta} = \mathbf{X}\boldsymbol{\beta}$$

where $\boldsymbol{\eta}$ is the $n \times 1$ vector of expected values for the response, $\mathbf{X}$ is the $n \times p$ matrix of independent variables, and $\boldsymbol{\beta}$ is the $p \times 1$ vector of parameters.

Since $\mathbf{x}_1, \mathbf{x}_2, \ldots, \mathbf{x}_p$ are the columns of $\mathbf{X}$, the normal equations

$$\mathbf{x}_1'(\mathbf{y} - \hat{\mathbf{y}}) = 0, \qquad \mathbf{x}_2'(\mathbf{y} - \hat{\mathbf{y}}) = 0, \qquad \ldots, \qquad \mathbf{x}_p'(\mathbf{y} - \hat{\mathbf{y}}) = 0 \qquad (14B.1)$$

can be written succinctly as

$$\mathbf{X}'(\mathbf{y} - \hat{\mathbf{y}}) = 0 \qquad (14B.2)$$

where the prime (′) means transpose. Upon substitution of $\hat{\mathbf{y}} = \mathbf{X}\mathbf{b}$

$$\mathbf{X}'(\mathbf{y} - \mathbf{X}\mathbf{b}) = 0 \qquad (14B.3)$$

or

$$\mathbf{X}'\mathbf{X}\mathbf{b} = \mathbf{X}'\mathbf{y} \qquad (14B.4)$$

Since we suppose that the columns of $\mathbf{X}$ are linearly independent, $\mathbf{X}'\mathbf{X}$ has an inverse, and

$$\mathbf{b} = [\mathbf{X}'\mathbf{X}]^{-1}\mathbf{X}'\mathbf{y} \qquad (14B.5)$$

In general it can be shown that the variance–covariance matrix for the estimates is

$$V(\mathbf{b}) = [\mathbf{X}'\mathbf{X}]^{-1}\sigma^2 \qquad (14B.6)$$

if σ^2 is known. Otherwise, assuming the model form is appropriate, we can substitute an estimate $s^2 = (\mathbf{y} - \hat{\mathbf{y}})'(\mathbf{y} - \hat{\mathbf{y}})/(n - p) = S_R/(n - p)$ for σ^2.

Example

To illustrate this use of matrices, consider the problem of fitting the quadratic model $\eta = \beta_0 + \beta_1 x + \beta_2 x^2$ (Equation 14.76) to the growth rate data in Table 14.10. The matrices needed are as follows:

$$
X = \begin{array}{c} x_0 \quad x \quad x^2 \\ \begin{bmatrix} 1 & 10 & 100 \\ 1 & 10 & 100 \\ 1 & 15 & 225 \\ 1 & 20 & 400 \\ 1 & 20 & 400 \\ 1 & 25 & 625 \\ 1 & 25 & 625 \\ 1 & 25 & 625 \\ 1 & 30 & 900 \\ 1 & 35 & 1225 \end{bmatrix} \end{array}, \quad y = \begin{bmatrix} 73 \\ 78 \\ 85 \\ 90 \\ 91 \\ 87 \\ 86 \\ 91 \\ 75 \\ 65 \end{bmatrix}, \quad b = \begin{bmatrix} b_0 \\ b_1 \\ b_2 \end{bmatrix} \qquad (14B.7)
$$

$$
X'X = \begin{bmatrix} n & \sum x & \sum x^2 \\ \sum x & \sum x^2 & \sum x^3 \\ \sum x^2 & \sum x^3 & \sum x^4 \end{bmatrix} = \begin{bmatrix} 10 & 215 & 5{,}225 \\ 215 & 5{,}225 & 138{,}125 \\ 5225 & 138{,}125 & 3{,}873{,}125 \end{bmatrix} \qquad (14B.8)
$$

$$
X'y = \begin{bmatrix} \sum y \\ \sum xy \\ \sum x^2 y \end{bmatrix} = \begin{bmatrix} 821 \\ 17{,}530 \\ 418{,}750 \end{bmatrix} \qquad (14B.9)
$$

Substituting these expressions into Equation 14B.4, we obtain the normal equations*

$$
\begin{bmatrix} 10 & 215 & 5{,}225 \\ 215 & 5{,}225 & 138{,}125 \\ 5225 & 138{,}125 & 3{,}873{,}125 \end{bmatrix} \begin{bmatrix} b_0 \\ b_1 \\ b_2 \end{bmatrix} = \begin{bmatrix} 821 \\ 17{,}530 \\ 418{,}750 \end{bmatrix} \qquad (14B.10)
$$

The solution is

$$
\begin{bmatrix} b_0 \\ b_1 \\ b_2 \end{bmatrix} = \begin{bmatrix} 10 & 215 & 5225 \\ 215 & 5{,}225 & 138{,}125 \\ 5225 & 138{,}125 & 3{,}873{,}125 \end{bmatrix}^{-1} \begin{bmatrix} 821 \\ 17{,}530 \\ 418{,}750 \end{bmatrix}
$$

$$
= \begin{bmatrix} 4.888449 & -0.468434 & 0.010111 \\ -0.468434 & 0.04823 & -0.001088 \\ 0.010111 & -0.001088 & 0.00002542 \end{bmatrix} \begin{bmatrix} 821 \\ 17{,}530 \\ 418{,}750 \end{bmatrix} = \begin{bmatrix} 35.66 \\ 5.26 \\ -0.128 \end{bmatrix}
$$

$$(14B.11)$$

that is, $b_0 = 35.66$, $b_1 = 5.26$, and $b_2 = -0.128$. Space is available in this book only for a very brief sketch of this topic. For practical computation various devices are available to simplify calculations and improve the "conditioning" of the normal equations. The reader should refer to Noble (1969) and to Draper and Smith (1966).

* Specifically the normal equations are

$$
\begin{aligned}
10b_0 + \quad 215b_1 + \quad\quad 5225b_2 &= \quad\quad 821 \\
215b_0 + \quad 5225b_1 + \quad 138{,}125b_2 &= \quad 17{,}530 \\
5225b_0 + 138{,}125b_1 + 3{,}873{,}125b_2 &= 418{,}750
\end{aligned}
$$

APPENDIX 14C. ANALYSIS OF FACTORIALS, BOTCHED AND OTHERWISE

The calculation of main effects and interactions for factorial and fractional factorial designs in the manner discussed in Chapters 10 and 12 is equivalent to the use of least squares. When the runs are not perfectly carried out according to plan, as in the example shown in Table 14C.1, analysis can still be performed using least squares.

Suppose that the first-order model

$$\eta = \beta_0 + \beta_1 x_1 + \beta_2 x_2 + \beta_3 x_3 \tag{14C.1}$$

is to be fitted to the data on resiliency given in Table 14C.1 as these runs were actually

TABLE 14C.1. Two versions of a factorial design: as planned and as executed

a. As planned

	natural units			coded units		
run	temperature (°C)	pH	water (grams)	x_1	x_2	x_3
1	100	6	30	-1	-1	-1
2	150	6	30	$+1$	-1	-1
3	100	8	30	-1	$+1$	-1
4	150	8	30	$+1$	$+1$	-1
5	100	6	60	-1	-1	$+1$
6	150	6	60	$+1$	-1	$+1$
7	100	8	60	-1	$+1$	$+1$
8	150	8	60	$+1$	$+1$	$+1$

b. As executed

	natural units			coded units			
run	temperature (°C)	pH	water (grams)	x_1	x_2	x_3	resiliency
1	100	6.3	30	-1	-0.7	-1	55.7
2	150	6.5	30	$+1$	-0.5	-1	56.7
3	100	8.3	30	-1	1.3	-1	61.4
4	150	7.9	30	$+1$	0.9	-1	64.2
5	110	6.1	60	-0.6	-0.9	$+1$	45.9
6	150	5.9	60	$+1$	-1.1	$+1$	48.7
7	100	8.0	60	-1	1.0	$+1$	53.9
8	150	8.6	60	$+1$	1.6	$+1$	59.0

carried out. Then the normal equations are

$$
\begin{aligned}
8.00b_0 + 0.40b_1 + 1.60b_2 + 0.00b_3 &= 445.50 \\
0.40b_0 + 7.36b_1 - 0.16b_2 - 0.40b_3 &= 30.06 \\
1.60b_0 - 0.16b_1 + 8.82b_2 + 0.40b_3 &= 123.68 \\
0.00b_0 - 0.40b_1 + 0.40b_2 + 8.00b_3 &= -30.50
\end{aligned}
\tag{14C.2}
$$

yielding

$$
b_0 = 54.831, \qquad b_1 = 1.390, \qquad b_2 = 3.934, \qquad b_3 = -3.685 \tag{14C.3}
$$

If the factorial design had been actually carried out as planned, the sums and cross products in Equation 14C.2 would have taken specially simple values as follows:

$$
\sum_{u=1}^{n} x_{iu} = 0, \qquad i = 1, 2, 3
$$

$$
\sum_{u=1}^{n} x_{iu} x_{ju} = 0, \qquad i \neq j \tag{14C.4}
$$

$$
\sum_{u=1}^{n} x_{iu}^2 = 8, \qquad i = 1, 2, 3
$$

All but the diagonal terms in the normal equations would have vanished, and the design would be said to be *orthogonal*. Suppose that the design in Table 14C.1a had been executed perfectly and the data of Table 14C.1b had been obtained. In that case we would have had equations and estimates as follows:

$$
8b_0 = 445.5, \qquad b_0 = \frac{445.5}{8} = 55.688
$$

$$
8b_1 = 11.7, \qquad b_1 = \frac{11.7}{8} = 1.463
$$

$$
8b_2 = 31.5, \qquad b_2 = \frac{31.5}{8} = 3.938
$$

$$
8b_3 = -30.5, \qquad b_3 = \frac{-30.5}{8} = -3.813
$$

These estimates differ somewhat from those obtained previously.

To obtain the usual main effect estimates E_1, E_2, and E_3, the least squares estimates b_1, b_2, and b_3 must be multiplied by 2 (i.e., $E_i = 2b_i$) because E_i measures the change in response as the input variable changes by *two* units from the low level ($x_i = -1$) to the high level ($x_i = +1$), whereas the slope b_i measures the change in response as x_i is changed by *one* unit from $x_i = 0$ to $x_i = 1$.

APPENDIX 14D. UNWEIGHTED AND WEIGHTED LEAST SQUARES

The ordinary method of least squares produces parameter estimates having smallest variance* if the (supposedly independent) observations have equal variance. In some instances, however, the variances may not be equal but may differ in some known manner. Suppose that the observations $y_1, y_2, \ldots, y_n$ have variances $\sigma^2/w_1, \sigma^2/w_2, \ldots, \sigma^2/w_n$, where σ^2 is unknown but the constants $w_1, w_2, \ldots, w_n$ that determine the *relative* accuracy of the observations are known. These constants w are then called the *weights* of the observations. They are proportional to the reciprocals of the variances of the individual observations. If (to an adequate approximation) the observations are distributed independently, it is possible to show that then the estimation procedure yielding estimates with the smallest variance is not ordinary least squares but *weighted least squares*.

If $\eta_u = \eta_u(\boldsymbol{\beta}, \mathbf{x})$ is the mean value of the response at the uth set of experimental conditions, the weighted least squares estimates are obtained by minimizing the weighted sum of squares

$$S_* = \sum w_u(y_u - \eta_u)^2 \tag{14D.1}$$

instead of the unweighted sum of squares $S = \sum (y_u - \eta_u)^2$. If the model is nonlinear in the parameters $\boldsymbol{\beta}$, S_* can be minimized just as in the unweighted case by iterative search methods. In the particular case of a model linear in the parameters the minimum can be found algebraically by substituting weighted sums of squares and products in the normal equations. For example, for the model

$$\eta_u = \beta_1 x_{1u} + \beta_2 x_{2u} \tag{14D.2}$$

the normal equations are

$$\beta_1 \sum w_u x_{1u}^2 \quad + \beta_2 \sum w_u x_{1u} x_{2u} = \sum w_u y_u x_{1u} \tag{14D.3}$$

$$\beta_1 \sum w_u x_{1u} x_{2u} + \quad \beta_2 \sum w_u x_{2u}^2 \quad = \sum w_u y_u x_{2u}$$

The variances and covariances of the estimates are also obtained by substituting weighted for unweighted sums of squares and products. For example,

$$V(b_1) = \left[\frac{\sum w_u x_{2u}^2}{\sum w_u x_{1u}^2 \sum w_u x_{2u}^2 - (\sum w_u x_{1u} x_{2u})^2} \right] \sigma^2 \tag{14D.4}$$

Finally, with p fitted parameters the estimate of σ^2 is provided by substituting the minimum weighted sum of squares in the formula for s^2. Thus

$$s^2 = \frac{S_{*\min}}{n - p} \tag{14D.5}$$

To provide some further insight into weighted least squares we consider some examples.

---

* Strictly, smallest variance among all unbiased estimates that are linear functions of the data.

Example 1: Replicated Data

Suppose that we had some data for which the variance was constant but in which certain observations were genuine replicates of each other, as in the following example.

	experimental conditions		
	1	2	3
observations	63 67 62	71 75	87 89 83 85
averages	64	73	86
number of observations	3	2	4

It need not concern us what form the model actually takes. The three experimental conditions could be three levels of temperature to which it was believed that the response was linearly related, or three different treatments. In either case, for any given values of the model parameters, the mean value η will be constant for *each* replicated set of experimental conditions. Let it be η_1 for the first set, η_2 for the second, and η_3 for the third. Then, (correctly) using ordinary least squares for the individual observations, we estimate the model parameters by minimizing

$$S = (63 - \eta_1)^2 + (67 - \eta_1)^2 + (62 - \eta_1)^2 + (71 - \eta_2)^2 + (75 - \eta_2)^2$$
$$+ (87 - \eta_3)^2 + (89 - \eta_3)^2 + (83 - \eta_3)^2 + (85 - \eta_3)^2 \qquad (14D.6)$$

Alternatively, we can regard the data as consisting of the three averages 64, 73, 86. Since these data have variances $\sigma^2/3$, $\sigma^2/2$, and $\sigma^2/4$, appropriate weights are $w_1 = 3$, $w_2 = 2$, $w_3 = 4$. Fitting these by weighted least squares, we minimize

$$S_* = 3(64 - \eta_1)^2 + 2(73 - \eta_2)^2 + 4(86 - \eta_3)^2 \qquad (14D.7)$$

For *any* fitted model this will give exactly the same parameter values as minimizing S with the individual data values.

Consider, for example, the first three terms in S; these have an average of 64, so that

$$(63 - \eta_1)^2 + (67 - \eta_1)^2 + (62 - \eta_1)^2$$
$$= [(63 - 64)^2 + (67 - 64)^2 + (62 - 64)^2] + 3(64 - \eta_1)^2 \quad (14D.8)$$

The right-hand side of the expression consists of a within-group sum of squares in brackets *not containing* η_1 (hence not containing the parameters of the model) and a second part that is the first term in S_*. If the pooled within-group sum of squares is denoted by S_W, the total sum of squares is therefore

$$S = S_w + S_* \qquad (14D.9)$$

Since S_w contains only deviations from the group averages that are independent of the model used, minimization of S implies minimization of S_*, and vice versa.

This example shows that weighted least squares are like ordinary least squares, in which observations are replicated in proportion to their weights, $w_1, w_2, \ldots, w_n$. Suppose, say, that a straight line is fitted to data $(x_1, y_1), (x_2, y_2), \ldots, (x_n, y_n)$ and the weights are $w_1, w_2, \ldots, w_n$. This equivalent to fitting by ordinary least squares to $\bar{y}$'s from runs replicated $w_1, w_2, \ldots, w_n$ times.

If the weights do not greatly differ, the weighted estimates will not be greatly different from the unweighted estimates. The following example illustrates what can be a contrary situation.

Example 2: Estimating the Slope of a Straight Line Passing through the Origin

Consider a model expressing the fact that the mean response is proportional to x:

$$y = \beta x + \epsilon \qquad (14D.10)$$

with the errors ϵ independently distributed. Then, if the variance of ϵ is constant, we know that the least squares estimate of β is $b = \sum xy / \sum x^2$. It quite often happens, however, that σ_ϵ^2 increases as the mean value $E(y) = \eta$ of the response increases. What difference should this make to the estimates? The following table shows three possibilities.

relationship between σ and η	weighting	estimate $b = \sum wxy / \sum wx^2$
I $\quad \sigma_\epsilon^2$ constant	$w = $ constant	$b_{\mathrm{I}} = \sum xy / \sum x^2$
II $\quad \sigma_\epsilon^2 \propto \eta \propto x$	$w \propto x^{-1}$	$b_{\mathrm{II}} = \bar{y}/\bar{x}$
III $\quad \sigma_\epsilon^2 \propto \eta^2 \propto x^2$	$w \propto x^{-2}$	$b_{\mathrm{III}} = \sum (y/x)/n$

The estimate b_{I} is the usual least squares estimate, appropriate if the variance is independent of the level of response. If, however, the variance is proportional to η (and hence to x), the appropriate estimate b_{II} is $\bar{y}/\bar{x}$. Thus the best fitting line is obtained by joining the origin to the point $(x = \bar{x}, y = \bar{y})$. Finally, if the standard deviation is proportional to η, so that the variance is proportional to η^2 and hence to x^2, then $b_{\mathrm{III}} = \sum (y/x)/n$. The best fitting straight line has a slope that is the average of all the slopes of the lines joining the individual points to the origin.

Example 3: Weighted Least Squares by the Use of Pseudovariates

Consider any model that is linear in the parameters

$$y_u = \sum_{i=1}^{p} \beta_i x_{iu} + \epsilon_u \qquad (14D.11)$$

and suppose the weight of the uth observation is w_u [i.e., $V(y_u) = \sigma^2/w_u$]. It is sometimes convenient to perform the weighted analysis by carrying out ordinary least squares on a constructed set of pseudovariates:

$$Y_u = \sqrt{w_u}\, y_u, \qquad X_{iu} = \sqrt{w_u}\, x_{iu} \qquad (14D.12)$$

Thus the model becomes

$$Y_u = \sum_{i=1}^{p} \beta_i X_{iu} + e_u$$

in which the errors e_u have constant variance since $e_u = \sqrt{w_u}\epsilon_u$ and $V(e_u) = \sigma^2$. An ordinary unweighted least squares analysis using Y_u and X_{iu} will now yield the correct *weighted* least squares estimates and their standard errors. The reader should ᴜnfirm this statement by substituting $Y_u = \sqrt{w_u}y_u$ and $X_{iu} = \sqrt{w_u}x_{iu}$ in the ordinary least squares formulas and noting that they correctly give the weighted least squares formulas.

REFERENCES AND FURTHER READING

For further information on the method of least squares, see:

Draper, N. R., and H. Smith (1966). *Applied Regression Analysis*, Wiley.
Daniel, C., and F. S. Wood (1971). *Fitting Equations to Data*, Wiley.
Plackett, R. L. (1960). *Regression Analysis*, Oxford University Press.
Box, G. E. P., and G. M. Jenkins (1970). *Time Series Analysis: Forecasting and Control*, Holden Day.

The data for the example on biochemical oxygen demand in Section 14.6 were taken from this article:

Marske, D. M., and L. B. Polkowski (1972). Evaluation of methods for estimating biochemical oxygen demand parameters, *J. Water Pollut. Control Fed.*, **44**, No. 10, 1987.

If errors are correlated, statistical methods based on the assumption of independence of errors can give very misleading results. A dramatic illustration of this point is provided by the following pair of papers:

Coen, P. G., E. D. Gomme, and M. G. Kendall (1969). Lagged relationships in economic forecasting, *J. Roy. Stat. Soc., Ser. A*, **132**, 133.
Box, G. E. P., and Paul Newbold (1971). Some comments on a Paper of Coen, Gomme, and Kendall, *J. Roy. Stat. Soc., Ser. A*, **134**, 229.

A survey of optimal design theory is given in this article:

St. John, R. C., and N. R. Draper (1975). D-optimality for regression designs: a review, *Technometrics*, **17**, 15.

The basics of matrices are very clearly explained in the following book:

Aitken, A. C. (1951). *Determinants and Matrices*, University Mathematical Texts, Interscience, New York.

The practice and theory of modeling and dynamic system operation under closed-loop control is described in the following papers:

Box, G. E. P., and J. F. MacGregor (1974). The analysis of closed-loop dynamic-stochastic systems, *Technometrics*, **16**, 391.

Box, G. E. P., and J. F. MacGregor (1976). Parameter estimation with closed loop operating data, *Technometrics*, **18**, 371.

An excellent text which explains the ill-conditioning problem in least squares analysis is

Noble, B., (1969). *Applied Linear Algebra*, Prentice Hall (First Edition), Section 8.8, p. 260.

QUESTIONS FOR CHAPTER 14

1. What is the method of least squares, and why is it useful?
2. There are two fundamentally different ways of using a fitted least squares equation. What are they?
3. What is the definition of a linear model?
4. Can the method of least squares be used for models that are nonlinear in the parameters? If not, why? If so, explain how.
5. Why should residuals always be calculated and examined?
6. What are normal equations? Is there any connection between them and the normal distribution? Why are they called *normal* equations?
7. If lurking variables are present, why is it essential that randomization be employed if the fitted equation is to be used as the basis for manipulating the setting of one or more variables in the hope of favorably changing the response?
8. Are there situations in which lurking variables are present but fitted equations may be properly used?
9. What are some of the common pitfalls associated with the use of regression analysis?

CHAPTER 15

Response Surface Methods

Response surface methodology consists of a group of techniques used in the empirical study of relationships between one or more measured responses such as yield, color index, and viscosity, on the one hand, and a number of input variables such as time, temperature, pressure, and concentration, on the other. The techniques have been used to answer questions of a number of different kinds, such as the following:

1. How is a particular response affected by a given set of input variables over some specified region of interest?
2. What settings, if any, of the inputs will give a product simultaneously satisfying desired specifications?
3. What values of the inputs will yield a maximum for a specific response, and what is the response surface like close to this maximum?

In what follows we present a summary account with references for further reading.

15.1. WEAKNESS OF CLASSICAL ONE-VARIABLE-AT-A-TIME STRATEGY: CHEMICAL EXAMPLE

We begin by discussing the third application mentioned above, namely, the determination of the approximate location and appearance of a maximum.* Specifically, suppose that a chemist wishes to maximize the yield of a chemical reaction by varying reaction time (t) and reaction temperature (T)

---

* If the response y were, for example, cost of production, one would be interested in a minimum, not a maximum. Similar principles apply, of course, in both situations. (For example, minimizing y is equivalent to maximizing $c - y$ where c is any constant.)

510

The Classical One-Variable-at-a-Time Strategy

If the chemist employed the classical one-variable-at-a-time approach, he might follow the course illustrated in Figure 15.1. In Figure 15.1a temperature has been fixed at $T = 225°C$, and reaction time t has been varied from about 60 to 180 minutes. A rough graph drawn through the points leads to the conclusion that for this fixed temperature the best reaction time is 130 minutes, at which point the yield is about 75 grams.

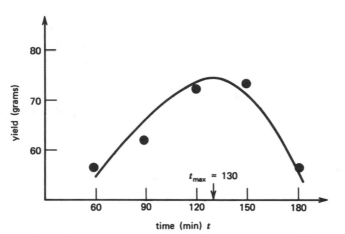

(a) First set of experiments: yield versus reaction time, temperature held fixed at 225°C.

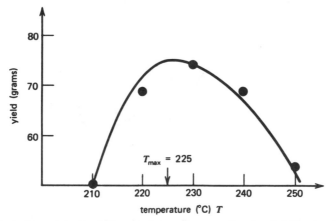

(b) Second set of experiments: yield versus temperature, reaction time held fixed at 130 minutes.

FIGURE 15.1. Hypothetical results from one-variable-at-a-time approach.

Following the classical one-variable-at-a-time strategy, the experimenter would now fix the reaction time t at this "best" value of 130 minutes and might vary the temperature T as in Figure 15.1b, leading to the conclusion that, at this "best" value of time, the best value of temperature is not far from the temperature of 225°C used for the first series of runs; again a yield of roughly 75 grams of product is obtained. Now the conclusion might seem justified that an *overall* maximum yield of about 75 grams is achieved with the reaction conditions $t = 130$ minutes and $T = 225$°C.

Certainly the graphs show that, if *either* time or temperature is *individually* increased or decreased from these conditions, a reduction in yield will occur. What they do not establish is what might happen if these variables were changed, not individually, but *together*. To understand the possible nature of the joint effect of time and temperature on yield we must think in terms of joint functional dependence of mean yield on time and temperature. This dependence can be conveniently represented by a yield contour diagram like that in Figure 15.2, which shows a commonly occurring kind of response

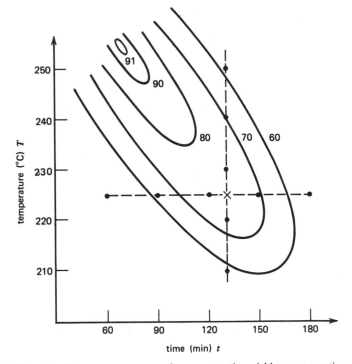

FIGURE 15.2. Possible true response surface representing yield versus reaction time and temperature, with points shown for one-variable-at-a-time approach.

function and is completely concordant with the graphs in Figure 15.1. However, if Figure 15.2 represented the true relationship, the actual maximum yield would be about 91 grams, not 75 grams. Also this yield would be achieved at conditions of time and temperature of about $t = 65$ minutes and $T = 255°C$, which are quite different from the conditions $t = 130$ minutes and $T = 225°C$ found by the one-variable-at-a-time method.

Study of Figures 15.1 and 15.2 shows that the one-variable-at-a-time strategy fails in this example because it tacitly assumes that the maximizing value of one variable is independent of the level of the other. Usually this is not true.

We now illustrate an alternative, usually better procedure, in which the experimenter first employs the method of steepest ascent to get close to the maximum and then approximates the surface in the vicinity of the maximum with a fitted second-degree equation.

15.2. ILLUSTRATION OF RESPONSE SURFACE METHODOLOGY: CHEMICAL EXAMPLE

The objective of the following laboratory investigation was to find settings of time (t) and temperature (T) that produced maximum yield subject to constraints discussed later. The best conditions known at the beginning of the work were $t = 75$ minutes, $T = 130°C$, and past experience had suggested that the experimental error standard deviation was about $\sigma = 1.5$.

A First-Order Design

Taking as a starting point the best conditions known, the investigator varied time from 70 to 80 minutes and temperature from 127.5 to 132.5°C according to the design shown in Table 15.1. The levels of the variables in coded units were

$$x_1 = \frac{\text{time} - 75 \text{ minutes}}{5 \text{ minutes}}, \qquad x_2 = \frac{\text{temperature} - 130°C}{2.5°C} \qquad (15.1)$$

The design employed was a 2^2 factorial with three center points; it is indicated by crosses in the bottom left corner of Figure 15.3. It is called a *first-order design* because it allows efficient fitting and checking of the first-degree polynomial model,

$$y = \beta_0 + \beta_1 x_1 + \beta_2 x_2 + \epsilon \qquad (15.2)$$

It was chosen because, at this stage of the investigation, the experimenter believed, but was not certain, that he was some distance away from the

TABLE 15.1. **Results from first factorial design with three center points, chemical example**

	variables in original units		variables in coded units		response: yield (grams)
run*	time (min)	temperature (°C)	x_1	x_2	y
1	70	127.5	-1	-1	54.3
2	80	127.5	$+1$	-1	60.3
3	70	132.5	-1	$+1$	64.6
4	80	132.5	$+1$	$+1$	68.0
5	75	130.0	0	0	60.3
6	75	130.0	0	0	64.3
7	75	130.0	0	0	62.3

* The random order in which runs were actually performed was 5, 4, 2, 6, 1, 7, 3.

maximum (some way down the smooth hillside that represents the true response surface). In these circumstances it was likely that the predominant local characteristics of the surface were its gradients and that the local surface could be roughly represented by the planar model (Equation 15.2) having slope β_1 in the x_1 direction and slope β_2 in the x_2 direction. If this idea was correct, by estimating β_1 and β_2 it should be possible to follow a direction of increasing yield up the hillside. The design chosen

1. allows the planar model to be efficiently *fitted*,
2. allows *checks* to be made to determine whether the planar model is representationally adequate, and
3. provides some estimate of experimental error.

Least Squares Fit

The least squares estimate* of β_1 is

$$b_1 = \tfrac{1}{4}[(-1 \times 54.3) + (1 \times 60.3) + (-1 \times 64.6) + (1 \times 68.0)] = 2.35$$

$$(15.3)$$

* The coefficient β_1 is the change that occurs in the response when x_1 is changed by one unit. The linear main effect in a factorial design is the change in response when x_1 is changed from -1 to $+1$, that is, by *two* units. Thus b_1 is half the linear main effect obtained from the difference of averages at the shorter and longer reaction times.

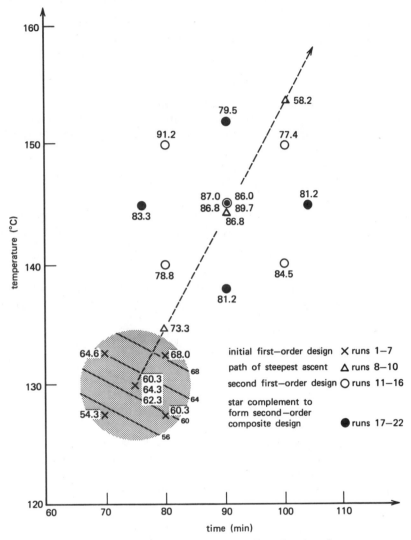

FIGURE 15.3. Data from first factorial design (runs 1–7), exploration of steepest ascent path (runs 8–10), second factorial design (runs 11–16), and added star (runs 17–22).

515

Similarly b_2 is 4.50. The least squares estimate of β_0 is the average of all seven observations, 62.01. We thus obtain the fitted equation

$$\hat{y} = 62.01 + 2.35x_1 + 4.50x_2$$
$$(\pm 0.57) \quad (\pm 0.75) \qquad (\pm 0.75) \tag{15.4}$$

where the standard errors shown beneath the coefficients of the equation are calculated on the assumption that $\sigma = 1.5$. Although least squares calculation tentatively assumes adequacy of the first-degree (planar) model, the design was chosen to allow checks of this assumption to be made.

Interaction Check

The planar model supposes that the effects of the variables are additive. Interaction between the variables would be measured by the coefficient β_{12} of an added cross-product term $x_1 x_2$ in the model.* This coefficient is estimated by

$$b_{12} = \tfrac{1}{4}[(1 \times 54.3) + (-1 \times 60.3) + (-1 \times 64.6) + (1 \times 68.0)]$$
$$= -0.65 \tag{15.5}$$

The standard error of this estimate is 0.75, the same as it is for b_1 and b_2 (we use $\sigma = 1.5$, as before).

Curvature Check

Another check of local planarity is supplied by comparing $\bar{y}_f$, the average of the four points of the 2^2 factorial, with $\bar{y}_c$, the average at the center of the design. By thinking of the design as sitting on a saucerlike surface, it is seen that $\bar{y}_f - \bar{y}_c$ is a measure of overall curvature of the surface. It can also be shown that, if β_{11} and β_{22} are coefficients of the terms x_1^2 and x_2^2, this curvature measure will be an estimate of $\beta_{11} + \beta_{22}$. Thus the estimate of the "overall curvature" is

$$b_{11} + b_{22} = \tfrac{1}{4}(54.3 + 60.3 + 64.6 + 68.0) - \tfrac{1}{3}(60.3 + 62.3 + 64.3)$$
$$= -0.50 \tag{15.6}$$

Using $\sigma = 1.5$, we obtain 1.15 for the standard error of this estimate.
   In summary, then, the planarity checking functions are

$$b_{12} = -0.65 \pm 0.75, \qquad b_{11} + b_{22} = -0.50 \pm 1.15 \tag{15.7}$$

and there is no reason to question the adequacy of the planar model in this example.

---

* The coefficient b_{12} is half the interaction effect calculated in the ordinary factorial design.

Exercise 15.1. Calculate the standard errors of b_{12} and $(b_{11} + b_{22})$, given that $\sigma = 1.5$. (*Hint*: Write these statistics as linear combinations of the observations, and use Equation 3A.9.) *Answer*: 0.75, 1.15.

Estimation of Error

An estimate of the experimental error variance, which is very approximate since it has only two degrees of freedom, is obtained from the replicate observations at the center:

$$s_1^2 = \frac{60.3^2 + 62.3^2 + 64.3^2 - (186.9)^2/3}{2} = 4.0 \tag{15.8}$$

Thus

$$s_1 = 2.0 \tag{15.9}$$

This agrees fairly well with the preliminary value of $\sigma = 1.5$ used above.

Fitted Equation and Contour Diagram

The data are shown graphically in Figure 15.3, together with the contours of the fitted first-degree equation:

$$\hat{y} = 62.01 + 2.35x_1 + 4.50x_2 \tag{15.10}$$

The equation of the $\hat{y}_0$ contour of the fitted plane is obtained by substituting $\hat{y}_0$ into Equation 15.10 to obtain $62.01 - \hat{y}_0 + 2.35x_1 + 4.50x_2 = 0$. Successively setting $\hat{y}_0 = 56, 60, 64, 68$ gives the set of parallel equally spaced straight line contours of Figure 15.3. These contour lines can be tentatively accepted as a rough geometrical representation of the underlying response function *over the experimental region explored thus far*.

Exercise 15.2. Do the necessary least squares calculations to obtain Equation 15.10.

Exercise 15.3. Add 10 units to each of the response values in Table 15.1, and obtain the least squares estimates of the parameters β_0, β_1, and β_2 in the first-order model, Equation 15.2. *Answer*: $b_0 = 72.01$, $b_1 = 2.35$, $b_2 = 4.50$.

Path of Steepest Ascent

The path of steepest ascent,* which is perpendicular to the contour lines, is indicated in Figure 15.3. It can be calculated without drawing contours of the fitted plane as follows: starting at the center of the experimental region, the

---

* This is the direction of *steepest* ascent if the space is measured in the relative units in which the experimenter currently has chosen to scale the design. These are the units in which the factorial design appears as a square.

TABLE 15.2. Points on the path of steepest ascent, chemical example

	coded conditions		time (min)	temperature (°C)		observed
	x_1	x_2	t	T	run	yield
center conditions	0	0	75	130.0	5, 6, 7	62.3 (average)
	1	1.91	80	134.8	8	73.3
	2	3.83	85	139.6		
path of steepest	3	5.74	90	144.4	10	86.8
ascent	4	7.66	95	149.1		
	5	9.57	100	153.9	9	58.2

path is followed by simultaneously moving $b_2 = +4.50$ units in x_2 for every $b_1 = +2.35$ units moved in x_1, or, equivalently, $4.50/2.35 = 1.91$ units in x_2 for every 1 unit in x_1. A convenient set of points on the path of steepest ascent is shown in Table 15.2.

Exercise 15.4. Suppose that the first response in Table 15.1 had been 50.3 instead of 54.3 and the fourth had been 72.0 instead of 68.0. Fit the first-order model, determine the standard errors for b_0 and b_1, and b_2, compute the statistics b_{12} and $b_{11} + b_{22}$ with their standard errors, determine whether there is evidence of lack of fit, and plot the path of steepest ascent. (Assume $\sigma = 1.5$).
Answer:

$$\hat{y} = 62.01 + 4.35x_1 + 6.50x_2, \quad b_{12} = -0.65 \pm 0.75, \quad b_{11} + b_{22} = -0.50 \pm 1.15$$
$$(\pm 0.57) \quad (\pm 0.75) \quad (\pm 0.75)$$

There is no evidence of lack of fit.

Runs made at points 8, 9, and 10 gave yields as indicated. Run 8 ($y = 73.3$) was encouraging and led to a large jump being taken to run 9. This yield ($y = 58.2$) was low, however, indicating that too large a move had been made. Run 10 ($y = 86.8$) at intermediate conditions provided a very substantial improvement over any results obtained so far. The steepest ascent points are indicated by triangles in Figure 15.3. A graph of these results suggests that subsequent experiments should be made in the neighborhood of run 10.

As the region of interest moves up the surface, it can be expected that first-order effects will become smaller. Since the blurring effects of experimental errors have to be faced, the opportunity may be taken at this time to broaden the second design by a factor of, say, two to increase the absolute magnitude of the effects in relation to the error. (Furthermore, as a region of interest moves up the surface, the possibility increases that a second-degree approximation will be needed. Expanding the design is also sensible in this eventuality because a second-degree approximation should provide an adequate representation over a larger region than a first-degree approximation.)

A Second Design

A new 2^2 factorial design with two center points situated close to run 10 had coded variables

$$x_1 = \frac{\text{time} - 90 \text{ minutes}}{10 \text{ minutes}}, \qquad x_2 = \frac{\text{temperature} - 145°C}{5°C} \qquad (15.11)$$

The design points are indicated by open circles in Figure 15.3. The data obtained from the six runs performed in random order are shown in Table 15.3 (runs 11 through 16). Analyzing this second first-order design, we obtain the following results.

TABLE 15.3. Second factorial design augmented with star design to form a second-order (composite) design, chemical example

run*	variables in original units		variables in coded units		response: yield (grams)	
	time (min)	temperature (°C)	x_1	x_2		
11	80	140	-1	-1	78.8	
12	100	140	$+1$	-1	84.5	second
13	80	150	-1	$+1$	91.2	first-order
14	100	150	$+1$	$+1$	77.4	design
15	90	145	0	0	89.7	
16	90	145	0	0	86.8	
17	76	145	$-\sqrt{2}$	0	83.3	runs
18	104	145	$+\sqrt{2}$	0	81.2	added to
19	90	138	0	$-\sqrt{2}$	81.2	form a
20	90	152	0	$+\sqrt{2}$	79.5	composite
21	90	145	0	0	87.0	design
22	90	145	0	0	86.0	

* The order of these runs was randomized. The first six were obtained in the order 11, 15, 13, 12, 16, 14, and the next six in the order 19, 20, 17, 21, 22, 18.

Least Squares Fit

On the assumption that a first-degree polynomial model is again applicable, least squares estimation yields the fitted equation

$$y = 84.73 - 2.025x_1 + 1.325x_2 \qquad (15.12)$$
$$(\pm 0.61) \quad (\pm 0.75) \qquad (\pm 0.75)$$

Exercise 15.5. Do the calculations necessary to obtain Equation 15.12.

Interaction and Curvature Checks

The checking functions yield the values

$$b'_{12} = -4.88 \pm 0.75, \qquad b'_{11} + b'_{22} = -5.28 \pm 1.15 \qquad (15.13)$$

from which it is evident that in the present region the first-degree equation is quite inadequate to represent the local response function.

Estimation of Error

An error estimate having one degree of freedom is provided by

$$s_2^2 = 89.7^2 + 86.8^2 - \frac{(176.5)^2}{2} = \frac{(89.7 - 86.8)^2}{2} = 4.21 \qquad (15.14)$$

Thus
$$s_2 = 2.1 \qquad (15.15)$$

By combining this estimate with the previous one (Equation 15.8), we obtain a pooled estimate or error variance having three degrees of freedom:

$$s^2 = \frac{(2 \times 4.00) + (1 \times 4.21)}{3} = 4.07 \qquad (15.16)$$

$$s = 2.02 \qquad (15.17)$$

which again agrees reasonably well with the prior estimate, $\sigma = 1.5$.

Augmenting the Design to Fit a Second-Order Model

Since the first-degree polynomial approximation had been shown to be quite inadequate in the new experimental region, a second-degree polynomial approximation,

$$y = \beta_0 + \beta_1 x_1 + \beta_2 x_2 + \beta_{11} x_1^2 + \beta_{22} x_2^2 + \beta_{12} x_1 x_2 + \epsilon \qquad (15.18)$$

was now contemplated.

To estimate efficiently all six coefficients in this model and to provide for appropriate checking and error determination, the second factorial group of points (runs 11 through 16) was augmented with a "star" design consisting of four axial points and two center points (runs 17 through 22 in Table 15.3, indicated by black dots in Figure 15.3). The two parts of the resulting second-order *composite design* are thus shown by open circles and black dots in Figure 15.3.

Least Squares Fit

The second-degree equation fitted by least squares to the 12 results from the composite design (runs 11 through 22) is

$$\hat{y} = 87.36 - 1.39x_1 + 0.37x_2 - 2.15x_1^2 - 3.12x_2^2 - 4.88x_1x_2 \quad (15.19)$$

The contours of this fitted equation are shown in Figure 15.4.

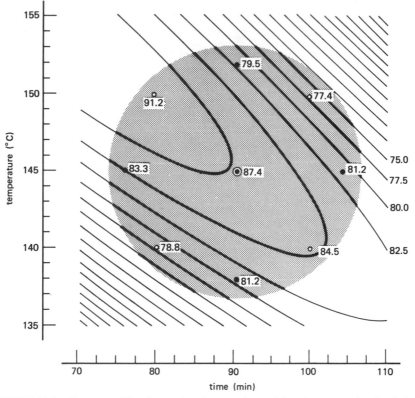

FIGURE 15.4. Contours of fitted second-order equation and data from second-order-design (runs 11–22).

Exercise 15.6.   Do the calculations necessary to obtain Equation 15.19.

Exercise 15.7.   Using Equation 15.19, obtain the predicted value $\hat{y}$ for time $t = 100$ minutes and temperature $T = 152°C$. Does this value agree with the one read from Figure 15.4?   *Answer:* $\hat{y} = 71.20$, yes.

Checks

We have seen that the first-order design was chosen so that estimates of selected second-order terms or combinations of them could be isolated and thereby supply checks on the adequacy of the first-degree equation. The composite design has been chosen in the same way so that selected combinations of third-order terms can be isolated to check the adequacy of the second-degree equation. For example, consider the distribution of points along the x_1 axis shown in Figure 15.5. If the surface is *exactly quadratic* in this direction,

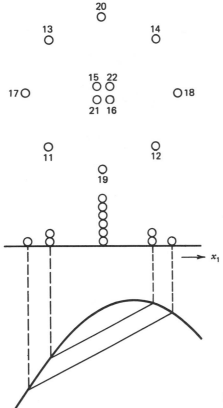

FIGURE 15.5.   Nature of lack of fit contrasts from second-order composite design.

it can be shown that the estimate of slope obtained from the axial points y_{17} and y_{18} will be the same as that obtained from the factorial points y_{11}, y_{12}, y_{13}, and y_{14}. A measure of the discrepancy in the two measures of slope is

$$\frac{1}{\sqrt{8}}(y_{18} - y_{17}) - \tfrac{1}{4}(y_{12} + y_{14} - y_{11} - y_{13})$$

We will denote this measure by $b_{111} - b_{122}$, since it can be shown that this expression supplies an estimate of $\beta_{111} - \beta_{122}$ (where β_{111} and β_{122} are coefficients of x_1^3 and of $x_1 x_2^2$, respectively, in a third-degree polynomial). Both of the latter coefficients would be zero if the surface were described by a second-degree equation. If we consider the distribution of points along the x_2 axis, a similar estimate $b_{222} - b_{112}$ of $\beta_{222} - \beta_{112}$ is obtained. For these data we find

$$b_{111} - b_{122} = 1.28 \pm 1.06 \qquad (15.20)$$

$$b_{222} - b_{112} = -1.93 \pm 1.06 \qquad (15.21)$$

There is indication of some slight model inadequacy in the x_2 direction, but we shall ignore it in this elementary account.*

Estimate of Error

A further error estimate having one degree of freedom is provided by

$$s_3^2 = \frac{(87.0 - 86.0)^2}{2} = 0.5 \qquad (15.22)$$

The pooled estimate with four degrees of freedom is now updated to

$$s^2 = \frac{(2 \times 4.0) + (1 \times 4.2) + (1 \times 0.5)}{4} = 3.18 \qquad (15.23)$$

$$s = 1.78 \qquad (15.24)$$

Blocking

The composite design used in this experiment has another interesting feature. Some days elapsed between the completion of the first and second parts of the design, thereby increasing the possibility that conditions might have

---

* When a particular model does not fit adequately, instead of immediately considering a higher-order polynomial model it is often advantageous to contemplate improvement by transformations of the variables and/or the response, or to consider some entirely different form of model, for example, a model nonlinear in the parameters. See, for instance, Box and Tidwell (1962), Box and Cox (1964), and Draper and Hunter (1969).

changed between the two sets of runs. The design used, however, has the remarkable property that its two parts are orthogonal blocks; this means that arbitrary constants added to the response in either block or in both do not change the estimates $b_1, b_2, b_{11}, b_{22}, b_{12}$ of the coefficients of the fitted second-degree polynomial. Also, an estimate of the possible change that occurred between the first and second blocks is provided for this design by the difference $\bar{y}_1 - \bar{y}_2$, where $\bar{y}_1$ is the arithmetic average of the first six results, and $\bar{y}_2$ the arithmetic average of the second six results. For these data we obtain

$$\text{block effect} = \bar{y}_1 - \bar{y}_2 = 84.73 - 83.03 = 1.70 \qquad (15.25)$$

where the standard error of this difference is given by $\sqrt{\sigma^2/6 + \sigma^2/6} = 1.5/\sqrt{3} = 0.87$. In this trial there is some evidence that somewhat lower yields were obtained in the second block of six runs.

Nature of the Fitted Surface

Before attempting to interpret the fitted surface, we need to consider whether or not it is estimated with sufficient precision. This can be done by a special application of the analysis of variance (see Box and Wetz, 1973). However, in this present short survey we approach the problem from a simpler but equivalent point of view.

It can be shown that, no matter what the design, the *average* variance of the fitted values $\hat{y}$ is

$$\bar{V}(\hat{y}) = \frac{1}{n} \sum_{i=1}^{n} V(\hat{y}_i) = \frac{p\sigma^2}{n} \qquad (15.26)$$

where p is the number of parameters fitted. In this example $p = 6$ and $n = 12$, so that the average variance of the fitted value is $6(1.5)^2/12 = 1.125$, and the corresponding average standard error of the $\hat{y}$'s is $\sqrt{1.125} = 1.06$. On the other hand, the fitted $\hat{y}$'s (see Figure 15.4) range from about 77 to 90. Thus in this example (1) we have failed to show any substantial lack of fit, and (2) the predicted change of $\hat{y}$ is 12 times the average standard error of $\hat{y}$. It therefore appears worthwhile* to interpret the fitted surface.

Exercise 15.8. Using Equations 15.10 and 15.26, determine the average variance for $\hat{y}$ in the first seven runs (Table 15.1). For those data what are the largest and smallest values for $\hat{y}$? Does this analysis indicate that the fitted Equation 15.10 approximates the true surface with sufficient precision to allow it to be interpreted and used?

Answer: 0.965, 66.86, 55.16, yes.

---

* A common misapplication of response surface methodology is to attempt to interpret an inadequately estimated response function.

We see from the contour plot of the fitted equation in Figure 15.4 that it represents an oblique rising ridge with yields increasing from the lower right to the top left corner of the diagram, that is, yield increases as we progressively *increase* temperature and simultaneously *reduce* reaction time. If yield had been the only response of importance, subsequent experimentation would have followed and further explored this rising ridge. In fact, in this application other considerations were of importance.

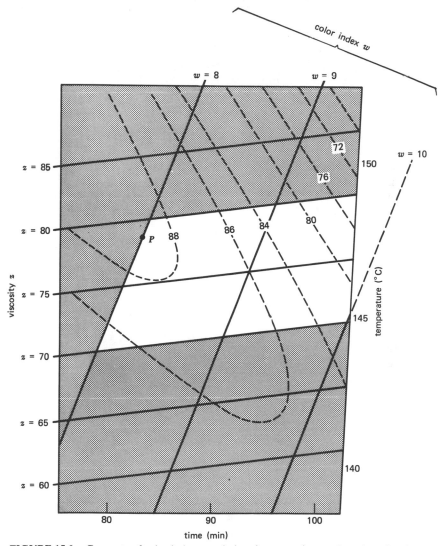

FIGURE 15.6. Contours of color index w and viscosity z superimposed on those for yield y.

15.3. A SPECIFICATION PROBLEM

In this investigation not only the yield y but also the color index w and the viscosity z of the product were of importance and were measured for each run. It was desired to maintain the color index at a value greater than 8 and to maintain the viscosity of the product between 70 and 80 units. For both color index w and viscosity z, linear response relationships were found to be adequate over the region of interest, and contours of the fitted first-degree equations for these responses are shown in Figure 15.6. The estimated contours suggest that at or near the point marked P a product is obtained, giving close to the highest possible yield (about 88 grams) consonant with satisfaction of the specifications on color index and viscosity.

The advantages of the "contour overlay" approach of Figure 15.6 are that it (1) clearly shows how (and whether) complex requirements may be met, (2) allows new conditions to be determined if specifications change, and (3) indicates in what direction process changes should be made if the characteristics of the manufactured products stray outside specification.

This approach also serves to spotlight problems that need to be solved. For example, suppose that the contour overlay had shown that the desired specifications could not be met simultaneously because the contour $w = 8$ and all higher w contours were below, and parallel to, the band represented by the inequality $70 < z < 80$. With the basic nature of the problem thus visualized and understandable, the experimenter could be led to consider what *new* variable might be employed to move the band of acceptable viscosity downward to overlap the color index constraint. The technique thus helps the investigator at the difficult *inductive* stage of scientific learning.

15.4. MAXIMA, RIDGES, AND CANONICAL ANALYSIS

The strategy we have outlined (1) uses first-order designs to determine the local slope of the surface, (2) employs steepest ascent to approach a maximal region, and (3) uses a second-order design to obtain a local representation of this maximal region. We now consider this fitted second-degree equation somewhat more carefully, using for illustration the case of just two variables. The fitted equation will be of the form

$$\hat{y} = b_0 + b_1 x_1 + b_2 x_2 + b_{11} x_1^2 + b_{22} x_2^2 + b_{12} x_1 x_2 \qquad (15.27)$$

Now, depending on the coefficients $b_0, b_1, b_2, b_{11}, b_{22}, b_{12}$, this equation can take a number of different forms. Figure 15.7 shows contours for four hypothetical fitted second-degree equations. The corresponding contour systems are obviously of very different kinds, but it is not easy to visualize what a

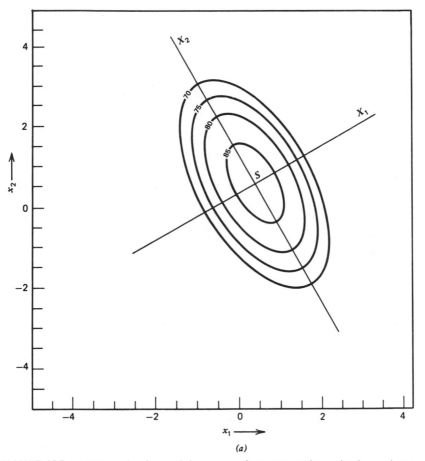

(a)

FIGURE 15.7. (a) Example of second-degree equations representing a simple maximum.

$$\hat{y} = 83.57 + 9.39x_1 + 7.12x_2 - 7.44x_1^2 - 3.71x_2^2 - 5.80x_1x_2$$
$$\hat{y} - 87.69 = -9.02X_1^2 - 2.13X_2^2$$

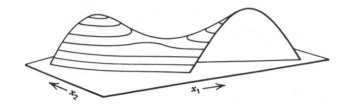

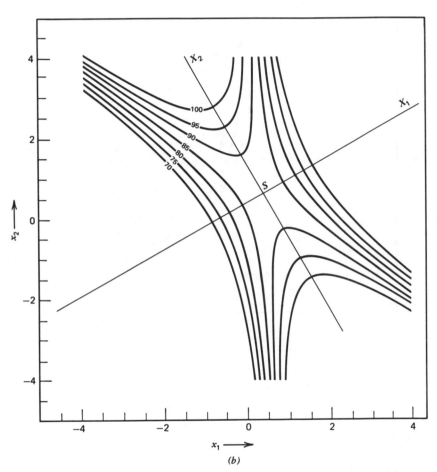

FIGURE 15.7. (*b*) Example of second-degree equations representing a minimax.

$$\hat{y} = 84.29 + 11.06x_1 + 4.05x_2 - 6.46x_1^2 - 0.43x_2^2 - 9.38x_1x_2$$
$$\hat{y} - 87.69 = -9.02X_1^2 + 2.13X_2^2$$

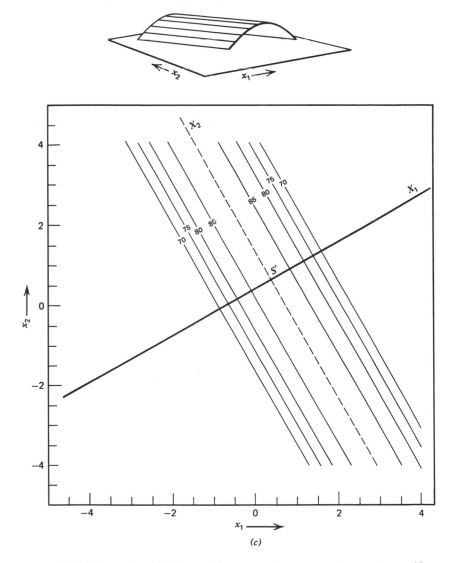

FIGURE 15.7. (c) Example of second-degree equations representing a stationary ridge.

$$\hat{y} = 83.93 + 10.23x_1 + 5.59x_2 - 6.95x_1^2 - 2.07x_2^2 - 7.59x_1x_2$$
$$\hat{y} - 87.69 = -9.02X_1^2 + 0.00X_2^2$$

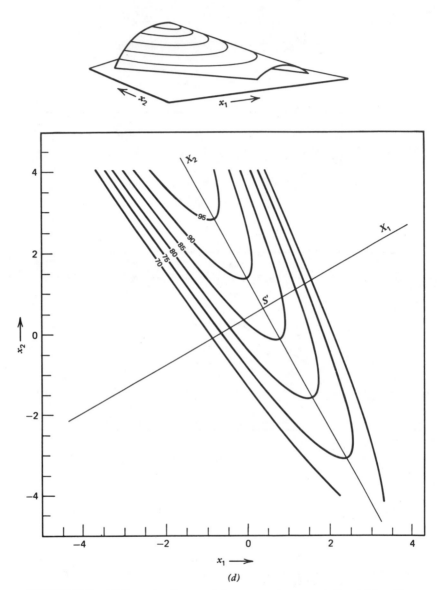

FIGURE 15.7. (d) Example of a second-degree equation representing a rising ridge.

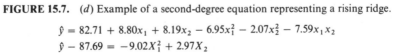

$$\hat{y} = 82.71 + 8.80x_1 + 8.19x_2 - 6.95x_1^2 - 2.07x_2^2 - 7.59x_1x_2$$
$$\hat{y} - 87.69 = -9.02X_1^2 + 2.97X_2$$

surface will look like simply by examining the magnitudes of the coefficients b_0, b_1, ..., b_{12} in Equation 15.27. By canonical analysis the equation is rewritten in such a way that its fundamental nature is clearly revealed.

Canonical Analysis

Canonical analysis consists of (1) shifting the origin to a new point S and (2) rotating the axes so that they correspond to the axes of the contours. The rotated axes with the new origin are labeled X_1 and X_2. When related to this new system of axes, the various forms of the second-order equations are greatly simplified and their geometrical nature becomes obvious. Our object here is to explain the ideas. Details of the necessary calculations will be found in the indicated references at the end of the chapter.

The forms in Figure 15.7 show the kinds of surfaces that can be generated by a second-degree equation in two variables. They repay careful study. The forms labeled (a), (c), and (d) are especially important since they approximate systems frequently met in practice. Consider Figure 15.7a. The origin S is set at the center of the system, and the canonical equation is then of the form

$$\hat{y} - \hat{y}_s = B_{11}X_1^2 + B_{22}X_2^2 \tag{15.28}$$

with B_{11} and B_{22} both negative. We immediately know therefore that movement in any direction from S results in a quadratic *loss* in response. Thus the fitted surface has a *maximum* at S. Since B_{11} is numerically larger than B_{22}, we know that the fall-off is greater in the X_1 direction than in the X_2 direction. The canonical equation in Figure 15.7b is also of the form of Equation 15.28, but B_{22} is now positive, opposite in sign to B_{11}. We thus have a minimax (saddle or col) in which movement away from S in the X_1 direction results in quadratic loss, but movement along X_2 results in a somewhat smaller quadratic gain. In practice a true minimax is a rarity. It is fairly common, however, to meet a surface that is locally approximated by the one in Figure 15.7c. Here there is not a single point maximum but a line of maxima. We have a stationary ridge. It is very important to detect and exploit this situation when it occurs, because it implies that a wide choice of optimal conditions exists (along the X_2 axis). The existence of this situation is announced by B_{22} in Equation 15.28 being equal to zero (in practice close to zero). Theoretically

$$\hat{y} - \hat{y}_s = B_{11}X_1^2.$$

There is then a line of centers, and S' may be taken anywhere on that line. (In practice it is taken as close as possible to the center of the design.)

Another surface that often approximates real situations is the one shown in Figure 15.7d. To visualize this surface imagine the stationary ridge of (c)

tipped up. Notice that the canonical equation is no longer like Equation 15.28 but is of the form

$$\hat{y} - \hat{y}_s = B_{11}X_1^2 + B_2 X_2 \qquad (15.29)$$

In this form B_{22} is zero, but there is now a linear term $B_2 X_2$ which is the contribution from the tilt of the ridge. Again the system, properly speaking, has no unique center, so we take a convenient origin S' at some point on the ridge close to the center of the design. As before, this form is not found exactly in practice but is commonly approached. In general, notice that the vanishing of B_{22} (in practice the fact that B_{22} is small compared with B_{11}) points to the existence of *some* kind of ridge. The existence of a linear term in X_2 implies a rising ridge.

Consider, for example, the fitted equation (Equation 15.19) graphed in Figure 15.4. It may be shown that we can calculate a new origin S' lying on the ridge at minimum distance from the design center and also calculate a rotation of axes so that X_2 lies along the ridge. The fitted equation then becomes

$$\hat{y} - 88.2 = -5.12X_1^2 + 2.4X_2 - 0.15X_2^2 \qquad (15.30)$$

In this equation the term $B_{22} = 0.15$ is negligible so that, by inspection of this form of the equation only, we know that the system is closely approximated by a rising ridge as in Figure 15.7d. It is also possible to calculate the equation of the ridge:

$$0.63x_1 + 0.77x_2 + 0.33 = 0$$

and thus, if we wish, conduct experiments along it.

In this example the canonical analysis merely confirms what we can see from a contour diagram. The importance of canonical analysis is that it enables us to analyze systems of maxima and minima in many dimensions and, in particular to identify complicated ridge systems, where direct geometric representation is not possible.

Ridges

The reason for the occurrence of ridge systems can be seen when it is remembered that factors like temperature, time, pressure, and concentration are regarded as "natural" variables only because they happen to be quantities that can be conveniently *measured* separately. A more *fundamental variable* that is not directly measured, but in terms of which the behavior of the system could be described more economically (in chemistry, e.g., the frequency of a particular type of molecular collision), will often be a function of two or more *natural variables* (e.g., temperature, concentration, pressure). For this reason many combinations of natural variables may correspond to the best level of a fundamental variable.

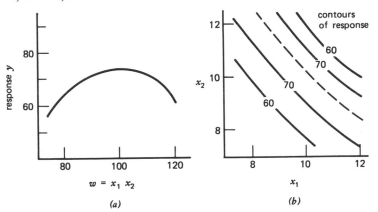

FIGURE 15.8. Example of redundancy. (a) Plot of y versus $w = x_1 x_2$. (b) Plot of y versus x_1 and x_2.

Suppose, for example, that in the region of interest the response was most economically described in terms of some fundamental variable the level of which was proportional to the product $w = x_1 x_2$ of two measured variables x_1 and x_2. Thus $w = x_1 x_2$ and $\eta = f(w)$. Suppose that the latter relationship was the one represented in Figure 15.8a. If the experimenter did not know that the system could adequately be described in terms of the compound variable w, and carried out experiments in which x_1 and x_2 were varied separately, he would be exploring a system for which the yield surface was the one shown in Figure 15.8b. Figure 15.8b, which describes a commonly occurring, practical situation, was constructed by drawing contour lines through the points giving a constant product $x_1 x_2$, the appropriate yield being read off from Figure 15.8a. For example, y might be the size of the kick of a frog's leg muscle when an electrical voltage w was applied. If the experimenter performed the experiment by varying electrical resistance x_1 in the circuit and also the current x_2 flowing through it, in ignorance of Ohm's law ($w = x_1 x_2$), we would have the situation illustrated.

Instead of the function $w = x_1 x_2$ we might have considered other functions $w = f(x_1, x_2)$. It will be found, for example, that the functions $w = a + bx_1 + cx_2$, $w = ax_1^b x_2^c$, $w = ax_1^b e^{-(c/x_2)}$, with a, b, and c positive, all produce diagonal ridge systems running from the top left to the bottom right of the diagram as in Figure 15.8b, while the functions $w = a + bx_1 - cx_2$, $w = ax_1^b/x_2^c$, $w = ax_1^b e^{-(cx_2)}$ all produce ridges running in the contrary sense. These ridge systems are of course associated with interaction, the former type with a negative two-factor interaction between x_1 and x_2 and the latter with a positive two-factor interaction.

The simplest type of ridge system is that produced by the linear relationship $w = a \pm bx_1 \pm cx_2$. The ridge system generated by such a function will have parallel straight line contours running in a direction determined by the relative magnitudes of b and c. A section at right angles to these contours will reproduce the original graph of η on w. Over limited ranges we should expect most systems to be capable of approximate representation by this simple type.

We could say in the examples described above that there is a "redundancy" of one variable. The apparently two-variable system can be expressed in terms of a single fundamental "compound" variable w. The physical and biological sciences abound with examples where, over suitable ranges of the variables, relationships exist similar to the above, often in many variables. Canonical analysis sometimes makes it possible to find out how many such compound variables are driving the system and what their natures are. See, for example, the paper by Box and Youle (1955). In particular the existence of a stationary ridge may point to a "scientific law."

15.5. APPLICATIONS OF RESPONSE SURFACE METHODS

In this brief introduction we have illustrated response surface techniques for only $k = 2$ variables. All the ideas, however, generalize to more variables. For example, Figure 15.9 shows a composite design for $k = 3$ variables.

The methods have been successfully used since about 1950 on a wide variety of problems in, for example, chemical engineering, agriculture, chemistry, and mechanical engineering. Hill and Hunter (1966) and Mead and Pike (1975) in survey articles give references to many such studies, and Meyer (1963) has discussed uses in education and psychology. Other references listed at the end of this chapter describe investigations on such widely different

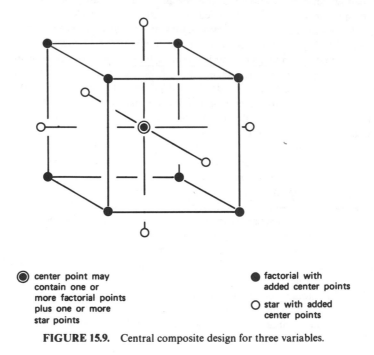

⊙ center point may contain one or more factorial points plus one or more star points

● factorial with added center points

○ star with added center points

FIGURE 15.9.   Central composite design for three variables.

activities as storing bovine spermatozoa, bleaching cotton, removing radio-nuclides, growing lettuce, dry ashing pulps, baking pie crusts, and com-pounding truck tires.

15.6. SUMMARY

Response surface methods can be used to answer a number of different questions:

1. How is a particular response affected by a set of variables over some specified region?
2. What settings of the variables will give a product or process satisfying desirable specifications?
3. What settings of the variables will yield a maximum (or minimum) response, and what is the local geography of the response surface near this maximal (or minimal) value?

Question 1 is concerned with obtaining a "snapshot" of the surface at a particular place in the space of the variables. After analysis has indicated the representational adequacy of the fitted surface, contour diagrams of interesting *sections* of a k-dimensional surface are often especially useful.

Question 2 arises because in many practical problems it is necessary to consider several responses simultaneously and to allow for constraints that these impose. The argument has been advanced that it should be possible to obtain some single function of many responses (e.g., profitability) that alone measures the desirability of a given setting of the variables. This approach has rather limited applicability in practice, and the technique illustrated in Section 15.3 of optimizing some principal response subject to restriction is usually better. In the manufacture of a drug, for example, what monetary value could be placed on impurities that might make the drug dangerous? For two, three, and even four input variables overlay plotting of contour lines and surfaces produces and dramatically illustrates solutions. If there are more variables, linear and nonlinear programming methods can sometimes be used to find desired conditions.

The shortcomings of using the classical one-variable-at-a-time approach to answer question 3 were explained in Section 15.1. The chemical example discussed in Section 15.2 showed a better way to tackle this problem.

Some attractive features of response surface methodology are that (a) it is a sequential approach, the results at each stage guiding the experimentation to be conducted at the next (at each step of the iteration only a modest number of

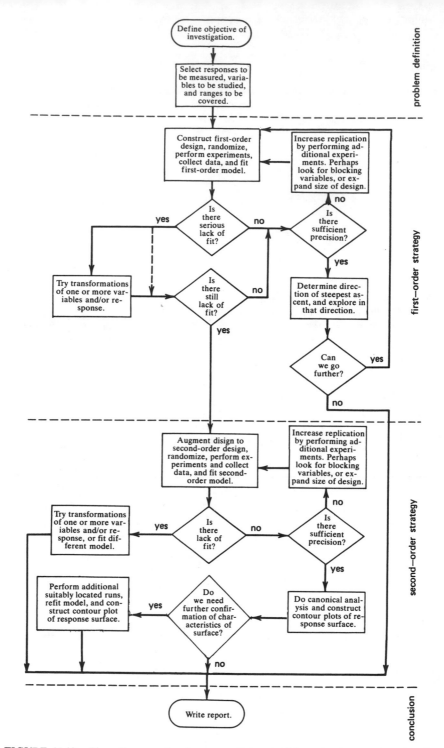

FIGURE 15.10. Flow diagram, showing some of the possible paths that can be taken in response surface studies.

experimental trials are run, thereby ensuring that the experimenter's resources are not squandered on unproductive trials); (b) it casts the experimental problem in readily understood geometric terms; and (c) it is applicable for any number of variables.

A response surface study typically ends with the writing of a report, in which the results should be presented in an understandable form. For this purpose graphical summaries and contour plots are most effective, being more digestible for most people than mathematical equations.

The highly adaptive characteristics of response surface methods are illustrated in Figure 15.10, which shows a somewhat formalized flow diagram for an investigation to find and explore a maximum. Even this rather complicated looking diagram by no means exhausts all the options. In particular the objective could change during the investigation (this could be a result of feedback produced by evolving knowledge of the system). The reader should notice the degree to which informed human judgment decides the final outcome.

REFERENCES AND FURTHER READING

For further information on response surface methods see:

Box, G. E. P., N. R. Draper, and J. S. Hunter, *Empirical Model-Building with Response Surfaces*, a book to be published.
Davies, O. L. (Ed.) (1954). *Design and Analysis of Industrial Experiments*, Hafner (Macmillan), Chapter 11.

Response surface methodology was first developed and described in these articles:

Box, G. E. P., and K. B. Wilson (1951). On the experimental attainment of optimum conditions, *J. Roy. Stat. Soc., Ser. B*, **13**, 1.
Box, G. E. P. (1954). The exploration and exploitation of response surfaces: some general consideration and examples, *Biometrics*, **10**, 16.
Box, G. E. P., and P. V. Youle (1955). The exploration and exploitation of response surfaces: An example of the link between the fitted surface and the basic mechanism of the system, *Biometrics*, **11**, 287.

Consult the above references for details of canonical analysis.

The following review articles contain references to applications of response surface methodology:

Hill, W. J., and W. G. Hunter (1966). A review of response surface methodology: A literature survey, *Technometrics*, **8**, 571.
Mead, R., and D. J. Pike (1975). A review of response surface methodology from a biometric point of view, *Biometrics*, **31**, 803.

Included in these review articles, for example, are the following references:

Cragle, R. G., R. M. Myers, R. K. Waugh, J. S. Hunter, and R. L. Anderson (1955). The effects of various levels of sodium citrate, glycerol, and equilibration time on survival of bovine spermatozoa after storage at −79°C, *J. Dairy Sci.*, **38**, 508.
Gaido, J. J. and H. D. Terhune (1961). Evaluation of variables in the pressure-kier bleaching of cotton, *Am. Dyestuff Reporter*, **50**, No. 21, 23.
Gardiner, D. A. and K. Cowser (1961). Optimization of radionuclide removal from low-level process wastes by the use of response surfaces, *Health Phys.*, **5**, 70.
Hader, R. J., M. E. Harward, D. D. Mason, and D. P. Moore (1957). An investigation of some of the relationships between copper, iron, and molybdenum in the growth and nutrition of lettuce. I. Experimental design and statistical methods for characterizing the response surface, *Proc. Soil Sci. Soc. Am.*, **21**, No. 1, 59.
Meyer, D. L. (1963). Response surface methodology in education and psychology, *J. Exp. Educ.*, **31**, 329.
Phifer, L. H., and J. B. Maginnis (1960). Dry ashing of pulp and factors which influence it, *TAPPI*, **53**, No. 1, 38.
Smith, H., and A. Rose (1963). Subjective responses in process investigation, *Ind. Eng. Chem.*, **55**, 25.
Weissert, F. C., and R. R. Cundiff (1963). Compounding Diene/NR blends for truck tires, *Rubber Age*, **92**, 881.

For methods to determine the precision with which a fitted response surface is estimated see:

Box, G. E. P., and J. Wetz (1973). *Criteria for Judging Adequacy of Estimation by an Approximating Response Function*, Technical Report 9, Department of Statistics, University of Wisconsin-Madison.

Sometimes in response surface investigations it is useful to transform one or more of the responses and variables. Transformations are discussed in these articles:

Box, G. E. P., and P. W. Tidwell (1962). Transformations of the independent variables, *Technometrics*, **4**, 531.

Box, G. E. P., and D. R. Cox (1964). An analysis of transformations, *J. Roy. Stat. Soc., Ser. B*, **26**, 211.

Draper, N. R., and W. G. Hunter (1969). Transformations: Some examples revisited, *Technometrics*, **11**, 23.

QUESTIONS FOR CHAPTER 15

1. What is response surface methodology? Name three kinds of problems for which it can be useful.

2. Can you think of a problem, preferably in your own field, for which response surface methodology might profitably be used?

3. When the object is to find and explore a maximum, what techniques may be useful when the model that locally approximates the response function is a first-degree polynomial? A second-degree polynomial?

4. At what points in a response surface study would the method of least squares be used?

5. What checks might be employed to detect (*a*) bad data values (*b*) an inadequate model?

6. Why is it useful to have replicate points?

7. What is the path of steepest ascent, and how is it calculated? Are there problems in which the path of steepest *descent* would be of more interest than the path of steepest *ascent*?

8. What is a composite design? In practice, why is it frequently built up sequentially?

9. Has response surface methodology been used in your field? Has the one-variable-at-a-time approach been used?

10. What is the essential nature of canonical analysis of the fitted second-degree equation? Why is it helpful?

11. What checks would you employ to ensure that the fitted equation was sufficiently well estimated to make further interpretation worthwhile?

CHAPTER 16

Mechanistic Model Building

The purpose of this chapter is to introduce mechanistically based models, to describe the model-building process, to explain the role of each of a number of techniques used in it, and to direct the reader to sources of further information.

16.1. EMPIRICAL AND MECHANISTIC MODELS

Empirical models are adequate for many purposes. Consider Old Faithful geyser in Yellowstone National Park. The park rangers need a model that will predict when the geyser will next erupt because that is what the tourists want to know. For this purpose an empirical model is perfectly satisfactory, and according to an information sheet supplied to tourists this is how the time of the next eruption can be predicted: if the previous eruption was short, lasting about 2 minutes, then 45 to 55 minutes will pass before the next eruption will begin; if the eruption was a long one, lasting 3 to 5 minutes, the next eruption will usually follow in 70 to 85 minutes. It is a rare tourist who wants a detailed explanation of the mechanism that governs this spectacular natural phenomenon.

A scientist who was making a study of geysers, however, might be interested in obtaining a better understanding of what makes geysers behave in the way they do. For this purpose he would prefer a model that actually described mathematically the principal factors involved in the process occurring below ground. It might be necessary to consider the rate of pressure buildup, the size and shape of the underground cavern, the temperature of the liquid involved, and so on. The scientist's model might yield more accurate prediction of eruptions, and it could do much more. It could produce basic understanding of the physical process involved and hence, for example, allow prediction of the behavior of other, rather different geysers. To see how a mechanistic model can be developed consider a simpler example.

540

An Example of a Mechanistic Model

Suppose that we have two similar tanks A and B as in Figure 16.1a. Initially A is full of water, and valve V is closed. Suppose that at time $x_1 = 0$ the valve is suddenly opened $x_2 = 0.2$ turns, and water runs from tank A to tank B. Observations of the height y of water in tank B are then made every minute. The resulting points are plotted in Figure 16.1b.

A *mechanistic* model for the dependence of $\eta = E(y)$ on time x_1 and valve opening x_2 would be a mathematical form derived from consideration of a supposed mechanism. For the present example such a mechanistic model is provided by the equation

$$\eta = \beta_0(1 - e^{-\beta_1 x_1 x_2}) \tag{16.1}$$

This equation may be derived as follows. Suppose that when *all* the water has run into tank B, its level is β_0. Assume that the rate of flow of water from A to B is proportional to (1) the number of turns x_2 by which the valve is opened and (2) the height of water in tank A (which is proportional to $\beta_0 - \eta$).

The stated assumptions imply that the rate of increase in η is

$$\frac{d\eta}{dx_1} = \beta_1 x_2(\beta_0 - \eta) \tag{16.2}$$

where β_1 is a constant and x_1 is the time elapsed since time zero. We also have the condition that η is zero when $x_1 = 0$. (The model is similar to the one for biochemical oxygen demand discussed in Section 14.6.) Equation 16.1 is the appropriate solution of the differential Equation 16.2.

If we set $x_2 = 0.2$ as in the experiment actually described

$$\eta = \beta_0(1 - e^{-0.2\beta_1 x_1}) \tag{16.3}$$

Figure 16.1c shows a graph of Equation 16.3 with $\beta_0 = 1$ and $\beta_1 = 2.5$. Comparison with Figure 16.1b suggests that, with suitably chosen values for the constants, the model should be adequate to describe the data. If we also allow x_2 (the amount we open the valve) to vary, then, for any given choice of the constants β_0 and β_1, we obtain a surface for η like that represented by the contours in Figure 16.1d.

Finally, if the observed level y differs from $E(y) = \eta$ by an experimental error ϵ so that $y - \eta = \epsilon$, the model for the observed level y is

$$y = \beta_0(1 - e^{-\beta_1 x_1 x_2}) + \epsilon \tag{16.4}$$

In what follows we assume that the errors ϵ are roughly IIDN(0, σ^2) and that the parameters are estimated by least squares.

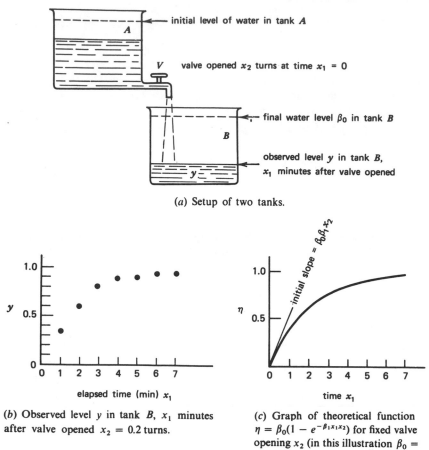

(a) Setup of two tanks.

(b) Observed level y in tank B, x_1 minutes after valve opened $x_2 = 0.2$ turns.

(c) Graph of theoretical function $\eta = \beta_0(1 - e^{-\beta_1 x_1 x_2})$ for fixed valve opening x_2 (in this illustration $\beta_0 = 1$, $\beta_1 = 2.5$, $x_2 = 0.2$).

FIGURE 16.1. Modeling hydraulic data.

In general, two technical problems exist in mechanistic model fitting.

1. Mechanistic models are typically formulated in terms of differential equations. In the simple example above, differential Equation 16.2 is immediately soluble in the form of Equation 16.1. More often, an explicit expression for η such as that illustrated in Equation 16.1 is not available for the solution. This creates no real difficulty, however, since all that is required for estimating the parameters are the numerical values of the responses for various specific settings of the parameters and the input variables. These can always be found by numerical methods and are quickly calculated with a computer.

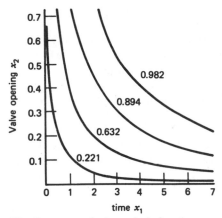
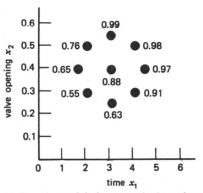
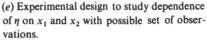

(d) Contours of theoretical function $\eta = \beta_0(1 - e^{-\beta_1 x_1 x_2})$ for different values of time x_1 and valve opening x_2 (in this illustration $\beta_0 = 1$, $\beta_1 = 2.5$).

(e) Experimental design to study dependence of η on x_1 and x_2 with possible set of observations.

FIGURE 16.1. (continued) Modeling hydraulic data.

2. Mechanistic models are usually nonlinear in the parameters. Estimates are therefore not immediately available from the linear least squares algorithm but must be obtained using iterative nonlinear least squares computer routines. Again effective routines are now generally available and their use normally presents no difficulty.

When and When Not To Use a Mechanistic Model

Judgment is needed in deciding when and when not to use mechanistic models. Suppose, for example, that the strength y of an extruded plastic sheet is known to depend on the rate of extrusion (x_1) and the rate of cooling immediately after extrusion (x_2) but that very little is known of the physics involved. A study to discover the appropriate mechanistic relation could be difficult and time consuming and might be improvident if all that was needed was to obtain a plastic sheet of greater strength. This limited objective might be achieved much more economically by empirical methods using factorials and response surface designs.

By contrast, a mechanistic approach is justified (a) whenever a basic understanding of the system is essential to progress or (b) when the state of the art is sufficiently advanced to make a useful mechanistic model easily available. Mechanistic modeling provides a great opportunity for human ingenuity. A gifted scientist can sometimes create a surprisingly effective model seemingly out of almost nothing.

16.2. POSSIBLE ADVANTAGES OF MECHANISTIC MODELS

Mechanistic models can (1) contribute to scientific understanding, (2) provide a basis for extrapolation, (3) provide a representation of the response function that is more parsimonious than the one attainable empirically.

For example, we might obtain a fair approximation locally to the surface of Figure 16.1d by fitting, say, a second-degree polynomial,

$$\eta = \beta_0 + \beta_1 x_1 + \beta_2 x_2 + \beta_{11} x_1^2 + \beta_{22} x_2^2 + \beta_{12} x_1 x_2 \qquad (16.5)$$

using a design like that shown in Figure 16.1e. Alternatively, we could use the data from this design (or preferably a design specifically chosen with this mechanistic model in mind) to fit and check the model of Equation 16.1.

If, however, this mechanistic model, believed to be supported by the basic physics of the system, could be verified, we would be in a much stronger position than would be attainable by mere empiricism. This is so because a well-tested mechanistic model does more than just graduate the data. It confirms that *our scientific understanding of the system* has been verified by the experiment. In addition, a polynomial equation, although it may be adequate to represent what is happening in the immediate region of study, provides only a very shaky basis for extrapolation. A mechanistic model, on the other hand, can suggest with greater certainty new sets of experimental conditions that are worthy of investigation. This better basis for extrapolation is provided because it is the mechanism, not a mere empirical curve, that is being supposed to apply more widely, and this mechanism is based on a partially verified understanding of the system itself. Of course, as we move in the space of the experimental variables, the mechanism may change or estimation errors may become serious, so *unchecked* extrapolation is *never* safe. Thus even a mechanistic model should preferably be used only to suggest regions where *further experimentation* might be fruitful.

If the mechanistic model is well founded, we can expect it to give a closer representation of the response over a wider region than is possible with a purely empirical function. Estimation of the response will then be better, because lack of fit (bias), measured by $E(\hat{y}) - \eta$, will tend to be less. In addition, since a mechanistic model is usually more parsimonious in the use of parameters, this causes less of the random error to be transmitted to $\hat{y}$. A general rule (strictly appropriate only for comparing linear models but providing an approximation for somewhat nonlinear models) is that the average variance of the estimated yields $\hat{y}$ over all the experimental points for *any* design is $p\sigma^2/n$, where p is the number of parameters fitted and n is the number of observations. Accordingly, the model given by Equation 16.1 (with $p = 2$) transmits to the estimated response only about one third as much of the variance as does the model given by Equation 16.5 (with $p = 6$).

TABLE 16.1. **Some experimental problems associated with studying a system represented by the model $\eta = f(\mathbf{x}, \boldsymbol{\beta})$**

problem type	objective	supposed known	supposed unknown
1. screening variables	Determine subset $\mathbf{x}$ of important variables from a given larger set $\mathbf{X}$ of potentially important variables (e.g., using a highly fractionated design).		$f, \mathbf{x}, \boldsymbol{\beta}$
2. empirical modeling	(a) Determine empirical effects of known input variables $\mathbf{x}$ (e.g., using a factorial design).	$\mathbf{x}$	$f, \boldsymbol{\beta}$
	(b) Determine a local interpolation approximation* $g(\mathbf{x}, \boldsymbol{\beta}')$ to $f(\mathbf{x}, \boldsymbol{\beta})$ (e.g., using response surface methods).	$\mathbf{x}$	$f, \boldsymbol{\beta}$
3. mechanistic modeling	(a) *parameter estimation* Design: Plan experiments for estimation of parameters. Analysis: Estimate the parameters.	$f, \mathbf{x}$	$\boldsymbol{\beta}$
	(b) *model discrimination* Design: Plan experiments for selection of form of f from m possibilities. Analysis: Determine the relative plausibility of m candidate models.	$f_1, f_2, \ldots, f_m, \mathbf{x}$	$f, \boldsymbol{\beta}$
	(c) *model testing* Design: Plan experiments to strain tentative model f_0. Analysis: Suitably modify f_0 to arrive at adequate form f.	f_0	$f, \mathbf{x}, \boldsymbol{\beta}$

* The parameters in the two functions g and f are different—hence the different designations, $\boldsymbol{\beta}'$ and $\boldsymbol{\beta}$.

Different Kinds of Experimental Investigations

Experimental investigations are usually concerned with elucidating certain aspects of a relationship $\eta = f(\mathbf{x}, \boldsymbol{\beta})$, where $\mathbf{x}$ is a set of experimental variables and $\boldsymbol{\beta}$ a group of empirical or theoretical parameters. The objectives and degrees of sophistication of such investigations can be widely different. To help place mechanistic model building in context, three particular types of study are distinguished in Table 16.1:

1. *Screening* studies aimed at discovering *which* of a large number of possible variables materially affect the response.
2. *Empirical* studies aimed at producing an *empirical model* that describes *how* the variables affect the response, the true nature of the function f being unknown.
3. *Mechanistic* studies aimed at producing a *mechanistic model* that can lead to the "correct" functional form f and can explain *why* the response is affected by the variables in the manner observed.

16.3.   TECHNIQUES FOR MECHANISTIC MODELING

We described screening studies and empirical modeling in earlier chapters. In the rest of this chapter we briefly review the methods, tactics, and strategy of mechanistic modeling.

Parameter Estimation, Single Response

Suppose that the functional form f is assumed known and the problem is to obtain estimates of the parameters $\beta_1, \beta_2, \ldots, \beta_p$. Given the data, and the assumption that the errors are IIDN(0, σ^2), appropriate estimates are obtained by using the method of least squares. Since most mechanistic models are nonlinear in the parameters, iterative algorithms are necessary to carry out these computations. See Section 14.6 and the references listed at the end of this chapter under the heading "Nonlinear Estimation."

Nonlinear Designs for Parameter Estimation

When fitting nonlinear models, one quickly realizes that experimental design is of paramount importance. The problem of choosing experimental conditions for nonlinear models to achieve reliable estimates of the parameters is discussed in the references listed at the end of this chapter under the heading "Nonlinear Design." It turns out that a sequential strategy is usually needed, since for models that are nonlinear in the parameters the best selection of the experimental conditions depends on the values of the parameters themselves. These values are initially unknown. As the investigation proceeds and better estimates become available, better experimental conditions can be determined.

Exercise 16.1. A chemist is studying a system in which a reactant A decomposes to form the desired product B, which in turn decomposes into an undesired by-product C according to the scheme:

$$A \overset{\beta_1}{\to} B \overset{\beta_2}{\to} C,$$

where β_1 and β_2 are reaction rate constants. On the assumption of first-order kinetics,

$$-\frac{d\eta_1}{dx} = \beta_1 \eta_1 \qquad \text{and} \qquad -\frac{d\eta_2}{dx} = \beta_2 \eta_2 - \beta_1 \eta_1$$

where η_1 is the concentration of A and η_2 is the concentration of B in gram moles per liter. If at time zero $(x = 0)$ the concentration of A is unity and the concentrations of B and C are zero, then the concentration of B at any time x is given by

$$\eta_2 = \frac{\beta_1}{\beta_1 - \beta_2} (e^{-\beta_2 x} - e^{-\beta_1 x})$$

A series of experiments in which the reaction is terminated after x minutes gives the following results:

time at which reaction is terminated	x	1	2	4	5	6	7
relative concentration of product B	y	0.22	0.51	0.48	0.29	0.20	0.12

Over a grid of values for β_1 and β_2, calculate sums of squares $S(\beta_1, \beta_2)$, plot rough contours, and obtain least squares estimates of the parameters β_1 and β_2 and an approximate 80% confidence region. If you have a computer and an iterative nonlinear least squares program, use it to confirm your estimates.

Exercise 16.2. Repeat Exercise 16.1 with the following data:

x	0.0625	0.125	0.25	0.5	1	2
y	0.01	0.02	0.08	0.15	0.22	0.51

Exercise 16.3. From the plotted data, fitted curves, and sums of squares contours for Exercises 16.1 and 16.2, comment on the suitability of the two designs, that is, the two sets of levels for x.

Parameter Estimation from Multiresponse Data

Scientists and engineers frequently need to estimate parameters from data in which each of a *number* of responses is related to one or more of the parameters by nonlinear equations. For example, in Exercise 16.1 suppose that, at each of the six different reaction times, data y_1, y_2, and y_3 were collected on the concentrations of each of the products A, B, and C.

As would be expected, if data on *all three* responses are available, one can get better checks on mechanism as well as more precise estimates of parameters. For further details consult the references listed at the end of this chapter under "Multiresponse Estimation."

Model Discrimination

We assumed above that a single adequate form of the function f was known. However, sometimes in the early stages of an investigation there are m candidate mechanisms, leading to m models $f_1, f_2, \ldots, f_m$. The problem then is to select the best form of f from among them. We call this the *model discrimination* problem. The analysis problem is, with a given set of data, to obtain the weight of evidence favoring each of the m models. This analysis can be made by appropriate use of Bayes's theorem. Once again, the design problem is also of great importance. The question is how to choose experiments that will permit the most efficient discrimination among the candidate models. For further information on this problem see the references listed at the end of this chapter under "Model Discrimination."

Model Testing

Experimentation designed to strain the model and so place it in jeopardy is called *model testing*. Procedures may be made diagnostic by applying strains in such a way that specific deficiencies can point to likely sources of inadequacy. Such procedures can lead to sequential model improvement. For further discussion of this topic see Section 16.5 and the references under "Model Testing."

16.4. THE MODEL-BUILDING PROCESS

The techniques listed above are of use at different stages of the overall model-building process. As mentioned in Chapter 1, models (empirical or mechanistic) almost always have to be arrived at by iteration, that is, by "trial and error." The iteration (see Figure 1.3) may involve several inferential sequences embedded in sequences of data gathering, using experimental design. As the investigation proceeds, more and more is learned about *which* variables we should be considering, the *scales* in which they should be measured, the *transformations* that should be employed, and the *degree of complexity* needed in the model. The designs performed toward the end of an investigation will usually be better adapted to the problem than those at the beginning. Initial experiments necessarily suffer from relative lack of knowledge.

Model building is typified by the iterative sequence:

specification → selection and fitting → diagnostic checking
(or identification)

The dashed line indicates that, when diagnostic checks point to model inadequacies, respecification may be needed, possibly after additional data have been collected, leading to a further cycle in the iteration.

Specification (or Identification)

Specification is an informal process in which graph drawing, preliminary data analysis, and interactive thinking about the fundamental elements of the system to be modeled are used to arrive at a class of models worthy to be entertained.

Selection and Fitting

This second stage involves the more formal application of techniques of parameter estimation and model discrimination with the object of selecting the best of the class of models so far entertained.

Diagnostic Checking

This is a process of criticism whereby the adequacy of the model selected in the previous stages is evaluated. Plotting of residuals is an important part of this process. Diagnostic checks can reveal not only that a model is inadequate but *how* it is inadequate, and so indicate modifications that should be made in later iterations.

An Adequate Model: Transformation of the Data to White Noise

What is a statistical model? We have an adequate statistical model if it is plausible that a set of parameters β exists such that

$$F(\mathbf{y}, \mathbf{x}, \boldsymbol{\beta}) = \varepsilon \qquad (16.6)$$

where $\boldsymbol{\varepsilon} = (\epsilon_1, \epsilon_2, \ldots, \epsilon_n)$ is a *white noise* sequence unrelated to any known variable. By a "white noise" sequence we mean that the ϵ's have zero mean and constant variance, and are distributed independently of each other. Thus Equation 16.6 provides a transformation F from the data $\mathbf{x}$ and $\mathbf{y}$ to white noise $\boldsymbol{\varepsilon}$.

For example, the regression equation

$$y_t^{1/2} = \beta_0 + \beta_1 x_t + \epsilon_t, \qquad t = 1, 2, \ldots, n \tag{16.7}$$

is an adequate statistical model if the ϵ_t (the elements of $\boldsymbol{\varepsilon}$) produced by the transformation

$$\left. \begin{array}{l} y_1^{1/2} - \beta_0 - \beta_1 x_1 = \epsilon_1 \\ y_2^{1/2} - \beta_0 - \beta_1 x_2 = \epsilon_2 \\ \vdots \\ y_n^{1/2} - \beta_0 - \beta_1 x_n = \epsilon_n \end{array} \right\} \tag{16.8}$$

are plausibly a white noise sequence unrelated to any other known variable.

Suppose, by contrast, that the ϵ_t were linearly related to the times t at which the experiments were performed. The appropriate model would then be

$$y_t^{1/2} = \beta_0 + \beta x_t + \gamma t + \epsilon_t' \tag{16.9}$$

The model $\eta = f(\mathbf{x}_1) + \epsilon$, often used to explain data, can be written more exactly as $y = f(\mathbf{x}_1) + \epsilon(\mathbf{x}_2)$, where the set of variables $\mathbf{x}_1$ in the model consists of the ones we think we know, and the set of variables $\mathbf{x}_2$ that are responsible for the "error" consists of the ones we currently do not know. If the ϵ component contains an important contribution from a particular variable, residuals from the initial model will contain a component from that variable which could suggest its identity.

Many statistical techniques (in particular the use of quality control charts, evolutionary operation, designed experiments, and normal plotting, as well as the analysis of residuals) are basically methods to move variables out of the ϵ "box" into the f "box." This substitution:

$$y = f(\overset{\frown}{\mathbf{x}_1}) + \epsilon(\overset{\frown}{\mathbf{x}_2})$$

is an important means by which the advancement of learning occurs.

16.5. MODEL TESTING WITH DIAGNOSTIC PARAMETERS

Models that are inadequate for a given purpose do not necessarily show their inadequacy with a particular set of data. To test a model it is important that the investigator run trials that put the model in jeopardy over important

TABLE 16.2. Estimated values of parameters from hydraulic data

valve opening (turns) x_2	$\hat{\beta}_0$	$\hat{\beta}_1$
0.2	1.016	2.61
0.4	0.993	2.52
0.6	1.002	2.55
0.8	1.005	2.50
1.0	0.991	2.48
1.2	1.009	2.39

ranges of variables. One device, useful not only in revealing inadequacy but also in pointing to its possible cause, employs the principle that *in an adequate model constants stay constant when variables are varied*. The method is often used when data appear as time plots. Consider again the hydraulic example of Section 16.1. Suppose that several series of measurements are made, one series for each of various choices of the valve opening x_2, to yield a series of graphs similar to the one in Figure 16.1b. For each series the amount of water in tank B is plotted against the time x_1. Suppose that β_0 and β_1 are now estimated for each curve and the estimates are plotted against x_2. Examination of these plots (see Figure 16.2) indicates that the "constant" β_1 does not, in fact, stay constant but decreases as the valve opening x_2 is increased. Why would x_2 and β_1 be related? A likely reason is that the valve has nonlinear characteristics, that is, the flow rate B is *not* directly

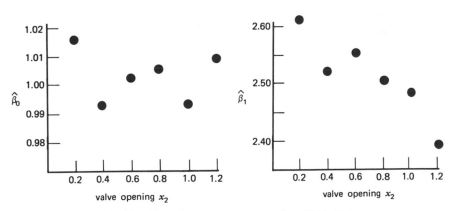

FIGURE 16.2. Plots of parameter estimates from Table 16.2.

proportional to the valve opening. This possibility could be checked directly by separate experiments. If it were confirmed, the model could be modified to allow appropriately for the valve nonlinearity.

Notice how this technique helps to advance scientific learning at the difficult *inductive* stage, leading to "appropriate modification of conjecture."

Importance of Placing the Model in Jeopardy

It is a common error to perform "confirmatory experimental runs" that in reality do not place a model in jeopardy. A conclusion that "there is no evidence of model inadequacy" based on such trials provides little justification for adopting and using the model. However, if a model passes *severe* tests, the investigator can feel greater confidence in its worth. If it fails, he is helped to see why, and is guided toward improving it.

16.6. IMPORTANCE OF PLOTTING DATA IN THE AGE OF COMPUTERS

Without computers most of the work done on nonlinear models would not be feasible. However, the more sophisticated the model and the more elaborate the techniques employed, the more important it is to submit complicated analyses to surveillance by data plots, residual plots, and other visual displays. The modern computer can make the plots itself, but graphs need not only be made but also to be carefully examined and thought about. The data analyst must "fondle" the data. Hand plotting used to be one of the ways this came about. The original data, as well as the various plots, whether made by hand or by the computer, should be mulled over. The experimenter's imagination, intuition, subject-matter knowledge, and experience must interact. This interaction will often lead to new ideas that may, in turn, lead to further analysis or experimentation.

The computer is fast and precise; the human mind is comparatively slow and inaccurate. But whereas the mind's powers of critical and inductive reasoning are very great, those of the computer are minuscule. Mind and machine therefore should each contribute what it does best, and data analysis should be organized so that such cooperation is encouraged.

16.7. SUMMARY

Mechanistic models, when appropriately employed, have several advantages over empirical models. They provide greater scientific insight, a better basis for extrapolation, and usually more parsimonious representation. Mechanistic models are typically based on differential equations and are usually nonlinear in the parameters. Numerical methods exist for solving the differential equations and fitting such models to data.

In building mechanistic models, a number of questions arise. How can experiments be devised that will test a mechanism and diagnose possible deficiencies? If there is more than one candidate model, how should experiments be planned so that the inadequate models are discarded? What experiments should be performed so that the parameters are estimated as precisely as possible? Statistical techniques have been developed to help answer such questions, and references to these are given at the end of the chapter. It is possible to start with a model based on imperfect theoretical understanding and be led along a path toward improvement as more information is gathered. It is hoped to discuss the topics of this chapter in much greater detail in a later book.

REFERENCES AND FURTHER READING

Modeling Philosophy

Box, G. E. P. (1960). Some general considerations in process optimization, *J. Basic Eng., Trans. ASME*, **113**.

Box, G. E. P., and W. G. Hunter (1965). The experimental study of physical mechanisms, *Technometrics*, **7**, 23.

Hunter, W. G., J. R. Kittrell, and P. Mezaki (1967). Experimental strategies for mechanistic modeling, *Trans. Inst. Chem. Eng.*, **45**, T146.

Box, G. E. P., and G. M. Jenkins (1970). *Time Series Analysis, Forecasting, and Control*, Holden Day.

Nonlinear Estimation

Levenberg, K. (1944). A method for the solution of certain non-linear problems in least squares, *Quart. Appl. Math.*, **2**, 164.

Box, G. E. P. (1958). Use of statistical methods in the elucidation of basic mechanism, *Bull. Inst. Int. Stat.*, **36**, 215.

Box, G. E. P. (1960). Fitting empirical data, *Ann. N.Y. Acad. Sci.*, **86**, 792.

Marquardt, D. W. (1963). An algorithm for least squares estimation of non-linear parameters, *J. Soc. Ind. Appl. Math.*, **2**, 431.

Hunter, J. S. (1966). Estimation and design for non-linear models, *Proc. 11th Conf. Des. Exp.*, Army Reserve Office, Report 66-2.

Draper, N. R. and H. Smith (1966). *Applied Regression Analysis*, Wiley.

Daniel, C., and F. S. Wood (1971). *Fitting Equations to Data*, Wiley.

Bard, Y. (1974). *Nonlinear Parameter Estimation*, Academic Press.

Beck, J. V., and K. J. Arnold (1977). *Parametric Estimation in Engineering and Science*, Wiley.

Nonlinear Design

Fisher, R. A. (1935). *The Design of Experiments*, Oliver and Boyd.

Box, G. E. P., and H. L. Lucas (1959). Design of experiments in nonlinear situations, *Biometrika*, **46**, 77.

Box, G. E. P., and W. G. Hunter (1965). Sequential design or experiments for nonlinear models, *I.B.M. Scientific Computing Symposium on Statistics*, 113.

Atkinson, A. C., and W. G. Hunter (1968). The design of experiments for parameter estimation, *Technometrics*, **10**, 271.

Draper, N. R., and W. G. Hunter (1967). Use of prior distributions in the design of experiments for parameter estimation in nonlinear situations, *Biometrika*, **54**, 147.

Graham, R. J., and F. D. Stevenson (1972). Kinetics of chlorination of niobium oxychloride by phosgene in a tube flow reactor: Application of sequential experimental design, *Ind. Eng. Chem. Process Des. Dev.*, **11**, 160.

Cochran, W. G. (1973). Experiments for nonlinear functions, *J. Am. Stat. Assoc.*, **68**, 771.

Multiresponse Estimation

Box, G. E. P., and N. R. Draper (1965). The Bayesian estimation of common parameters from several responses, *Biometrika*, **52**, 355.

Draper, N. R., and W. G. Hunter (1966). Design of experiments for parameter estimation in multiresponse situations, *Biometrika*, **53**, 525.

Hunter, W. G. (1967). Estimation of unknown constants from multiresponse data, *Ind. Eng. Chem. Fundam.*, **6**, 461.

Box, G. E. P., W. G. Hunter, J. F. MacGregor, and J. Erjavec (1973). Some problems associated with the analysis of multiresponse data, *Technometrics*, **15**, 33.

Model Discrimination

Hunter, W. G., and A. M. Reiner (1965). Designs for discriminating between two rival models, *Technometrics*, **7**, 307.

Box, G. E. P., and W. J. Hill (1967). Discrimination among mechanistic models, *Technometrics*, **9**, 57.

Hunter, W. G., and Mezaki, R. (1967). An experimental design strategy for distinguishing among rival models, *Can. J. Chem. Eng.*, **45**, 247.

Atkinson, A. C., and D. R. Cox (1974). Planning experiments for discriminating between models, *J. Roy. Stat. Soc.*, Ser. B, **36**, 321.

Model Testing

Box, G. E. P., and W. G. Hunter (1962). A useful method for model-building, *Technometrics*, **4**, 301.

Hunter, W. G., and R. Mezaki (1964). A model building technique for chemical engineering kinetics. *Am. Inst. Chem. Eng. J.*, **10**, 315.

Kittrell, J. R. and R. Mezaki (1967). Discrimination among rival Hougen–Watson models through intrinsic parameters, *Am. Inst. Chem. Eng. J.*, **13**(2), 389.

QUESTIONS FOR CHAPTER 16

1. What is a mechanistic model?

2. Why is a mechanistic model sometimes preferred to an empirical model? In some situations why do investigators (correctly) settle for an empirical model? Give examples, preferably from your own field.

3. What are some of the difficulties encountered in using mechanistic models? How can they be surmounted?

4. Why are mechanistic models often nonlinear in the parameters?

5. Can the method of least squares be used to fit mechanistic models?

6. What are diagnostic parameters, and how can they be used in model-building projects?

CHAPTER 17

Study of Variation

It is frequently important to locate, estimate, and control major sources of variation. In this chapter we survey three techniques directed to these ends: (1) control charts, (2) error transmission studies, and (3) estimation of variance components. The first can detect assignable causes of variation and can provide surveillance and feedback for the control of operating systems. The second can show how errors injected in various parts of a system will *synthesize* in their effects on the final result. The third allows final variation to be *analyzed* into components associated with individual causes.

17.1. GRAPHS AND CONTROL CHARTS: IMPURITY DETERMINATION EXAMPLE

The Value of Graphs in the Study of Variation

Of all the devices for analysis of data, perhaps the most valuable is the simple graph. Understanding comes as a result of information *properly communicated*; the existence of information is not enough. Whether the objective is the control of a manufacturing plant, the management of a testing laboratory, or the care of a patient, it is important not only that appropriate information be collected, but also that this information be fed back in a readily understood form to those responsible for taking action.

Unfortunately it is rather common that the major effort of data collection is made and the minor additional effort of communication is not. Information can remain buried in notebooks, tables, process record sheets, or computer tapes, or embalmed in abstruse mathematical equations. To be informative data must be displayed so that present and past experience can be readily compared, and concomitant variation in two or more impinging responses can be simultaneously considered. Whatever may be done later in the way of more formal analysis, appropriate plotting of data is never a waste of time

556

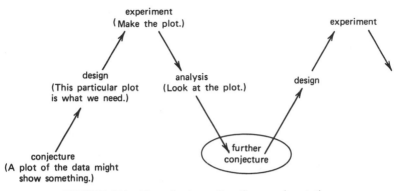

FIGURE 17.1. Data plotting as iterative experimentation.

and frequently reveals unexpected characteristics that might otherwise be overlooked.

Figure 17.1 illustrates that the making of plots is a particular example of iterative experimentation. Perhaps a first plot reveals copying errors, unsuspected trends, or cycles. If so, these phenomena will need to be examined further. Possibly this may be done by additional plots or more formal analysis of the same data; possibly it may be necessary to collect additional data. In any case such plots can sometimes change the course of the whole study. It is an unnatural but frequently committed crime to force the square peg of inappropriate data into the round hole of computerized statistical method.

Exercise 17.1. Plot the following sets of data. Do you think they ought to be treated as random samples from roughly normal populations? Say what may be wrong for any set you are doubtful about.

(a) 269	264	275	268	227	270	267	272	274	269
(b) 261	266	261	262	264	266	276	272	272	278
(c) 267	261	263	268	262	264	266	260	261	269

Answer: (a) Copying error? (b) Trend? (c) Suspicious pattern (in groups of three)?

Exercise 17.2. Draw a square peg in a round hole (plan view), and plot residuals at 15° angular intervals. How does a plot of residuals suggest the nature of the lack of fit?

The Shewhart Chart

One important application of data plotting is the Shewhart control chart, which is used for surveillance of repetitive operations in industrial plants and scientific laboratories. The control chart consists of a plot of data in time order

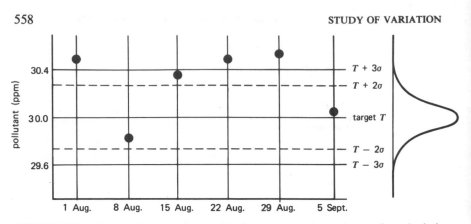

FIGURE 17.2. Control chart, showing weekly determinations by analyst A of standard air sample with 30 ppm of pollutant.

like that in Figure 17.2. Its single most important merit is to ensure that data are displayed and not buried, but the Shewhart chart has the additional feature of indicating how much variation about the target value T can be expected to occur by chance. This helps to ensure that both overreaction and underreaction to apparent peculiarities in the data are avoided. In effect, one judges the plotted data against a reference distribution representing expected performance when the plant is operating correctly. Rather than show the reference distribution itself, as is done in the right-hand margin of Figure 17.2, it is customary to show only control lines drawn in such a way that when the process is operating correctly the lines include, say, 99.9% of the observations.

If the measured quality characteristic y is assumed to vary normally with standard deviation σ, the control lines are usually centered on the target value T and are drawn at $T \pm 3\sigma$. These are approximate 99.9% limit lines. Often approximate 95% lines at $T \pm 2\sigma$ are added also. The value of σ used in drawing the control lines is the standard deviation of the process when it is "in control," that is, when it is varying free of known disturbances. Details of routine construction and use of quality control charts will be found in the references at the end of this chapter. The following example illustrates the versatility of these plotting methods.

Example: Determination of Air Pollutant

Figure 17.2 is one section of a Shewhart chart kept in an analytical laboratory for the routine determination of an air pollutant for which $\sigma = 0.13$ ppm. It was important to keep the concentration of this particular pollutant below 30 ppm, and to achieve this many dozens of work samples were analyzed each week. As a control, a standard air mixture containing exactly 30 ppm of this pollutant was also run each week. The chart shows the determination of the *standard* obtained by one analyst over a particular period of 6 weeks.

There is little doubt that during the period studied the analyses of the standard mixture were markedly out of control. No less than three of the recorded points fell outside the 3σ limits. The most obvious (but, as it turned out, incorrect) explanation was that the analyst had failed to perform the analyses with the same care that was taken in the runs used to establish the value of σ.

A Conjecture Coming from the Inspection of Two Control Charts

Look at Figure 17.3, which shows control charts over the same period for analyst *A* and analyst *B*. Similar changes in level occurred from one week to the next for *both* analysts. Evidently some factor was affecting the results of

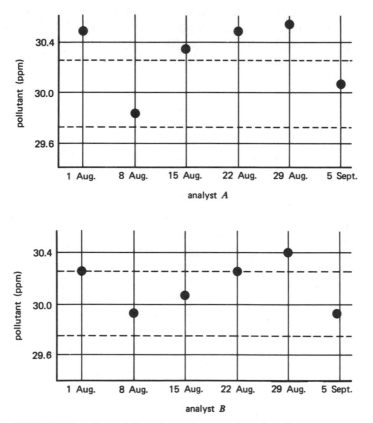

FIGURE 17.3. Control charts for analysts *A* and *B*, air pollutant example.

both analysts together. The relative humidity of the laboratory received early study, and the analysts' results were found to be highly correlated with this variable. A small number of special experiments subsequently showed that the relative humidity did indeed affect the measured response, and the analytical test procedure was later modified to eliminate the influence of this factor.

Exercise 17.3. Sketch the iteration that occurred in this investigation. (Use a diagram like Figure 17.1.)

Plots of the Average and Range

When variation between contemporaneous observations is large, charts of greater sensitivity are obtained by plotting averages of n results. Thus samples of five bobbins of nylon yarn randomly drawn from 50 possible positions on a draw-twist machine might be tested for strength at hourly intervals. Mean strength could then be monitored by plotting the sample averages on a chart having appropriate control lines. At the same time *variability* in strength could also be checked by plotting the sample range with suitable control limits obtained from tables given in the references at the end of the chapter.

Making Feedback Effective

Statistical quality control charts ensure continuous comparison of current results with a reference distribution indicating the capability of the process when behaving properly. The charts encourage feedback leading to appropriate action. However, even when charts are prepared and displayed, they are not always seen and worried over by the appropriate people.

To ensure effective feedback one company we know of conducts weekly chart meetings, which are attended by the plant manager and process superintendents. The charts are made especially large and are hung so as to be visible to everyone. The quality control manager then goes through the charts, pointing out where they show abnormalities. The abnormalities from a particular process are then publicly discussed with the person responsible for that part of the process. This person explains the deviations for which he can account and describes what action is being taken to set things right. Feedback of valuable ideas and generation of new hypotheses can occur as various explanations of the historical evidence are offered.

The plotting of data is a valuable tool for surveillance of processes. Such plots can ensure rapid feedback of information when something is going wrong and, in the words of Walter Shewhart, can lead to the discovery of "assignable causes."

TABLE 17.1. Quality data with calculation of cumulative sum

observation number t	1	2	3	4	5	6	7	8	9	10	11	12	13	14	15	16	17	18	19	20
original data y_t	63.1	59.0	58.4	62.4	57.6	60.9	59.9	59.7	61.1	60.0	62.0	62.6	64.6	62.9	62.5	58.6	63.6	58.9	63.6	59.4
original data minus 60 $y_t - 60$	3.1	−1.0	−1.6	2.4	−2.4	0.9	−0.1	−0.3	1.1	0.0	2.0	2.6	4.6	2.9	2.5	−1.4	3.6	−1.1	3.6	−0.6
cumulative sum $S_t = \sum (y - 60)$	3.1	2.1	0.5	2.9	0.5	1.4	1.3	1.0	2.1	2.1	4.1	6.7	11.3	14.2	16.7	15.3	18.9	17.8	21.4	20.8

Action resulting from study of data plots is one important way in which the transfer of variables occurs from the unknown noise component of the model to the known deterministic component–a fundamental means by which the advancement of knowledge occurs.

Cusum Charts

The charts devised by Shewhart are not the only way of plotting data for process surveillance, although they are the most *generally* useful in the sense that they provide an overall picture of what is happening, whatever it may be.

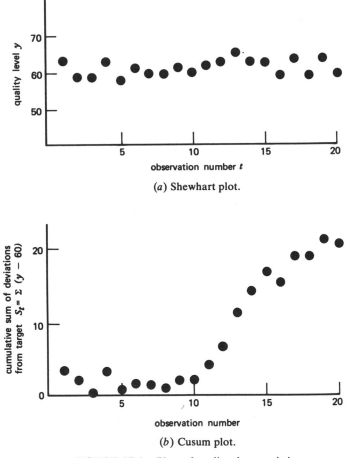

(*a*) Shewhart plot.

(*b*) Cusum plot.

FIGURE 17.4.   Plots of quality characteristic.

When changes of a specific kind are expected, additional charts, most sensitive to this particular kind of change, may be employed. In particular, for detecting small changes in mean level (of an approximately normal process), the cumulative sum (cusum) chart is more sensitive than the Shewhart chart.

For illustration consider the data of Table 17.1, plotted in Figure 17.4a. This records the level of a quality characteristic that, when in control, varies about a target value $T = 60$. Figure 17.4b shows a cusum plot of the same data in which the cumulative sum of the deviations $\sum (y - T)$ is plotted at each stage.

If, when the process is on target, $y_j - 60 = \epsilon_j$ is a random error with zero mean, then $S_t = \sum_{j=1}^{t} \epsilon_j$ is the sum of t such random errors. Suppose, however, that at some intermediate time, $j = n$, the mean of the process shifted to a new value $60 + \delta$; then, at time t, $S_t = \sum_{j=1}^{t} \epsilon_j + (t - n)\delta$. Besides the random sum there is now a systematic component $(t - n)\delta$ increasing steadily with each new observation. A change in mean in the process will thus be indicated by a change in *slope* of the cusum plot. Although it is not at all obvious from the Shewhart plot in Figure 17.4a, an increase in mean (of about two units) appears, from the cusum plot in Figure 17.4b, to have occurred at or near observation 10. For more information on cumulative sum charts, see the references at the end of this chapter.

Exercise 17.4. The following 20 numbers y are random drawings from a normal distribution with mean $\eta = 10$ and standard deviation $\sigma = 1$:

  9.9, 9.8, 10.3, 10.1, 9.9, 9.7, 8.4, 10.4, 8.2, 10.3, 11.0, 9.0, 10.4, 11.4, 10.4, 11.3, 9.5, 9.4, 9.3, 9.8

(*a*) Plot the values y in time order.
(*b*) Plot the cumulative sum $S = \Sigma (y - 10)$.
(*c*) Subtract 0.5 $(= \sigma/2)$ from the last 10 numbers, and denote the 20 new values by y'.
(*d*) Plot these values y' in time order.
(*e*) Plot the cumulative sum $S' = \Sigma (y' - 10)$. What conclusions do you draw?

17.2. TRANSMISSION OF ERROR

In discussing the central limit effect (Section 2.4), we mentioned that the error, in, say, a measured yield is typically an aggregate of many errors. When we know the functional relationship connecting the final result with the variable components, as well as the error structure for these components, we can study how the variation in the components is *transmitted* to the final result. In particular we can decide which components are the major causes of variability in the final result.

Consider a linear function of random variables

$$Y = \theta_0 + \theta_1 y_1 + \theta_2 y_2 + \theta_3 y_3 \tag{17.1}$$

where $\theta_0, \theta_1, \theta_2,$ and θ_3 are constants, and $y_1, y_2,$ and y_3 are random variables. From Equation 3A.2 we have

$$V(Y) = \theta_1^2 V(y_1) + \theta_2^2 V(y_2) + \theta_3^2 V(y_3) + 2\theta_1 \theta_2 \, \text{Cov}(y_1 y_2)$$
$$+ 2\theta_1 \theta_3 \, \text{Cov}(y_1 y_3) + 2\theta_2 \theta_3 \, \text{Cov}(y_2 y_3) \tag{17.2}$$

where $V(y_u)$ and $\text{Cov}(y_u y_v)$ stand for the variance of y_u and covariance of y_u and y_v, respectively. If $y_1, y_2,$ and y_3 are uncorrelated,

$$V(Y) = \theta_1^2 V(y_1) + \theta_2^2 V(y_2) + \theta_3^2 V(y_3) \tag{17.3}$$

The terms on the right-hand side of this expression represent the separate contributions to the overall variance. For *linear* functions of random variables, therefore, the problem of error transmission is solved. But what if Y is a *nonlinear* function of the y's? In the rest of this section we consider this question.

Error in a Nonlinear Function of One Variable: Volume of a Bubble

Suppose that the volume Y of a spherical bubble is calculated by measuring its diameter y from a photograph and applying the formula

$$Y = \frac{\pi}{6} y^3 = 0.524 y^3 \tag{17.4}$$

Suppose that the diameter of a particular bubble is measured as $y_0 = 1.80$ mm, and the error of measurement is known to have a standard deviation of $\sigma_y = 0.02$ mm. Approximately how big an error will be induced in the derived volume Y by the measurement error in the diameter y? Figure 17.5 shows a plot of Y against y over the range $y = 1.74$ to $y = 1.86$, that is, over the range $y_0 \pm 3\sigma_y$, which is (arbitrarily) chosen so as to include most of the error distribution.

Now y^3 is certainly a nonlinear function of y, yet *over this narrow range* the plot is essentially linear. Also, since Y increases by about 5.1 units per unit increase in y and the graph passes through the point ($y = 1.80$, $Y = 3.05$), the approximate linear relation is

$$Y = 3.05 + 5.1(y - 1.80) \qquad (1.74 < y < 1.86) \tag{17.5}$$

or, equivalently,

$$Y = -6.13 + 5.1y \qquad (1.74 < y < 1.86) \tag{17.6}$$

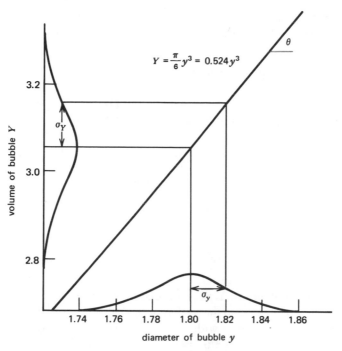

FIGURE 17.5. Transmission of error from measured diameter reading to reported volume, bubble example.

But for a linear relation

$$Y = \theta_0 + \theta y \qquad (17.7)$$

we have

$$V(Y) = \theta^2 V(y) \qquad (17.8)$$

Taking square roots on both sides of this equation gives:

$$\sigma_Y = \theta \sigma_y \qquad (17.9)$$

The gradient or slope θ of the approximating straight line is thus the magnifying factor that converts σ_y to σ_Y.

For this example θ is about 5.1. Thus from Equation 17.9

$$\sigma_Y \simeq 5.1 \times 0.02 = 0.102 \qquad (17.10)$$

The diameter measurement of 1.80 ± 0.02 therefore corresponds to a volume measurement of about 3.05 ± 0.10. Notice that the *percentage* error in measurement of volume is about three times that for the diameter.

Exercise 17.5. Given $V(y) = 1.0$, find $V(Y)$ for these two cases:
(a) $Y = 10y$ and (b) $Y = 100y$.   *Answers*: (a) 100, (b) 10,000.

Checking Linearity and Estimating the Slope

A check on linearity and an estimate of the gradient are conveniently obtained numerically from a table of the following kind:

y	Y	differences of Y
$y_0 - 3\sigma = 1.74$	2.76	
		0.30
$y_0 \quad\;\; = 1.80$	3.06	
		0.31
$y_0 + 3\sigma = 1.86$	3.37	

In this table the values of Y are recorded at y_0, $y_0 - 3\sigma$, and $y_0 + 3\sigma$. The close agreement between the differences $3.06 - 2.76 = 0.30$ and $3.37 - 3.06 = 0.31$ indicates (even without the graph) that a straight line provides a good approximation in this particular example. Also, a good average value for θ is provided by the expression

$$\frac{\text{overall change in } Y}{\text{overall change in } y} = \frac{3.37 - 2.76}{1.86 - 1.74} = \frac{0.61}{0.12} = 5.1 \qquad (17.11)$$

which is also the value we read from the graph. Readers familiar with calculus will know that this slope can also be obtained by differentiating the function $Y = 0.524y^3$ with respect to y. Notice, however, that the numerical procedure offers two advantages over the calculus method:

1. It can be applied even when the function $Y = f(y)$ is not known explicitly; for example, Y might be the output from a complex simulation, and y one of the variables in the system.
2. It provides a convenient check on linearity.

Error in a Nonlinear Function of Several Variables

Figure 17.6 shows how a nonlinear function of two variables

$$Y = f(y_1, y_2) \qquad (17.12)$$

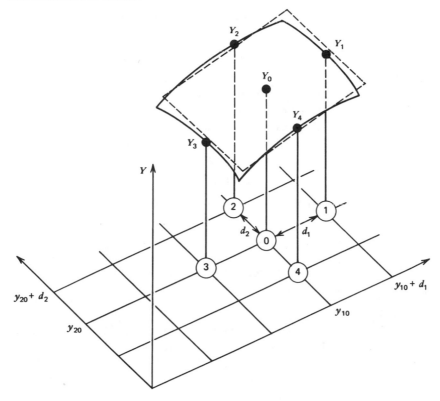

FIGURE 17.6. Local linearization of the function $Y = f(y_1, y_2)$.

(such as $Y = 0.366y_1^{1/2}/y_2$) can be represented graphically by a curved surface. Over a sufficiently small region in the neighborhood of some point 0 of interest (y_{10}, y_{20}), it may then be possible to approximate this surface by a plane

$$Y = Y_0 + \theta_1(y_1 - y_{10}) + \theta_2(y_2 - y_{20}) \tag{17.13}$$

where θ_1 and θ_2 are gradients in the directions of axes y_1 and y_2. Equation 17.13 is referred to as a "local linearization" of the function of Equation 17.12.

The gradients θ_1, θ_2 may be determined using calculus, but for our purpose it is better to obtain them numerically. In Figure 17.6 consider the points labeled 1, 2, 3, 4. The gradient in the direction of axis y_1 is approximated by

$$\theta_1 \simeq \frac{Y_1 - Y_3}{2d_1} \tag{17.14}$$

and the gradient in the direction of axis y_2 is approximated by

$$\theta_2 \simeq \frac{Y_2 - Y_4}{2d_2} \tag{17.15}$$

Also, linearity can be checked in the y_1 direction by comparing $Y_1 - Y_0$ with $Y_0 - Y_3$, and in the y_2 direction by comparing $Y_2 - Y_0$ with $Y_0 - Y_4$.

The approximating plane will cover the region of interest if d_1 and d_2 are set equal to $3\sigma_1$ and $3\sigma_2$, respectively, where σ_1^2 and σ_2^2 are the error variances of y_1 and y_2. If, then, the linear approximation is adequate, and if y_1 and y_2 vary independently, the variance of Y is approximated as follows:

$$V(Y) \simeq \theta_1^2 \sigma_1^2 + \theta_2^2 \sigma_2^2$$

The method is readily generalized to more variables, as is illustrated in the following example.

An Example with Four Variables: Molecular Weight of Sulfur

The following problem appeared in the St. Catherine's College, Cambridge, Scholarship Papers for 1932:*

"A sample of pure prismatic sulphur melted initially at 119.25°C, but in the course of a few moments the melting point fell to 114.50°C. When the sulphur had completely melted at this temperature, the liquid sulphur was plunged into iced water; 3.6 per cent of the resultant solid sulphur was then found to be insoluble in carbon disulphide. Deduce the molecular formula of the type of sulphur insoluble in carbon disulphide. The latent heat of fusion of sulphur is 9 cals. per gram."

Denote the initial melting point of the pure sulfur in degrees Kelvin by y_1, the value in degrees Kelvin to which the melting point fell by y_2, the percentage in the result of the allotropic form (the part insoluble in carbon disulfide) by y_3, and the latent heat of fusion of sulfur (in calories per gram) by y_4. The molecular weight Y of sulfur in the allotropic form is then given by the expression

$$Y = f(y_1, y_2, y_3, y_4) = \frac{0.02 y_3 (100 - y_3) y_1^2}{100 y_4 (y_1 - y_2)} \tag{17.16}$$

Substituting the values

$$y_{10} = 392.25, \qquad y_{20} = 387.5, \qquad y_{30} = 3.6, \qquad y_{40} = 9 \tag{17.17}$$

we obtain

$$Y_0 = \frac{0.02 \times (392.25)^2 \times 3.6 \times 96.4}{9 \times 4.75 \times 100} = 249.8 \tag{17.18}$$

* The question is quoted by permission of the Cambridge University Press.

from which the answer required for this particular examination question can be deduced.

Equation 17.16 is typical of the nonlinear functions that occur in practical work, and we use it to illustrate the transmission of error formula, supposing the standard deviations of y_1, y_2, y_3, and y_4 to be

$$\sigma_1 = 0.1, \qquad \sigma_2 = 0.1, \qquad \sigma_3 = 0.2, \qquad \sigma_4 = 0.5 \qquad (17.19)$$

Appropriate calculations are shown in Table 17.2. There is evidence of some nonlinearity in this example. Checking linearity over so wide a range as $\pm 3\sigma$, however, is a conservative procedure, and linearization should yield a reasonably good approximation for $V(Y)$ even in this example.

TABLE 17.2. **Numerical computation of gradients and check on linearity for four variables, molecular weight example**

	value of Y	difference	numerical value for gradient
$y_{10} + 3\sigma_1$	392.55	235.3	
		-14.5	
y_{10}	392.25	249.8	$\theta_1 \simeq -51.5$
		-16.4	
$y_{10} - 3\sigma_1$	391.95	266.2	
$y_{20} + 3\sigma_2$	387.8	266.6	
		16.8	
y_{20}	387.5	249.8	$\theta_2 \simeq 52.7$
		14.8	
$y_{20} - 3\sigma_2$	387.2	235.0	
$y_{30} + 3\sigma_3$	4.2	289.6	
		39.8	
y_{30}	3.6	249.8	$\theta_3 \simeq 66.8$
		40.3	
$y_{30} - 3\sigma_3$	3.0	209.5	
$y_{40} + 3\sigma_4$	10.5	214.1	
		-35.7	
y_{40}	9.0	249.8	$\theta_4 \simeq -28.6$
		-50.0	
$y_{40} - 3\sigma_4$	7.5	299.8	

Assuming, as might be expected, that y_1, y_2, y_3, and y_4 vary independently, we have

$$V(Y) \simeq (-51.5)^2(0.01) + (52.7)^2(0.01) + (66.8)^2(0.04) + (-28.6)^2(0.25)$$
$$= 26.52 + 27.77 + 178.48 + 204.50$$
$$= 437.27 \tag{17.20}$$

whence

$$\sigma(Y) = 20.9$$

Thus the estimate of the molecular weight Y of 249.8 would be subject to a standard deviation of about 21.

It is seen from the size of the individual contributions in Equation 17.20 that the greater part of the variation arises because of uncertainty in y_3 and y_4. If substantial improvement in the estimation of Y were required, it would be essential to find ways of obtaining more accurate measurements of these quantities. No great improvement in accuracy can be expected by determining temperatures y_1 and y_2 more precisely, since their total contribution to the variance is only $26.52 + 27.77 = 54.29$.

Exercise 17.6. With appropriate assumptions the number of moles of oxygen in a fixed volume of air is given by the formula $Y = 0.336 y_1/y_2$, where $Y =$ moles of oxygen, $y_1 =$ atmospheric pressure (centimeters of mercury), and $y_2 =$ atmospheric temperature (degrees Kelvin). Suppose that average laboratory conditions are $y_{10} = 76.0$ centimeters of mercury and $y_{20} = 295°K$.
(a) Obtain a linear approximation for Y at average laboratory conditions. (b) Assuming $\sigma_1 = 0.2$ centimeters and $\sigma_2 = 0.5°K$, check the linearity assumption over the ranges $y_{10} \pm 3\sigma_1$, $y_{20} \pm 3\sigma_2$. (c) Obtain an approximate value for the standard deviation of Y at average laboratory conditions.
$$\text{Answer: (a) } Y = 0.08656 + 0.001139(y_1 - 76.0) - 0.000293(y_2 - 295).$$
$$\text{(c) } \sigma_Y = 0.00027.$$

Notice that the main contributions to $V(Y)$ are not necessarily from variables having large variances. Error transmission also depends critically on the values of the gradients.

Misuse of Formulas for Error Transmission

Attempts are sometimes made to find the *overall* error of a method by direct use of Equation 17.3. Such calculations may seriously underestimate the total error. There are two reasons for this:

1. Errors to which the individual measurements are subject tend to be underestimated.
2. The investigator necessarily omits unknown components of error.

17.3. VARIANCE COMPONENTS: PIGMENT PASTE EXAMPLE

The topic of transmission of error might be called variance component *synthesis*. We now discuss its converse, namely, variance component *analysis*, the objective of which is to deduce the values of component variances that cannot be measured directly. Consider, for example, a batch process used in the manufacture of a pigment paste. One measurement of interest is the *moisture content* of this product. Suppose that a batch of pigment has been made, a chemical sample has been taken to the laboratory, and a portion of the sample has been tested to yield a moisture content y. If each of a number of batches is sampled and tested once, the variation of the resulting moisture content y will reflect some combination of the variability arising from

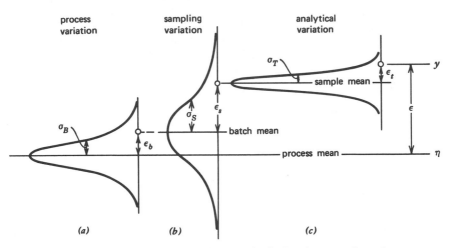

FIGURE 17.7. Three components of variance in the final moisture reading, pigment paste example. (*a*) Distribution of batch means about the process mean η. (*b*) Distribution of sample means about the batch mean. (*c*) Distribution of analytical test results about the sample mean.

analytical test variation, variation of chemical sampling, and batch-to-batch variation. To separate this total variation into parts assignable to these various sources, we set down a model that can explain the generation of the observations.

If η is the long-run process mean for the moisture content, then, as is illustrated in Figure 17.7, the overall error $\epsilon = y - \eta$ will contain three separate error components: $\epsilon = \epsilon_t + \epsilon_s + \epsilon_b$, where ϵ_t is the analytical test error, ϵ_s the error made in taking the chemical samples, and ϵ_b the batch-to-batch error.

A Model

The deviation of a particular analytical result y from the chemical sample mean (i.e., from the mean of the conceptual population of test analyses that might have been obtained *from this one chemical sample*) we call the *analytical test error*, and we denote it by ϵ_t. Similarly, the deviation of the mean of this one chemical sample from the batch mean (i.e., from the mean of the conceptual population of means that might have been obtained for different chemical samples taken *from this one batch*) we call the *chemical sampling error*, and we denote it by ϵ_s. Finally, the deviation of the batch mean from the process mean (i.e., from the mean of the conceptual population of batch means that might be obtained for different batches coming from this process) we call the *batch error*, and we denote it by ϵ_b.

According to these definitions, ϵ_t, ϵ_s, and ϵ_b all have zero means. We assume that they can be represented by *random* (independent) drawings from normal distributions having fixed variances σ_T^2, σ_S^2, and σ_B^2. Thus, if a single test were performed on a single sample taken from each batch, the variance of the test results would be the sum of these three variances. We now consider how an experiment could be run that would allow these components of the overall variance to be separately estimated.

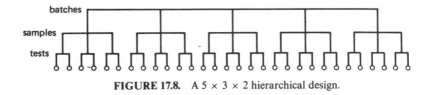

FIGURE 17.8.   A $5 \times 3 \times 2$ hierarchical design.

The Nested (Hierarchical) Design

The design usually employed to obtain the required estimates of the components of variance is called a *nested* or *hierarchical* design. For the chemical example we would choose B batches of product. Each batch would be sampled S times, and for each of these samples T replicate analyses would be performed. In all we would have $B \times S \times T$ individual results. Figure 17.8 shows diagramatically a nested design with $B = 5$, $S = 3$, and $T = 2$. Obviously, branching designs of this kind can be elaborated to accommodate any number of factors. We shall continue our discussion, however, in terms of the three factors—batches, samples, and tests.

An Investigation To Estimate Variance Components

In the chemical process under study each batch of the pigment was routinely tested for moisture content by performing a single test on a single sample. It is convenient to take a unit of measured moisture content to be one tenth of 1%. In terms of these units the moisture content thus determined varied about a mean of approximately 25 with a standard deviation of about 6. This variation was regarded as excessive. The analytical chemists believed the large variation occurred because of lack of control of the process by the engineers. The process engineers thought it occurred because of inaccuracies in the chemists' analytical procedures.

The data of Table 17.3 were collected from an experiment intended to resolve the issue, in which $B = 15$, $S = 2$, and $T = 2$. Thus each of 15 batches of product was independently sampled twice, yielding 30 samples. Each sample was then thoroughly mixed and split into two parts, yielding 60 subsamples. These subsamples were numbered and randomly introduced into the stream of routine analyses going to the analytical laboratory.

Calculation of V_T

For the first sample of the first batch there are two test results, 40 and 39. An estimate of the testing variance obtained from this one pair of results is given by (difference)$^2/2 = (-1)^2/2 = 0.5$. This estimate has one degree of freedom. Pooling the 30 estimates of this kind, we have

$$V_T = \frac{(-1)^2 + 0^2 + \cdots + 2^2}{(2)(30)} = 0.92 \qquad (17.21)$$

an estimate of σ_T^2 with 30 degrees of freedom.

In general, suppose that there are T replicated observations on each sample and that $y_{bs1}, y_{bs2}, y_{bs3}, \ldots, y_{bsT}$ are the T replicated analytical tests made on the sth sample of the bth batch. From these results we can calculate the individual sample averages $\bar{y}_{bs}$, each based on T observations, and the estimated variance $[\sum_{t=1}^{T} (y_{bst} - \bar{y}_{bs})^2]/(T - 1)$, which is an estimate of σ_T^2 with $T - 1$ degrees of freedom. Now, if all such estimates from the $B \times S$ samples are pooled, we obtain

$$V_T = \frac{\sum_b^B \sum_s^S \sum_t^T (y_{bst} - \bar{y}_{bs})^2}{BS(T - 1)} \qquad (17.22)$$

an estimate of σ_T^2 having $BS(T - 1)$ degrees of freedom.

TABLE 17.3. Moisture contents from duplicate tests on duplicate samples of 15 batches of pigment, pigment paste example

		tests		samples			batches	
batch	sample	subsample results (1) (2)	difference (2) − (1) (3)	sum (1) + (2) (4)	average (4) ÷ 2 (5)	difference between entries in (5) (6)	sum of entries in (4) (7)	average (7) ÷ 4 (8)
1	1 2	40 39 30 30	−1 0	79 60	39.5 30.0	−9.5	139	34.75
2	3 4	26 28 25 26	2 1	54 51	27.0 25.5	−1.5	105	26.25
3	5 6	29 28 14 15	−1 1	57 29	28.5 14.5	−14.0	86	21.50
4	7 8	30 31 24 24	1 0	61 48	30.5 24.0	−6.5	109	27.25
5	9 10	19 20 17 17	1 0	39 34	19.5 17.0	−2.5	73	18.25
6	11 12	33 32 26 24	−1 −2	65 50	32.5 25.0	−7.5	115	28.75

7	13 14	23 24 / 32 33	1 1	47 65	23.5 32.5	9.0	112	28.00
8	15 16	34 34 / 29 29	0 0	68 58	34.0 29.0	−5.0	126	31.50
9	17 18	27 27 / 31 31	0 0	54 62	27.0 31.0	4.0	116	29.00
10	19 20	13 16 / 27 24	3 −3	29 51	14.5 25.5	11.0	80	20.00
11	21 22	25 23 / 25 27	−2 2	48 52	24.0 26.0	2.0	100	25.00
12	23 24	29 29 / 31 32	0 1	58 63	29.0 31.5	2.5	121	30.25
13	25 26	19 20 / 29 30	1 1	39 59	19.5 29.5	10.0	98	24.50
14	27 28	23 24 / 25 25	1 0	47 50	23.5 25.0	1.5	97	24.25
15	29 30	39 37 / 26 28	−2 2	76 54	38.0 27.0	−11.0	130	32.50

units: tenth of 1%   grand average = 26.783

575

Calculation of V_S

For the first batch there are two sample averages, 39.5 and 30.0. An estimate having one degree of freedom of the variance of the sample averages is therefore given by $(39.5 - 30.0)^2/2 = 45.125$. Pooling the 15 estimates of this kind, we get $V_S = 28.99$ with 15 degrees of freedom.

In general, there would be S sample averages, $\bar{y}_{b1}, \bar{y}_{b2}, \ldots, \bar{y}_{bS}$, for the bth batch. From these results we can calculate the average $\bar{y}_b$ and the variance $[\sum_s^S (\bar{y}_{bs} - \bar{y}_b)^2]/(S - 1)$, which has $S - 1$ degrees of freedom. Pooling the variances for all the batches, we obtain

$$V_S = \frac{\sum_b^B \sum_s^S (\bar{y}_{bs} - \bar{y}_b)^2}{B(S - 1)} \tag{17.23}$$

which has $B(S - 1)$ degrees of freedom.

Now V_S is *not* an estimate of σ_S^2 alone, because each sample is represented by an *average* of T test analyses. Thus, supposing the chemical sampling and testing errors to be independently distributed, V_S provides an estimate of $\sigma_S^2 + (\sigma_T^2/T)$.

For this example $V_S = 28.99$ is an estimate of $\sigma_S^2 + (\sigma_T^2/2)$.

Calculation of V_B

Consider now the 15 batch averages 34.75, 26.25, ..., 32.50. Their overall average is 26.783, and their variance is

$$V_B = \frac{(34.75 - 26.783)^2 + (26.25 - 26.783)^2 + \cdots + (32.50 - 26.783)^2}{14}$$

$$= 21.64* \tag{17.24}$$

which has 14 degrees of freedom.

In general, there would be B batch averages, $\bar{y}_1, \bar{y}_2, \ldots, \bar{y}_B$, each of which would be an average over S samples and ST analytical tests. The quantity

$$V_B = \frac{\sum_{b=1}^B (\bar{y}_b - \bar{y})^2}{B - 1} \tag{17.25}$$

where $\bar{y}$ is the grand average, is thus an estimate of $\sigma_B^2 + (\sigma_S^2/S) + (\sigma_T^2/ST)$.

For this example $V_B = 21.64$ is an estimate of $\sigma_B^2 + (\sigma_S^2/2) + (\sigma_T^2/4)$.

* Or, equivalently,

$$V_B = \frac{(34.75)^2 + (26.25)^2 + \cdots + (32.50)^2 - 15(26.278)^2}{14} = 21.64$$

Estimates of Components of Variance

Collecting the results, we have:

estimate	quantity estimated
$V_B = 21.6$	$\sigma_B^2 + \dfrac{\sigma_S^2}{2} + \dfrac{\sigma_T^2}{4}$
$V_S = 29.0$	$\sigma_S^2 + \dfrac{\sigma_T^2}{2}$
$V_T = 0.92$	σ_T^2

Thus estimates of the desired components are:

$$\hat{\sigma}_T^2 = 0.92$$

$$\hat{\sigma}_S^2 = 29.0 - \frac{0.92}{2} = 28.5 \qquad\qquad (17.26)$$

$$\hat{\sigma}_B^2 = 21.6 - \frac{28.5}{2} - \frac{0.92}{4} = 6.9$$

Appendix 17A shows a more commonly used method for calculating the components by means of an analysis of variance table.

Discussion of Results

Figure 17.9 shows normal distributions with standard deviations equal to those estimated for the three error components. The experiments suggested that the largest individual source of variation was the error arising in chemical sampling. Here there were two possible explanations:

1. The standard sampling procedure was satisfactory *but was not being carried out.*
2. The standard sampling procedure was not satisfactory.

A batch of pigment paste consisted of approximately 80 drums of material. The recommended (and supposedly standard) procedure for taking a sample consisted of (1) randomly selecting five drums of the material from a batch, (2) carefully taking a certain amount of material from each selected drum with a special, vertically inserted, sampling tube designed to obtain representative material from all layers in the drum, (3) thoroughly mixing the five portions of material, and finally (4) taking a sample of the aggregate mixture.

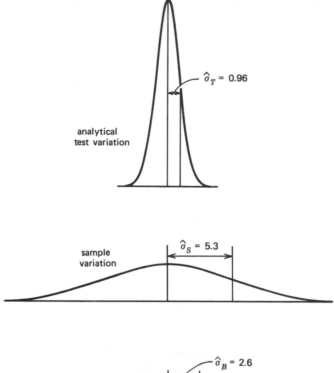

$\hat{\sigma}_T = 0.96$

analytical
test variation

$\hat{\sigma}_S = 5.3$

sample
variation

$\hat{\sigma}_B = 2.6$

process
variation

FIGURE 17.9. Diagrammatic summary of results of experiment to determine components of variance, pigment paste example.

By walking down to the plant and talking to the operators, the investigators discovered that this procedure was not being followed and that the operators responsible for conducting the sampling were unaware that there *was* any such procedure. They believed that all that was required was to take a sample from one drum and put it directly into the sampling bottle. The special sampling tube was nowhere to be found. Over a period of time poor communication, laxness in supervision, and changes in personnel had brought about this situation. When a test was conducted with the operators correctly using the recommended sampling procedure, a dramatic reduction in variance was found. The action taken as a result of the study was

1. to explain to the operators what was expected of them,
2. to introduce adequate supervision in the taking of samples, and
3. to initiate studies to discover whether further improvement in sampling was possible.

Choice of Numbers of Samples and Tests

Another part of the follow-up investigation consisted of performing a second hierarchical design similar to that just discussed, with the correct sampling procedure being followed except that each sample was a single mixed specimen from a sampling tube taken from *one* drum rather than from five. We denote the corresponding variance component by σ_U^2. The experiment yielded these estimates:

$$\hat{\sigma}_B^2 = 4.3, \qquad \hat{\sigma}_U^2 = 16.0, \qquad \hat{\sigma}_T^2 = 1.3 \qquad (17.27)$$

Deciding on a Routine Sampling Scheme

The question next considered was: How can the results of this special experiment be used to set up a good routine sampling procedure? Suppose that, for each batch, the procedure consisted of taking one mixed specimen from each of U randomly selected drums and performing a *total* of A analyses on these U samples. (If T tests were performed on each of the U samples, $A = UT$. If all U samples were aggregated and then a single test made, $A = 1$.) The average of the T tests performed on the U samples would be distributed about the true batch mean with variance

$$V = \frac{\sigma_U^2}{U} + \frac{\sigma_T^2}{A} \qquad (17.28)$$

Some alternative schemes are listed in Table 17.4. (The fourth column refers to a cost analysis to be discussed later.) The third column of this table shows values of the variance V for various choices of U and A. It should be noted that in this example, because of the large relative size of the analytical sampling error, little is gained by replicate testing. Large reduction in variance, however, can be obtained by replicate sampling.

Schemes in which A, the total number of analyses, is less than the number of samples can be physically realized by using the device of the "bulked" sample. For example, the scheme $U = 10$ and $A = 1$ is accomplished by taking 10 samples, mixing them together, and performing one test on the aggregate mixture.

TABLE 17.4. **Alternative sampling and testing schemes, pigment paste example**

number of samples U	total number of test observations $A = TU$	variance of estimate of batch mean $V = \dfrac{16.0}{U} + \dfrac{1.3}{A}$	cost of procedure ($)
1	1	17.3	21
1	10	16.1	201
1	100	16.0	2001
10	1	2.9	30
10	10	1.7	210
10	100	1.6	2010
100	1	1.5	120
100	10	0.3	300
100	100	0.2	2100

Minimum Cost Sampling Schemes

The fourth column in Table 17.4 shows the cost of each alternative procedure on the assumption that it costs $1 to take a sample and $20 to make an analytical test.

Let C_U be the cost of taking one sample, and C_T the cost of making one test. The cost of the sampling scheme is

$$C = UC_U + AC_T \qquad (17.29)$$

Minimization of C for fixed V leads to the conclusion that any desired value of the variance can be achieved for minimum cost by choosing the ratio U/A of samples to tests so that

$$\frac{U}{A} = \sqrt{\frac{\sigma_U^2}{\sigma_T^2} \times \frac{C_T}{C_U}} \qquad (17.30)$$

If, for instance, we substitute in Equation 17.30 the estimates $\hat{\sigma}_U^2 = 16.0$, $\hat{\sigma}_T = 1.3$, and $C_T/C_U = 20$, we obtain

$$\frac{U}{A} = \sqrt{\frac{16.0}{1.3} \times 20} = 15.7$$

In this case, therefore, there is no point in making more than a single analytical test unless more than 16 samples are to be taken.

TABLE 17.5. Variances and costs associated with alternative sampling schemes, pigment paste example

scheme	number of samples U	total number of tests A	variance V	cost ($/batch)
1	1	1	17.3	21
2	5	1	4.5	25
3	8	1	3.3	28
4	16	1	2.3	36
5	32	2	1.1	72

Results of this kind of study are usually best presented in terms of a series of alternative possible schemes with associated variances and costs. Five schemes for the present case are shown in Table 17.5.

The table makes it clear that the previously recommended standard procedure (scheme 2 with $U = 5$ and $A = 1$) is good if properly executed. It also appears, however, that by bulking more samples some further reduction in variance could be achieved at little extra expense. In fact, scheme 3 with $U = 8$ and $A = 1$ was eventually chosen for routine sampling and testing purposes.

Exercise 17.7. Suppose that $\sigma_U^2 = 1$, $\sigma_T^2 = 5$, $C_U = \$10$, and $C_T = \$1$. Discuss possible alternative testing schemes that use no more than five samples and five tests.

APPENDIX 17A. CALCULATING VARIANCE COMPONENTS FROM AN ANALYSIS OF VARIANCE TABLE

It is customary to set out variance component calculations in an analysis of variance table. The general algebraic forms are shown in Table 17A.1, and numerical values for the various quantities in the pigment paste example are given in Table 17A.2.

The analysis of variance for the three-stage hierarchy here discussed is based on the algebraic identity

$$\sum_b^B \sum_s^S \sum_t^T y_{bst}^2 = BST\bar{y}^2 + ST \sum_b^B (\bar{y}_b - \bar{y})^2$$

$$+ T \sum_b^B \sum_s^S (\bar{y}_{bs} - \bar{y}_b)^2 + \sum_b^B \sum_s^S \sum_t^T (y_{bst} - \bar{y}_{bs})^2$$

The mean squares m_B, m_S, and m_T in the analysis of variance table are simply related to the quantities calculated in our previous calculations; in fact,

$$m_B = STV_B, \qquad m_S = TV_S, \qquad m_T = V_T$$

TABLE 17A.1. **General layout of analysis of variance table for separation of variance components associated with batches, samples, and tests**

source of variation	sum of squares	degrees of freedom	mean square	expected value of mean square
average	$BST\bar{y}^2$	1		
batches	$ST \displaystyle\sum_{b}^{B}(\bar{y}_b - \bar{y})^2$	$B - 1$	m_B	$ST\sigma_B^2 + T\sigma_S^2 + \sigma_T^2$
samples	$T \displaystyle\sum_{b}^{B}\sum_{s}^{S}(\bar{y}_{bs} - \bar{y}_b)^2$	$B(S - 1)$	m_S	$T\sigma_S^2 + \sigma_T^2$
tests	$\displaystyle\sum_{b}^{B}\sum_{s}^{S}\sum_{t}^{T}(y_{bst} - \bar{y}_{bs})^2$	$BS(T - 1)$	m_T	σ_T^2
total	$\displaystyle\sum_{b}^{B}\sum_{s}^{S}\sum_{t}^{T} y_{bst}^2$	BST		

TABLE 17A.2. **Analysis of variance table, pigment paste example**

source of variation	sum of squares	degrees of freedom	mean square	expected value of mean square
average	43,040.8	1		
batches	1,211.0	14	86.6	$4\sigma_B^2 + 2\sigma_S^2 + \sigma_T^2$
samples	869.7	15	58.0	$2\sigma_S^2 + \sigma_T^2$
tests	27.5	30	0.9	σ_T^2
total	45,149.0	60		

The expected values of these mean squares are appropriate multiples of quantities we have previously denoted as being estimated by V_B, V_S, and V_T. The two approaches are equivalent. The quantities involved are the same, but the multipliers in the analysis of variance table are such that, if there were no variation due to batches or to samples, each mean square would provide a separate estimate of σ_T^2.

Equating the mean squares to their expectations and solving the resulting equations gives the same estimates as before.

REFERENCES AND FURTHER READING

The following books contain further information on quality control:

Grant, Eugene L., and R. S. Leavenworth (1972). *Statistical Quality Control*, 4th ed., McGraw-Hill.

Duncan, Acheson J. (1974). *Quality Control and Industrial Statistics*, 4th ed., Irwin.
Juran, J. M. (Ed.) (1974). *Quality Control Handbook*, 3rd ed., McGraw-Hill.
Wetherill, G. B. (1969). *Sampling Inspection and Quality Control*, Methuen.

Cumulative sum charts were first proposed in these papers:

Page, E. S. (1954). Continuous inspection schemes, *Biometrika*, **41**, 100.
Barnard, G. A. (1959). Control charts and stochastic processes, *J. Roy. Stat. Soc., Ser. B*, **21**, 239.

Information on cumulative sum charts is contained in most modern books on quality control. In addition see:

Ewan, W. D. (1963). When and how to use cu-sum charts, *Technometrics*, **5**, 1.
Woodward, R. H. and P. L. Goldsmith (1964). *Cumulative Sum Techniques*, Oliver and Boyd.

For more on variance components and their uses see:

Kempthorne, O., and J. L. Folks (1971). *Probability, Statistics and Data Analysis*, Iowa State University Press.
Davies, O. L. (Ed.) (1960). *Design and Analysis of Industrial Experiments*, Hafner (Macmillan).
Box, G. E. P., and G. C. Tiao (1973). *Bayesian Inference in Statistical Analysis*, Addison Wesley.

QUESTIONS FOR CHAPTER 17

1. Describe (*a*) a Shewhart control chart, and (*b*) a cumulative sum chart. How might such charts be useful (i) in mass production industry, (ii) in a testing laboratory, (iii) in a hospital? How would you ensure that information supplied by such charts was fed back to appropriate people? Who do you think "appropriate people" might be?

2. Using an example from a field with which you are familiar, describe how an error transmission study might be made and how it could be useful. How could you check the assumptions (*a*) that function linearization was justified and (*b*) that errors in two variables were uncorrelated? What circumstances would *predispose* you to believe or to doubt these assumptions?

3. Write a report to the chief chemist of an analytical laboratory, explaining how and why variance component experiments could supply him with useful information. Illustrate with hypothetical examples.

4. Using your knowledge of variance components, write a report for a lawyer in the Environmental Protection Agency, explaining difficulties that might be anticipated in setting and enforcing environmental standards and in drafting requirements for sampling river water and testing the samples.

CHAPTER 18

Modeling Dependence: Time Series

In the early chapters of this book we saw that many standard statistical procedures are based on the hypothesis of random sampling. In particular, statistical methods for comparing means using the t and F distributions could be precisely justified provided it could be assumed that the errors were *independently* as well as normally distributed. For real data the independence assumption, whose violation produces serious inaccuracies, is often invalid. If appropriate randomization is introduced into the conduct of the experiment, however, the standard procedures can be justified in the sense that they yield sampling distributions that roughly approximate corresponding randomization distributions.

When it can be adopted, the randomization approach has the advantage that it does not require specificity about the *nature* of the dependence which is feared. Situations arise, however, where randomization is not used (in some cases it cannot be used). One way to deal with such problems is to use a specific model for dependence. If such a model can be relied upon, it is possible to develop more sensitive procedures than those that depend only on randomization. Two questions thus arise:

1. How may dependence be modeled?
2. How can we take advantage of such modeling?

To answer these questions comprehensively would take us beyond the intended scope of this book. We will be content here to illustrate by means of elementary examples. Rather unexpectedly, perhaps, we will find that the ability to model dependence can lead to the solution of problems in forecasting, feedback control, and intervention analysis.

18.1. THE INDUSTRIAL DATA OF CHAPTER 2 RECONSIDERED AS A TIME SERIES

The industrial data first presented in Table 2.2 are a time series of 210 successive batch yields produced by a chemical process. The data were used to provide the reference distribution of Figure 2.6 appropriate to assess the difference $\bar{y}_B - \bar{y}_A = 1.30$ units obtained in a plant-scale trial. In this trial 10 successive batches had been produced by a modified process (B), and these were compared with 10 successive batches produced by the standard process (A) which had immediately preceded them. An appropriate reference distribution, shown in Figure 2.6, was obtained by calculating the 191 differences between averages of adjacent sets of 10 observations in the time series. By referring the observed difference 1.30 to this distribution, it was possible to find out how often a difference this large occurred in successive runs of 10 batches when *no modification* had been made. The procedure assumed, not that the observations were independent, but only that groups of observations were similarly related in probability *if they were similarly spaced in time.*

Autocorrelation

To see how we can explicitly model serial dependence, first recall the discussion of Section 3.1. Figure 18.1 shows a serial plot of the industrial data for observations one and two steps apart. Figure 18.1*a*, introduced earlier as Figure 3.1, shows a plot of y_{t+1} against y_t. To make the plot, the second observation (81.7) is plotted against the first (85.5), the third (80.6) against the second (81.7), and so on. The plot suggests that successive observations are negatively correlated. Similarly Figure 18.1*b* suggests that there is very little correlation between observations two intervals apart.

The correlation between observations k intervals apart is quantitatively measured by the *lag k sample autocorrelation coefficient*, given by

$$r_k = \frac{\sum (y_i - \bar{y})(y_{i+k} - \bar{y})}{\sum (y_i - \bar{y})^2} \tag{18.1}$$

where $\bar{y}$ is the average, 84.1, and summations are taken over all available products and squares. For example, for $k = 2$ we have:

$$r_2 = \frac{(85.5 - 84.1)(80.6 - 84.1) + (81.7 - 84.1)(84.7 - 84.1) + \cdots + (82.3 - 84.1)(80.2 - 84.1)}{(85.5 - 84.1)^2 + (81.7 - 84.1)^2 + \cdots + (80.2 - 84.1)^2}$$

$$= \frac{-231.1}{1734.6} = -0.13 \tag{18.2}$$

By a similar calculation it will be found that the value of r_1 is -0.29.

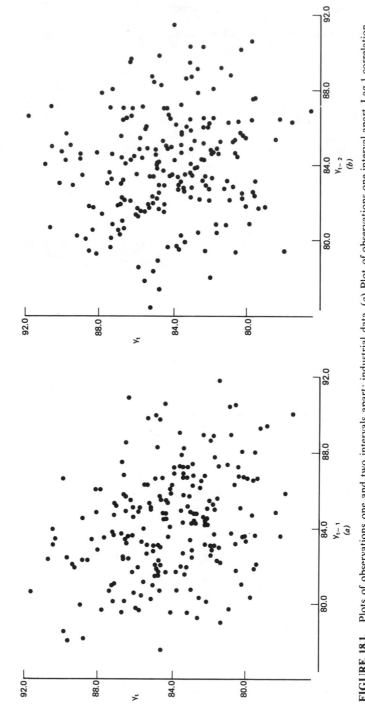

FIGURE 18.1. Plots of observations one and two intervals apart: industrial data. (*a*) Plot of observations one interval apart. Lag 1 correlation $r_1 = -0.29$. (*b*) Plot of observations two intervals apart. Lag 2 correlation $r_2 = -0.13$.

An autocorrelation cannot be larger than plus 1 or smaller than minus 1. Obviously $r_0 = \sum (y_i - \bar{y})^2 / \sum (y_i - \bar{y})^2$ is always equal to 1; this makes sense because the series is perfectly correlated with its (unlagged) self.

When the autocorrelations for the batch data at lags 0, 1, 2, ... are plotted against k, the value of the lag, we obtain the *sample autocorrelation function* shown in Figure 18.2. If it is true (as it is likely to be in this example) that the statistical dependence of one observation on another is a function only of their distance apart, the series is said to be *stationary*, and in particular it will have a *fixed mean*. If for such a series we imagine the number of observations in the sample to be increased without limit, r_k will converge to some value ρ_k, which we will call the *lag k theoretical* autocorrelation. The sample autocorrelations r_k that are actually available from the existing body of data may be regarded as estimates of the theoretical autocorrelations ρ_k of a hypothetical infinite sequence.

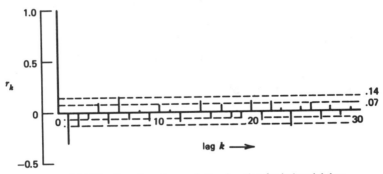

FIGURE 18.2. Sample autocorrelation function for industrial data.

The standard error of the sample autocorrelations r_k, if the theoretical autocorrelations are all zero, is approximately $1/\sqrt{n}$. For the industrial example $1/\sqrt{210} \simeq 0.07$. To give some idea of the statistical significance of the sample autocorrelation, therefore, dashed lines at ± 0.07 and at ± 0.14 are shown in Figure 18.2.

Serial correlation is, naturally, most expected at low lags. Inspection of the sample autocorrelation function of Figure 18.2 suggests that for this system there exists a negative autocorrelation ρ_1 between adjacent observations, estimated by $r_1 = -0.29$. Beyond this point the correlation pattern seems to die out. If we accept this representation of the serial dependence in the data (specifically, that $\rho_2, \rho_3, \ldots$ are all zero), we can use it to generate an appropriate reference distribution for $\bar{y}_B - \bar{y}_A$.

A Reference Distribution for $\bar{y}_B - \bar{y}_A$ That Takes Account of Serial Correlation in the Observations

We are familiar with the result that σ^2/n is the variance of the average of n independent observations drawn from a population with variance σ^2. It is shown in Appendix 3A that, when observations are drawn from a stationary time series with theoretical autocorrelations all zero except ρ_1, this variance is modified and becomes

$$V(\bar{y}) = \frac{\sigma^2}{n}\left[1 + \frac{2(n-1)}{n}\rho_1\right] \tag{18.3}$$

Also (see Appendix 18A), the variance of the difference $\bar{y}_B - \bar{y}_A$ of two successive averages of n observations is not $2\sigma^2/n$ but

$$V(\bar{y}_B - \bar{y}_A) = \frac{2\sigma^2}{n}\left[1 + \frac{2n-3}{n}\rho_1\right] \tag{18.4}$$

In the industrial example there were two successive groups of $n = 10$ observations with $\bar{y}_B - \bar{y}_A = 1.30$, and for this large group of 210 observations the variance was 8.29. Assuming the observations to be independent (so that $\rho_1 = 0$) and substituting 8.29 for σ^2, we get from Equation 18.4 that the estimated variance of $\bar{y}_B - \bar{y}_A$ is $2 \times 8.29/10 = 1.67$. As we saw in Chapter 3, by referring $1.30/\sqrt{1.67} = 1.01$ to the normal table, we obtain a significance level of 0.16. If, on the other hand, we suppose that ρ_1 is nonzero and substitute for it the estimate $r_1 = -0.29$, we obtain for the estimated variance of $\bar{y}_B - \bar{y}_A$

$$1.67[1 + 1.7(-0.29)] = 0.84 \tag{18.5}$$

A significance level of 0.08 is then obtained by referring $1.30/\sqrt{0.84} = 1.41$ to the normal table. This is much closer to the value 0.05 obtained in Chapter 2 by using the reference set of 191 differences between successive averages.

18.2. STATISTICAL MODELING REVISITED

Earlier chapters have included discussions of statistical *models* appropriate to the various problems examined. For example, for the randomized block design we considered the model

$$y_{ji} = \eta + \beta_i + \tau_j + \epsilon_{ji} \tag{18.6}$$

where τ_j represented the jth treatment effect, and β_i the ith block effect.

Again, for a simple linear regression we considered the straight line model passing through the origin:

$$y_j = \beta x_j + \epsilon_j \tag{18.7}$$

where β was the slope of the regression line. In both instances the ϵ's were a sequence of errors that were assumed to vary *independently* about zero with some fixed variance σ^2. We called such a sequence *white noise*, and we noted that white noise is informationless in the sense that knowledge of the value of one such error supplies no additional knowledge of any other.

The two models above can be rearranged as follows:

$$\epsilon_{ji} = y_{ji} - \eta - \beta_i - \tau_j \tag{18.8}$$

$$\epsilon_j = y_j - \beta x_j \tag{18.9}$$

As we saw earlier, each model is a recipe for generating white noise. All the models in this book can be similarly viewed. Each model is a prescription for transforming the data to white noise, uncorrelated with itself *or with any other known phenomenon.* If this prescription succeeds, we have a fully efficient model, for if it transforms informative data to informationless white noise; the model contains within itself all the information in the data. Looked at in this way, studies of residuals are seen as checks to ensure that there is no leak of relevant information from the model. This is analogous to checking the efficiency of an extraction process by testing the filtrate to determine whether it contains any of the material that was supposed to be filtered out. How can we obtain a similar model for a time series? What can we do with it when we have it? We ponder the first question first.

A Time Series Model for the Plant Trial Data

Consider the industrial data consisting of 210 successive yield observations. We know that the manufacturer, so far as he was able, performed precisely the same operation throughout this period. It is reasonable, then, to represent the tth result y_t as the sum of a fixed mean μ plus an error z_t:

$$y_t = \mu + z_t \tag{18.10}$$

Since μ is the mean of the y's, the mean of the errors z_t is zero. In practice these errors are unlikely to be independent because uncontrollable influences happen over periods of time, and when z_t is unusually high it may be that z_{t+1} will also tend to be unusually high (or, in some examples, unusually low). The actual data (Figure 18.2) suggest that errors were (negatively) correlated one lag apart but not at greater lags. How can we write a model producing autocorrelations of this kind?

Suppose that $\epsilon_1, \epsilon_2, \ldots, \epsilon_t, \ldots$ is a *white noise* sequence with variance σ_ϵ^2. Then with θ, a parameter whose absolute magnitude is less than 1, let us write

$$
\begin{aligned}
z_t &= \epsilon_t - \theta \; \epsilon_{t-1} \\
z_{t-1} &= \epsilon_{t-1} - \theta\epsilon_{t-2} \\
z_{t-2} &= \epsilon_{t-2} - \theta\epsilon_{t-3} \\
z_{t-3} &= \epsilon_{t-3} - \theta\epsilon_{t-4} \\
\text{etc.}
\end{aligned}
\tag{18.11}
$$

Then, exactly as required, z_t will be correlated with z_{t-1} because they both contain the random variable ϵ_{t-1}. But z_t will not be correlated, for example, with z_{t-2} because the ϵ's that they contain, none of which are common, are distributed independently of each other. Clearly we can change the degree of lag 1 correlation by changing θ. Thus a time series model of the required form is

$$
y_t = \mu + \epsilon_t - \theta\epsilon_{t-1} \tag{18.12}
$$

Just like other statistical models, this is an equation containing a systematic part (μ) and a part expressible in terms of a white noise sequence (ϵ_t). The elements ϵ_t are sometimes referred to as the "random shocks" that "drive" the system. If we write $\tilde{y}_t$ for the deviation from the true mean ($\tilde{y}_t = y_t - \mu$), the model is

$$
\tilde{y}_t = \epsilon_t - \theta\epsilon_{t-1} \tag{18.13}
$$

and is called a *first-order moving average process*. It is easy to show that for this model

$$
\sigma_y^2 = (1 + \theta^2)\sigma_\epsilon^2 \tag{18.14}
$$

$$
\rho_1 = \frac{-\theta}{1 + \theta^2} \tag{18.15}
$$

and all other autocorrelations are zero. Now, for the plant data, the sample average $\bar{y}$, sample variance s^2, and sample autocorrelation at lag 1 are $\bar{y} = 84.1$, $s^2 = 8.29$, $r_1 = -0.29$, from which we can obtain approximate values* for μ, σ_ϵ^2, and θ by substituting $\bar{y}$ for μ and choosing σ_ϵ^2 and θ so that, in Equation 18.14, $\sigma_y^2 = s^2 = 8.29$ and, in Equation 18.15, $\rho_1 = r_1 = -0.29$. The parameter estimates thus obtained are $\hat{\sigma}_\epsilon^2 = 7.5$ and $\hat{\theta} = 0.32$. The estimated *time series model* is then

$$
y_t = 84.1 + \epsilon_t - 0.32\epsilon_{t-1}, \qquad \hat{\sigma}_\epsilon^2 = 7.5 \tag{18.16}
$$

---

* More precise maximum likelihood estimates of the parameters would be preferable, but their calculation is outside the scope of the present discussion.

What Can We Do with a Time Series Model Once We Have It?

We see that it is possible to build time series models representing observations that are serially dependent. But of what use are such models?

In the next three sections we illustrate three different applications. Although by no means exhaustive, these elementary examples give some idea of the breadth and power of the times series approach. The applications are to forecasting, feedback control, and intervention analysis.

18.3. FORECASTING: REFRIGERATOR SALES EXAMPLE

Figure 18.3a shows a plot of monthly sales figures for a certain type of refrigerator. The data extend from September 1963 to August 1972. Suppose that we are in August 1972 and have this record. How can we best forecast the value for next month, September 1972? The answer is that if we can *build a model* for the series it is a simple matter to obtain a best* forecast.

The Model

As is usually true for sales data, the series of Figure 18.3a does not appear to be stationary and, in particular, does not appear to vary about a *fixed* mean. However, the first differences $d_t = y_t - y_{t-1}$, that is, the monthly *changes* in sales plotted in Figure 18.3b, do appear to be stationary and to vary about a fixed mean that is close to zero. We therefore model $d_t = y_t - y_{t-1}$ rather than modeling the nonstationary y_t directly.

Figure 18.4 shows the autocorrelation function for d_t. There is a fairly large autocorrelation at lag 1, $r_1 = -0.48$, followed by autocorrelation at higher lags whose existence can be readily accounted for by sampling error. This suggests a first-order moving average model for d_t of the kind discussed in Section 18.2. Now the sample variance of d_t is 602.4. Once more, employing Equations 18.14 and 18.15, we obtain an approximate model

$$d_t = \epsilon_t - 0.7\epsilon_{t-1}, \qquad \hat{\sigma}_\epsilon^2 = 404.3 \qquad (18.17)$$

Equivalently, the model expressed in terms of the original y_t's is

$$y_t - y_{t-1} = \epsilon_t - 0.7\epsilon_{t-1}, \qquad \hat{\sigma}_\epsilon^2 = 404.3 \qquad (18.18)$$

Using the Model to Make a Forecast

We have built a time series model for sales based on past data. We now want to use it at time n (August 1972) to forecast the value at time $n + 1$ (September

---

* The forecast is best in the sense that it has the smallest mean square error.

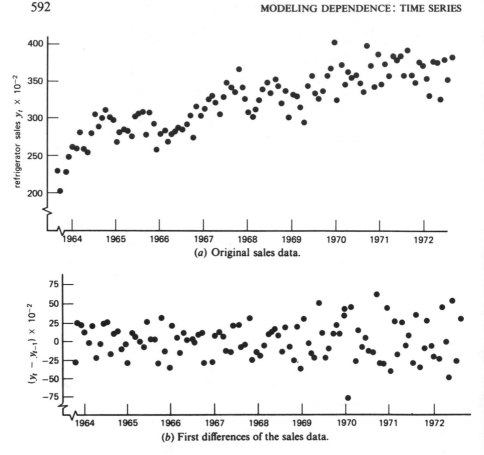

(a) Original sales data.

(b) First differences of the sales data.

FIGURE 18.3. Monthly sales of refrigerators (September 1963–August 1972).

1972). To do this we first write the model for the month to be forecast, that is, we write the model for y_{n+1}. This is*

$$y_{n+1} = y_n + \epsilon_{n+1} - 0.7\epsilon_n \qquad (18.19)$$

It turns out that the forecast of y_{n+1} giving a forecast error with smallest standard deviation is obtained by substituting on the right-hand side of Equation 18.19 the best values we know *at the time the forecast is made*. Now at time n (in August 1972) we know y_n, the August sales figure. We do not know ϵ_{n+1}, the shock that will enter the system in September, but we

---

* Adding y_{t-1} to both sides of Equation 18.18, we get $y_t = y_{t-1} + \epsilon_t - 0.7\epsilon_{t-1}$. Advancing the index by unity gives $y_{t+1} = y_t + \epsilon_{t+1} - 0.7\epsilon_t$. Putting $t = n$, we obtain Equation 18.19.

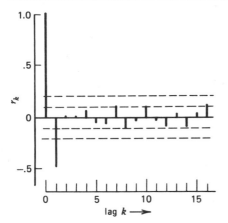

FIGURE 18.4. Sample autocorrelation function for the first differences of the sales data.

do know it has mean zero and is uncorrelated with any data that are so far available. The best value to use for ϵ_{n+1}, therefore, is zero, and the best forecast $\hat{y}_{n+1}$ is

$$\hat{y}_{n+1} = y_n - 0.7\epsilon_n \qquad (18.20)$$

But what is ϵ_n? Subtracting Equation 18.20 from Equation 18.19 gives

$$y_{n+1} - \hat{y}_{n+1} = \epsilon_{n+1} \qquad (18.21)$$

Thus ϵ_{n+1} is simply the difference between y_{n+1}, the actual sales figure due next month, and $\hat{y}_{n+1}$, its forecast made now. But if

$$\epsilon_{n+1} = y_{n+1} - \hat{y}_{n+1} \qquad (18.22)$$

then

$$\epsilon_n = y_n - \hat{y}_n \qquad (18.23)$$

Thus the shocks ..., ϵ_{n+1}, ϵ_n, ... turn out to be just the one-step-ahead forecast errors. They measure differences between what actually happened in a given month and what the previous month's forecast *said* would happen. It is no surprise that, when best forecasts are used, the one-step-ahead forecast errors ..., ϵ_{n+1}, ϵ_n, ... must be uncorrelated. If they were correlated, it would be possible to forecast the next month's error from previous forecast errors and thus improve the forecast.

If now we substitute Equation 18.23 into 18.20, we get

$$\hat{y}_{n+1} = y_n - 0.7(y_n - \hat{y}_n) \qquad (18.24)$$

or

$$\hat{y}_{n+1} = 0.3y_n + 0.7\hat{y}_n \qquad (18.25)$$

and, therefore, since $y_n = 385$ and $\hat{y}_n = 359.4$, the desired forecast is $\hat{y}_{n+1} = 367.1$. This is an interesting formula. It tells us how we may obtain our forecast $\hat{y}_{n+1}$ by updating the previous forecast $\hat{y}_n$. This is done by taking a weighted average of the new value y_n actually observed in August and the old forecast $\hat{y}_n$ made in July. In general terms Equation 18.25 is

$$\hat{y}_{n+1} = (1 - \theta)y_n + \theta\hat{y}_n \qquad (18.26)$$

The Forecast as a Weighted Average of Previous Data Values

Equation 18.26 is an extremely convenient formula for successively updating the forecast once an initial forecast has been obtained. However, it does not make immediately apparent how the forecast $\hat{y}_{n+1}$ uses the data $y_n, y_{n-1}, y_{n-2}, \dots$

According to Equation 18.26, at stage n we have

$$\hat{y}_{n+1} = (1 - \theta)y_n \quad + \theta\hat{y}_n$$

so that at stage $n - 1$

$$\hat{y}_n = (1 - \theta)y_{n-1} + \theta\hat{y}_{n-1} \qquad (18.27)$$

and at stage $n - 2$

$$\hat{y}_{n-1} = (1 - \theta)y_{n-2} + \theta\hat{y}_{n-2} \qquad (18.28)$$

etc.

If we multiply Equation 18.27 by θ, Equation 18.28 by θ^2, and so on, we have

$$
\begin{aligned}
\hat{y}_{n+1} &= (1 - \theta)y_n &&+ \theta\hat{y}_n \\
\theta\hat{y}_n &= (1 - \theta)\theta y_{n-1} &&+ \theta^2\hat{y}_{n-1} \qquad (18.29) \\
\theta^2\hat{y}_{n-1} &= (1 - \theta)\theta^2 y_{n-2} &&+ \theta^3\hat{y}_{n-2}
\end{aligned}
$$

etc.

Adding, we get

$$\hat{y}_{n+1} = (1 - \theta)(y_n + \theta y_{n-1} + \theta^2 y_{n-2} + \cdots) \qquad (18.30)$$

or, for our specific example with $\theta = 0.7$,

$$\hat{y}_{n+1} = 0.30y_n + 0.21y_{n-1} + 0.15y_{n-2} + 0.10y_{n-3} + 0.07y_{n-4} + \cdots \qquad (18.31)$$

The forecast thus turns out to be a *weighted average* of the current and previous values of the series. This is illustrated in Figure 18.5, where the

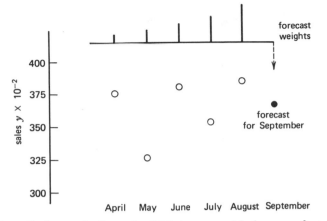

FIGURE 18.5. The forecast for September 1972, shown as weighed average of previous values of the series, refrigerator sales example.

last few values of the series, the forecast for September 1972, and the weight function are shown.

Any linear aggregate $w_1 y_n + w_2 y_{n-1} + w_3 y_{n-2} + \cdots$ of previous observations is an *average* if the weights $w_1, w_2, w_3, \ldots$ *add up to unity* ($\sum w = 1$). This property is necessary to ensure that the average is a measure of the *location* of the series. Why? Suppose that we add some constant k to all the values in the series. Then only if $\sum w = 1$ will the same constant k be added to the average, because

$$w_1(y_n + k) + w_2(y_{n-1} + k) + \cdots = (w_1 y_n + w_2 y_{n-1} + \cdots) + k \sum w$$
$$= (w_1 y_n + w_2 y_{n-1} + \cdots) + k$$

(18.32)

only if $\sum w = 1$.

The usual arithmetic average

$$\bar{y} = \frac{\sum y}{n} = \frac{1}{n} y_n + \frac{1}{n} y_{n-1} + \cdots + \frac{1}{n} y_1$$

(18.33)

is of this form with weights all equal to $1/n$. But the average of Equation 18.30 is "exponentially weighted." The current observation receives greatest weight, and the weights applied to previous observations fall off exponentially with a "damping factor" θ (see Figure 18.6). For data like those we are considering this makes a great deal of sense. Indeed, *exponentially weighted averages* were used for forecasting before their theoretical justification was understood.

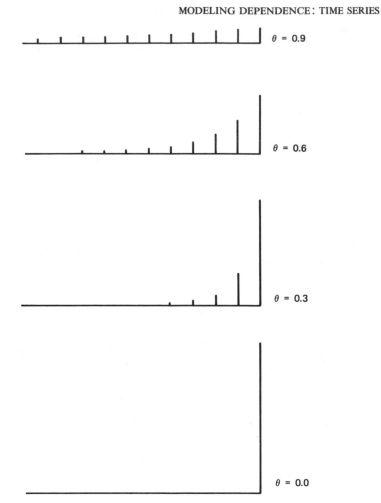

FIGURE 18.6.   Weights for exponentially weighted averages for various values of the parameter θ.

Nature of the Forecast

Notice a number of interesting facts:

1. Once we get the forecasting process started in some way, for example, by directly using Equation 18.31, we can update the forecast as each new piece of data becomes available simply by using Equation 18.26.
2. At time n, all relevent information about the series is stored in the *single* number $\hat{y}_n$. This fact is particularly valuable in inventory control, where

the signal for reordering is given when the forecast of stocks available reaches a sufficiently low level. Suppose that we have 1000 different items to keep track of. If the above model is appropriate, we need only store the 1000 values $\hat{y}_n$ in a computer and program it so as to update these quantities appropriately as soon as the new values y_{n+1} become available.

3. The exponentially weighted average is an example of the compromise (which we all must make) between the conflicting demands of enterprise, on the one hand, and conservatism, on the other. Suppose that we have a variable nonstationary series, and we want to estimate its level during the next time interval. Because the series is variable, caution would tell us to take an average that stretched back into the series over a large number of data values. This could be done by setting θ close to 1 (see $\theta = 0.9$ in Figure 18.6). Unfortunately this procedure would also guarantee that the forecast would be slow to respond to real changes in level. The most rapid response will be obtained when $\theta = 0$. In that case $\hat{y}_{n+1} = y_n$, and response is immediate because the forecast uses only the latest value of the series and ignores all previous values. A compromise must be made between the calls of immediacy and inertia. The best compromise employs the value of θ determined by the modeling process for the particular series under study.

Summary and Interim Discussion

Let us recapitulate. Time series modeling has been illustrated in a simple case where it appeared that the lag 1 autocorrelation ρ_1 was nonzero but higher autocorrelations were zero. The first difference of the refrigerator sales data appeared to be represented by a model of this form. The forecast directly following from this model was examined and was found to agree with common sense. We should add the following at this point:

1. Had the correlation structure been different, the appropriate model and the resulting forecast would also have been different. In particular, the forecast would not have taken the form of an exponentially weighted average.
2. To meet the need for conciseness an inadequate picture of the modeling process has been given. We have talked only of what may be called the *identification* (or *specification*) stage of time series modeling. We have not discussed the accurate *fitting* and *diagnostic checking* of the model. The method of choosing the constant θ by matching the sample autocorrelation r_1 of the data to the theoretical autocorrelation $\rho_1 = -\theta/(1 + \theta^2)$, which we have used, does not take full advantage of the data. In practice more efficient maximum likelihood methods would

be used for the fitting process. Furthermore the fitted model would have been submitted to a process of diagnostic checking in which the residuals (the ϵ's obtained when the best estimate was substituted for θ) would have been examined to ensure that no apparent inadequacy of the model had been overlooked. See Box and Jenkins (1970) for further details.

We now illustrate how a time series model may be used to devise a feedback control scheme.

18.4. FEEDBACK CONTROL: DYE LEVEL EXAMPLE

Consider Figure 18.7a. A polymer was colored by adding dye at the inlet of a continuous reactor. At the outlet the color index y was checked every 15 minutes, and, if it deviated from the target value of $T = 9$, the rate of

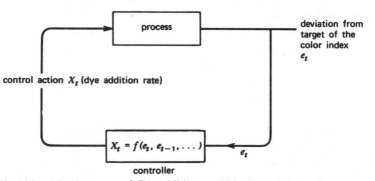

(a) Closed loop, feedback control. Dye addition rate X, is changed, depending on past and present values of the deviation e, of the color index from its target value T.

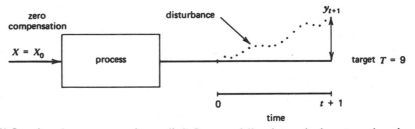

(b) Open loop (zero compensation applied). Because of disturbances in the system, the color index would have wandered from target to some value y_{t+1} at time $t + 1$.

FIGURE 18.7.   Diagrammatic representation of dyeing operation.

addition of dye X at the inlet could be increased or decreased. This is a feedback control scheme where at time t past and present deviations at the output $e_t, e_{t-1}, \ldots$, determine the level X_t to which the input is adjusted. Thus $X_t = f(e_t, e_{t-1}, \ldots.)$ The problem is to choose this *control function* or *controller* so that the deviations e_t from target have the smallest standard deviation. In what follows we denote the incremental adjustment by x_t. Thus $x_t = X_t - X_{t-1}$.

Suppose that the following relevant facts were known about the dyeing process:

1. If *no control action was taken* at all (see Figure 18.7b), the color index would drift away from the target value $T = 9$ in a time series represented approximately by the model (see Equation 18.18)

$$y_t - y_{t-1} = \epsilon_t - 0.7\epsilon_{t-1} \qquad (18.34)$$

2. The dye addition rate X at the input had to be increased by 16.6 units to eventually increase the color index at the output by 1 unit.
3. Any change in dye addition rate took one time interval (15 minutes) to become effective at the output.

Suppose that at some time "zero" the dye addition rate X_0 was such that no deviation from target occurred at the output. At some later time t the dye rate will be $X_t = X_0 + x_1 + x_2 + \cdots + x_t$ (the initial rate plus the sum of all the adjustments). Now, if no adjustments had been made (Figure 18.7b), by time $t + 1$ the output color index would have strayed to some value y_{t+1} and would be off target by an amount $y_{t+1} - T$. We call $y_{t+1} - T$ the "disturbance" at time $t + 1$.

Bearing in mind that control action computed at time t could not be effective until time $t + 1$, we would like to choose X_t so that it eliminated as much as possible of the disturbance $y_{t+1} - T$. Our best estimate at time t of the disturbance is $\hat{y}_{t+1} - T$, where $\hat{y}_{t+1}$ is the optimal forecast. For the time series model of Equation 18.34 we know this forecast to be the exponentially weighted average of past values

$$\hat{y}_{t+1} = 0.3(y_t + 0.7y_{t-1} + 0.49y_{t-2} + \cdots) \qquad (18.35)$$

The compensatory action to be applied at time t that will just cancel the forecasted disturbance $\hat{y}_{t+1} - T$ is therefore

$$X_t = -16.6(\hat{y}_{t+1} - T) \qquad (18.36)$$

If this action is taken, the error e_{t+1} at the output is equal to ϵ_{t+1} for

$$(y_{t+1} - T) - (\hat{y}_{t+1} - T) = y_{t+1} - \hat{y}_{t+1} = \epsilon_{t+1} \qquad (18.37)$$

Similarly at time $t - 1$ the action should have been

$$X_{t-1} = -16.6(\hat{y}_t - T) \tag{18.38}$$

Subtracting Equation 18.38 from 18.36, we find that the *adjustment* of the dye addition rate at time t should be

$$x_t = X_t - X_{t-1} = -16.6(\hat{y}_{t+1} - \hat{y}_t) \tag{18.39}$$

Now, since $\hat{y}_{t+1} = y_{t+1} - \epsilon_{t+1}$, we can write

$$x_t = X_t - X_{t-1} = -16.6[(y_{t+1} - y_t) - (\epsilon_{t+1} - \epsilon_t)] \tag{18.40}$$

But the disturbance is modeled by

$$y_{t+1} - y_t = \epsilon_{t+1} - 0.7\epsilon_t \tag{18.41}$$

Substituting Equation 18.41 into 18.40 yields

$$x_t = X_t - X_{t-1} = -16.6 \times 0.3\epsilon_t \tag{18.42}$$

Thus finally the adjustment should be, at each stage,

$$x_t = X_t - X_{t-1} = -5\epsilon_t = -5e_t \tag{18.43}$$

and therefore the dye level should be adjusted so that

$$X_t = X_{t-1} - 5e_t \tag{18.44}$$

The last equation says that, if the dye level at the output was too high by an amount e_t, the dye addition rate at the input should be reduced by an amount $5e_t$. This is the optimal control equation (or "controller"). It can be simply put into effect by using the chart of Figure 18.8, whereby the operator simply plots the current value of the color index given on the right-hand side of the chart and reads off the action he should take on the left. For example, in the situation depicted he has just observed a deviation from target of -0.14 unit and should therefore increase the rate of dye addition by 0.7 unit.

The optimal scheme requires that we take *some* action at every opportunity, that is, every 15 minutes. In practice, usually little is lost if the "rounded" control chart indicated at the extreme left of Figure 18.8 is used. In this simplification, if a point falls in the center band, no action is taken. Correspondingly, if a point falls in one of the other bands, the rate of dye addition is raised or lowered by one or two units.

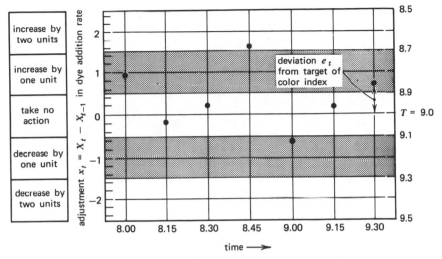

FIGURE 18.8. A feedback control chart for dye addition.

The Control Action in Terms of the Disturbances $y_t - T$ That Would Have Occurred Had No Action Been Taken

We see from the argument of Section 18.3 that the control action is the same *as if* at each stage we made a forecast of the level of the disturbance in the next time interval by taking an exponentially weighted average of the disturbances $y_t - T$, $y_{t-1} - T$, ... and applying the degree of compensation X_t that would exactly cancel the forecasted deviation. Notice carefully, however, that, although our action is equivalent to doing this, we do not actually observe the disturbance and we do not allow it to develop. Because of feedback action we observe only the errors $\epsilon_t = y_t - \hat{y}_t$. In particular, what is done is *not* equivalent to taking an exponentially weighted average of the *deviations on the chart*.

The Control Action in Terms of the Observed Errors When Feedback Action Is Taken

The adjustment at time t is $X_t - X_{t-1} = -5e_t$, that at time $t-1$ was $X_{t-1} - X_{t-2} = -5e_{t-1}$, and so on. Adding together such equations, we obtain

$$X_t = -5(e_t + e_{t-1} + \cdots + e_1) \tag{18.45}$$

Which is the control equation in the form $X_t = f(e_t, e_{t-1}, \ldots)$

The total adjustment applied is thus proportional to the cumulative sum*
of all the errors at the output for this particular set up.

The method illustrated above for obtaining optimal feedback control is
readily extended to more general situations where the disturbance and also
the dynamic properties of the system may be of a much more complex nature.
In more complex cases and those in which several controllers must be handled
simultaneously, "discrete" control of this kind, that is, control at discrete
times, is ideally suited to the digital minicomputer.

In the final example the effect of an intervention is estimated using a
time series model.

18.5. INTERVENTION ANALYSIS: LOS ANGELES AIR POLLUTION EXAMPLE

An indication of the level of smog in downtown Los Angeles is the oxidant
concentration, denoted by O_3. Monthly averages of O_3 for January 1955
to December 1965 are shown in Figure 18.9. In early 1960 two events
occurred: (1) the opening of the Golden State throughway and (2) the coming
into effect of a new law (rule 63) that reduced the allowable proportion of
reactive hydrocarbons in gasoline sold locally. Let us call these joint events
an intervention. Did this intervention coincide with any difference in the O_3
concentration?

The oxidant levels in Figure 18.9 vary a great deal; moreover there are
strong seasonal effects. It would be totally inappropriate, therefore, to
apply, for example, a t test to the difference in means before and after January
1960 because of the obvious dependence between the observations. Furthermore no question of randomization can possibly arise since the ordering
of the observations cannot be changed.

The problem can be tackled as follows. Denote the O_3 concentration at
time t by y_t. Then write

$$y_t = \beta x_t + z_t \qquad (18.46)$$

where x_t has the value zero before January 1960 and the value unity after-
wards, and z_t is represented by a time series model that takes account of the
local and seasonal dependence. With this arrangement β will represent the
possible change in concentration of oxidant that occurred as a result of
the intervention.

---

* The control action for this case is, therefore, a direct analog of the integral control action
used in continuous control schemes, where $X_t = k \int_0^t e_t \, dt$. Notice that the feedback chart is
not the same as a cumulative sum chart and has quite a different purpose.

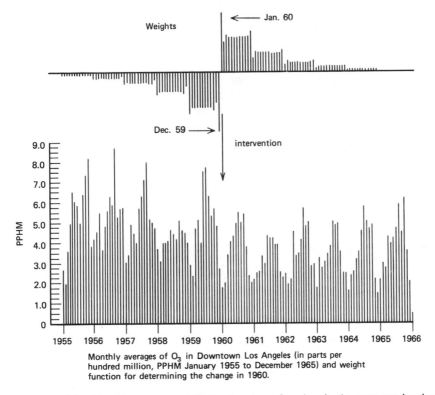

Monthly averages of O_3 in Downtown Los Angeles (in parts per hundred million, PPHM January 1955 to December 1965) and weight function for determining the change in 1960.

FIGURE 18.9. Monthly averages of O_3 in downtown Los Angeles in parts per hundred million (pphm) January 1955 to December 1965; with weight function for determining the change in 1960.

For these data the following time series model was found appropriate for z_t:

$$z_t - z_{t-12} = \epsilon_t + 0.2\epsilon_{t-1} - 0.8\epsilon_{t-12} - 0.16\epsilon_{t-13} \qquad (18.47)$$

This model is somewhat more complicated than those considered earlier since it allows for seasonal characteristics of the data. It is, however, of the same general form, in which z_t is linearly related to a white noise sequence ϵ_t.

It is possible to show (see Box and Tiao, 1975) that with this model the maximum likelihood estimate of the change is

$$\hat{\beta} = -1.1 \pm 0.1 \qquad (18.48)$$

where the number following $\pm$ is the standard error of $\hat{\beta}$. The analysis strongly suggests, therefore, that the intervention (or something else that occurred at the same time) produced an estimated reduction in oxidant level of about 25%.

It turns out that $\hat{\beta}$ is estimated by the difference of two weighted averages before and after the event. In fact, the weight function for $\hat{\beta}$ is shown at the top of Figure 18.9. Notice that the greatest weight is given to data immediately before and immediately after the intervention and that "steps" in the exponential discounting allow for averaging over 12-month periods. Intervention methods of this kind help the investigator to find out what happened when, for example, the new law was passed, the nuclear power station was started up, the teaching method was changed, or free contraceptives were distributed. Notice that investigations of this kind do not *prove* causality. Rather they establish whether or not there is any reason to believe that a change of a kind that could imply causation has really occurred, and they estimate the size of that change.

APPENDIX 18A. DERIVATION OF EQUATION 18.4

Variance of the Difference of Two Successive Averages of n Observations Serially Correlated at Lag 1 (Equation 18.4)

The variance of the difference of two averages $\bar{y}_B$ and $\bar{y}_A$ is

$$V(\bar{y}_B - \bar{y}_A) = V(\bar{y}_B) + V(\bar{y}_A) - 2\,\text{Cov}(\bar{y}_B, \bar{y}_A) \tag{18A.1}$$

It was shown in Appendix 3A that for n observations with common mean and variance serially correlated only at lag 1

$$V(\bar{y}_B) = V(\bar{y}_A) = \frac{\sigma^2}{n}\left[1 + \frac{2(n-1)}{n}\rho_1\right] \tag{18A.2}$$

If $\tilde{y}_j$ is the deviation of y_j from its mean value,

$$\text{Cov}(\bar{y}_A, \bar{y}_B) = E\left[\frac{1}{n^2}(\tilde{y}_1 + \tilde{y}_2 + \cdots + \tilde{y}_n)(\tilde{y}_{n+1} + \tilde{y}_{n+2} + \cdots + \tilde{y}_{2n})\right]$$

$$= \frac{\sigma^2}{n^2}\rho_1 \tag{18A.3}$$

so that

$$V(\bar{y}_B - \bar{y}_A) = \frac{2\sigma^2}{n}\left(1 + \frac{2n-3}{n}\rho_1\right)$$

This is Equation 18.4.

REFERENCES AND FURTHER READING

For more details about applications of time series analysis see:

Box, G. E. P., and G. M. Jenkins (1968). Some recent advances in forecasting and control, part I, *Appl. Stat.*, **17**, 91.

Box, G. E. P., G. M. Jenkins, and J. F. MacGregor (1974). Some recent advances in forecasting and control, part II, *Appl. Stat.* **23**, 158.

Box, G. E. P., and G. M. Jenkins (1970). *Time Series Analysis: Forecasting and Control*, Holden-Day.

Box, G. E. P., and G. C. Tiao (1975). Intervention analysis with applications to economic and environmental problems, *J. Am. Stat. Assoc.* (*Theory and Methods Section*), **70**, 70.

QUESTIONS FOR CHAPTER 18

1. Sometimes an investigator must work with data that are autocorrelated in circumstances where it is impossible to perform a randomized experiment. Can you think of examples, preferably from your own field?

2. How can time series data be modeled?

3. Why does it make sense to define an adequate model for a set of data as a filter that reduces these data to white noise?

4. Once a model for some time series data has been established, how can it be used?

5. Explain how a time series model can be used to forecast future values of the series.

6. How can a time series model be used in designing a feedback control scheme?

7. What is intervention analysis? For the Los Angeles O_3 data, why should we not assess the intervention using the familiar t test described in Chapters 3 and 4?

Problems for Part IV

Whether or not specifically asked, the reader should always (1) plot the data in any potentially useful way, (2) state the assumptions made, (3) comment on the appropriateness of these assumptions, and (4) consider alternative analyses.

1. In a physics laboratory experiment on thermal conductivity a student collected the following data:

x = time (sec)	$y = \log I$
300	0.79
420	0.77
540	0.76
660	0.73

(a) Plot these data, fit a straight line by eye, and determine its slope and intercept.

(b) By least squares fit the model $y = \beta_0 + \beta x + \epsilon$ to these data. Plot the least squares line $\hat{y} = b_0 + bx$ with the data.

(c) Compare the answers to parts (a) and (b).

2. (a) An electrical engineer obtained the following data from six randomized experiments:

dial setting x	measured voltage y
1	31.2
2	32.4
3	33.4
4	34.0
5	34.6
6	35.0

Suggest a simple empirical formula to relate y to x, assuming that the standard deviation of an individual y value is $\sigma = 0.5$.

(b) Would your answer change if $\sigma = 0.05$? Why?

3. A metallurgist has collected the following data from work on a new alloy.

x	temperature (°F)	amount of iron oxidized (mg)
1	1100	6
2	1200	6
3	1300	10
4	1400	30

The first column gives

$$x = \frac{\text{temperature} - 1000}{100}$$

If the prediction equation $\hat{y} = bx^2$ is used (where $\hat{y}$ is the predicted value of the amount of iron oxidized), what is the least squares estimate for the amount of iron oxidized at 1500°F? Comment on the assumptions necessary to validate this estimate.

4. (a) Using the method of least squares, fit a straight line to the following data. What are the least squares estimates of the slope and intercept of the line?

x	10	20	30	40	50	60
y	2.7	3.6	5.2	6.1	6.0	4.9

(b) Calculate 99% confidence intervals for the slope and intercept.
(c) Comment on the data and analysis, and carry out any further analysis you think is appropriate.

5. Fit the model $y = \beta_1 x + \beta_2 x^2 + \epsilon$ to these data, which were collected in random order:

x	1	1	2	2	3	3	4	4
y	15	21	36	32	38	49	33	30

(a) Plot the data.
(b) Obtain least squares estimates b_1 and b_2 for the parameters β_1 and β_2.
(c) Draw the curve $\hat{y} = b_1 x + b_2 x^2$.
(d) Do you think the model is adequate? Explain your answer.

6. Using the method of least squares, an experimenter fitted the model

$$\eta = \beta_0 + \beta_1 x \tag{I}$$

to the data below. It is known that σ is about 0.2. A friend suggested it would be better to fit the following model instead:

$$\eta = \alpha + \beta(x - \bar{x}) \tag{II}$$

where $\bar{x}$ is the average value of the x's

(a) Is $\hat{\alpha} = \hat{\beta}_0$? Explain your answer. (The caret indicates the least squares estimate of the parameter.)
(b) Is $\hat{\beta} = \hat{\beta}_1$? Explain your answer.
(c) Are $\hat{y}_I$ and $\hat{y}_{II}$, the predicted values of the responses for the two models, identical at $x = 40$? Explain your answer.
(d) Considering the two models above, which would you recommend the experimenter use: I or II or both or neither? Explain your answer.

x	10	12	14	16	18	20	22	24	26	28	30	32
y	80.0	83.5	84.5	84.8	84.2	83.3	82.8	82.8	83.3	84.2	85.3	86.0

7. The residual mean square is often used to estimate σ^2. There are circumstances, however, in which it can produce (a) a serious underestimate of σ^2 or (b) a serious overestimate of σ^2. Can you give examples of such circumstances?

8. The following data, which are given in coded units, were obtained from a tool-life testing investigation in which x_1 = measure of the feed, x_2 = measure of the speed, y = measure of the tool life:

x_1	x_2	y
-1	-1	61
1	-1	42
-1	1	58
1	1	38
0	0	50
0	0	51

(a) Obtain the least squares estimates of the parameters β_0, β_1, and β_2 in the model

$$\eta = \beta_0 + \beta_1 x_1 + \beta_2 x_2$$

where η = the mean value of y.

(b) Which variable influences tool life more over the region of experimentation, feed or speed?

(c) Obtain a 95% confidence interval for β_2, assuming that the standard deviation of the observations $\sigma = 2$.

(d) Carry out any further analysis that you think is appropriate.

(e) Is there a simpler model that will adequately fit the data?

9. Using the data below, determine the least squares estimates of the coefficients in this model:

$$\eta = \beta_0 + \beta_1 x_1 + \beta_2 x_2 + \beta_3 x_3$$

where η is the mean value of the drying rate.

temperature (°C)	feed rate (kg/min)	humidity (%)	drying rate (coded units)
300	15	40	3.2
400	15	40	5.7
300	20	40	3.8
400	20	40	5.9
300	15	50	3.0
400	15	50	5.4
300	20	50	3.3
400	20	50	5.5

Coding for the variables is as follows:

x_i	$i = 1$ temperature (°C)	$i = 2$ feed rate (kg/min)	$i = 3$ humidity (%)
-1	300	15	40
$+1$	400	20	50

The coding equation for temperature is

$$x_1 = \frac{(\text{temperature in units of °C}) - 350}{50}$$

(a) What are the coding equations for feed rate and humidity?

(b) The design is a 2^3 factorial. What are the relations between b_0, b_1, b_2, b_3 and the factorial effects calculated in the usual way?

10. Suppose that the first seven runs specified in problem 9 had been performed with the results given, but in the eighth run the temperature had

been set at 350°C, the feed rate at 20 kg/min, and the humidity at 45%, and the observed drying rate had been 3.9.

(a) Analyze these data.

(b) Compare the answers to those obtained for problem 9.

11. An experimenter intended to perform a full 2^3 factorial design, but two values were lost as follows:

x_1	x_2	x_3	y
-1	-1	-1	3
$+1$	-1	-1	5
-1	$+1$	-1	4
$+1$	$+1$	-1	6
-1	-1	$+1$	3
$+1$	-1	$+1$	data missing
-1	$+1$	$+1$	data missing
$+1$	$+1$	$+1$	6

He now wants to estimate the parameters in the model

$$\eta = \beta_0 + \beta_1 x_1 + \beta_2 x_2 + \beta_3 x_3$$

With the data above, is this possible? Explain your answer. If your answer is yes, set up the equation(s) he would have to solve. If your answer is no, what is the minimum amount of additional information that would be required?

12. Mr. A and Ms. B, both newly employed, pack crates of material. Let x_1 and x_2 be indicator variables showing which packer is on duty, and y be the number of crates packed. On 9 successive working days these data were collected:

day t	A x_{1t}	B x_{2t}	y_t
1	1	0	48
2	1	1	51
3	1	0	39
4	0	1	24
5	0	1	24
6	1	1	27
7	0	1	12
8	1	0	27
9	1	1	24

Denote by β_1 and β_2 the number of crates packed per day by A and B, respectively. First entertain the model

$$\eta = \beta_1 x_1 + \beta_2 x_2$$

Carefully stating what assumptions you make, do the following:
(a) Obtain least squares estimates of β_1 and β_2.
(b) Compute the residuals $y_t - \hat{y}_t$.
(c) Calculate $\sum_{t=1}^9 x_{1t}(y_t - \hat{y}_t)$ and $\sum_{t=1}^9 x_{2t}(y_t - \hat{y}_t)$.
(d) Show an analysis of variance appropriate for contemplation of the hypothesis that A and B pack at a rate of 30 crates per day, the rate laid down by the packers' union.
(e) Split the residual sum of squares into a lack-of-fit term, $S_L = \sum_{u=1}^3 3(\hat{y}_u - \bar{y}_u)^2$, and an "error" term $\sum_{u=1}^3 \sum_{i=1}^3 (y_{ui} - \bar{y}_u)^2$.
(*Note*: The above design consists of three distinct sets of conditions, $(x_1, x_2) = (0, 1), (1, 0), (1, 1)$, which are designated by the subscripts $u = 1, 2, 3$, respectively. The subscripts $i = 1, 2, 3$ are used to designate the three individual observations included in each set.)
(f) Make a table of the following quantities, each having nine values: $x_1, x_2, y_{ui}, \hat{y}_u, \bar{y}_u, \hat{y}_u - \bar{y}_u$, and $y_{ui} - \bar{y}_u$, and make relevant plots. Look for a time trend in the residuals.
(g) At this point it was suggested that the presence of B might stimulate A to do more (or less) work by an amount β_{12}, while the presence of A might stimulate B to do more (or less) work by an amount β_{21}. Furthermore, there might be a time trend. A new model was therefore tentatively entertained:

$$\eta = (\beta_1 + \beta_{12}x_2)x_1 + (\beta_2 + \beta_{21}x_1)x_2 + \beta_4 t$$

that is,

$$\eta = \beta_1 x_1 + \beta_2 x_2 + (\beta_{12} + \beta_{21})x_1 x_2 + \beta_4 t$$

The β_{12} and β_{21} terms cannot be separately estimated, but the combined effect $\beta_{12} + \beta_{21} = \beta_3$ can be estimated from this model:

$$\eta = \beta_1 x_1 + \beta_2 x_2 + \beta_3 x_3 + \beta_4 t$$

where $x_1 x_2 = x_3$. Fit this model, examine the residuals, and, if the model appears adequate, report least squares estimates of the parameter together with their standard errors.
(h) What do you think is going on? Consider other ways in which the original model could have been wrong.

13. (a) Fit a first-order model to the following data on the strength of welds.
(b) Plot contours of the fitted surface in the region of experimentation.

(c) Stating assumptions, say how ultimate tensile stress might be increased.

ambient temperature (°C) x_1	wind velocity (km/hr) x_2	ultimate tensile stress (coded units) y
5	2	83.4
20	2	93.5
5	30	79.0
20	30	87.1

The standard deviation of an individual y reading is approximately one unit.

14. Using least squares, fit a straight line to the following data:

x	12	13	14	15	16
y	404	412	413	415	422

Carefully stating *all* assumptions you make, obtain a 95% confidence interval for the mean value of the response at $x = 10$.

15. The following results were obtained by a chemist (where all the data are given in coded units):

temperature x_1	pH x_2	yield of chemical reaction y
−1	−1	6
+1	−1	14
−1	+1	13
+1	+1	7
−1	−1	4
+1	−1	14
−1	+1	10
+1	+1	8

(a) Fit the model $\eta = \beta_0 + \beta_1 x_1 + \beta_2 x_2$ by the method of least squares (η is the mean value of the yield).

(b) Obtain an estimate of the experimental error variance of an individual yield reading, assuming this model is adequate.

(c) Fit the model $\eta = \beta_0 + \beta_1 x_1 + \beta_2 x_2 + \beta_{11} x_1^2 + \beta_{22} x_2^2 + \beta_{12} x_1 x_2$. What difficulties arise? Explain these. What parameters and linear combinations of parameters can be estimated with this design?

(d) Consider what experimental design is being used here, and make an analysis in terms of factorial effects. Relate the two analyses.

16. An engineer wants to fit the model $\eta = \beta x$ to the following data by the method of least squares. The tests were made in the random order indicated below. Do this work for him. First make a plot of the data and then:

(a) Obtain the least squares estimate for the parameter β.

(b) Find the variance of this estimate (assuming that the variances of the individual y readings are constant and equal to 16).

(c) Obtain a 90% confidence interval for β.

(d) Make any further analysis you think appropriate, and comment on your results.

test order u	temperature (°C) x_u	surface finish y
1	51	61
2	38	52
3	40	58
4	67	57
5	21	41
6	13	21

State any assumptions you make.

17. Which of the following models from physics, as written, are linear in the parameters?

(a) $s = \frac{1}{2} a t^2$ (a is the parameter).

(b) $d = (1/Y)(FL^3/12\pi r^4)$ ($1/Y$ is the parameter).

(c) $d = CL^n$ (C and n are parameters).

(d) $\ln d = \ln C + n \ln L$ ($\ln C$ and n are parameters).

(e) $\log I = \log I_0 - (KAL/2.303MLS)$ ($\log I_0$ and K are parameters).

(f) $\Delta I_p = g_m \Delta E_g$ (g_m is the parameter).

(g) $N = N_0 e^{-\lambda t}$ (λ is the parameter).

(h) $\log T = -kn$ (k is the parameter).

18. Which of the following models are linear in the parameters? Explain your answer.

(a) $\eta = \beta_1 x_1 + \beta_2 \log x_2$.

(b) $\eta = \beta_1 \beta_2 x_1 + \beta_2 \log x_2$.

(c) $\eta = \beta_1 \sin x_2 + \beta_2 \cos x_2$.

(d) $\eta = \beta_1 \log \beta_1 x_1 + \beta_2 x_1 x_2$.

(e) $\eta = \beta_1 (\log x_1) e^{-x_2} + \beta_2 e^{x_3}$.

19. State whether each of the following models is linear in β_1, linear in β_2, and so on. Explain your answers.

(a) $\eta = \beta_1 e^{x_1} + \beta_2 e^{x_2}$.

(b) $\eta = \beta_1 \beta_2 x_1 + \beta_3 x_2$.

(c) $\eta = \beta_1 + \beta_2 x + \beta_3 x^2$.

(d) $\eta = \beta_1 (1 - e^{-\beta_2 x})$.

(e) $\eta = \beta_1 x_1^{\beta_2} x_2^{\beta_3}$.

20. Suppose that a chemical engineer uses the method of least squares with the data given below to estimate the parameters β_0, β_1, and β_2 in the model

$$\eta = \beta_0 + \beta_1 x_1 + \beta_2 x_2$$

where η is the mean response (peak gas temperature, °R). He obtains the following fitted equation:

$$\hat{y} = 1425.8 + 123.3 x_1 + 96.7 x_2$$

where $\hat{y}$ is the predicted value of the response, and x_1 and x_2 are given by the equations

$$x_1 = \frac{\text{compression ratio} - 12.0}{2}$$

and

$$x_2 = \frac{\text{cooling water temperature} - 550°R}{10°R}$$

test	compression ratio	cooling water temperature (°R)	peak gas temperature (°R) y	residual $y - \hat{y}$
1	10	540	1220	14
2	14	540	1500	48
3	10	560	1430	41
4	14	560	1650	4
5	8	550	1210	31
6	16	550	1700	29
7	12	530	1200	−32
8	12	570	1600	−29
9	12	550	1440	14
10	12	550	1450	24
11	12	550	1350	−76
12	12	550	1360	−66

The tests were run in random order. Check the engineer's calculations. Do you think the model form is adequate?

21. (a) Obtain least squares estimates of the parameters in this model:

$$y = \beta_1(1 - e^{-\beta_2 x}) + \epsilon$$

given these data:

x (day)	1	2	3	4	5	6
y (average BOD, mg/liter)	4.3	8.2	9.5	10.35	12.1	13.1

(b) Construct an approximate 95% confidence region for the parameters. [Source: L. B. Polkowski and D. M. Marske (1972), J. Water Pollut. Control Fed., 44, 1987.]

22. Using the data from Example 2 in Chapter 13, fit the model $\eta = \beta_0 + \beta_1 x_1 + \beta_2 x_2$ and draw an appropriate contour diagram in the experimental region, where x_1 = coded level of acid concentration and x_2 = coded level of catalyst concentration. Plot the data on this contour diagram.

23. The following data were obtained from a study on a chemical reaction system:

trial	temperature (°F)	concentration (%)	pH	yield		
1	150	40	6	73	70	
2	160	40	6	75	74	
3	150	50	6	78	80	
4	160	50	6	82	82	
5	150	40	7	75		
6	160	40	7	76	79	
7	150	50	7	87	85	82
8	160	50	7	89	88	

The 16 runs were performed in random order, trial 7 being run three times, trial 5 once, and all the rest twice. (The intention was to run each test twice, but a mistake was made in setting the concentration level on one of the tests.)

(a) Analyze these data. One thing you might wish to consider is fitting the following model to them: $\eta = \beta_0 + \beta_1 x_1 + \beta_2 x_2 + \beta_3 x_3$, where x_1, x_2, and x_3 (preferably, though not necessarily, in coded units) refer to the variables temperature, concentration, and pH, respectively, and η is the mean value of the yield. Alternatively, a good indication of how elaborate a model is needed will be obtained by averaging the available data and making a preliminary factorial analysis on the results.

(b) Compare these two approaches.

24. If suitable assumptions are made, the following model can be derived from chemical kinetic theory for a consecutive reaction with η the yield of intermediate product, x the reaction time, and β_1 and β_2 the rate constants to be estimated from the experimental data:

$$\eta = \frac{\beta_1}{\beta_1 - \beta_2} (e^{-\beta_2 x} - e^{-\beta_1 x})$$

(a) A chemist wants to obtain the most precise estimates possible of β_1 and β_2 from the following data and asks you what he should do. What would you ask him? What might you tell him?

reaction time (min) x	concentration of product (gram-mole/liter) y
10	0.20
20	0.52
30	0.69
40	0.64
50	0.57
60	0.48

The reaction is this type: reactant $\rightarrow$ product $\rightarrow$ by-product; β_1 is the rate constant of the first reaction, and β_2 of the second reaction.

(b) Obtain the least squares estimates for β_1 and β_2.

25. As part of a pollution study, a sanitary engineer has collected some biochemical oxygen demand (BOD) data on a particular sample of water taken from a certain stream in Wisconsin. The BOD equation is

$$\eta = L_a(1 - e^{-kt})$$

where η = mean value of BOD, t = time, and L_a and k are two parameters to be estimated. His data consist of one BOD reading for each of several times t. He wants to estimate the parameters by the method of least squares.

(a) Why might it be inappropriate to use the ordinary method of least squares in this situation?

(b) Under what circumstances would it be appropriate to use the method of least squares?

(c) Assuming that appropriate circumstances exist, explain in your own words how the method of least squares could be used to solve this problem. Your explanation should be understandable to an engineer who has had no experience with statistical methods.

26. (a) Fit a suitable model to the following data.
    (b) Draw any conclusions that you consider appropriate. State assumptions

coded values of three adjustable variables on prototype engine			measured noise (coded units)
x_1	x_2	x_2	
−1	−1	−1	10
+1	−1	−1	11
−1	+1	−1	4
+1	+1	−1	4
−1	−1	+1	9
+1	−1	+1	10
−1	+1	+1	5
+1	+1	+1	3
0	0	0	6
0	0	0	7
0	0	0	8
0	0	0	7

27. As a result of a statistically designed experiment the fitted model

$$\hat{y} = 89.30 + 16.48x_1 + 3.38x_2 - 16.50x_1^2 - 17.20x_2^2 - 6.99x_1x_2$$

was found to adequately represent the response function in the neighborhood of interest. In this expression

$$x_1 = \frac{(\text{temperature of reaction in } °F) - 110°F}{10°F}$$

$$x_2 = \frac{(\text{pressure in atmospheres}) - 2 \text{ atmospheres}}{2/\sqrt{3} \text{ atmospheres}}$$

and y is the observed yield. The process cannot be operated on a production scale at a temperature above 110°F or a pressure above 2 atmospheres for reasons of safety. On the basis of these data what settings of temperature and pressure would you recommend to maximize yield?

28. Write down the design matrix for a 32-run composite design having five variables with six center points.

29. The data given below were obtained when working on a lathe with a new alloy. From these data, which are in coded units, obtain the least squares estimates of β_0, β_1, and β_2 in the following model:

$$\eta = \beta_0 + \beta_1 x_1 + \beta_2 x_2$$

Criticize this analysis, and carry out any further analysis that you think is appropriate.

speed x_1	feed x_2	tool life y
−1	−1	1.3
0	−1	0.4
+1	−1	1.0
−1	0	1.2
0	0	1.0
+1	0	0.9
−1	+1	1.1
0	+1	1.0
+1	+1	0.8

30. The following data were collected by an experimenter using a composite design:

x_1	x_2	y
−1	−1	2
+1	−1	4
−1	+1	3
+1	+1	5
−2	0	1
+2	0	4
0	−2	1
0	+2	5
0	0	3

Assume that a second-degree equation is to be fitted to these data using the method of least squares.

(a) What is the matrix of independent variables?

(b) Set up the normal equations that would have to be solved to obtain the least squares estimates b_0, b_1, b_2, b_{12}, b_{11}, and b_{22}.

(c) Solve these normal equations.

(d) Criticize the analysis.

31. A chemist obtained the following data from eight experiments on a particular chemical reaction:

	variable		
temperature (°C)	pH	catalyst concentration (%)	response: yield (grams)
160	6	4	42
180	6	4	50
160	7	4	39
180	7	4	50
160	6	6	60
180	6	6	69
160	7	6	60
180	7	6	70

The chemist claims that from past experience there is reason to believe that the standard deviation for an individual yield reading is approximately 1 gram.

(a) Compute the main effects and interactions for these data, and interpret the results.

(b) Comment on the hypothesis that changing the pH for this reaction over the range used in the above experiment has no measurable effect on the yield.

(c) Use the method of least squares to fit the model

$$y = \beta_0 + \beta_1 x_1 + \beta_2 x_2 + \epsilon$$

where y = yield (in grams)

$$x_1 = \frac{\text{temperature (in °C)} - 170°C}{10°C}$$

$$x_2 = \frac{\text{catalyst concentration (in \%)} - 5\%}{1\%}$$

ϵ = random error

(d) Does this model fit the data adequately? Explain your answer.

(e) Do you think the chemist's estimate of the standard deviation is plausible?

(f) Draw a contour map for the fitted equation.

(g) Determine the settings (in units of °C for temperature and % for concentration) for two further experiments that you would recommend the chemist try next to increase yield.

32. The following data were collected by a mechanical engineer studying the performance of an industrial process for the manufacture of cement blocks, where x_1 is travel, x_2 is vibration, and y is a measure of operating efficiency. All values are in coded units. High values of y are desired. Do the following:

(a) Fit a first-order model to these data, using the method of least squares.

(b) Draw contour lines of the fitted response surface.

(c) Draw the direction of steepest ascent on this surface.

(d) Criticize your analysis.

travel x_1	vibration x_2	efficiency y
−1	−1	74
+1	−1	74
−1	+1	72
+1	+1	73
−2	0	71
+2	0	72
0	−2	75
0	+2	71
0	0	73
0	0	75

33. Suppose that an experimenter uses the following design and wants to fit a second-degree polynomial model to the data, using the method of least squares.

(a) Write the 16×10 matrix of independent variables $\mathbf{X}$.

(b) Is this a linear or nonlinear model?

(c) How many normal equations are there?

(d) What equation(s) would have to be solved to obtain the least squares estimate for β_{13}?

run	x_1	x_2	x_3	run	x_1	x_2	x_3
1	−1	−1	−1	9	−2	0	0
2	+1	−1	−1	10	+2	0	0
3	−1	+1	−1	11	0	−2	0
4	+1	+1	−1	12	0	+2	0
5	−1	−1	+1	13	0	0	−2
6	+1	−1	+1	14	0	0	+2
7	−1	+1	+1	15	0	0	0
8	+1	+1	+1	16	0	0	0

34. The object of an experiment, the results from which are set out below, was to determine the best settings for x_1 and x_2 to ensure a high yield and a low filtration time.

(a) Analyze the data, and draw contour diagrams for yield and filtration time.

(b) Find the settings of x_1 and x_2 that give (i) the highest predicted yield and (ii) the lowest predicted filtration time.

(c) Specify *one* set of conditions for x_1 and x_2 that will simultaneously give high yield and low filtration time. At this set of conditions what field color and how much crystal growth would you expect?

trial	x_1	x_2	yield (grams)	field color	crystal growth	filtration time (sec)
1	−	−	21.1	blue	none	150
2	−	0	23.7	blue	none	10
3	−	+	20.7	red	none	8
4	0	−	21.1	slightly red	none	35
5	0	0	24.1	blue	very slight	8
6	0	+	22.2	*unobserved*	slight	7
7	+	−	18.4	slightly red	slight	18
8	+	0	23.4	red	much	8
9	+	+	21.9	very red	much	10

	variable level		
variable	−	0	+
x_1 = condensation temperature (°C)	90	100	110
x_2 = amount of B (cc)	24.4	29.3	34.2

(*Source*: W. J. Hill and W. R. Demler, *Ind. Eng. Chem.*, October 1970, 60–65.)

35. The following data are from a tire radial runout study. Low runout values are desirable.

	x_1	x_2	x_3	x_4	y
1	9.6	40	305	4.5	0.51
2	32.5	40	305	4.5	0.28
3	9.6	70	305	4.5	0.65
4	32.5	70	305	4.5	0.51
5	9.6	40	335	4.5	0.24
6	32.5	40	335	4.5	0.38
7	9.6	70	335	4.5	0.45
8	32.5	70	335	4.5	0.49
9	9.6	40	305	7.5	0.30
10	32.5	40	305	7.5	0.35
11	9.6	70	305	7.5	0.45
12	32.5	70	305	7.5	0.82
13	9.6	40	335	7.5	0.24
14	32.5	40	335	7.5	0.54
15	9.6	70	335	7.5	0.35
16	32.5	70	335	7.5	0.51
17	60.0	55	320	6.0	0.53
18	5.0	55	320	6.0	0.56
19	17.5	85	320	6.0	0.67
20	17.5	25	320	6.0	0.45
21	17.5	25	350	6.0	0.41
22	17.5	25	290	6.0	0.23
23	17.5	25	320	9.0	0.41
24	17.5	25	320	3.0	0.47
25	17.5	25	320	6.0	0.32

Here x_1 = postinflation time (minutes), x_2 = postinflation pressure (psi), x_3 = cure temperature (°F), x_4 = rim size (inches), and y = radial runout (mils).

(a) Make what you think is an appropriate analysis of the data.

(b) Make three separate two-dimensional contour plots of $\hat{y}$ versus x_1 and x_3 over the ranges $5 \le x_1 \le 60$ and $290 \le x_3 \le 350$, one for each of the following sets of conditions:

   (i) $x_2 = 40$, $x_4 = 3$;   (ii) $x_2 = 55$, $x_4 = 9$;   (iii) $x_2 = 55$, $x_4 = 4.5$

   Conditions (iii) correspond to the conditions used in the plant.

(c) Comment on this conclusion: "It was surprising to observe that either very wide (9-inch) or very narrow (3-inch) rims could be used to reach low radial runout levels."

(d) Comment on any aspect of the design, the data, or the analysis which merits attention.

(Source: K. R. Williams, Rubber Age, August 1968, 65–71.)

36. The following data were collected by a chemist:

	variable		response: yield (%)	order in which experiments were performed
	---	---	---	---
trial	temperature (°C)	concentration (%)		
1	70	21	78.0	8
2	110	21	79.7	3
3	70	25	79.3	11
4	110	25	79.1	4
5	60	23	78.5	6
6	120	23	80.2	10
7	90	20	78.2	2
8	90	26	80.2	7
9	90	23	80.0	1
10	90	23	80.3	5
11	90	23	80.2	9

Using these data, do the following:
(a) Fit the model $\eta = \beta_0 + \beta_1 x_1 + \beta_2 x_2 + \beta_{11} x_1^2 + \beta_{22} x_2^2 + \beta_{12} x_1 x_2$, using the method of least squares (first code the data so that the levels of the variables are -1.5, -1.0, 0, $+1.0$, $+1.5$).
(b) Check the adequacy of fit of this model.
(c) Map the surface in the vicinity of the experiments.
(d) Suppose that the last three data values listed above (for trials 9, 10, and 11, which were the 1st, 5th, and 9th experiments performed) had been different, as follows:

trial	temperature (°C)	concentration (%)	yield (%)	order in which experiments were performed
9	90	23	79.5	1
10	90	23	80.8	5
11	90	23	80.2	9

Would your answers for (a), (b), and (c) be different with these new data?

37. For a chemical reaction of the type

$$A + B \rightarrow F, \qquad A + F \rightarrow G$$

the following data were collected:

	design				concentration of F (moles $\times 10^2$/liter) t (min) =				
run	A	B	C	D	80	160	320	640	1280
1	−	−	−	−	3.17	5.39	8.66	15.9	22.6
2	+	−	−	−	14.7	23.4	34.3	34.6	20.3
3	−	+	−	−	4.80	10.8	22.5	34.6	42.0
4	+	+	−	−	23.2	39.0	55.6	63.4	41.6
5	−	−	+	−	3.72	3.81	17.2	20.0	23.9
6	+	−	+	−	17.9	28.3	40.5	34.2	21.6
7	−	+	+	−	8.60	13.3	25.9	39.8	50.8
8	+	+	+	−	30.9	51.4	72.2	76.4	38.9
9	−	−	−	+	7.48	9.93	20.0	30.9	24.9
10	+	−	−	+	25.3	35.3	39.1	28.4	7.50
11	−	+	−	+	13.3	27.1	43.0	58.0	49.4
12	+	+	−	+	50.8	75.6	84.2	57.0	11.5
13	−	−	+	+	9.15	15.8	27.5	33.9	23.0
14	+	−	+	+	30.8	44.4	46.7	24.9	2.94
15	−	+	+	+	22.8	37.2	57.9	69.1	53.9
16	+	+	+	+	62.6	88.0	89.5	43.4	5.80

	A (moles/liter)	B (moles/liter)	C (mmoles/liter)	D (°C)
+	40	2	1.0	175
−	20	1	0.5	165

The experimenter proposes the following mechanistic model:

$$\eta = \frac{B\beta_1}{\beta_1 - \beta_2} (e^{-\beta_2 t} - e^{-\beta_1 t})$$

where η is the mean value of F, the desired product: B is the concentration of reactant B (either 1 or 2 moles per liter, depending on the run); t is the reaction time (80, 160, ..., or 1280 minutes); and β_1 and β_2 are rate constants for the two reactions.

(a) For each set of 16 observations corresponding to one of the five values of time t (80, 160, ..., 1280), calculate the effects using Yates's algorithm. What conclusions can you draw from this way of analyzing the data?

From these data least squares estimates ln $\hat{\beta}_1$ and ln $\hat{\beta}_2$ were obtained for each of the 16 runs as follows:

run	1	2	3	4	5	6	7	8	9	10	11	12	13	14	15	16
$-\ln \hat{\beta}_1$	8.0	6.2	7.7	6.5	7.5	6.0	7.6	6.1	6.9	5.6	6.9	5.5	6.6	5.2	6.5	5.2
$-\ln \hat{\beta}_2$	7.3	6.2	6.8	6.2	6.7	6.2	6.9	6.2	6.5	5.8	6.4	5.8	6.3	5.8	6.4	5.7

(b) From the 16 estimates of ln β_1 calculate main effects and interactions, and repeat, using the 16 estimates of ln β_2. Hence suggest a modification of the model that might explain all the data.

[*Source*: G. E. P. Box, and W. G. Hunter, (1962), *Technometrics*, **4**, 301.]

38. (a) The estimated volume Y of a container is obtained by measuring the length y_1, width y_2, and depth y_3 and then using this formula:

$$Y = y_1 y_2 y_3$$

If each of the three measurements is subject to the same errors of measurement with standard deviation 0.3 millimeter, estimate the standard error of the computed volume for containers whose mean length, width, and depth are 4.75, 16.4, and 8.7 millimeters, respectively.

(b) Repeat part (a) for containers whose mean length, width, and depth are 547.5, 516.4, and 508.7 millimeters.

(c) Repeat parts (a) and (b), assuming the standard deviation to be 0.1 mm.

39. (a) Analyze these results from a hierarchical design:

batch	sample	tests		
1	1	41	40	38
	2	32	31	33
	3	35	36	35
2	4	26	27	30
	5	25	24	23
	6	35	33	32
3	7	30	29	29
	8	14	16	19
	9	24	23	25
4	10	30	31	31
	11	25	24	24
	12	24	26	25
5	13	19	18	21
	14	16	17	17
	15	27	26	29

(b) Criticize the experiment.

Data for the Next Six Problems

The following data were collected from a sewage treatment plant on 140 consecutive days. For each day four quantities were measured: the 5-day biochemical oxygen demand for the influent stream, in milligrams per liter (series *A*); the temperature, in °F (series *B*); the suspended solids, in milligrams per liter (series *C*); and the 5-day BOD for the effluent stream, in milligrams per liter (series *D*). Read the data from left to right.

Series A

202	202	141	173	233	214	208	227	227	201
182	199	182	222	206	206	147	178	204	194
287	185	120	120	189	193	172	197	193	193
144	188	225	183	221	219	219	127	195	219
192	239	177	177	166	190	171	201	188	174
174	136	159	216	204	250	233	233	199	209
185	203	225	225	225	119	174	190	238	225
257	257	121	163	144	172	226	184	184	184
177	160	183	144	159	159	118	159	147	154
179	-222	222	162	196	212	205	192	170	170
106	140	143	161	181	156	156	139	167	155
172	212	253	253	134	177	192	179	193	216
216	137	185	156	160	186	200	200	166	166
158	213	205	171	171	194	191	169	173	183

Series B

53	53	51	48	44	48	49	48	53	51
50	50	48	49	48	52	50	44	44	46
50	53	53	45	48	47	43	40	44	44
43	46	44	45	52	52	47	48	45	45
46	50	47	50	48	50	48	50	53	60
50	50	42	44	51	47	54	53	50	46
46	47	50	52	51	50	46	52	50	50
54	50	51	50	50	49	48	48	51	49
48	47	50	53	53	50	52	47	52	54
56	47	48	50	49	53	56	55	53	52
54	53	50	52	54	54	57	56	54	56
54	55	58	53	55	53	52	55	56	58
56	56	55	58	58	60	59	58	58	57
61	56	58	58	62	62	59	64	64	59

Series C

1100	1100	1100	1100	1120	1090	1130	1180	1540	1540
1240	1200	1300	1210	1250	1320	1320	1160	1330	1340
1200	1880	1330	1330	1280	1250	1300	1260	1290	1380
1380	1250	1110	1130	1160	1380	1410	1410	1380	1340
1380	1320	1330	1350	1350	1310	1360	1310	1300	1230
1330	1330	1470	1400	1440	1310	1370	1330	1330	1360
1310	1170	1360	1190	1290	1290	1220	1280	1290	1240
1250	1030	1030	1170	1300	1250	1190	1260	1280	1280
1170	1190	1230	1070	1090	1130	1130	1110	1080	1110
1020	1040	1180	1180	1150	1220	1210	1210	1310	1330
1330	1020	1310	1240	1110	1120	1180	1180	1160	1220
1210	1120	1060	1140	1140	1100	1180	1250	1170	1120
1140	1140	1140	1210	1250	1120	1210	1310	1310	1200
1250	1270	1300	1300	1430	1430	1660	1610	2010	1480

Series D

37	37	16	27	29	23	26	26	26	12
27	28	22	24	25	25	10	20	18	14
28	26	12	12	22	25	18	27	24	24
16	36	30	27	31	23	23	8	26	20
19	24	18	18	6	16	19	13	19	19
19	7	14	18	13	21	22	22	8	16
15	17	28	23	23	10	16	18	22	24
24	24	10	14	15	10	18	11	11	8
17	18	17	25	18	18	8	19	14	16
14	15	15	10	22	25	18	15	9	9
9	27	25	19	11	15	15	6	17	13
11	15	13	13	6	14	13	22	13	10
10	8	18	13	8	14	14	14	8	13
15	15	14	11	11	17	16	19	19	21

The following six problems can be done using any one of the above series of data.

40. Plot the data. What conclusions do you reach?
41. Make a serial plot of the data for y_{t+1} against y_t. What conclusions do you reach?

42. Make a serial plot of the data for y_{t+2} against y_t. What conclusions do you reach?

43. Plot the autocorrelation function. Comment on its appearance.

44. Can the data be regarded as a first-order moving average process?

45. Compute the first differences $w_t = y_t - y_{t-1}$. Using the w_t data, repeat problems 40 through 46.

46. Describe three areas of application for time series models. How do you think such models might be applied in your own field?

APPENDIX: TABLES

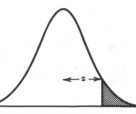

TABLE A. Tail area of unit normal distribution

z	0.00	0.01	0.02	0.03	0.04	0.05	0.06	0.07	0.08	0.09
0.0	0.5000	0.4960	0.4920	0.4880	0.4840	0.4801	0.4761	0.4721	0.4681	0.4641
0.1	0.4602	0.4562	0.4522	0.4483	0.4443	0.4404	0.4364	0.4325	0.4286	0.4247
0.2	0.4207	0.4168	0.4129	0.4090	0.4052	0.4013	0.3974	0.3936	0.3897	0.3859
0.3	0.3821	0.3783	0.3745	0.3707	0.3669	0.3632	0.3594	0.3557	0.3520	0.3483
0.4	0.3446	0.3409	0.3372	0.3336	0.3300	0.3264	0.3228	0.3192	0.3156	0.3121
0.5	0.3085	0.3050	0.3015	0.2981	0.2946	0.2912	0.2877	0.2843	0.2810	0.2776
0.6	0.2743	0.2709	0.2676	0.2643	0.2611	0.2578	0.2546	0.2514	0.2483	0.2451
0.7	0.2420	0.2389	0.2358	0.2327	0.2296	0.2266	0.2236	0.2206	0.2177	0.2148
0.8	0.2119	0.2090	0.2061	0.2033	0.2005	0.1977	0.1949	0.1922	0.1894	0.1867
0.9	0.1841	0.1814	0.1788	0.1762	0.1736	0.1711	0.1685	0.1660	0.1635	0.1611
1.0	0.1587	0.1562	0.1539	0.1515	0.1492	0.1469	0.1446	0.1423	0.1401	0.1379
1.1	0.1357	0.1335	0.1314	0.1292	0.1271	0.1251	0.1230	0.1210	0.1190	0.1170
1.2	0.1151	0.1131	0.1112	0.1093	0.1075	0.1056	0.1038	0.1020	0.1003	0.0985
1.3	0.0968	0.0951	0.0934	0.0918	0.0901	0.0885	0.0869	0.0853	0.0838	0.0823
1.4	0.0808	0.0793	0.0778	0.0764	0.0749	0.0735	0.0721	0.0708	0.0694	0.0681
1.5	0.0668	0.0655	0.0643	0.0630	0.0618	0.0606	0.0594	0.0582	0.0571	0.0559
1.6	0.0548	0.0537	0.0526	0.0516	0.0505	0.0495	0.0485	0.0475	0.0465	0.0455
1.7	0.0446	0.0436	0.0427	0.0418	0.0409	0.0401	0.0392	0.0384	0.0375	0.0367
1.8	0.0359	0.0351	0.0344	0.0336	0.0329	0.0322	0.0314	0.0307	0.0301	0.0294
1.9	0.0287	0.0281	0.0274	0.0268	0.0262	0.0256	0.0250	0.0244	0.0239	0.0233
2.0	0.0228	0.0222	0.0217	0.0212	0.0207	0.0202	0.0197	0.0192	0.0188	0.0183
2.1	0.0179	0.0174	0.0170	0.0166	0.0162	0.0158	0.0154	0.0150	0.0146	0.0143
2.2	0.0139	0.0136	0.0132	0.0129	0.0125	0.0122	0.0119	0.0116	0.0113	0.0110
2.3	0.0107	0.0104	0.0102	0.0099	0.0096	0.0094	0.0091	0.0089	0.0087	0.0084
2.4	0.0082	0.0080	0.0078	0.0075	0.0073	0.0071	0.0069	0.0068	0.0066	0.0064
2.5	0.0062	0.0060	0.0059	0.0057	0.0055	0.0054	0.0052	0.0051	0.0049	0.0048
2.6	0.0047	0.0045	0.0044	0.0043	0.0041	0.0040	0.0039	0.0038	0.0037	0.0036
2.7	0.0035	0.0034	0.0033	0.0032	0.0031	0.0030	0.0029	0.0028	0.0027	0.0026
2.8	0.0026	0.0025	0.0024	0.0023	0.0023	0.0022	0.0021	0.0021	0.0020	0.0019
2.9	0.0019	0.0018	0.0018	0.0017	0.0016	0.0016	0.0015	0.0015	0.0014	0.0014
3.0	0.0013	0.0013	0.0013	0.0012	0.0012	0.0011	0.0011	0.0011	0.0010	0.0010
3.1	0.0010	0.0009	0.0009	0.0009	0.0008	0.0008	0.0008	0.0008	0.0007	0.0007
3.2	0.0007	0.0007	0.0006	0.0006	0.0006	0.0006	0.0006	0.0005	0.0005	0.0005
3.3	0.0005	0.0005	0.0005	0.0004	0.0004	0.0004	0.0004	0.0004	0.0004	0.0003
3.4	0.0003	0.0003	0.0003	0.0003	0.0003	0.0003	0.0003	0.0003	0.0003	0.0002
3.5	0.0002	0.0002	0.0002	0.0002	0.0002	0.0002	0.0002	0.0002	0.0002	0.0002
3.6	0.0002	0.0002	0.0001	0.0001	0.0001	0.0001	0.0001	0.0001	0.0001	0.0001
3.7	0.0001	0.0001	0.0001	0.0001	0.0001	0.0001	0.0001	0.0001	0.0001	0.0001
3.8	0.0001	0.0001	0.0001	0.0001	0.0001	0.0001	0.0001	0.0001	0.0001	0.0001
3.9	0.0000	0.0000	0.0000	0.0000	0.0000	0.0000	0.0000	0.0000	0.0000	0.0000

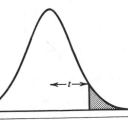

TABLE B1. Probability points of the
t **distribution with ν degrees of freedom**

					tail area probability					
ν	0.4	0.25	0.1	0.05	0.025	0.01	0.005	0.0025	0.001	0.0005
1	0.325	1.000	3.078	6.314	12.706	31.821	63.657	127.32	318.31	636.62
2	0.289	0.816	1.886	2.920	4.303	6.965	9.925	14.089	22.326	31.598
3	0.277	0.765	1.638	2.353	3.182	4.541	5.841	7.453	10.213	12.924
4	0.271	0.741	1.533	2.132	2.776	3.747	4.604	5.598	7.173	8.610
5	0.267	0.727	1.476	2.015	2.571	3.365	4.032	4.773	5.893	6.869
6	0.265	0.718	1.440	1.943	2.447	3.143	3.707	4.317	5.208	5.959
7	0.263	0.711	1.415	1.895	2.365	2.998	3.499	4.029	4.785	5.408
8	0.262	0.706	1.397	1.860	2.306	2.896	3.355	3.833	4.501	5.041
9	0.261	0.703	1.383	1.833	2.262	2.821	3.250	3.690	4.297	4.781
10	0.260	0.700	1.372	1.812	2.228	2.764	3.169	3.581	4.144	4.587
11	0.260	0.697	1.363	1.796	2.201	2.718	3.106	3.497	4.025	4.437
12	0.259	0.695	1.356	1.782	2.179	2.681	3.055	3.428	3.930	4.318
13	0.259	0.694	1.350	1.771	2.160	2.650	3.012	3.372	3.852	4.221
14	0.258	0.692	1.345	1.761	2.145	2.624	2.977	3.326	3.787	4.140
15	0.258	0.691	1.341	1.753	2.131	2.602	2.947	3.286	3.733	4.073
16	0.258	0.690	1.337	1.746	2.120	2.583	2.921	3.252	3.686	4.015
17	0.257	0.689	1.333	1.740	2.110	2.567	2.898	3.222	3.646	3.965
18	0.257	0.688	1.330	1.734	2.101	2.552	2.878	3.197	3.610	3.922
19	0.257	0.688	1.328	1.729	2.093	2.539	2.861	3.174	3.579	3.883
20	0.257	0.687	1.325	1.725	2.086	2.528	2.845	3.153	3.552	3.850
21	0.257	0.686	1.323	1.721	2.080	2.518	2.831	3.135	3.527	3.819
22	0.256	0.686	1.321	1.717	2.074	2.508	2.819	3.119	3.505	3.792
23	0.256	0.685	1.319	1.714	2.069	2.500	2.807	3.104	3.485	3.767
24	0.256	0.685	1.318	1.711	2.064	2.492	2.797	3.091	3.467	3.745
25	0.256	0.684	1.316	1.708	2.060	2.485	2.787	3.078	3.450	3.725
26	0.256	0.684	1.315	1.706	2.056	2.479	2.779	3.067	3.435	3.707
27	0.256	0.684	1.314	1.703	2.052	2.473	2.771	3.057	3.421	3.690
28	0.256	0.683	1.313	1.701	2.048	2.467	2.763	3.047	3.408	3.674
29	0.256	0.683	1.311	1.699	2.045	2.462	2.756	3.038	3.396	3.659
30	0.256	0.683	1.310	1.697	2.042	2.457	2.750	3.030	3.385	3.646
40	0.255	0.681	1.303	1.684	2.021	2.423	2.704	2.971	3.307	3.551
60	0.254	0.679	1.296	1.671	2.000	2.390	2.660	2.915	3.232	3.460
120	0.254	0.677	1.289	1.658	1.980	2.358	2.617	2.860	3.160	3.373
∞	0.253	0.674	1.282	1.645	1.960	2.326	2.576	2.807	3.090	3.291

Source: Taken with permission from E. S. Pearson and H. O. Hartley (Eds.) (1958), *Biometrika Tables for Statisticians*, Vol. 1, Cambridge University Press.
    Parts of the table are also taken from Table III of Fisher and Yates: *Statistical Tables for Biological, Agricultural and Medical Research*, published by Longman Group Ltd., London (previously published by Oliver and Boyd, Edinburgh), by permission of the authors and publishers.

631

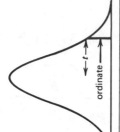

TABLE B2. Ordinates of t distribution with v degrees of freedom

							value of t							
v	0.00	0.25	0.50	0.75	1.00	1.25	1.50	1.75	2.00	2.25	2.50	2.75	3.00
1	0.318	0.300	0.255	0.204	0.159	0.124	0.098	0.078	0.064	0.053	0.044	0.037	0.032
2	0.354	0.338	0.296	0.244	0.193	0.149	0.114	0.088	0.068	0.053	0.042	0.034	0.027
3	0.368	0.353	0.313	0.261	0.207	0.159	0.120	0.090	0.068	0.051	0.039	0.030	0.023
4	0.375	0.361	0.322	0.270	0.215	0.164	0.123	0.091	0.066	0.049	0.036	0.026	0.020
5	0.380	0.366	0.328	0.276	0.220	0.168	0.125	0.091	0.065	0.047	0.033	0.024	0.017
6	0.383	0.369	0.332	0.280	0.223	0.170	0.126	0.090	0.064	0.045	0.032	0.022	0.016
7	0.385	0.372	0.335	0.283	0.226	0.172	0.126	0.090	0.063	0.044	0.030	0.021	0.014
8	0.387	0.373	0.337	0.285	0.228	0.173	0.127	0.090	0.062	0.043	0.029	0.019	0.013
9	0.388	0.375	0.338	0.287	0.229	0.174	0.127	0.090	0.062	0.042	0.028	0.018	0.012
10	0.389	0.376	0.340	0.288	0.230	0.175	0.127	0.090	0.061	0.041	0.027	0.018	0.011
11	0.390	0.377	0.341	0.289	0.231	0.176	0.128	0.089	0.061	0.040	0.026	0.017	0.011
12	0.391	0.378	0.342	0.290	0.232	0.176	0.128	0.089	0.060	0.040	0.026	0.016	0.010
13	0.391	0.378	0.343	0.291	0.233	0.177	0.128	0.089	0.060	0.039	0.025	0.016	0.010
14	0.392	0.379	0.343	0.292	0.234	0.177	0.128	0.089	0.060	0.039	0.025	0.015	0.010
15	0.392	0.380	0.344	0.292	0.234	0.177	0.128	0.089	0.059	0.038	0.024	0.015	0.009

16	0.393	0.380	0.344	0.293	0.235	0.178	0.128	0.089	0.059	0.038	0.024	0.015	0.009
17	0.393	0.380	0.345	0.293	0.235	0.178	0.128	0.089	0.059	0.038	0.024	0.014	0.009
18	0.393	0.381	0.345	0.294	0.235	0.178	0.129	0.088	0.059	0.037	0.023	0.014	0.008
19	0.394	0.381	0.346	0.294	0.236	0.179	0.129	0.088	0.058	0.037	0.023	0.014	0.008
20	0.394	0.381	0.346	0.294	0.236	0.179	0.129	0.088	0.058	0.037	0.023	0.014	0.008
22	0.394	0.382	0.346	0.295	0.237	0.179	0.129	0.088	0.058	0.036	0.022	0.013	0.008
24	0.395	0.382	0.347	0.296	0.237	0.179	0.129	0.088	0.057	0.036	0.022	0.013	0.007
26	0.395	0.383	0.347	0.296	0.237	0.180	0.129	0.088	0.057	0.036	0.022	0.013	0.007
28	0.395	0.383	0.348	0.296	0.238	0.180	0.129	0.088	0.057	0.036	0.021	0.012	0.007
30	0.396	0.383	0.348	0.297	0.238	0.180	0.129	0.088	0.057	0.035	0.021	0.012	0.007
35	0.396	0.384	0.348	0.297	0.239	0.180	0.129	0.088	0.056	0.035	0.021	0.012	0.006
40	0.396	0.384	0.349	0.298	0.239	0.181	0.129	0.087	0.056	0.035	0.020	0.011	0.006
45	0.397	0.384	0.349	0.298	0.239	0.181	0.129	0.087	0.056	0.034	0.020	0.011	0.006
50	0.397	0.385	0.350	0.298	0.240	0.181	0.129	0.087	0.056	0.034	0.020	0.011	0.006
∞	0.399	0.387	0.352	0.301	0.242	0.183	0.130	0.086	0.054	0.032	0.018	0.009	0.004

TABLE C. Probability points of the χ^2 distribution with v degrees of freedom

tail area probability

v	0.995	0.99	0.975	0.95	0.9	0.75	0.5	0.25	0.1	0.05	0.025	0.01	0.005	0.001
1	—	—	—	—	0.016	0.102	0.455	1.32	2.71	3.84	5.02	6.63	7.88	10.8
2	0.010	0.020	0.051	0.103	0.211	0.575	1.39	2.77	4.61	5.99	7.38	9.21	10.6	13.8
3	0.072	0.115	0.216	0.352	0.584	1.21	2.37	4.11	6.25	7.81	9.35	11.3	12.8	16.3
4	0.207	0.297	0.484	0.711	1.06	1.92	3.36	5.39	7.78	9.49	11.1	13.3	14.9	18.5
5	0.412	0.554	0.831	1.15	1.61	2.67	4.35	6.63	9.24	11.1	12.8	15.1	16.7	20.5
6	0.676	0.872	1.24	1.64	2.20	3.45	5.35	7.84	10.6	12.6	14.4	16.8	18.5	22.5
7	0.989	1.24	1.69	2.17	2.83	4.25	6.35	9.04	12.0	14.1	16.0	18.5	20.3	24.3
8	1.34	1.65	2.18	2.73	3.49	5.07	7.34	10.2	13.4	15.5	17.5	20.1	22.0	26.1
9	1.73	2.09	2.70	3.33	4.17	5.90	8.34	11.4	14.7	16.9	19.0	21.7	23.6	27.9
10	2.16	2.56	3.25	3.94	4.87	6.74	9.34	12.5	16.0	18.3	20.5	23.2	25.2	29.6
11	2.60	3.05	3.82	4.57	5.58	7.58	10.3	13.7	17.3	19.7	21.9	24.7	26.8	31.3
12	3.07	3.57	4.40	5.23	6.30	8.44	11.3	14.8	18.5	21.0	23.3	26.2	28.3	32.9
13	3.57	4.11	5.01	5.89	7.04	9.30	12.3	16.0	19.8	22.4	24.7	27.7	29.8	34.5
14	4.07	4.66	5.63	6.57	7.79	10.2	13.3	17.1	21.1	23.7	26.1	29.1	31.3	36.1
15	4.60	5.23	6.26	7.26	8.55	11.0	14.3	18.2	22.3	25.0	27.5	30.6	32.8	37.7

16	5.14	5.81	6.91	7.96	9.31	11.9	15.3	19.4	23.5	26.3	28.8	32.0	34.3	39.3
17	5.70	6.41	7.56	8.67	10.1	12.8	16.3	20.5	24.8	27.6	30.2	33.4	35.7	40.8
18	6.26	7.01	8.23	9.39	10.9	13.7	17.3	21.6	26.0	28.9	31.5	34.8	37.2	42.3
19	6.84	7.63	8.91	10.1	11.7	14.6	18.3	22.7	27.2	30.1	32.9	36.2	38.6	43.8
20	7.43	8.26	9.59	10.9	12.4	15.5	19.3	23.8	28.4	31.4	34.2	37.6	40.0	45.3
21	8.03	8.90	10.3	11.6	13.2	16.3	20.3	24.9	29.6	32.7	35.5	38.9	41.4	46.8
22	8.64	9.54	11.0	12.3	14.0	17.2	21.3	26.0	30.8	33.9	36.8	40.3	42.8	48.3
23	9.26	10.2	11.7	13.1	14.8	18.1	22.3	27.1	32.0	35.2	38.1	41.6	44.2	49.7
24	9.89	10.9	12.4	13.8	15.7	19.0	23.3	28.2	33.2	36.4	39.4	43.0	45.6	51.2
25	10.5	11.5	13.1	14.6	16.5	19.9	24.3	29.3	34.4	37.7	40.6	44.3	46.9	52.6
26	11.2	12.2	13.8	15.4	17.3	20.8	25.3	30.4	35.6	38.9	41.9	45.6	48.3	54.1
27	11.8	12.9	14.6	16.2	18.1	21.7	26.3	31.5	36.7	40.1	43.2	47.0	49.6	55.5
28	12.5	13.6	15.3	16.9	18.9	22.7	27.3	32.6	37.9	41.3	44.5	48.3	51.0	56.9
29	13.1	14.3	16.0	17.7	19.8	23.6	28.3	33.7	39.1	42.6	45.7	49.6	52.3	58.3
30	13.8	15.0	16.8	18.5	20.6	24.5	29.3	34.8	40.3	43.8	47.0	50.9	53.7	59.7

Source: Adapted from Table 8, Percentage points of the χ^2 distribution, in E. S. Pearson and H. O. Hartley (Eds.) (1966). *Biometrika Tables for Statisticians*, Vol. 1, 3rd ed., Cambridge University Press. Used by permission.

TABLE D. Percentage points of the F distribution: upper 25% points

v_2 \ v_1	1	2	3	4	5	6	7	8	9	10	12	15	20	24	30	40	60	120	∞
1	5.83	7.50	8.20	8.58	8.82	8.98	9.10	9.19	9.26	9.32	9.41	9.49	9.58	9.63	9.67	9.71	9.76	9.80	9.85
2	2.57	3.00	3.15	3.23	3.28	3.31	3.34	3.35	3.37	3.38	3.39	3.41	3.43	3.43	3.44	3.45	3.46	3.47	3.48
3	2.02	2.28	2.36	2.39	2.41	2.42	2.43	2.44	2.44	2.44	2.45	2.46	2.46	2.46	2.47	2.47	2.47	2.47	2.47
4	1.81	2.00	2.05	2.06	2.07	2.08	2.08	2.08	2.08	2.08	2.08	2.08	2.08	2.08	2.08	2.08	2.08	2.08	2.08
5	1.69	1.85	1.88	1.89	1.89	1.89	1.89	1.89	1.89	1.89	1.89	1.89	1.88	1.88	1.88	1.88	1.87	1.87	1.87
6	1.62	1.76	1.78	1.79	1.79	1.78	1.78	1.78	1.77	1.77	1.77	1.76	1.76	1.75	1.75	1.75	1.74	1.74	1.74
7	1.57	1.70	1.72	1.72	1.71	1.71	1.70	1.70	1.69	1.69	1.68	1.68	1.67	1.67	1.66	1.66	1.65	1.65	1.65
8	1.54	1.66	1.67	1.66	1.66	1.65	1.64	1.64	1.63	1.63	1.62	1.62	1.61	1.60	1.60	1.59	1.59	1.58	1.58
9	1.51	1.62	1.63	1.63	1.62	1.61	1.60	1.60	1.59	1.59	1.58	1.57	1.56	1.56	1.55	1.54	1.54	1.53	1.53
10	1.49	1.60	1.60	1.59	1.59	1.58	1.57	1.56	1.56	1.55	1.54	1.53	1.52	1.52	1.51	1.51	1.50	1.49	1.48
11	1.47	1.58	1.58	1.57	1.56	1.55	1.54	1.53	1.53	1.52	1.51	1.50	1.49	1.49	1.48	1.47	1.47	1.46	1.45
12	1.46	1.56	1.56	1.55	1.54	1.53	1.52	1.51	1.51	1.50	1.49	1.48	1.47	1.46	1.45	1.45	1.44	1.43	1.42
13	1.45	1.55	1.55	1.53	1.52	1.51	1.50	1.49	1.49	1.48	1.47	1.46	1.45	1.44	1.43	1.42	1.42	1.41	1.40
14	1.44	1.53	1.53	1.52	1.51	1.50	1.49	1.48	1.47	1.46	1.45	1.44	1.43	1.42	1.41	1.41	1.40	1.39	1.38
15	1.43	1.52	1.52	1.51	1.49	1.48	1.47	1.46	1.46	1.45	1.44	1.43	1.41	1.41	1.40	1.39	1.38	1.37	1.36
16	1.42	1.51	1.51	1.50	1.48	1.47	1.46	1.45	1.44	1.44	1.43	1.41	1.40	1.39	1.38	1.37	1.36	1.35	1.34
17	1.42	1.51	1.50	1.49	1.47	1.46	1.45	1.44	1.43	1.43	1.41	1.40	1.39	1.38	1.37	1.36	1.35	1.34	1.33
18	1.41	1.50	1.49	1.48	1.46	1.45	1.44	1.43	1.42	1.42	1.40	1.39	1.38	1.37	1.36	1.35	1.34	1.33	1.32
19	1.41	1.49	1.49	1.47	1.46	1.44	1.43	1.42	1.41	1.41	1.40	1.38	1.37	1.36	1.35	1.34	1.33	1.32	1.30
20	1.40	1.49	1.48	1.47	1.45	1.44	1.43	1.42	1.41	1.40	1.39	1.37	1.36	1.35	1.34	1.33	1.32	1.31	1.29
21	1.40	1.48	1.48	1.46	1.44	1.43	1.42	1.41	1.40	1.39	1.38	1.37	1.35	1.34	1.33	1.32	1.31	1.30	1.28
22	1.40	1.48	1.47	1.45	1.44	1.42	1.41	1.40	1.39	1.39	1.37	1.36	1.34	1.33	1.32	1.31	1.30	1.29	1.28
23	1.39	1.47	1.47	1.45	1.43	1.42	1.41	1.40	1.39	1.38	1.37	1.35	1.34	1.33	1.32	1.31	1.30	1.28	1.27
24	1.39	1.47	1.46	1.44	1.43	1.41	1.40	1.39	1.38	1.38	1.36	1.35	1.33	1.32	1.31	1.30	1.29	1.28	1.26
25	1.39	1.47	1.46	1.44	1.42	1.41	1.40	1.39	1.38	1.37	1.36	1.34	1.33	1.32	1.31	1.29	1.28	1.27	1.25
26	1.38	1.46	1.45	1.44	1.42	1.41	1.39	1.38	1.37	1.37	1.35	1.34	1.32	1.31	1.30	1.29	1.28	1.26	1.25
27	1.38	1.46	1.45	1.43	1.42	1.40	1.39	1.38	1.37	1.36	1.35	1.33	1.32	1.31	1.30	1.28	1.27	1.26	1.24
28	1.38	1.46	1.45	1.43	1.41	1.40	1.39	1.38	1.37	1.36	1.34	1.33	1.31	1.30	1.29	1.28	1.27	1.25	1.24
29	1.38	1.45	1.45	1.43	1.41	1.40	1.38	1.37	1.36	1.35	1.34	1.32	1.31	1.30	1.29	1.27	1.26	1.25	1.23
30	1.38	1.45	1.44	1.42	1.41	1.39	1.38	1.37	1.36	1.35	1.34	1.32	1.30	1.29	1.28	1.27	1.26	1.24	1.23
40	1.36	1.44	1.42	1.40	1.39	1.37	1.36	1.35	1.34	1.33	1.31	1.30	1.28	1.26	1.25	1.24	1.22	1.21	1.19
60	1.35	1.42	1.41	1.38	1.37	1.35	1.33	1.32	1.31	1.30	1.29	1.27	1.25	1.24	1.22	1.21	1.19	1.17	1.15
120	1.34	1.40	1.39	1.37	1.35	1.33	1.31	1.30	1.29	1.28	1.26	1.24	1.22	1.21	1.19	1.18	1.16	1.13	1.10
∞	1.32	1.39	1.37	1.35	1.33	1.31	1.29	1.28	1.27	1.25	1.24	1.22	1.19	1.18	1.16	1.14	1.12	1.08	1.00

Source: M. Merrington and C. M. Thompson (1943). Tables of percentage points of the inverted beta (F) distribution, Biometrika, **33**, 73. Used by permission.

TABLE D (continued). Percentage points of the F distribution: upper 10% points

v_2 \ v_1	1	2	3	4	5	6	7	8	9	10	12	15	20	24	30	40	60	120	∞
1	39.86	49.50	53.59	55.83	57.24	58.20	58.91	59.44	59.86	60.19	60.71	61.22	61.74	62.00	62.26	62.53	62.79	63.06	63.33
2	8.53	9.00	9.16	9.24	9.29	9.33	9.35	9.37	9.38	9.39	9.41	9.42	9.44	9.45	9.46	9.47	9.47	9.48	9.49
3	5.54	5.46	5.39	5.34	5.31	5.28	5.27	5.25	5.24	5.23	5.22	5.20	5.18	5.18	5.17	5.16	5.15	5.14	5.13
4	4.54	4.32	4.19	4.11	4.05	4.01	3.98	3.95	3.94	3.92	3.90	3.87	3.84	3.83	3.82	3.80	3.79	3.78	3.76
5	4.06	3.78	3.62	3.52	3.45	3.40	3.37	3.34	3.32	3.30	3.27	3.24	3.21	3.19	3.17	3.16	3.14	3.12	3.10
6	3.78	3.46	3.29	3.18	3.11	3.05	3.01	2.98	2.96	2.94	2.90	2.87	2.84	2.82	2.80	2.78	2.76	2.74	2.72
7	3.59	3.26	3.07	2.96	2.88	2.83	2.78	2.75	2.72	2.70	2.67	2.63	2.59	2.58	2.56	2.54	2.51	2.49	2.47
8	3.46	3.11	2.92	2.81	2.73	2.67	2.62	2.59	2.56	2.54	2.50	2.46	2.42	2.40	2.38	2.36	2.34	2.32	2.29
9	3.36	3.01	2.81	2.69	2.61	2.55	2.51	2.47	2.44	2.42	2.38	2.34	2.30	2.28	2.25	2.23	2.21	2.18	2.16
10	3.29	2.92	2.73	2.61	2.52	2.46	2.41	2.38	2.35	2.32	2.28	2.24	2.20	2.18	2.16	2.13	2.11	2.08	2.06
11	3.23	2.86	2.66	2.54	2.45	2.39	2.34	2.30	2.27	2.25	2.21	2.17	2.12	2.10	2.08	2.05	2.03	2.00	1.97
12	3.18	2.81	2.61	2.48	2.39	2.33	2.28	2.24	2.21	2.19	2.15	2.10	2.06	2.04	2.01	1.99	1.96	1.93	1.90
13	3.14	2.76	2.56	2.43	2.35	2.28	2.23	2.20	2.16	2.14	2.10	2.05	2.01	1.98	1.96	1.93	1.90	1.88	1.85
14	3.10	2.73	2.52	2.39	2.31	2.24	2.19	2.15	2.12	2.10	2.05	2.01	1.96	1.94	1.91	1.89	1.86	1.83	1.80
15	3.07	2.70	2.49	2.36	2.27	2.21	2.16	2.12	2.09	2.06	2.02	1.97	1.92	1.90	1.87	1.85	1.82	1.79	1.76
16	3.05	2.67	2.46	2.33	2.24	2.18	2.13	2.09	2.06	2.03	1.99	1.94	1.89	1.87	1.84	1.81	1.78	1.75	1.72
17	3.03	2.64	2.44	2.31	2.22	2.15	2.10	2.06	2.03	2.00	1.96	1.91	1.86	1.84	1.81	1.78	1.75	1.72	1.69
18	3.01	2.62	2.42	2.29	2.20	2.13	2.08	2.04	2.00	1.98	1.93	1.89	1.84	1.81	1.78	1.75	1.72	1.69	1.66
19	2.99	2.61	2.40	2.27	2.18	2.11	2.06	2.02	1.98	1.96	1.91	1.86	1.81	1.79	1.76	1.73	1.70	1.67	1.63
20	2.97	2.59	2.38	2.25	2.16	2.09	2.04	2.00	1.96	1.94	1.89	1.84	1.79	1.77	1.74	1.71	1.68	1.64	1.61
21	2.96	2.57	2.36	2.23	2.14	2.08	2.02	1.98	1.95	1.92	1.87	1.83	1.78	1.75	1.72	1.69	1.66	1.62	1.59
22	2.95	2.56	2.35	2.22	2.13	2.06	2.01	1.97	1.93	1.90	1.86	1.81	1.76	1.73	1.70	1.67	1.64	1.60	1.57
23	2.94	2.55	2.34	2.21	2.11	2.05	1.99	1.95	1.92	1.89	1.84	1.80	1.74	1.72	1.69	1.66	1.62	1.59	1.55
24	2.93	2.54	2.33	2.19	2.10	2.04	1.98	1.94	1.91	1.88	1.83	1.78	1.73	1.70	1.67	1.64	1.61	1.57	1.53
25	2.92	2.53	2.32	2.18	2.09	2.02	1.97	1.93	1.89	1.87	1.82	1.77	1.72	1.69	1.66	1.63	1.59	1.56	1.52
26	2.91	2.52	2.31	2.17	2.08	2.01	1.96	1.92	1.88	1.86	1.81	1.76	1.71	1.68	1.65	1.61	1.58	1.54	1.50
27	2.90	2.51	2.30	2.17	2.07	2.00	1.95	1.91	1.87	1.85	1.80	1.75	1.70	1.67	1.64	1.60	1.57	1.53	1.49
28	2.89	2.50	2.29	2.16	2.06	2.00	1.94	1.90	1.87	1.84	1.79	1.74	1.69	1.66	1.63	1.59	1.56	1.52	1.48
29	2.89	2.50	2.28	2.15	2.06	1.99	1.93	1.89	1.86	1.83	1.78	1.73	1.68	1.65	1.62	1.58	1.55	1.51	1.47
30	2.88	2.49	2.28	2.14	2.05	1.98	1.93	1.88	1.85	1.82	1.77	1.72	1.67	1.64	1.61	1.57	1.54	1.50	1.46
40	2.84	2.44	2.23	2.09	2.00	1.93	1.87	1.83	1.79	1.76	1.71	1.66	1.61	1.57	1.54	1.51	1.47	1.42	1.38
60	2.79	2.39	2.18	2.04	1.95	1.87	1.82	1.77	1.74	1.71	1.66	1.60	1.54	1.51	1.48	1.44	1.40	1.35	1.29
120	2.75	2.35	2.13	1.99	1.90	1.82	1.77	1.72	1.68	1.65	1.60	1.55	1.48	1.45	1.41	1.37	1.32	1.26	1.19
∞	2.71	2.30	2.08	1.94	1.85	1.77	1.72	1.67	1.63	1.60	1.55	1.49	1.42	1.38	1.34	1.30	1.24	1.17	1.00

637

TABLE D (*continued*). Percentage points of the *F* distribution: upper 5% points

v_2 \ v_1	1	2	3	4	5	6	7	8	9	10	12	15	20	24	30	40	60	120	∞
1	161.4	199.5	215.7	224.6	230.2	234.0	236.8	238.9	240.5	241.9	243.9	245.9	248.0	249.1	250.1	251.1	252.2	253.3	254.3
2	18.51	19.00	19.16	19.25	19.30	19.33	19.35	19.37	19.38	19.40	19.41	19.43	19.45	19.45	19.46	19.47	19.48	19.49	19.50
3	10.13	9.55	9.28	9.12	9.01	8.94	8.89	8.85	8.81	8.79	8.74	8.70	8.66	8.64	8.62	8.59	8.57	8.55	8.53
4	7.71	6.94	6.59	6.39	6.26	6.16	6.09	6.04	6.00	5.96	5.91	5.86	5.80	5.77	5.75	5.72	5.69	5.66	5.63
5	6.61	5.79	5.41	5.19	5.05	4.95	4.88	4.82	4.77	4.74	4.68	4.62	4.56	4.53	4.50	4.46	4.43	4.40	4.36
6	5.99	5.14	4.76	4.53	4.39	4.28	4.21	4.15	4.10	4.06	4.00	3.94	3.87	3.84	3.81	3.77	3.74	3.70	3.67
7	5.59	4.74	4.35	4.12	3.97	3.87	3.79	3.73	3.68	3.64	3.57	3.51	3.44	3.41	3.38	3.34	3.30	3.27	3.23
8	5.32	4.46	4.07	3.84	3.69	3.58	3.50	3.44	3.39	3.35	3.28	3.22	3.15	3.12	3.08	3.04	3.01	2.97	2.93
9	5.12	4.26	3.86	3.63	3.48	3.37	3.29	3.23	3.18	3.14	3.07	3.01	2.94	2.90	2.86	2.83	2.79	2.75	2.71
10	4.96	4.10	3.71	3.48	3.33	3.22	3.14	3.07	3.02	2.98	2.91	2.85	2.77	2.74	2.70	2.66	2.62	2.58	2.54
11	4.84	3.98	3.59	3.36	3.20	3.09	3.01	2.95	2.90	2.85	2.79	2.72	2.65	2.61	2.57	2.53	2.49	2.45	2.40
12	4.75	3.89	3.49	3.26	3.11	3.00	2.91	2.85	2.80	2.75	2.69	2.62	2.54	2.51	2.47	2.43	2.38	2.34	2.30
13	4.67	3.81	3.41	3.18	3.03	2.92	2.83	2.77	2.71	2.67	2.60	2.53	2.46	2.42	2.38	2.34	2.30	2.25	2.21
14	4.60	3.74	3.34	3.11	2.96	2.85	2.76	2.70	2.65	2.60	2.53	2.46	2.39	2.35	2.31	2.27	2.22	2.18	2.13
15	4.54	3.68	3.29	3.06	2.90	2.79	2.71	2.64	2.59	2.54	2.48	2.40	2.33	2.29	2.25	2.20	2.16	2.11	2.07
16	4.49	3.63	3.24	3.01	2.85	2.74	2.66	2.59	2.54	2.49	2.42	2.35	2.28	2.24	2.19	2.15	2.11	2.06	2.01
17	4.45	3.59	3.20	2.96	2.81	2.70	2.61	2.55	2.49	2.45	2.38	2.31	2.23	2.19	2.15	2.10	2.06	2.01	1.96
18	4.41	3.55	3.16	2.93	2.77	2.66	2.58	2.51	2.46	2.41	2.34	2.27	2.19	2.15	2.11	2.06	2.02	1.97	1.92
19	4.38	3.52	3.13	2.90	2.74	2.63	2.54	2.48	2.42	2.38	2.31	2.23	2.16	2.11	2.07	2.03	1.98	1.93	1.88
20	4.35	3.49	3.10	2.87	2.71	2.60	2.51	2.45	2.39	2.35	2.28	2.20	2.12	2.08	2.04	1.99	1.95	1.90	1.84
21	4.32	3.47	3.07	2.84	2.68	2.57	2.49	2.42	2.37	2.32	2.25	2.18	2.10	2.05	2.01	1.96	1.92	1.87	1.81
22	4.30	3.44	3.05	2.82	2.66	2.55	2.46	2.40	2.34	2.30	2.23	2.15	2.07	2.03	1.98	1.94	1.89	1.84	1.78
23	4.28	3.42	3.03	2.80	2.64	2.53	2.44	2.37	2.32	2.27	2.20	2.13	2.05	2.01	1.96	1.91	1.86	1.81	1.76
24	4.26	3.40	3.01	2.78	2.62	2.51	2.42	2.36	2.30	2.25	2.18	2.11	2.03	1.98	1.94	1.89	1.84	1.79	1.73
25	4.24	3.39	2.99	2.76	2.60	2.49	2.40	2.34	2.28	2.24	2.16	2.09	2.01	1.96	1.92	1.87	1.82	1.77	1.71
26	4.23	3.37	2.98	2.74	2.59	2.47	2.39	2.32	2.27	2.22	2.15	2.07	1.99	1.95	1.90	1.85	1.80	1.75	1.69
27	4.21	3.35	2.96	2.73	2.57	2.46	2.37	2.31	2.25	2.20	2.13	2.06	1.97	1.93	1.88	1.84	1.79	1.73	1.67
28	4.20	3.34	2.95	2.71	2.56	2.45	2.36	2.29	2.24	2.19	2.12	2.04	1.96	1.91	1.87	1.82	1.77	1.71	1.65
29	4.18	3.33	2.93	2.70	2.55	2.43	2.35	2.28	2.22	2.18	2.10	2.03	1.94	1.90	1.85	1.81	1.75	1.70	1.64
30	4.17	3.32	2.92	2.69	2.53	2.42	2.33	2.27	2.21	2.16	2.09	2.01	1.93	1.89	1.84	1.79	1.74	1.68	1.62
40	4.08	3.23	2.84	2.61	2.45	2.34	2.25	2.18	2.12	2.08	2.00	1.92	1.84	1.79	1.74	1.69	1.64	1.58	1.51
60	4.00	3.15	2.76	2.53	2.37	2.25	2.17	2.10	2.04	1.99	1.92	1.84	1.75	1.70	1.65	1.59	1.53	1.47	1.39
120	3.92	3.07	2.68	2.45	2.29	2.17	2.09	2.02	1.96	1.91	1.83	1.75	1.66	1.61	1.55	1.50	1.43	1.35	1.25
∞	3.84	3.00	2.60	2.37	2.21	2.10	2.01	1.94	1.88	1.83	1.75	1.67	1.57	1.52	1.46	1.39	1.32	1.22	1.00

TABLE D (*continued*), Percentage points of the F distribution: upper 1% points

ν_2 \ ν_1	1	2	3	4	5	6	7	8	9	10	12	15	20	24	30	40	60	120	∞
1	4052	4999.50	5403	5625	5764	5859	5928	5982	6022	6056	6106	6157	6209	6235	6261	6287	6313	6339	6366
2	98.50	99.00	99.17	99.25	99.30	99.33	99.36	99.37	99.39	99.40	99.42	99.43	99.45	99.46	99.47	99.47	99.48	99.49	99.50
3	34.12	30.82	29.46	28.71	28.24	27.91	27.67	27.49	27.35	27.23	27.05	26.87	26.69	26.60	26.50	26.41	26.32	26.22	26.13
4	21.20	18.00	16.69	15.98	15.52	15.21	14.98	14.80	14.66	14.55	14.37	14.20	14.02	13.93	13.84	13.75	13.65	13.56	13.46
5	16.26	13.27	12.06	11.39	10.97	10.67	10.46	10.29	10.16	10.05	9.89	9.72	9.55	9.47	9.38	9.29	9.20	9.11	9.02
6	13.75	10.92	9.78	9.15	8.75	8.47	8.26	8.10	7.98	7.87	7.72	7.56	7.40	7.31	7.23	7.14	7.06	6.97	6.88
7	12.25	9.55	8.45	7.85	7.46	7.19	6.99	6.84	6.72	6.62	6.47	6.31	6.16	6.07	5.99	5.91	5.82	5.74	5.65
8	11.26	8.65	7.59	7.01	6.63	6.37	6.18	6.03	5.91	5.81	5.67	5.52	5.36	5.28	5.20	5.12	5.03	4.95	4.86
9	10.56	8.02	6.99	6.42	6.06	5.80	5.61	5.47	5.35	5.26	5.11	4.96	4.81	4.73	4.65	4.57	4.48	4.40	4.31
10	10.04	7.56	6.55	5.99	5.64	5.39	5.20	5.06	4.94	4.85	4.71	4.56	4.41	4.33	4.25	4.17	4.08	4.00	3.91
11	9.65	7.21	6.22	5.67	5.32	5.07	4.89	4.74	4.63	4.54	4.40	4.25	4.10	4.02	3.94	3.86	3.78	3.69	3.60
12	9.33	6.93	5.95	5.41	5.06	4.82	4.64	4.50	4.39	4.30	4.16	4.01	3.86	3.78	3.70	3.62	3.54	3.45	3.36
13	9.07	6.70	5.74	5.21	4.86	4.62	4.44	4.30	4.19	4.10	3.96	3.82	3.66	3.59	3.51	3.43	3.34	3.25	3.17
14	8.86	6.51	5.56	5.04	4.69	4.46	4.28	4.14	4.03	3.94	3.80	3.66	3.51	3.43	3.35	3.27	3.18	3.09	3.00
15	8.68	6.36	5.42	4.89	4.56	4.32	4.14	4.00	3.89	3.80	3.67	3.52	3.37	3.29	3.21	3.13	3.05	2.96	2.87
16	8.53	6.23	5.29	4.77	4.44	4.20	4.03	3.89	3.78	3.69	3.55	3.41	3.26	3.18	3.10	3.02	2.93	2.84	2.75
17	8.40	6.11	5.18	4.67	4.34	4.10	3.93	3.79	3.68	3.59	3.46	3.31	3.16	3.08	3.00	2.92	2.83	2.75	2.65
18	8.29	6.01	5.09	4.58	4.25	4.01	3.84	3.71	3.60	3.51	3.37	3.23	3.08	3.00	2.92	2.84	2.75	2.66	2.57
19	8.18	5.93	5.01	4.50	4.17	3.94	3.77	3.63	3.52	3.43	3.30	3.15	3.00	2.92	2.84	2.76	2.67	2.58	2.49
20	8.10	5.85	4.94	4.43	4.10	3.87	3.70	3.56	3.46	3.37	3.23	3.09	2.94	2.86	2.78	2.69	2.61	2.52	2.42
21	8.02	5.78	4.87	4.37	4.04	3.81	3.64	3.51	3.40	3.31	3.17	3.03	2.88	2.80	2.72	2.64	2.55	2.46	2.36
22	7.95	5.72	4.82	4.31	3.99	3.76	3.59	3.45	3.35	3.26	3.12	2.98	2.83	2.75	2.67	2.58	2.50	2.40	2.31
23	7.88	5.66	4.76	4.26	3.94	3.71	3.54	3.41	3.30	3.21	3.07	2.93	2.78	2.70	2.62	2.54	2.45	2.35	2.26
24	7.82	5.61	4.72	4.22	3.90	3.67	3.50	3.36	3.26	3.17	3.03	2.89	2.74	2.66	2.58	2.49	2.40	2.31	2.21
25	7.77	5.57	4.68	4.18	3.85	3.63	3.46	3.32	3.22	3.13	2.99	2.85	2.70	2.62	2.54	2.45	2.36	2.27	2.17
26	7.72	5.53	4.64	4.14	3.82	3.59	3.42	3.29	3.18	3.09	2.96	2.81	2.66	2.58	2.50	2.42	2.33	2.23	2.13
27	7.68	5.49	4.60	4.11	3.78	3.56	3.39	3.26	3.15	3.06	2.93	2.78	2.63	2.55	2.47	2.38	2.29	2.20	2.10
28	7.64	5.45	4.57	4.07	3.75	3.53	3.36	3.23	3.12	3.03	2.90	2.75	2.60	2.52	2.44	2.35	2.26	2.17	2.06
29	7.60	5.42	4.54	4.04	3.73	3.50	3.33	3.20	3.09	3.00	2.87	2.73	2.57	2.49	2.41	2.33	2.23	2.14	2.03
30	7.56	5.39	4.51	4.02	3.70	3.47	3.30	3.17	3.07	2.98	2.84	2.70	2.55	2.47	2.39	2.30	2.21	2.11	2.01
40	7.31	5.18	4.31	3.83	3.51	3.29	3.12	2.99	2.89	2.80	2.66	2.52	2.37	2.29	2.20	2.11	2.02	1.92	1.80
60	7.08	4.98	4.13	3.65	3.34	3.12	2.95	2.82	2.72	2.63	2.50	2.35	2.20	2.12	2.03	1.94	1.84	1.73	1.60
120	6.85	4.79	3.95	3.48	3.17	2.96	2.79	2.66	2.56	2.47	2.34	2.19	2.03	1.95	1.86	1.76	1.66	1.53	1.38
∞	6.63	4.61	3.78	3.32	3.02	2.80	2.64	2.51	2.41	2.32	2.18	2.04	1.88	1.79	1.70	1.59	1.47	1.32	1.00

TABLE D (continued), Percentage points of the F distribution: upper 0.1 % points

v_2 \ v_1	1	2	3	4	5	6	7	8	9	10	12	15	20	24	30	40	60	120	∞
1	4053*	5000*	5404*	5625*	5764*	5859*	5929*	5981*	6023*	6056*	6107*	6158*	6209*	6235*	6261*	6287*	6313*	6340*	6366*
2	998.5	999.0	999.2	999.2	999.3	999.3	999.4	999.4	999.4	999.4	999.4	999.4	999.4	999.5	999.5	999.5	999.5	999.5	999.5
3	167.0	148.5	141.1	137.1	134.6	132.8	131.6	130.6	129.9	129.2	128.3	127.4	126.4	125.9	125.9	125.0	124.5	124.0	123.5
4	74.14	61.25	56.18	53.44	51.71	50.53	49.66	49.00	48.47	48.05	47.41	46.76	46.10	45.77	45.43	45.09	44.75	44.40	44.05
5	47.18	37.12	33.20	31.09	29.75	28.84	28.16	27.64	27.24	26.92	26.42	25.91	25.39	25.14	24.87	24.60	24.33	24.06	23.79
6	35.51	27.00	23.70	21.92	20.81	20.03	19.46	19.03	18.69	18.41	17.99	17.56	17.12	16.89	16.67	16.44	16.21	15.99	15.75
7	29.25	21.69	18.77	17.19	16.21	15.52	15.02	14.63	14.33	14.08	13.71	13.32	12.93	12.73	12.53	12.33	12.12	11.91	11.70
8	25.42	18.49	15.83	14.39	13.49	12.86	12.40	12.04	11.77	11.54	11.19	10.84	10.48	10.30	10.11	9.92	9.73	9.53	9.33
9	22.86	16.39	13.90	12.56	11.71	11.13	10.70	10.37	10.11	9.89	9.57	9.24	8.90	8.72	8.55	8.37	8.19	8.00	7.81
10	21.04	14.91	12.55	11.28	10.48	9.92	9.52	9.20	8.96	8.75	8.45	8.13	7.80	7.64	7.47	7.30	7.12	6.94	6.76
11	19.69	13.81	11.56	10.35	9.58	9.05	8.66	8.35	8.12	7.92	7.63	7.32	7.01	6.85	6.68	6.52	6.35	6.17	6.00
12	18.64	12.97	10.80	9.63	8.89	8.38	8.00	7.71	7.48	7.29	7.00	6.71	6.40	6.25	6.09	5.93	5.76	5.59	5.42
13	17.81	12.31	10.21	9.07	8.35	7.86	7.49	7.21	6.98	6.80	6.52	6.23	5.93	5.78	5.63	5.47	5.30	5.14	4.97
14	17.14	11.78	9.73	8.62	7.92	7.43	7.08	6.80	6.58	6.40	6.13	5.85	5.56	5.41	5.25	5.10	4.94	4.77	4.60
15	16.59	11.34	9.34	8.25	7.57	7.09	6.74	6.47	6.26	6.08	5.81	5.54	5.25	5.10	4.95	4.80	4.64	4.47	4.31
16	16.12	10.97	9.00	7.94	7.27	6.81	6.46	6.19	5.98	5.81	5.55	5.27	4.99	4.85	4.70	4.54	4.39	4.23	4.06
17	15.72	10.66	8.73	7.68	7.02	6.56	6.22	5.96	5.75	5.58	5.32	5.05	4.78	4.63	4.48	4.33	4.18	4.02	3.85
18	15.38	10.39	8.49	7.46	6.81	6.35	6.02	5.76	5.56	5.39	5.13	4.87	4.59	4.45	4.30	4.15	4.00	3.84	3.67
19	15.08	10.16	8.28	7.26	6.62	6.18	5.85	5.59	5.39	5.22	4.97	4.70	4.43	4.29	4.14	3.99	3.84	3.68	3.51
20	14.82	9.95	8.10	7.10	6.46	6.02	5.69	5.44	5.24	5.08	4.82	4.56	4.29	4.15	4.00	3.86	3.70	3.54	3.38
21	14.59	9.77	7.94	6.95	6.32	5.88	5.56	5.31	5.11	4.95	4.70	4.44	4.17	4.03	3.88	3.74	3.58	3.42	3.26
22	14.38	9.61	7.80	6.81	6.19	5.76	5.44	5.19	4.99	4.83	4.58	4.33	4.06	3.92	3.78	3.63	3.48	3.32	3.15
23	14.19	9.47	7.67	6.69	6.08	5.65	5.33	5.09	4.89	4.73	4.48	4.23	3.96	3.82	3.68	3.53	3.38	3.22	3.05
24	14.03	9.34	7.55	6.59	5.98	5.55	5.23	4.99	4.80	4.64	4.39	4.14	3.87	3.74	3.59	3.45	3.29	3.14	2.97
25	13.88	9.22	7.45	6.49	5.88	5.46	5.15	4.91	4.71	4.56	4.31	4.06	3.79	3.66	3.52	3.37	3.22	3.06	2.89
26	13.74	9.12	7.36	6.41	5.80	5.38	5.07	4.83	4.64	4.48	4.24	3.99	3.72	3.59	3.44	3.30	3.15	2.99	2.82
27	13.61	9.02	7.27	6.33	5.73	5.31	5.00	4.76	4.57	4.41	4.17	3.92	3.66	3.52	3.38	3.23	3.08	2.92	2.75
28	13.50	8.93	7.19	6.25	5.66	5.24	4.93	4.69	4.50	4.35	4.11	3.86	3.60	3.46	3.32	3.18	3.02	2.86	2.69
29	13.39	8.85	7.12	6.19	5.59	5.18	4.87	4.64	4.45	4.29	4.05	3.80	3.54	3.41	3.27	3.12	2.97	2.81	2.64
30	13.29	8.77	7.05	6.12	5.53	5.12	4.82	4.58	4.39	4.24	4.00	3.75	3.49	3.36	3.22	3.07	2.92	2.76	2.59
40	12.61	8.25	6.60	5.70	5.13	4.73	4.44	4.21	4.02	3.87	3.64	3.40	3.15	3.01	2.87	2.73	2.57	2.41	2.23
60	11.97	7.76	6.17	5.31	4.76	4.37	4.09	3.87	3.69	3.54	3.31	3.08	2.83	2.69	2.55	2.41	2.25	2.08	1.89
120	11.38	7.32	5.79	4.95	4.42	4.04	3.77	3.55	3.38	3.24	3.02	2.78	2.53	2.40	2.26	2.11	1.95	1.76	1.54
∞	10.83	6.91	5.42	4.62	4.10	3.74	3.47	3.27	3.10	2.96	2.74	2.51	2.27	2.13	1.99	1.84	1.66	1.45	1.00

* Multiply these entries by 100.

TABLE E. Scales for normal plots

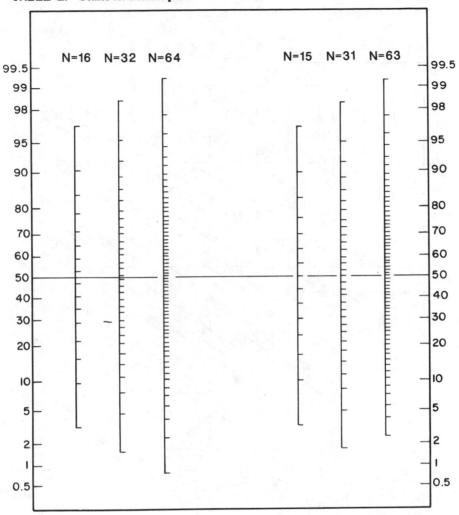

TABLE F. Confidence limits for p in binomial sampling, given a sample fraction $\hat{p} = y/n$: confidence coefficient $= 0.95$*

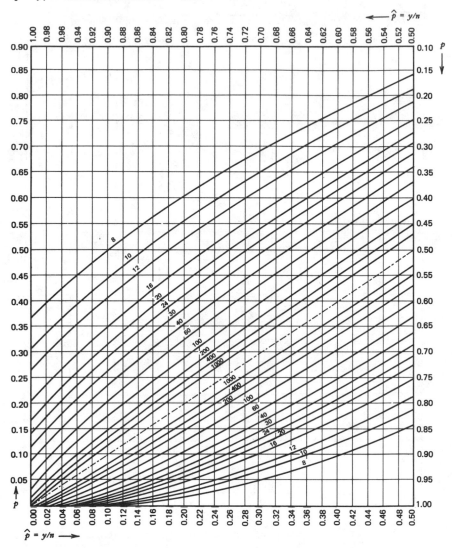

* The numbers printed along the curves indicate the sample size n. If, for a given value of the abscissa y/n, p_A and p_B are the ordinates read from (or interpolated between) the appropriate lower and upper curves, then $\Pr(p_A \leqslant p \leqslant p_B) \leqslant 0.95$.

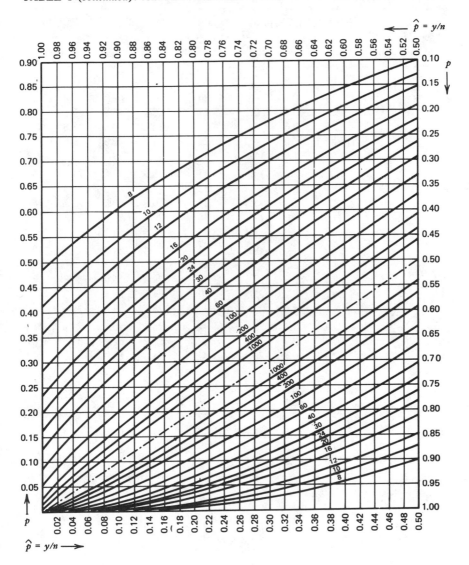

TABLE G. Confidence limits for the mean value of a Poisson variable

observed frequency	99% limit lower	99% limit upper	95% limit lower	95% limit upper	observed frequency	99% limit lower	99% limit upper	95% limit lower	95% limit upper
0	0.0	5.3	0.0	3.7					
1	0.0	7.4	0.1	5.6	26	14.7	42.2	17.0	38.0
2	0.1	9.3	0.2	7.2	27	15.4	43.5	17.8	39.2
3	0.3	11.0	0.6	8.8	28	16.2	44.8	18.6	40.4
4	0.6	12.6	1.0	10.2	29	17.0	46.0	19.4	41.6
5	1.0	14.1	1.6	11.7	30	17.7	47.2	20.2	42.8
6	1.5	15.6	2.2	13.1	31	18.5	48.4	21.0	44.0
7	2.0	17.1	2.8	14.4	32	19.3	49.6	21.8	45.1
8	2.5	18.5	3.4	15.8	33	20.0	50.8	22.7	46.3
9	3.1	20.0	4.0	17.1	34	20.8	52.1	23.5	47.5
10	3.7	21.3	4.7	18.4	35	21.6	53.3	24.3	48.7
11	4.3	22.6	5.4	19.7	36	22.4	54.5	25.1	49.8
12	4.9	24.0	6.2	21.0	37	23.2	55.7	26.0	51.0
13	5.5	25.4	6.9	22.3	38	24.0	56.9	26.8	52.2
14	6.2	26.7	7.7	23.5	39	24.8	58.1	27.7	53.3
15	6.8	28.1	8.4	24.8	40	25.6	59.3	28.6	54.5
16	7.5	29.4	9.4	26.0	41	26.4	60.5	29.4	55.6
17	8.2	30.7	9.9	27.2	42	27.2	61.7	30.3	56.8
18	8.9	32.0	10.7	28.4	43	28.0	62.9	31.1	57.9
19	9.6	33.3	11.5	29.6	44	28.8	64.1	32.0	59.0
20	10.3	34.6	12.2	30.8	45	29.6	65.3	32.8	60.2
21	11.0	35.9	13.0	32.0	46	30.4	66.5	33.6	61.3
22	11.8	37.2	13.8	33.2	47	31.2	67.7	34.5	62.5
23	12.5	38.4	14.6	34.4	48	32.0	68.9	35.3	63.6
24	13.2	39.7	15.4	35.6	49	32.8	70.1	36.1	64.8
25	14.0	41.0	16.2	36.8	50	33.6	71.3	37.0	65.9

AUTHOR INDEX

SUBJECT INDEX

647